Angiogenesis and Vascularisation

Józef Dulak • Alicja Józkowicz • Agnieszka Łoboda
Editors

Angiogenesis and Vascularisation

Cellular and Molecular Mechanisms in Health and Diseases

Springer

Editors
Józef Dulak
Alicja Józkowicz
Agnieszka Łoboda
Department of Medical Biotechnology
Faculty of Biochemistry, Biophysics
and Biotechnology
Jagiellonian University
Kraków
Poland

ISBN 978-3-7091-1427-8 ISBN 978-3-7091-1428-5 (eBook)
DOI 10.1007/978-3-7091-1428-5
Springer Wien Heidelberg New York Dordrecht London

Library of Congress Control Number: 2014931375

Printed on acid-free paper

Springer is part of Springer Science+Business Media (www.springer.com)

Preface

There is no vertebrate life without blood vessels and blood circulation. Therefore, formation of blood vessels is strictly connected with the continuous well-being of human during development and adult life, and finally, disturbed—impaired or exaggerated—angiogenesis leads to development or aggravation of disease.

Understanding the mechanisms of physiological and pathological vascularization is thus really a matter of life and death. The discoveries of the last 25 years have identified plethora of angiogenic mediators including transcription factors, growth factors, cytokines, or extracellular matrix components. Recently, numerous studies have highlighted the role of specific microRNAs in modulating angiogenic signaling. Finally, the conditions in which cells are growing, like oxidative stress, pro-inflammatory state, or hypoxic microenvironment, affect vessel formation and may influence the effectiveness of therapies used for the treatment of angiogenesis-dependent diseases.

Due to such a large area of research we, the editors and authors, are aware that the chapters presented offer only the view on some limited aspects of vascular biology. However, we do believe that this volume can fulfill its aim of presentation of the field from a little bit different perspective than some other books on similar subjects. We hope our wishes will meet the expectations of the readers.

The book is organized in three sections. Part I describes the major components of the angiogenic process with the special emphasis on understanding the significance of different populations of endothelial progenitor cells and the role of specific mediators. A discussion on the mechanisms of lymphangiogenesis is also presented in this part, together with the chapters devoted to endothelial cells and their precursors. Part II continues by discussing the molecular mechanisms of the angiogenic process, concentrating on the role of hypoxia, antioxidant genes, and microRNAs. The significance of gaseous mediators, like carbon monoxide and nitric oxide, is also discussed, presented with some broader perspectives of involvement of Nrf2/heme oxygenase-1 pathway, nitric oxide synthases, and the modulators of their activities. Accordingly, besides stimulators, like ETS transcription factors or immediate early genes, the role of some novel angiogenic inhibitors is also highlighted. Part III concentrates on the pathological aspects of vascularization process in diabetes, rheumatoid arthritis, and cancer, discussing also some new modalities for potential therapies and presenting novel models for investigation of angiogenesis.

There is no reason to describe in detail each chapter and its content which, we hope, in total will be of interest for every reader of this book. In this preface only some aspects of the book will be highlighted. No doubt, the reader can recognize that this volume content reflects the scientific interests of book editors. The significant part is devoted to the nature and role of potential endothelial cell precursors as well as the involvement of hypoxia and gaseous mediators, like carbon monoxide and nitric oxide, in angiogenesis. The obvious links to inflammatory processes underlying the angiogenesis-related diseases, such as cancer, diabetes, and rheumatoid arthritis, are indicated. The molecular mechanisms are discussed with special emphasis on the role of microRNAs, as well as the new emerging models of angiogenesis are presented.

The authors of the chapters did not avoid to address the controversies in vascular biology. Among them are so-called endothelial progenitor cells (EPCs). Described for the first time by Asahara and coworkers in 1997, EPCs have quite dominated the field of angiogenesis translational research for several years. It appears, however, that the nature of those putative progenitors was not clearly specified and their blood-forming properties were exaggerated. Those issues are discussed in several chapters in this book, and although not claiming to be comprehensive, the presented views may help to clarify this continuous controversy. The initial hypothesis on the plasticity, heterogeneity, or both of the bone marrow cells, claimed to be able to differentiate into the endothelial cells, or even heart or brain cells, appears to be misleading. Although still of course requiring the independent confirmation, a hypothesis on the local stem or progenitor precursors in organs appears to be attractive and may be more biologically valid than widely acclaimed, but not really confirmed, presumptions on the miracle nature of bone marrow cells.

Another issue we want to highlight in this preface are microRNAs. Their involvement in regulation of gene expression in angiogenesis is broadly discussed in this book. Beside the separate chapter, Chap. 9, devoted solely to those regulatory molecules, microRNAs are also described in Chap. 8 on hypoxic regulation of gene expression, in Chap. 14 presenting the zebrafish model of angiogenesis, as well as in Chap. 17, in which the correlation between microRNAs and tumor development is shortly discussed. Also, hypoxic regulation of angiogenesis is described in Chaps. 3, 8, and 9 and is addressed from different points in many others. The significance of the lowered oxygen level for regulation of angiogenesis is being recognized both in physiological and particularly pathological conditions. It links also to the role being played by nitric oxide synthase and nitric oxide itself, as well as to antioxidant systems, such as that dependent on Nrf2 and heme oxygenase-1. The role of those obviously cytoprotective pathways is being now considered also in the context of blood vessel formation, adding more noncanonical functions to these interesting, widely interconnected systems.

We do hope that this book can be of some help for all who would like with one volume to get a broader perspective on various aspects of blood vessel formation. The book may be, as we believe, of help for graduate and PhD students and scientists starting their research in vascular biology. We are grateful to our colleagues, coauthors of all the chapters, for their willingness and efforts to contribute to the work we present now to the scientific community.

Kraków, Poland

Józef Dulak
Alicja Józkowicz
Agnieszka Łoboda

Contents

Part I Cells and Mediators

1 **Endothelial Cell Origin, Differentiation, Heterogeneity and Function** . . . 3
Anna Grochot-Przęczek, Magdalena Kozakowska, Józef Dulak, and Alicja Józkowicz

2 **Developmental and Pathological Lymphangiogenesis** . . . 27
Angelika Chachaj and Andrzej Szuba

3 **Vascular-Resident Endothelial Side Population Cells** . . . 67
Hisamichi Naito and Nobuyuki Takakura

4 **Multiple Roles of Proteases in Angiogenesis** . . . 85
Ester M. Weijers and Victor W.M. van Hinsbergh

5 **Serpins in Angiogenesis** . . . 101
Czeslaw S. Cierniewski and Joanna Boncela

Part II Molecular Mechanisms

6 **Endothelial Transcriptional Networks in the Control of Angiogenesis: The ETS Factor** . . . 121
Anna M. Randi

7 **Insights into Roles of Immediate-Early Genes in Angiogenesis** . . . 145
Leonel Prado-Lourenço, Ahmad M.N. Alhendi, and Levon M. Khachigian

8 **Molecular Mechanisms of Hypoxia-Regulated Angiogenesis** . . . 163
Agnieszka Łoboda, Alicja Józkowicz, and Józef Dulak

9 **MicroRNA Regulation of Angiogenesis** . . . 187
Chandan K. Sen and Yuk Cheung Cyrus Chan

10 **Nrf2 Transcription Factor and Heme Oxygenase-1 as Modulators of Vascular Injury and Angiogenesis** . . . 213
Urszula Florczyk, Alicja Józkowicz, and Józef Dulak

11 Metabolism of Methylarginines and Angiogenesis 241
Hilda Tsang, Lucio Iannone, and Beata Wojciak-Stothard

12 The Vasohibin Family: Novel Regulators of Angiogenesis 261
Yasufumi Sato

Part III Diseases and Therapy

13 Hypoxia-Induced Pathological Angiogenesis in Zebrafish 271
Lasse D. Jensen, Pegah Rouhi, and Yihai Cao

14 Tumor Angiogenesis: *Fishing* for Screening Models 293
Dafne Gays, Vera Mugoni, and Massimo M. Santoro

15 Vascular Complications in Diabetes 313
Louise L. Dunn, Kim Hoe Chan, Martin K.C. Ng,
and Roland Stocker

16 Angiogenesis in Rheumatoid Arthritis 339
Sofia Konisti, Serafim Kiriakidis, and Ewa M. Paleolog

17 Angiogenic Signaling and Structural Abnormalities in Tumors 367
Magdalena Tertil, Klaudia Skrzypek, and Agnieszka Łoboda

18 Endothelium in Pathologic Angiogenesis and Angiogenesis-Mediated Therapies 389
Maria Paprocka, Catherine Grillon, Danuta Duś,
and Claudine Kieda

Index 407

Part I

Cells and Mediators

Endothelial Cell Origin, Differentiation, Heterogeneity and Function

1

Anna Grochot-Przęczek, Magdalena Kozakowska, Józef Dulak, and Alicja Józkowicz

Contents

1.1 Introduction 4
1.2 New Blood Vessel Formation: Basic Information 4
1.3 Endothelial Cell Origin and Differentiation 5
1.4 Arterial–Venous–Lymphatic Identity 9
1.5 Angiogenesis: An Overview 11
1.6 Endothelial Cell Heterogeneity and Function 13
1.6.1 Heterogeneity in Structure of Endothelium 14
1.6.2 Heterogeneity in Expression of Markers and Phenotypes of ECs 15
1.6.3 Heterogeneity of ECs Function 17
References 21

Abstract

The endothelium forms a cellular lining of blood vessels in the circulatory system, but its function goes far beyond the creation of a structural barrier between blood and tissues. It is a multifunctional, complex organ, engaged in numerous physiological and pathological processes. This chapter focuses on several basic aspects related to endothelium and tries to answer the following questions: what is the precursor of endothelial cell, how the endothelial cell

A. Grochot-Przęczek • M. Kozakowska
Department of Medical Biotechnology, Faculty of Biochemistry, Biophysics and Biotechnology, Jagiellonian University, Gronostajowa 7, 30-387 Kraków, Poland

Jagiellonian Centre for Experimental Therapeutics, Jagiellonian University, Kraków, Poland

J. Dulak • A. Józkowicz (✉)
Department of Medical Biotechnology, Faculty of Biochemistry, Biophysics and Biotechnology, Jagiellonian University, Gronostajowa 7, 30-387 Kraków, Poland
e-mail: alicja.jozkowicz@uj.edu.pl

J. Dulak et al. (eds.), *Angiogenesis and Vascularisation*,
DOI 10.1007/978-3-7091-1428-5_1, © Springer-Verlag Wien 2013

gains venous, arterial or lymphatic fate, what are the main mechanisms responsible for blood vessel formation and finally, is there any variety in structure and function of endothelium throughout the vascular system?

Keywords
Endothelium • Hemangioblast • Endothelial cell differentiation • Angiogenesis

1.1 Introduction

The fascinating story of a vascular system and endothelium began almost 400 years ago, when William Harvey discovered that blood circulates within a closed unit [1]. Shortly after that, Marcello Malpighi described a network of blood vessels in lungs of a frog and 200 years later, these structures were found to be lined by cells, which were called endothelium [2]. Further studies in the twentieth century undeniably proved the inner cell monolayer lining the vessels to be a widespread (it is estimated that adult blood vessels would double the circuit of the Earth, when laid end to end) [3], complex, multifunctional organ, composed of 1–6 × 10^{13} cells [4], playing a crucial role in physiological and pathological states [1].

Endothelium of circulatory system comprises many different types of endothelial cells (blood and lymphatic, arterial and vein, capillary and large vessel endothelium), which makes it a heterogeneous organ both structurally and functionally. This chapter focuses on endothelial cell origin and function and provides an overview of molecular mechanisms regulating endothelial cell differentiation and blood vessel formation.

1.2 New Blood Vessel Formation: Basic Information

The existence of a functional circulatory system is a prerequisite for survival and growth of the developing embryo. Therefore blood vessels and blood cells, appearing in a parallel manner, form the functional cardiovascular system at the earliest during embryogenesis. The first endothelial and blood cells develop in the yolk sac hematopoietic islands on day 7 of mouse embryogenesis and over the next several hours the circulatory system is greatly enhanced [5]. During embryonic development blood vessels can be formed on the way of vasculogenesis or angiogenesis [6]. Vasculogenesis is the process of the formation of capillaries from primitive stem cells and occurs mainly in early embryogenesis. It involves the creation of first blood vessels: capillaries in the yolk sac and in the body of the embryo. In contrast, angiogenesis is the formation of blood vessels from pre-existing capillaries by proliferation and migration of fully differentiated endothelial cells. In embryonic development it plays a leading role in vascularisation of organ rudiments [7].

Three factors play a key role in the process of blood vessel formation. The first is the basic fibroblast growth factor (FGF2, bFGF), which induces the formation of angioblasts from mesenchymal cells [8, 9]. It was shown that bFGF receptor knockout (FGF-R1) mice die in early pregnancy. Studies using antisense RNA sequence against bFGF or dominant-negative form of FGF-R1 demonstrated strong vascular disorders in both embryo and yolk sac, evidencing the importance of the route bFGF–FGF-R1 in the early vasculogenesis [10, 11]. The second most important protein is a vascular endothelial growth factor (VEGF). VEGF protein family includes VEGF-A, VEGF-B, VEGF-C, VEGF-D and placental growth factor (PlGF). Among these, VEGF-A is the most well-characterised protein and plays a key role in the migration and proliferation of endothelial cells. VEGF-A is expressed in several isoforms, all of which transmit the signal through two receptors with the function of tyrosine kinases VEGF-R1 (Flt-1) and VEGF-R2 (Flk-1, KDR) [12]. During embryogenesis VEGF is released by mesenchymal cells located near the blood islands [13]. The great importance of VEGF was demonstrated by the fact that the exclusion of its single allele leads to the death of the embryo around day 11 [14, 15]. Full VEGF or Flk-1 gene knockout leads to the death of the embryo about 9 days after fertilisation, due to the weakening of the differentiation of endothelial cells and the lack of formation of blood islands [14–16]. Exclusion of both alleles of Flt-1 gene is also lethal, due to uncontrolled proliferation of angioblasts, leading to abnormal vasculature [17]. The third crucial factor is angiopoietin-1 (Ang-1). Ang-1, by binding to the receptor Tie-2, stimulates recruitment of pericytes, which cover endothelial sprouts and stabilise them. Mutations in the genes of Ang-1 and Tie-2 cause vessel malformations [18, 19].

1.3 Endothelial Cell Origin and Differentiation

There is now an established body of knowledge about the origin of endothelial cells and the mechanisms of their differentiation. New research and hitherto long-term investigations have consistently adjudged the hypothesis, originally proposed almost a century ago, of a common endothelial and hematopoietic precursor cell of mesodermal origin, a hemangioblast [20–22]. This concept has been supported by in vitro investigations of embryonic stem cell (ESC)-derived blast colony forming cells (BF-CFC), an in vitro equivalent of a hemangioblast, which undergo a bilineage differentiation into endothelial and hematopoietic cells when exposed to VEGF-A and bone morphogenic protein-4 (BMP-4) [23–25]. Subsequent in vivo studies trucked the potential hemangioblasts in mouse and zebrafish embryos [26, 27]. In contrary, a few studies suggested that endothelial and hematopoietic cells arise independently from mesoderm [28, 29]. Therefore the existence of a hemangioblast is still under debate, although majority of both in vitro and in vivo studies support the concept of a common cell at the divergent point in endothelial-hematopoietic differentiation [22].

Embryonic hemangioblasts are identified by coexpression of a mesodermal marker, transcription factor Brachyury (Bry) and KDR, an antigen characteristic for endothelial cells, but not specific (Fig. 1.1) [26, 30]. Importantly, a hemangioblast might be regarded as multipotent cell, as $Bry^{+}KDR^{+}$ cells can generate, apart from hematopoietic and endothelial cells, also vascular smooth

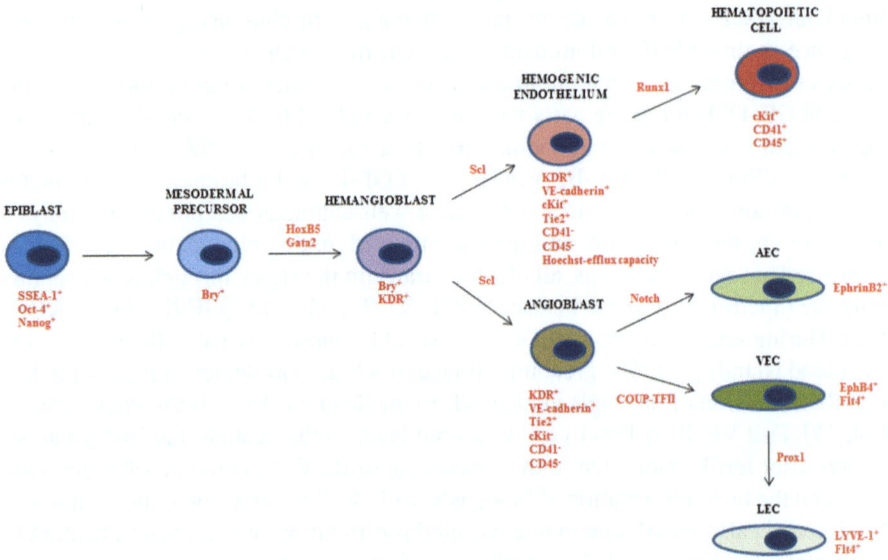

Fig. 1.1 Endothelial cell origin and differentiation. Hemangioblast Bry^+ KDR^+, a common cell at the divergent point in endothelial-hematopoietic differentiation, derives from mesodermal Bry^+ precursor cell. Subsequently, it differentiates either into a hemogenic endothelial cell, capable of hematopoietic cell generation or into an angioblast, which is a precursor for arterial and venous endothelial cells. The specific phenotype of cells and regulators of their differentiation are indicated

muscle cells [26] and embryonic KDR^+ cells give rise also to skeletal and cardiac muscle cells [31]. VEGF signalling is crucial for endothelial and hematopoietic cells to arise, as genetic KDR deficiency results in the absence of blood island in developing embryo, and its lethality [16]. In vitro differentiation of KDR^+ mesodermal cells showed that 10 % of seeded precursor cells give rise to hematopoietic colony in the absence of VEGF. When KDR^+ cells were exposed to VEGF, 10 % of them formed endothelial colonies and 5 % hematopoietic ones [32]. The upstream mechanisms activating KDR expression in mesodermal precursor cells involve HoxB5 transcription factor [33]. Generation of the hemangioblast from mesoderm is activated by BMP4 [34] and is mediated also by Gata2 transcription factor, which drives the expression of KDR and Scl (stem cell leukaemia), two downstream genes critical in hemangioblast differentiation [35].

Generally, the notion of a hemangioblast is rather related to embryonic development, but there are also several indications for the existence of such a cell in a postnatal life. The study by Pelosi et al. identified hemangioblast-like cells within $CD34^+KDR^+$ subpopulation from bone marrow and cord blood. Such cells were capable of bilineage differentiation as indicated by hemato-endothelial colony formation. Endothelial identity of the differentiated cells was confirmed by VE-cadherin expression, lack of CD45 and other hematopoietic markers, LDL uptake and capillary-like structure formation. Of note, long-term proliferative capacity of these hemangioblasts was confirmed [36]. Another evidence for the adult life hemangioblast concept came from leukaemia model, where the BCR/ABL fusion gene was found also in endothelial cells. This indicates that the rearrangement had occurred in a common precursor cell [37].

The fact that both endothelial and the bone marrow-resident hematopoietic cells originate from KDR$^+$ precursor cells [38] may suggest that hemangioblast may differentiate into hematopoietic cell indirectly, through an endothelial intermediate cell: a hemogenic endothelium (Fig. 1.1). Such concept is supported by the evidence that hematopoietic cells generated in aorta–gonad–mesonephrons (AGM) region, one of the sites of definitive hematopoiesis, localise very closely to endothelial surface of the aorta [39, 40], and hematopoietic clusters form tight junctions with aortic endothelial cells [41, 42]. The study investigating the phenotype of cells lining the chicken aortic lumen showed, that before the hematopoietic cluster emergence, cells of the aortic endothelium were uniformly positive for KDR. The appearance of the hematopoietic clusters was parallel with increase in CD45 expression and KDR downregulation, indicating that hematopoietic cells derive from a kind of endothelial precursor [43]. Moreover, first steps of the hemangioblast differentiation coincide with upregulation of endothelial markers, such as Tie2, VE-cadherin (vascular-endothelial cadherin), CD31 and KDR. Functional assays showed that such differentiated cells possess endothelial features and can finally acquire CD41 marker and generate primitive and definitive hematopoietic cells (Fig. 1.1) [44]. Accordingly, CD34$^+$CD45$^-$ and CD31$^+$CD45$^-$ endothelial cells isolated from human yolk sac, embryonic AGM region, embryonic and foetal liver and foetal bone marrow, co-cultured on MS-5 cell line (a mouse bone marrow stroma that supports the multilineage, long-term development of human hematopoietic stem cells) generated hematopoietic cells [45]. Other studies revealed that hematopoietic cells can be obtained from a single hemogenic endothelial cell in vitro [44, 46] and hematopoietic stem cells were shown to be generated from hemogenic endothelium in vivo [47, 48]. Molecular mechanism responsible for generation of hemogenic endothelium from hemangioblasts relies on the expression of Scl, as this intermediate endothelial cell is not formed in the absence of Scl [44]. On the other hand, endothelial-hematopoietic transition depends on Runx-1 (AML1) transcription factor [49–52]. The hemogenic endothelium can be distinguished from the angioblasts by cKit expression and also the capacity to extrude Hoechst dye, which is the characteristic for side population phenotype (Fig. 1.1) [53, 54].

A quest for an endothelial precursor cell in adult organisms led to the discovery of so-called endothelial progenitor cells (EPCs). The EPC concept, although at the beginning very promising, during several years of research was evolving and turned out to be rather misleading [55, 56]. The EPC story started in 1997, when Asahara and co-workers for the first time isolated human peripheral blood circulating CD34$^+$ cells able to home to ischemic tissue and to facilitate angiogenesis [57]. Of note, more than 30 years earlier, it had been demonstrated that cells circulating in peripheral blood can endothelialise artificial grafts sewn in blood vessels [58]. Confirmation of this study was published in 1994, describing the circulating endothelial cells as endothelialisation fallout [59], which included CD34$^+$ cells [60]. Subsequent studies showed that EPCs are mobilised from their niches, such as bone marrow, into peripheral blood, and migrate into hypoxic tissues where they participate in neovascularisation [61–64], mainly in paracrine manner [55]. Usually, EPCs are defined as the peripheral blood or bone marrow mononuclear cells, with the capability of adhesion to the extracellular matrix proteins, acLDL uptake and lectin binding. In addition, EPCs should express endothelial antigens, such as endothelial nitric oxide synthase (eNOS), CD31, VE-cadherin, von Willebrand factor (vWF) and VEGF-R2 [65, 66]. The literature provides neither one definition of EPCs nor a specific EPC phenotype, as well as presents several protocols of EPC isolation and culture. And this was the main problem, the consequences of which became apparent later on.

First paper, which drew attention to the heterogeneity of EPCs was published in 2004 [67]. It describes two subpopulations: early and late EPCs. Early EPCs are spindle shaped, proliferate extensively in the second and third week after isolation and can be maintained in the culture for 4 weeks. Late EPCs have cobblestone morphology, characteristic for endothelial cells. In culture, they begin to appear about 2–3 weeks after isolation, proliferate the fastest between the fourth and eighth week and survive in vitro for about 12 weeks. Importantly, both cell subpopulations have different expression of CD45, VEGF-R1, VEGF-R2 and VE-cadherin. Early EPCs produce larger quantities of pro-angiogenic factors such as VEGF and IL-8, which increase their paracrine potential for stimulation of blood vessel formation. Late EPCs produce more nitric oxide and efficiently create the capillary-like structures on Matrigel and effectively incorporate into the network formed by mature endothelium. Unexpectedly, in vivo both populations have similar vasculogenic potential [67]. Another example of the heterogeneity of EPCs was described by Yoder et al. Subpopulation of CFU-EC (colony forming unit-endothelial cells) was obtained by negative selection of cells on fibronectin. CFU-EC showed a high expression of both hematopoietic and endothelial markers. As explained by the authors, CFU-EC derived from hematopoietic stem cells, had properties of myeloid progenitors, differentiated into phagocytic macrophages and did not form functional blood vessels in vivo. ECFC subpopulation (endothelial colony-forming cells) was isolated by positive selection on collagen I. These cells appeared to be phenotypically closer to endothelium, because they did not exhibit hematopoietic CD45, CD14 and CD115 antigen expression, and almost 100 % expressed endothelial markers. Furthermore, in contrast to the CFU-EC, ECFC showed a high proliferative potential and most importantly, in vivo formed fully perfused vessels [68]. Finally, other papers revealed the true face of majority of so-called EPCs, which are in fact not endothelial precursors but can be better described as myeloid lineage-derived cells with pro-angiogenic properties. This problem was raised in many experiments. For example, culture of monocytes in endothelial-dedicated media resulted in a loss of CD14 and CD45 hematopoietic markers [69]; primary monocytes demonstrated a high level of most endothelial genes [69]; the presence of monocytes was indispensable for CFU-EC to form a network on Matrigel [69]; CD31$^+$F4/80$^+$ monocytes could be recruited to the damaged tissues and were directly involved in the formation of blood vessels [70]; proteome analysis showed that the presence of endothelial markers in early EPCs was a result of the contamination of EPCs with platelet fragments and was not related to endogenous synthesis [71]. Moreover, it was proven that EPCs are capable of antigen presentation at a level similar to monocytes and much stronger than mature endothelial cells and can co-stimulate naive lymphocytes. Thus, they are professional antigen-presenting cells [72]. Due to the fact of monocyte-endothelial mimicry of EPCs, all papers regarding these cells should be considered very carefully. Of note, ECFC seem to the most constantly comply the rules to be called genuine endothelial progenitor cells.

Our understanding of endothelial cell precursor biology is improving. Identification of the hemangioblast and its derivative, a hemogenic endothelium, outlined the molecular pathways regulating the emergence of endothelial and hematopoietic cells and revealed that endothelium is in fact precedent to hematopoiesis on the ladder of differentiation. Once a hemangioblast gains its identity towards an endothelial cell, then a nascent endothelial precursor must choose whether to become a lymphatic or blood endothelial cell, and if so, a vein or arterial endothelium.

1.4 Arterial–Venous–Lymphatic Identity

The blood vascular system is a bipolar network of arteries and veins presenting morphological, physiological, functional and molecular differences. Interestingly, the first functional distinction between arteries and veins was described by William Harvey, who observed that some vessels carry blood away from and some towards the heart [73]. The research of the past two decades definitely showed that the functional and structural differences between arteries and veins result from a genetic designation and can be not attributed to physiological factors. The arterio-venous fate of endothelial cell is appointed early during embryogenesis by genetic mechanisms. Majority of the studies investigating the arterial and venous differentiation were carried out on a zebrafish model, which is regarded to be an excellent organism to study vascular development (see also Chaps. 13 and 14). The development of the main trunk blood vessels in a zebrafish embryo starts when the angioblast, which arise in a lateral mesoderm, migrate to the midline of a trunk, aggregate and form a dorsal aorta and posterior cardinal vein [74]. On a cross section, these two blood vessels are localised below the notochord (Fig. 1.2a). The molecular pattern suggests that the endothelial cell arising from hemangioblast has a default venous identity, until arterial-promoting cascade involving Notch signalling is turn on and the cell gains arterial fate. Indeed, *flt4*, a venous marker, is initially expressed in both arterial and venous primordia [75]. The master mediator of a venous fate is chicken ovoalbumin upstream promoter-transcription factor II (COUP-TFII), a nuclear receptor, which downregulates arterial Notch signalling, releasing EphB4 and Flt4 from Notch-mediated repression (Fig. 1.2b) [76]. Studies in a zebrafish model convincingly showed how the arterial differentiation is turn on

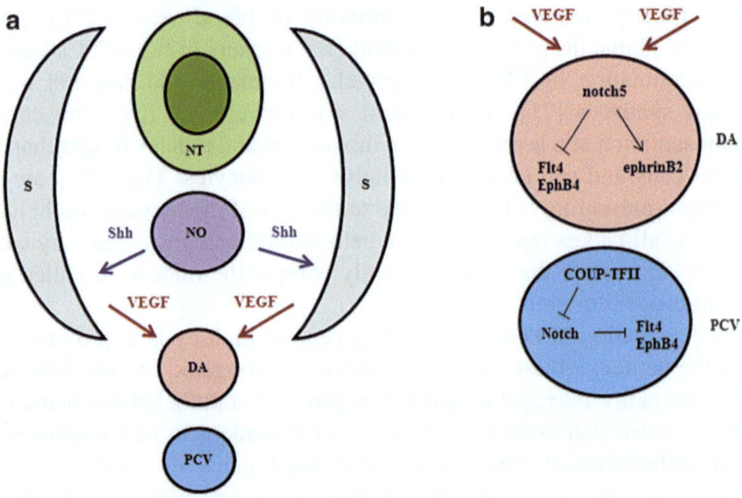

Fig. 1.2 Establishment of arteriovenous cell identity. (**a**) The arteriovenous fate of endothelial cell is appointed early during embryogenesis by genetic mechanisms, involving Shh-VEGF signalling. (**b**) Notch and COUP-TFII are the key mediators of genes characteristic for arterial and venous endothelial cells, respectively. *NT* neural tube, *NO* notochord, *S* somites, *DA* dorsal aorta, *PCV* posterior cardinal vein

and driven (Fig. 1.2a, b). Sonic hedgehog (Shh) protein was identified to act as a master upstream regulator of the arterial identity. Shh is a diffusible protein, which is produced in a notochord [77], and mediates the development of adjacent structures, for example organisation and patterning of somites [78] and neural tube [79]. Of note, mutation in *sonic- you* (*syu*) gene, which encodes Shh in zebrafish [80], causes the failure of dorsal aorta and posterior cardinal vein development and single-lumenized structure, positive for *flt4*, is formed instead [81]. It was demonstrated that Shh drives the expression of somitic VEGF, which directly provides a signal for arterial differentiation (Fig. 1.2a). Somitic VEGF activates Notch signalling in endothelial cells, in zebrafish particularly *notch5* [81], which is expressed exclusively in dorsal aorta in the vasculature during embryonic development [82]. Notch signalling drives arterial differentiation of endothelial cells by induction of arterial-specific genes, such as ephrin-B2 (*efnb2a*) and downregulation of venous markers, such as *flt4 or ephb4a* [81] (Fig. 1.2b). The repression of venous genes is achieved by gridlock protein (Grl), a transcriptional repressor acting downstream of Notch. Accordingly, disorders in the Notch-Grl signalling in zebrafish selectively perturbs assembly of the artery with the venous expansion [83]. The molecular pathway orchestrating arterial differentiation is already known, but the issue why some cells choose arterial characteristic and why others remain venous cells, demands further elucidation. In zebrafish embryos, the precursor cells for the trunk artery and vein are mixed together in a lateral posterior mesoderm. Later on their progeny is assigned either to dorsal aorta or to posterior cardinal vein [83]. The cells found at the position of dorsal aorta respond to VEGF more strongly than those placed at posterior cardinal vein, so apparently some angioblasts are predifferentiated and that is why they are more sensitive to VEGF signalling [81] and subsequently differentiate into arterial endothelium. Although zebrafish studies showed that arteriovenous cell fate is accomplished before the blood flow onset, on the other hand, studies conducted in avian embryos demonstrated the ability of embryonic endothelial cells to transdifferentiate from venous to arterial cell and vice versa after the circulation is already initiated, but before the embryonic day 7, when such arteriovenous lability does not occur any more [74, 84].

Determination of the arterial and venous fate of endothelial cell is one of the initial steps needed to be resolved during formation of circulatory system. Soon after that, another network of vessels is started to be established and this is a lymphatic vasculature, which has a venous origin (Fig. 1.1). The expression of a lymphatic markers such as lymphatic vessel endothelial hyaluronian 1 (Lyve-1) and Prox-1 is initiated on day 9 of embryonic development on a subpopulation of cardinal vein endothelial cells. COUP-TFII and Sox18 transcription factors are the key regulators of lymphatic cell identity. Then the lymphatically committed venous endothelial cells start to sprout and bud off, forming the primary lymphatic sacs [85]. The exact molecular pathways regulating establishment of lymphatic system are comprehensively described in the Chap. 2.

1.5 Angiogenesis: An Overview

Designation of the arterial, venous and lymphatic identity of endothelial cell is a process parallel to formation of blood vessels de novo. Subsequent development of vascular system relies on the expansion of the vessel network. This is achieved mostly in a process of angiogenesis. Physiologically, apart from embryogenesis, it occurs rarely and is limited to female reproductive system, growth of organs or takes place in the response to injuries. Contrary to pathological angiogenesis, which may last for years, physiological formation of blood vessels is a process strictly restricted in time [86, 87].

Angiogenesis involves series of events, with a key position of endothelial cells (ECs) responding to angiogenic factors produced either by endothelium or stromal cells [88] (Fig. 1.3). Initiation of angiogenesis is completed in response to hypoxia, to overcome oxygen depletion and starvation. Hypoxia-inducible factor-1α (HIF-1α), which is stable and active under low oxygen tension, drives expression of strong pro-angiogenic growth factors—among others VEFG, PDGF, Ang1 and Ang2 [89–91]. They are prerequisite for angiogenesis and activate ECs for proliferation, survival and migration via endothelial receptors: VEGF-R2, VEGF-R3 (VEGF and PDGF) and Tie2 receptor (Ang1) [87, 88, 92, 93], whereas VEGF-R1 is a decoy for VEGF [94].

Sprouting of capillaries from pre-existing blood vessels requires selection of a specific site on a vessel, where ECs start to invade extracellular matrix (ECM). In response to gradient of VEGF, one of the endothelial cells is induced into a tip cell phenotype—proliferates slowly, becomes motile, protrudes filopodia and expresses

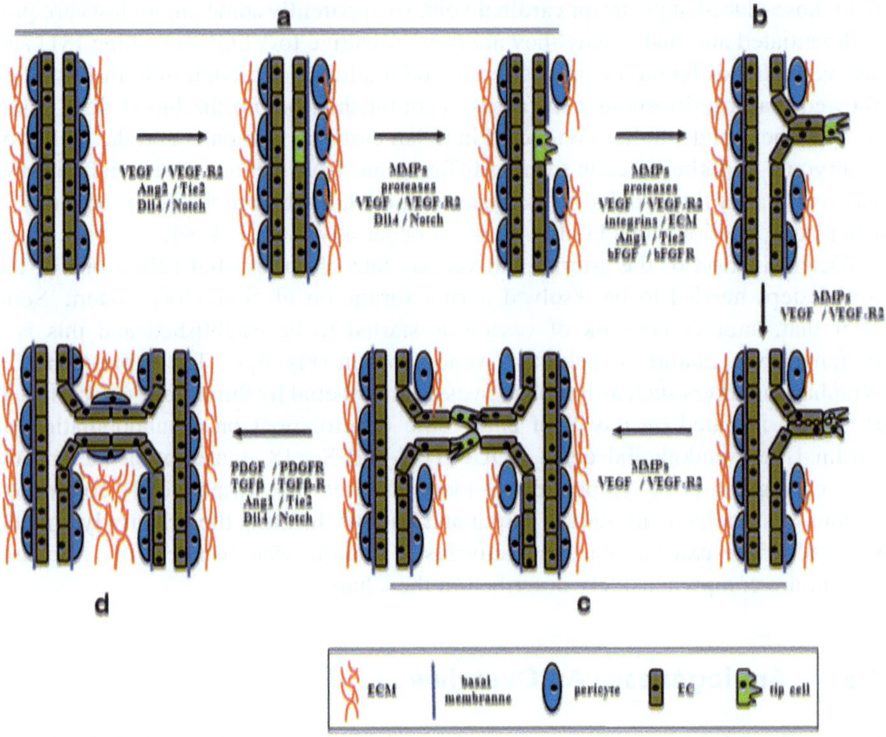

Fig. 1.3 Main stages of angiogenesis and their mediators. (**a**) Initial steps of angiogenesis involve conversion of a quiescent EC of a vessel wall into a tip cell phenotype and dissociation of pericytes. MMPs and other proteases degrade basal membrane and ECM components. (**b**) Proliferation and migration of both tip and stalk cells occur in response to pro-angiogenic factors. (**c**) Formation of lumen and anastomosis between two capillaries requires VEGF signalling and MMPs action and involves rearrangement in junctions between ECs. (**d**) Finally, nascent vessels are stabilised by adhesion of pericytes

high levels of proteases to facilitate digestion of ECM during migration. Stalk cells follow a tip cell and proliferate extensively to elongate a new vessel. Determination between tip and stalk identity is controlled by Notch and VEGF signalling in a negative feedback loop. Hypoxia and VEGF induces expression of Notch ligand (Dll4) in tip cells, which stimulates Notch pathway in neighbouring cells [86, 88, 91, 94]. Activated Notch signalling and high expression of VEGF-R1 with low level of VEGF-R2 and VEGF-R3 in stalk cells limit filopodia formation, VEGF responsiveness and therefore exclude them from a tip cell position [86, 88, 91].

Since capillaries are surrounded by basal membrane composed mainly of lamins, collagens and proteoglycans, liberation of ECs requires firstly degradation of ECM. This process is accomplished mainly by matrix metalloproteinases (MMPs), although other proteases such as plasmin, uPA and tPA also contribute to ECM degradation. Angiogenic factors can modulate their proteolytic activity, whereas proteases in turn are responsible for releasing pro-angiogenic and anti-angiogenic factors from ECM reservoir, as well as can inactivate membrane receptors for them [86, 95, 96]. Subsequent migration of ECs is necessary for new blood vessels to be formed. It involves reversible binding of ECs to ECM components as well as assembly and disassembly of cytoskeletal elements [92, 97]. ECM is a modulator of angiogenesis as it mediates the attachment of ECs and in basal conditions promotes their quiescence. However it is also a scaffold for growth factors which bind to it and become released during initial, proteolytic steps of angiogenesis [95–97]. ECs are connected to ECM at focal adhesions, which together with actin stress fibres are aligned in direction of migration [97]. Migration of endothelial cells can be guided both towards the gradient of soluble chemoattractants (VEGF, Ang and bFGF) and ECM components (collagen I, fibronectin), which activate small GTPase of the Rho family to induce filopodia formation. Integrins, which are heterodimeric, transmembrane receptors for extracellular components, link the cell with ECM at focal adhesions [95, 97]. This in turn facilitates protrusion of lamellopodia and afterwards—formation of new focal adhesion as well as assembly of actin stress fibres, enabling the movement [97].

Proliferation and migration of ECs create network of tubes, unable to convey blood, unless lumen is formed [86, 88]. Two main models of lumen establishment are proposed—intracellular and intercellular. Intracellular mechanism involves vacuoles generation, which subsequently fuse into transcellular lumen compartment, that is secreted, so the multicellular lumen between neighbouring ECs can be formed [86, 95, 98]. Intercellular model of lumen formation involves membrane invagination between two cells of defined apical–basal polarity due to rearrangement of tight junctions. It seems that both processes are dependent on VEGF-induced small GTPase proteins, whereas lumen expansion involves MMPs activity. Finally, two luminal segments integrate to create anastomosis. The fusion is established by creating VE-cadherin junctions which consolidate connection, and is facilitated by macrophages [86, 98, 99].

The final stage of angiogenesis is frequently termed arteriogenesis. It involves coverage of primary tubes composed of ECs with mural cells—pericytes, which directly envelop endothelium, and vascular smooth muscle cells (VSMC), which are separated from ECs by a layers of ECM. By secreting components of ECM and basal membrane they stabilise nascent vessels and regulate permeability. It is proposed, that while VEGF is necessary for initial formation of vascular tree, vessel maturation is mediated by Ang1 and only cooperation of these factors results in a proper angiogenesis. Once incorporated, pericytes and VSMC mediate quiescence of ECs, as contact with pericytes decreases endothelial proliferation, downregulates responsiveness for VEGF and finally reduces branching and sprouting [86, 88, 100].

1.6 Endothelial Cell Heterogeneity and Function

For a long time endothelium was recognised as a monolayer lining blood and lymphatic vessels, built up from a homogenous population of ECs, resting on a basement membrane and functioning as a static barrier between the circulatory system and adjacent tissue. However, as it is exposed to a variety of tissue environmental cues, affected by different metabolites, growth factors, cytokines, oxygen tension and mechanical forces, not surprisingly it varies much throughout the circulatory system. First of all, ECs of different organs are morphologically distinct and contain specific subcellular structures. Secondly, it is now well established that endothelium forms a metabolically active and dynamic structure of multiply functions—regulating blood flow, vessel permeability, thrombosis and thrombolysis, angiogenesis and immunity. Finally, endothelial heterogeneity has been also described on the level of gene expression and antigen composition [92, 101–104]. Moreover, nowadays even anatomical criteria seems to be not enough for the universal description of endothelium, since the phenomenon of vascular mimicry by other cell types was described [105]. Thus, finding an appropriate definition of this versatile and multifunctional organ remains challenging.

1.6.1 Heterogeneity in Structure of Endothelium

Endothelium has been classically defined as continuous and fenestrated, depending on the presence of specific structures and microdomains: fenestrae (with or without diaphragms), caveolae (with or without stomatal diaphragms) as well as transendothelial channels and vesiculo-vacuolar organelles [106].

The continuous, non-fenestrated endothelium forms an uninterrupted barrier between blood and tissue. It lines large blood and lymph vessels (arteries, arterioles and veins), capillaries (of myocardium, skin, muscles, lungs, kidney, vasa recta and brain) and chambers of the heart. In continuous endothelium there are almost no pores (neither fenestraes nor transendothelial channels), which are abundant in other types of endothelium. Instead, ECs are tightly connected to each other, surrounded by a basement membrane and characterised by numerous caveolae, as well as vesiculo-vacuolar organelles (VVOs). In this manner polarity between luminal and abluminal side of ECs is maintained and transcytosis enabled [101, 106]. Caveolae are recognised as single plasmolemmal vesicles with caveolin-1 as major protein component, whereas VVOs are morphologically defined as chains of interconnected vesicles spanning through cytoplasm of the endothelial cells. Although VVOs were also regarded as clusters of caveolae, they are found in mice lacking caveolin-1, what supports their status as independent and novel structure [106, 107]. Caveolae and VVOs are frequently connected with stomatal diaphragm, which may be found also in the neck of the caveola pore. This is a thin protein barrier, the major component of which is plasmalemma vesicle-associated protein (PV-1) [106].

Fenestrated endothelia are characterised by the presence of fenestrae microdomains, which can be described as pores of regular size (~70 nm) in endothelium, organised in planar clusters. Endothelium with fenestrae, transendothelial channels (pores spanning in ECs from luminal to abluminal side) and few caveolae is characteristic for capillaries of endocrine glands, choroid plexus, digestive tract mucosa and peritubular part of kidney. These structures possess diaphragm containing PV-1 glycoproteins. In this manner selective permeability and transendothelial transport, necessary for efficient absorption and secretion, are accomplished. Similar structures, but without diaphragms, can be found in kidney glomerulus, enabling blood filtration [92, 101, 106].

Finally, discontinuous or sinusoidal endothelium lines sinusoids of liver, spleen and bone marrow. It lacks basement membranes and instead of caveolae it develops clathrin-coated pits to accomplish high level of clathrin-mediated endocytosis. Its fenestrate lack diaphragms, they are of larger diameter (~100–200 nm) and heterogeneous in size, often referred as gaps, enabling cell trafficking. Interestingly, it seems that fenestrae subtended with diaphragms precede glomerular or sinusoidal fenestraes, as they are developed in glomerulus and liver in embryonic life and are replaced by non-diaphragm types later on [101, 106].

1.6.2 Heterogeneity in Expression of Markers and Phenotypes of ECs

The common and major function of vessels lined with endothelium is providing blood and lymph flow. However, the major conducting systems (arteries, veins and lymph vessels) vary a lot in relation to their structure and properties. It is then not surprising, that populations of ECs forming the innermost layer of vasculature are distinct as well, depending on which part of circulatory system they function in.

Arteries have thick wall, able to pulsate and are devoid of valves. Arterial endothelium has reduced permeability in comparison to the venous one due to tighter junctions between ECs and decreased ability to mediate an inflammatory response. There are however areas of arterial tree exposed to disturbed blood flow and thus ECs in these sites are especially designated for activation, inflammation and coagulation, due to expression of genes related to inflammation, oxidation and lipid metabolism [2, 108, 109]. Arterial ECs express high level of actin and deposit a thick layer of extracellular matrix, covered with vascular smooth muscle cells. The main function of arteriolar ECs is to regulate vascular tone and they are longer and narrower, but thicker, than their venous counterparts. They align according to the blood flow and preferentially express Delta-like 4, activin-receptor-like kinase 1, endothelial PAS domain protein, Hey1, Hey2 and neuropilin 1 [2, 108, 110].

Veins are larger in diameter but have thin and more flexible wall that is not pulsating. Their endothelium is more permeable then arterial, with less extracellular matrix, and possess a greater ability to mediate inflammatory response due to expression of selectins and integrins [108, 110]. Postcapillary venules which have prominent transcellular machinery and low level of tight junctions are predominantly well suited for this purpose and they are preferred site for leukocyte trafficking and attachment of platelets. Venous ECs are short and wide, with the exception of ECs of high endothelial vessels in lymph nodes, which have unique tall and cuboidal shape and express specific adhesion molecules for constitutive lymphocytes trafficking between blood and lymph [2, 110]. Genes preferentially expressed in ECs on the venal side include ephrinB4, neuropilin 2, COUP-TFII, vWF and eNOS [2, 101, 111].

Capillaries are the major exchange vessels of circulation system. Due to their low diameter these vessels comprise the majority of the endothelial surface of circulation. To minimise the length of diffusion path, their wall is extraordinary thin and made of flat ECs occasionally covered with pericytes [2]. They express protein components of basement membrane, to which microvascular ECs are tightly associated [108]. Microcirculation presents anti-coagulant phenotype, as synthesises among others tissue factor pathway inhibitor (TFPI) and thrombomodulin, whereas is devoid of tissue factor (TF) [112].

Lymphatic ECs make up a highly permeable structure devoid of basement membrane, where ECs are connected by button-like junctions, which open and close in response to interstitial pressure. Only major lymphatic vessels are supported by smooth muscle cells and pericytes and their endothelium is continuous and expressing higher level of NOS comparing to lymphatic capillaries [108, 113]. Lymphatic ECs synthesise also specifically membrane glycoprotein podoplanin, which level is regulated by another protein characteristic for lymphatic ECs—transcription factor Prox1 [104, 114]. LYVE-1 is an important component of the extracellular matrix in capillary lymphatic endothelium but not in collective lymph ducts, mediating cell migration in tissues [104, 113]. Its expression is not however limited to lymphatic system, as can be also detected in hepatic sinusoidal ECs [113].

Additionally to above differences, ECs significantly vary in structure and gene expression pattern depending on an organ of origin and even vascular bed within a given organ [102, 113]. For instance pulmonary capillary ECs contains large caveolae, while bronchial microvascular endothelium provides increased permeability [2]. Lung endothelial cell adhesion molecule-1 (Lu-ECAM-1) is specific for pulmonary postcapillary ECs, bronchial circulation expresses E-selectin consecutively in contrast to other vascular beds, tPA is the highest in brain and present in bronchial but not in pulmonary vessels, whereas uPA is confined to renal ECs. Hepatic sinusoidal ECs are characterised with high fenestration as well as endocytosis machinery and do not express selectins [2, 92]. Glomerular ECs lack expression of PV-1, which is frequent in other types of endothelia, and actively synthesise glycocalyx and basement membrane instead [2, 104, 106]. In contrast to other vascular beds, glomerular, sinusoidal and lymphatic ECs are not positive for vWF [2, 104]. Endocardial ECs are larger than ECs of other origins, express specific components of gap junctions (e.g. connexin 43) and have elevated level of vWF and eNOS in comparison to myocardial microvessels [2]. Thrombomodulin and P-selectin levels are lowest in endothelium of brain vessels, which are characterised by abundance of tight junctions and small amounts of caveolae [101, 112]. Indeed, there are almost none markers, which could be specific and uniform for ECs. Even the most frequently used platelet endothelial cell adhesion molecule-1 (PECAM-1, also known as CD31), playing role in stabilisation of endothelial monolayer, is also expressed on trophoblast or monocytes, whereas in liver it can be found in periportal vessels, but not in sinusoidal intrahepatic ECs [92, 104]. Furthermore, VE-cadherin despite its seemingly uniform expression seems to be restricted to arteries, arterioles and capillaries, with extremely low expression in venous ECs [101, 115].

1.6.3 Heterogeneity of ECs Function

The location of ECs in the interface between the circulatory system and surrounding tissue gives it an essential role in regulation a diversity of crucial properties of cardiovascular system—vessels permeability, immune response, hemostasis and blood tone. Surprisingly, the way these processes are modulated, strongly depends on the condition of ECs. Quiescent endothelium has anticoagulant, anti-adhesive and vasodilatory function with moderate level of permeability, whereas activated ECs express pro-coagulant, pro-adhesive and vasoconstricting properties, with elevated transmigration of immune cells [105] (Fig. 1.4). Such a broad heterogeneity of function makes ECs unique and indispensable for organisms homeostasis.

The major property of the interface created by ECs between blood and surrounded tissue is semipermeability, enabling transport of fluids, ions, macromolecules and cells in and out of blood. Under constitutive conditions, continuous and physiologically controllable flux takes place in capillaries, the major exchange vessels of circulatory system [101, 107]. After penetrating the extracellular glycocalyx surface, covering ECs, fluids and small solutes move passively via the paracellular route (between ECs), which is regulated by interendothelial junctions [107, 116]. Adhere junctions which are composed of cadherins and catenins, becoming phosphorylated–dephosphorylated in response to changes in Ca^{2+} level, provide vessel permeability. Tight junctions, with occludin and claudin as major protein components, are responsible for maintaining endothelial barrier and apical–basal polarity [107]. Macromolecules use energy dependent on transcellular route (via ECs itself), including VVOs, transcytosis channels and caveolae, that may be either mediated by specific receptors or receptor independent (fluid-phase transcytosis) [101, 107]. The level of permeability depends directly on the presence of junctions, fenestrae and transcytosis machinery. For instance, VEGF stimulation leads to changes in level of PV-1 as well as regulates phosphorylation–dephosphorylation of protein components of junctions and as a consequences—increases permeability of ECs. Similarly, pro-inflammatory mediators such as histamine and thrombin increase permeability by opening adhere junctions [107, 117]. Such inducible inflammatory permeability takes place predominantly in postcapillary venules. This type of blood vessels contains numerous VVOs, low level of tight junction and expresses preferentially receptors for histamine, bradykinin and serotonin, which induce ECs retraction and formation of intercellular gaps. Noteworthy, severe inflammatory state may cause permeability dysfunction, often leading to oedema, in other segments of circulatory system. This can be mediated by TNFα or VEGF which induce gaps and fenestrae formation [101, 117].

By forming a physical barrier for immune cell infiltration, ECs exerts anti-inflammatory function. However, upon viral or bacterial infection, ECs play an

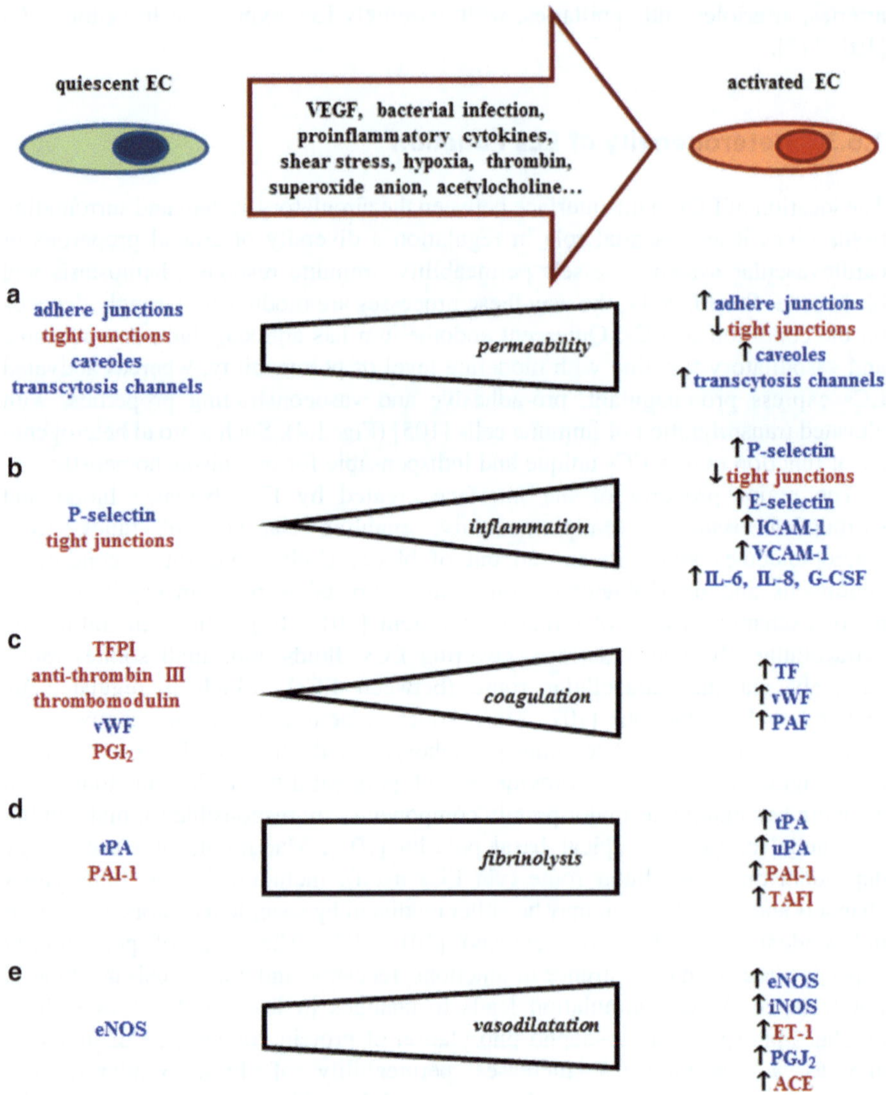

Fig. 1.4 Major functions of ECs in basal conditions and upon activation. Mediators of (**a**) endothelial cell permeability, (**b**) inflammation, (**c**) coagulation, (**d**) fibrinolysis and (**e**) vasodilatation are indicated (in blue – activators, in red – inhibitors)

important role in development both adaptive and innate immune response. Due to expression of MHC I and II, they are able to present antigen to memory T cells and when stimulated with LPS, TNFα or IL-1β, ECs form a passage for leukocytes to recruit immune cells into a tissue, as a first line of innate defence against pathogens. Canonical multistep cascade of events leading to transmigration of immune cells starts with releasing chemoattracting cytokines (IL-6, IL-8, G-CSF). Afterwards, it involves initial attachment and rolling of leukocytes, which is mediated by

P-selectin interaction with oligosaccharide ligands on immune cells. The second important component of immune-cells transmigration, E-selectin, is expressed on ECs only after their activation [92, 101, 118–120]. Subsequently, firm adhesion occurs due to interaction between integrins (LFA-1 and VLA-1) on leukocytes and endothelial adhesion molecules—ICAM-1 and VCAM-1, induced on activated ECs [92, 101, 103, 119]. Finally, leukocytes transmigrate through endothelium either by transcellular passage or by paracellular route owing to disassembly of VE-cadherin complex at tight junctions. Additionally, endothelium forms dome-like membrane structures surrounding passage for leukocytes to limit the loss of barrier function [101, 119]. Although such a close interaction between ECs and immune cells predominantly occurs in postcapillary venules, where shear stress is the lowest, it may also happen in other parts of circulatory system (large veins, arterioles, capillaries) due to changes in adhesion protein and cytokines expression [92, 101, 118–120]. Leukocytes migration in basal conditions, involving L-selectin interaction with its receptor on ECs, takes place in high endothelial venules of secondary lymphoid organs (lymph nodes, appendix, tonsils) [101, 102], owing to the production of potent chemoattractants for T cells and dendritic cells by ECs [118].

Importantly, ECs affect also hemostasis by influencing coagulation cascade, platelet adhesion and promoting limited clot formation in case of haemorrhage. ECs in basal conditions inhibit thrombin—a major pro-coagulant factor, activating platelets and several coagulation enzymes. TFPI synthesised by ECs prevents thrombin formation, while anti-thrombin III bound to heparin sulphate and glycosaminoglycans on endothelium surface inhibits thrombin and counteracts coagulation [92, 103, 112]. Additionally, ECs express predominantly thrombomodulin, which after forming complex with thrombin, reduces its ability to activate platelets, factors V and XIII and to cleave fibrinogen, facilitating instead enzymatic activation of protein C. This in turn induces its anticoagulant properties—ability to inactivate Va and VIIIa factors of coagulation cascade [103, 112]. However, ECs stimulated with thrombin or other factors (e.g. by bacterial infection and pro-inflammatory cytokines, shear stress, hypoxia, thrombin) may be activated to express TF and therefore to convert their phenotype to pro-coagulant type. TF stimulates activation of factors IX and X of coagulation cascade and prothrombinase activity resulting in fibrin formation, thus its expression in ECs occurs after vascular injury. ECs are also major source of vWF, which is normally stored in Weibel–Palade bodies and secreted upon activation, e.g. by thrombin. vWF promotes coagulation by stabilisation of factor VIII and together with ECs-derived platelet activating factor (PAF)—platelet activation and aggregation [92, 103, 112]. Similarly to leukocytes, platelets at first attach and then adhere to ECs to perform clot formation during haemorrhage [92]. However in basal conditions their association to endothelial surface is prevented by negatively charged glycosyaminoglycans and prostacyclin (PGI_2) produced by ECs also inhibits platelet aggregation [111, 112]. Finally, ECs influence the final effect of coagulation process—fibrin and its fibrinolysis. Endothelial-derived urokinase-type plasminogen activator (uPA) is expressed upon activation by proper stimuli (e.g. plasmin), whereas tissue plasminogen activator (tPA) is produced constitutively but is being cleared from circulation unless encounter its target—fibrin. Constitutively, ECs produce also small amount of tPA inhibitor, plasminogen activator inhibitor-1 (PAI-1), and its synthesis is upregulated potently due to thrombin and cytokines [92, 112]. Furthermore, complexes of thrombomodulin and thrombin induce expression of thrombin activatable fibrinolysis inhibitor (TAFI) which cleaves binding sites of plasminogen/plasmin and tPA on fibrin and therefore impairs fibrin degradation [92, 103].

Finally, ECs are potent regulators of vasomotor tone and blood pressure, by realising both vasodilators and vasoconstrictions, that occurs primarily at the level of arterioles [110]. The dilation of blood vessels in response to acetylcholine, leading to reduced blood pressure, was shown to be dependent on integrity of endothelium [111, 121]. NO is a mediator of this effect as its relaxes vascular smooth muscle cells due to activation of cGMP, as well as reduces their migration and proliferation [92, 111, 122]. NO is constitutively produced in ECs from L-arginine, by endothelial nitric oxide synthase (eNOS). However, thrombin, bradykinin, shear stress and acetylcholine stimulate eNOS, while cytokines upregulate inducible NOS (iNOS) expression [92, 111]. Surprisingly, ECs in response to superoxide anion, shear stress, thrombin or hypoxia synthesise also the most potent endogenous vasoconstrictor, endothelin-1 (ET-1), which affects smooth muscle cells by binding specific receptors on their surface [92, 111, 121]. In addition to these major endothelium-derived vasomotor regulators, activated ECs produce also PGI_2, PAF and express angiotensin-converting enzyme (ACE). PGI_2 by direct interaction with receptors on vascular smooth muscle cells induces its relaxation, while PAF may cause either vasodilating or vasoconstricting effects, depending on a concentration, time and vascular bed examined [92, 111, 121]. The product of ACE action, angiotensin II, is in turn vasoconstriction agent [111].

Concluding Remarks

Endothelium is a multifunctional, widespread and dynamic organ, influencing the function of the whole organism. Therefore, a clear understanding of the molecular mechanisms regulating endothelial differentiation, function and physiology is a great challenge in the light of pro- and anti-angiogenic therapeutic strategies. Particular attention is nowadays paid to application of precursor cells capable of stimulation of neovascularisation in pathological conditions, such as myocardial infarction, heart failure, impaired wound healing or peripheral artery disease. No less important is gene therapy targeted on endothelial cell response or application of vascularisation promoting/inhibiting compounds and macromolecules.

Acknowledgments The Faculty of Biochemistry, Biophysics and Biotechnology of the Jagiellonian University is a beneficiary of the structural funds from the European Union and the Polish Ministry of Science and Higher Education (grants no: POIG.02.01.00-12 064/08, POIG 01.01.02-00-109/09, POIG.02.02.00-014/08 and 01.01.02-00-069/09).

References

1. Fishman AP (1982) Endothelium: a distributed organ of diverse capabilities. Ann N Y Acad Sci 401:1–8
2. Aird WC (2007) Phenotypic heterogeneity of the endothelium: I. Structure, function, and mechanisms. Circ Res 100(2):158–173
3. DeWitt N (2005) Angiogenesis. Nature 438(931)
4. Augustin HG, Kozian DH, Johnson RC (1994) Differentiation of endothelial cells: analysis of the constitutive and activated endothelial cell phenotypes. Bioessays 16(12):901–906
5. McGrath KE, Koniski AD, Malik J, Palis J (2003) Circulation is established in a stepwise pattern in the mammalian embryo. Blood 101(5):1669–1676
6. Risau W, Sariola H, Zerwes HG, Sasse J, Ekblom P, Kemler R, Doetschman T (1988) Vasculogenesis and angiogenesis in embryonic-stem-cell-derived embryoid bodies. Development 102(3):471–478
7. Weinstein BM (1999) What guides early embryonic blood vessel formation? Dev Dyn 215 (1):2–11
8. Cox CM, Poole TJ (2000) Angioblast differentiation is influenced by the local environment: FGF-2 induces angioblasts and patterns vessel formation in the quail embryo. Dev Dyn 218 (2):371–382
9. Pardanaud L, Dieterlen-Lievre F (1999) Manipulation of the angiopoietic/hemangiopoietic commitment in the avian embryo. Development 126(4):617–627
10. Leconte I, Fox JC, Baldwin HS, Buck CA, Swain JL (1998) Adenoviral-mediated expression of antisense RNA to fibroblast growth factors disrupts murine vascular development. Dev Dyn 213(4):421–430
11. Lee SH, Schloss DJ, Swain JL (2000) Maintenance of vascular integrity in the embryo requires signaling through the fibroblast growth factor receptor. J Biol Chem 275(43): 33679–33687
12. Kanno S, Oda N, Abe M, Terai Y, Ito M, Shitara K, Tabayashi K, Shibuya M, Sato Y (2000) Roles of two VEGF receptors, Flt-1 and KDR, in the signal transduction of VEGF effects in human vascular endothelial cells. Oncogene 19(17):2138–2146
13. Millauer B, Wizigmann-Voos S, Schnurch H, Martinez R, Moller NP, Risau W, Ullrich A (1993) High affinity VEGF binding and developmental expression suggest Flk-1 as a major regulator of vasculogenesis and angiogenesis. Cell 72(6):835–846
14. Carmeliet P, Ferreira V, Breier G, Pollefeyt S, Kieckens L, Gertsenstein M, Fahrig M, Vandenhoeck A, Harpal K, Eberhardt C, Declercq C, Pawling J, Moons L, Collen D, Risau W, Nagy A (1996) Abnormal blood vessel development and lethality in embryos lacking a single VEGF allele. Nature 380(6573):435–439
15. Ferrara N, Carver-Moore K, Chen H, Dowd M, Lu L, O'Shea KS, Powell-Braxton L, Hillan KJ, Moore MW (1996) Heterozygous embryonic lethality induced by targeted inactivation of the VEGF gene. Nature 380(6573):439–442
16. Shalaby F, Rossant J, Yamaguchi TP, Gertsenstein M, Wu XF, Breitman ML, Schuh AC (1995) Failure of blood-island formation and vasculogenesis in Flk-1-deficient mice. Nature 376(6535):62–66
17. Kearney JB, Ambler CA, Monaco KA, Johnson N, Rapoport RG, Bautch VL (2002) Vascular endothelial growth factor receptor Flt-1 negatively regulates developmental blood vessel formation by modulating endothelial cell division. Blood 99(7):2397–2407
18. Suri C, Jones PF, Patan S, Bartunkova S, Maisonpierre PC, Davis S, Sato TN, Yancopoulos GD (1996) Requisite role of angiopoietin-1, a ligand for the TIE2 receptor, during embryonic angiogenesis. Cell 87(7):1171–1180
19. Vikkula M, Boon LM, Carraway KL III, Calvert JT, Diamonti AJ, Goumnerov B, Pasyk KA, Marchuk DA, Warman ML, Cantley LC, Mulliken JB, Olsen BR (1996) Vascular dysmorphogenesis caused by an activating mutation in the receptor tyrosine kinase TIE2. Cell 87(7):1181–1190

20. Sabin FR (1920) Studies on the origin of blood vessels and of red corpuscules as seen in the living blastoderm of the chick during the second day of incubation: contributions to embryology. Contrib Embryol 9:213–262
21. Murray PDF (1932) The development in vitro of the blood of the early chick embryo. Proc R Soc Lond 11:497–521
22. Hirschi KK (2012) Hemogenic endothelium during development and beyond. Blood 119(21): 4823–4827
23. Kennedy M, Firpo M, Choi K, Wall C, Robertson S, Kabrun N, Keller G (1997) A common precursor for primitive erythropoiesis and definitive haematopoiesis. Nature 386(6624): 488–493
24. Choi K, Kennedy M, Kazarov A, Papadimitriou JC, Keller G (1998) A common precursor for hematopoietic and endothelial cells. Development 125(4):725–732
25. Nishikawa SI, Nishikawa S, Hirashima M, Matsuyoshi N, Kodama H (1998) Progressive lineage analysis by cell sorting and culture identifies FLK1 + VE-cadherin + cells at a diverging point of endothelial and hemopoietic lineages. Development 125(9):1747–1757
26. Huber TL, Kouskoff V, Fehling HJ, Palis J, Keller G (2004) Haemangioblast commitment is initiated in the primitive streak of the mouse embryo. Nature 432(7017):625–630
27. Vogeli KM, Jin SW, Martin GR, Stainier DY (2006) A common progenitor for haematopoietic and endothelial lineages in the zebrafish gastrula. Nature 443(7109):337–339
28. Kinder SJ, Tsang TE, Quinlan GA, Hadjantonakis AK, Nagy A, Tam PP (1999) The orderly allocation of mesodermal cells to the extraembryonic structures and the anteroposterior axis during gastrulation of the mouse embryo. Development 126(21):4691–4701
29. Furuta C, Ema H, Takayanagi S, Ogaeri T, Okamura D, Matsui Y, Nakauchi H (2006) Discordant developmental waves of angioblasts and hemangioblasts in the early gastrulating mouse embryo. Development 133(14):2771–2779
30. Fehling HJ, Lacaud G, Kubo A, Kennedy M, Robertson S, Keller G, Kouskoff V (2003) Tracking mesoderm induction and its specification to the hemangioblast during embryonic stem cell differentiation. Development 130(17):4217–4227
31. Motoike T, Markham DW, Rossant J, Sato TN (2003) Evidence for novel fate of Flk1+ progenitor: contribution to muscle lineage. Genesis 35(3):153–159
32. Eichmann A, Corbel C, Nataf V, Vaigot P, Breant C, Le Douarin NM (1997) Ligand-dependent development of the endothelial and hemopoietic lineages from embryonic mesodermal cells expressing vascular endothelial growth factor receptor 2. Proc Natl Acad Sci U S A 94(10):5141–5146
33. Wu Y, Moser M, Bautch VL, Patterson C (2003) HoxB5 is an upstream transcriptional switch for differentiation of the vascular endothelium from precursor cells. Mol Cell Biol 23(16): 5680–5691
34. Park C, Afrikanova I, Chung YS, Zhang WJ, Arentson E, Fong Gh G, Rosendahl A, Choi K (2004) A hierarchical order of factors in the generation of FLK1- and SCL-expressing hematopoietic and endothelial progenitors from embryonic stem cells. Development 131 (11):2749–2762
35. Lugus JJ, Chung YS, Mills JC, Kim SI, Grass J, Kyba M, Doherty JM, Bresnick EH, Choi K (2007) GATA2 functions at multiple steps in hemangioblast development and differentiation. Development 134(2):393–405
36. Pelosi E, Valtieri M, Coppola S, Botta R, Gabbianelli M, Lulli V, Marziali G, Masella B, Muller R, Sgadari C, Testa U, Bonanno G, Peschle C (2002) Identification of the hemangioblast in postnatal life. Blood 100(9):3203–3208
37. Gunsilius E, Duba HC, Petzer AL, Kahler CM, Grunewald K, Stockhammer G, Gabl C, Dirnhofer S, Clausen J, Gastl G (2000) Evidence from a leukaemia model for maintenance of vascular endothelium by bone-marrow-derived endothelial cells. Lancet 355(9216): 1688–1691
38. Lugus JJ, Park C, Ma YD, Choi K (2009) Both primitive and definitive blood cells are derived from Flk-1+ mesoderm. Blood 113(3):563–566

39. de Bruijn MF, Speck NA, Peeters MC, Dzierzak E (2000) Definitive hematopoietic stem cells first develop within the major arterial regions of the mouse embryo. EMBO J 19(11): 2465–2474
40. Tavian M, Coulombel L, Luton D, Clemente HS, Dieterlen-Lievre F, Peault B (1996) Aorta-associated CD34+ hematopoietic cells in the early human embryo. Blood 87(1):67–72
41. Marshall CJ, Thrasher AJ (2001) The embryonic origins of human haematopoiesis. Br J Haematol 112(4):838–850
42. Tavian M, Cortes F, Charbord P, Labastie MC, Peault B (1999) Emergence of the haematopoietic system in the human embryo and foetus. Haematologica 84(Suppl EHA-4):1–3
43. Jaffredo T, Gautier R, Eichmann A, Dieterlen-Lievre F (1998) Intraaortic hemopoietic cells are derived from endothelial cells during ontogeny. Development 125(22):4575–4583
44. Lancrin C, Sroczynska P, Stephenson C, Allen T, Kouskoff V, Lacaud G (2009) The haemangioblast generates haematopoietic cells through a haemogenic endothelium stage. Nature 457(7231):892–895
45. Oberlin E, Tavian M, Blazsek I, Peault B (2002) Blood-forming potential of vascular endothelium in the human embryo. Development 129(17):4147–4157
46. Eilken HM, Nishikawa S, Schroeder T (2009) Continuous single-cell imaging of blood generation from haemogenic endothelium. Nature 457(7231):896–900
47. Chen MJ, Yokomizo T, Zeigler BM, Dzierzak E, Speck NA (2009) Runx1 is required for the endothelial to haematopoietic cell transition but not thereafter. Nature 457(7231):887–891
48. Zovein AC, Hofmann JJ, Lynch M, French WJ, Turlo KA, Yang Y, Becker MS, Zanetta L, Dejana E, Gasson JC, Tallquist MD, Iruela-Arispe ML (2008) Fate tracing reveals the endothelial origin of hematopoietic stem cells. Cell Stem Cell 3(6):625–636
49. Sroczynska P, Lancrin C, Kouskoff V, Lacaud G (2009) The differential activities of Runx1 promoters define milestones during embryonic hematopoiesis. Blood 114(26):5279–5289
50. Yokomizo T, Ogawa M, Osato M, Kanno T, Yoshida H, Fujimoto T, Fraser S, Nishikawa S, Okada H, Satake M, Noda T, Ito Y (2001) Requirement of Runx1/AML1/PEBP2alphaB for the generation of haematopoietic cells from endothelial cells. Genes Cells 6(1):13–23
51. Mukouyama Y, Chiba N, Hara T, Okada H, Ito Y, Kanamaru R, Miyajima A, Satake M, Watanabe T (2000) The AML1 transcription factor functions to develop and maintain hematogenic precursor cells in the embryonic aorta-gonad-mesonephros region. Dev Biol 220(1):27–36
52. North TE, de Bruijn MF, Stacy T, Talebian L, Lind E, Robin C, Binder M, Dzierzak E, Speck NA (2002) Runx1 expression marks long-term repopulating hematopoietic stem cells in the midgestation mouse embryo. Immunity 16(5):661–672
53. Nadin BM, Goodell MA, Hirschi KK (2003) Phenotype and hematopoietic potential of side population cells throughout embryonic development. Blood 102(7):2436–2443
54. Goldie LC, Lucitti JL, Dickinson ME, Hirschi KK (2008) Cell signaling directing the formation and function of hemogenic endothelium during murine embryogenesis. Blood 112(8):3194–3204
55. Pearson JD (2010) Endothelial progenitor cells—an evolving story. Microvasc Res 79(3): 162–168
56. Grochot-Przeczek A, Dulak J, Jozkowicz A (2013) Therapeutic angiogenesis for revascularization in peripheral artery disease. Gene 525(2):220–228
57. Asahara T, Murohara T, Sullivan A, Silver M, van der Zee R, Li T, Witzenbichler B, Schatteman G, Isner JM (1997) Isolation of putative progenitor endothelial cells for angiogenesis. Science 275(5302):964–967
58. Stump MM, Jordan GL Jr, Debakey ME, Halpert B (1963) Endothelium grown from circulating blood on isolated intravascular Dacron hub. Am J Pathol 43:361–367
59. Shi Q, Wu MH, Hayashida N, Wechezak AR, Clowes AW, Sauvage LR (1994) Proof of fallout endothelialization of impervious Dacron grafts in the aorta and inferior vena cava of the dog. J Vasc Surg 20(4):546–556, discussion 556–547

60. Shi Q, Rafii S, Wu MH, Wijelath ES, Yu C, Ishida A, Fujita Y, Kothari S, Mohle R, Sauvage LR, Moore MA, Storb RF, Hammond WP (1998) Evidence for circulating bone marrow-derived endothelial cells. Blood 92(2):362–367
61. Asahara T, Masuda H, Takahashi T, Kalka C, Pastore C, Silver M, Kearne M, Magner M, Isner JM (1999) Bone marrow origin of endothelial progenitor cells responsible for postnatal vasculogenesis in physiological and pathological neovascularization. Circ Res 85 (3):221–228
62. Asahara T, Takahashi T, Masuda H, Kalka C, Chen D, Iwaguro H, Inai Y, Silver M, Isner JM (1999) VEGF contributes to postnatal neovascularization by mobilizing bone marrow-derived endothelial progenitor cells. EMBO J 18(14):3964–3972
63. Takahashi T, Kalka C, Masuda H, Chen D, Silver M, Kearney M, Magner M, Isner JM, Asahara T (1999) Ischemia- and cytokine-induced mobilization of bone marrow-derived endothelial progenitor cells for neovascularization. Nat Med 5(4):434–438
64. Pearson JD (2009) Endothelial progenitor cells – hype or hope? J Thromb Haemost 7(2): 255–262
65. Loomans CJ, Wan H, de Crom R, van Haperen R, de Boer HC, Leenen PJ, Drexhage HA, Rabelink TJ, van Zonneveld AJ, Staal FJ (2006) Angiogenic murine endothelial progenitor cells are derived from a myeloid bone marrow fraction and can be identified by endothelial NO synthase expression. Arterioscler Thromb Vasc Biol 26(8):1760–1767
66. Hristov M, Erl W, Weber PC (2003) Endothelial progenitor cells: mobilization, differentiation, and homing. Arterioscler Thromb Vasc Biol 23(7):1185–1189
67. Hur J, Yoon CH, Kim HS, Choi JH, Kang HJ, Hwang KK, Oh BH, Lee MM, Park YB (2004) Characterization of two types of endothelial progenitor cells and their different contributions to neovasculogenesis. Arterioscler Thromb Vasc Biol 24(2):288–293
68. Yoder MC, Mead LE, Prater D, Krier TR, Mroueh KN, Li F, Krasich R, Temm CJ, Prchal JT, Ingram DA (2007) Redefining endothelial progenitor cells via clonal analysis and hematopoietic stem/progenitor cell principals. Blood 109(5):1801–1809
69. Rohde E, Malischnik C, Thaler D, Maierhofer T, Linkesch W, Lanzer G, Guelly C, Strunk D (2006) Blood monocytes mimic endothelial progenitor cells. Stem Cells 24(2):357–367
70. Kim SJ, Kim JS, Papadopoulos J, Wook Kim S, Maya M, Zhang F, He J, Fan D, Langley R, Fidler IJ (2009) Circulating monocytes expressing CD31: implications for acute and chronic angiogenesis. Am J Pathol 174(5):1972–1980
71. Prokopi M, Pula G, Mayr U, Devue C, Gallagher J, Xiao Q, Boulanger CM, Westwood N, Urbich C, Willeit J, Steiner M, Breuss J, Xu Q, Kiechl S, Mayr M (2009) Proteomic analysis reveals presence of platelet microparticles in endothelial progenitor cell cultures. Blood 114 (3):723–732
72. Raemer PC, Haemmerling S, Giese T, Canaday DH, Katus HA, Dengler TJ, Sivanandam VG (2009) Endothelial progenitor cells possess monocyte-like antigen-presenting and T-cell-co-stimulatory capacity. Transplantation 87(3):340–349
73. Harvey W (1957) Movement of the heart and blood in animals: an anatomical essay. Blackwell Scientific, Oxford, UK
74. Lawson ND, Weinstein BM (2002) Arteries and veins: making a difference with zebrafish. Nat Rev Genet 3(9):674–682
75. Thompson MA, Ransom DG, Pratt SJ, MacLennan H, Kieran MW, Detrich HW III, Vail B, Huber TL, Paw B, Brownlie AJ, Oates AC, Fritz A, Gates MA, Amores A, Bahary N, Talbot WS, Her H, Beier DR, Postlethwait JH, Zon LI (1998) The cloche and spadetail genes differentially affect hematopoiesis and vasculogenesis. Dev Biol 197(2):248–269
76. You LR, Lin FJ, Lee CT, DeMayo FJ, Tsai MJ, Tsai SY (2005) Suppression of Notch signalling by the COUP-TFII transcription factor regulates vein identity. Nature 435(7038): 98–104
77. Odenthal J, Haffter P, Vogelsang E, Brand M, van Eeden FJ, Furutani-Seiki M, Granato M, Hammerschmidt M, Heisenberg CP, Jiang YJ, Kane DA, Kelsh RN, Mullins MC, Warga RM,

Allende ML, Weinberg ES, Nusslein-Volhard C (1996) Mutations affecting the formation of the notochord in the zebrafish, *Danio rerio*. Development 123:103–115
78. Fan CM, Tessier-Lavigne M (1994) Patterning of mammalian somites by surface ectoderm and notochord: evidence for sclerotome induction by a hedgehog homolog. Cell 79(7): 1175–1186
79. Ericson J, Morton S, Kawakami A, Roelink H, Jessell TM (1996) Two critical periods of Sonic Hedgehog signaling required for the specification of motor neuron identity. Cell 87(4): 661–673
80. Schauerte HE, van Eeden FJ, Fricke C, Odenthal J, Strahle U, Haffter P (1998) Sonic hedgehog is not required for the induction of medial floor plate cells in the zebrafish. Development 125(15):2983–2993
81. Lawson ND, Vogel AM, Weinstein BM (2002) sonic hedgehog and vascular endothelial growth factor act upstream of the Notch pathway during arterial endothelial differentiation. Dev Cell 3(1):127–136
82. Lawson ND, Scheer N, Pham VN, Kim CH, Chitnis AB, Campos-Ortega JA, Weinstein BM (2001) Notch signaling is required for arterial-venous differentiation during embryonic vascular development. Development 128(19):3675–3683
83. Zhong TP, Childs S, Leu JP, Fishman MC (2001) Gridlock signalling pathway fashions the first embryonic artery. Nature 414(6860):216–220
84. Moyon D, Pardanaud L, Yuan L, Breant C, Eichmann A (2001) Plasticity of endothelial cells during arterial-venous differentiation in the avian embryo. Development 128(17):3359–3370
85. Albrecht I, Christofori G (2011) Molecular mechanisms of lymphangiogenesis in development and cancer. Int J Dev Biol 55(4–5):483–494
86. Potente M, Gerhardt H, Carmeliet P (2011) Basic and therapeutic aspects of angiogenesis. Cell 146(6):873–887
87. Folkman J (2007) Angiogenesis: an organizing principle for drug discovery? Nat Rev 6(4): 273–286
88. Fagiani E, Christofori G (2013) Angiopoietins in angiogenesis. Cancer Lett 328(1):18–26
89. Rey S, Semenza GL (2010) Hypoxia-inducible factor-1-dependent mechanisms of vascularization and vascular remodelling. Cardiovasc Res 86(2):236–242
90. Semenza GL (2007) Vasculogenesis, angiogenesis, and arteriogenesis: mechanisms of blood vessel formation and remodeling. J Cell Biochem 102(4):840–847
91. Al Haj Zen A, Madeddu P (2009) Notch signalling in ischaemia-induced angiogenesis. Biochem Soc Trans 37(Pt 6):1221–1227
92. Cines DB, Pollak ES, Buck CA, Loscalzo J, Zimmerman GA, McEver RP, Pober JS, Wick TM, Konkle BA, Schwartz BS, Barnathan ES, McCrae KR, Hug BA, Schmidt AM, Stern DM (1998) Endothelial cells in physiology and in the pathophysiology of vascular disorders. Blood 91(10):3527–3561
93. Ferrara N (2009) VEGF-A: a critical regulator of blood vessel growth. Eur Cytokine Netw 20 (4):158–163
94. Jakobsson L, Bentley K, Gerhardt H (2009) VEGFRs and Notch: a dynamic collaboration in vascular patterning. Biochem Soc Trans 37(Pt 6):1233–1236
95. Ucuzian AA, Gassman AA, East AT, Greisler HP (2010) Molecular mediators of angiogenesis. J Burn Care Res 31(1):158–175
96. Davis GE, Senger DR (2008) Extracellular matrix mediates a molecular balance between vascular morphogenesis and regression. Curr Opin Hematol 15(3):197–203
97. Lamalice L, Le Boeuf F, Huot J (2007) Endothelial cell migration during angiogenesis. Circ Res 100(6):782–794
98. Davis GE, Stratman AN, Sacharidou A, Koh W (2011) Molecular basis for endothelial lumen formation and tubulogenesis during vasculogenesis and angiogenic sprouting. Int Rev Cell Mol Biol 288:101–165
99. Domigan CK, Iruela-Arispe ML (2012) Recent advances in vascular development. Curr Opin Hematol 19(3):176–183

100. Ribatti D, Nico B, Crivellato E (2011) The role of pericytes in angiogenesis. Int J Dev Biol 55 (3):261–268
101. Aird WC (2007) Phenotypic heterogeneity of the endothelium: II. Representative vascular beds. Circ Res 100(2):174–190
102. Aird WC (2012) Endothelial cell heterogeneity. Cold Spring Harb Perspect Med 2(1): a006429
103. Sumpio BE, Riley JT, Dardik A (2002) Cells in focus: endothelial cell. Int J Biochem Cell Biol 34(12):1508–1512
104. Ordonez NG (2012) Immunohistochemical endothelial markers: a review. Adv Anat Pathol 19(5):281–295
105. Aird WC (2008) Endothelium in health and disease. Pharmacol Rep 60(1):139–143
106. Tse D, Stan RV (2010) Morphological heterogeneity of endothelium. Semin Thromb Hemost 36(3):236–245
107. Chavez A, Smith M, Mehta D (2011) New insights into the regulation of vascular permeability. Int Rev Cell Mol Biol 290:205–248
108. Dyer LA, Patterson C (2010) Development of the endothelium: an emphasis on heterogeneity. Semin Thromb Hemost 36(3):227–235
109. Davies PF, Civelek M, Fang Y, Guerraty MA, Passerini AG (2010) Endothelial heterogeneity associated with regional athero-susceptibility and adaptation to disturbed blood flow in vivo. Semin Thromb Hemost 36(3):265–275
110. dela Paz NG, D'Amore PA (2009) Arterial versus venous endothelial cells. Cell Tissue Res 335(1):5–16
111. Triggle CR, Samuel SM, Ravishankar S, Marei I, Arunachalam G, Ding H (2012) The endothelium: influencing vascular smooth muscle in many ways. Can J Physiol Pharmacol 90(6):713–738
112. van Hinsbergh VW (2011) Endothelium – role in regulation of coagulation and inflammation. Semin Immunopathol 34(1):93–106
113. Lee S, Choi I, Hong YK (2010) Heterogeneity and plasticity of lymphatic endothelial cells. Semin Thromb Hemost 36(3):352–361
114. Kume T (2010) Specification of arterial, venous, and lymphatic endothelial cells during embryonic development. Histol Histopathol 25(5):637–646
115. Quadri SK (2012) Cross talk between focal adhesion kinase and cadherins: role in regulating endothelial barrier function. Microvasc Res 83(1):3–11
116. Curry FE, Adamson RH (2012) Endothelial glycocalyx: permeability barrier and mechanosensor. Ann Biomed Eng 40(4):828–839
117. Bates DO (2010) Vascular endothelial growth factors and vascular permeability. Cardiovasc Res 87(2):262–271
118. Young MR (2012) Endothelial cells in the eyes of an immunologist. Cancer Immunol Immunother 61(10):1609–1616
119. Williams MR, Azcutia V, Newton G, Alcaide P, Luscinskas FW (2011) Emerging mechanisms of neutrophil recruitment across endothelium. Trends Immunol 32(10):461–469
120. Molema G (2010) Heterogeneity in endothelial responsiveness to cytokines, molecular causes, and pharmacological consequences. Semin Thromb Hemost 36(3):246–264
121. Barton M (2011) The discovery of endothelium-dependent contraction: the legacy of Paul M. Vanhoutte. Pharmacol Res 63(6):455–462
122. Schwartz BG, Economides C, Mayeda GS, Burstein S, Kloner RA (2010) The endothelial cell in health and disease: its function, dysfunction, measurement and therapy. Int J Impot Res 22 (2):77–90

Developmental and Pathological Lymphangiogenesis

2

Angelika Chachaj and Andrzej Szuba

Contents

2.1 Developmental and Physiological Lymphangiogenesis 28
2.2 Molecular Mechanisms of Lymphangiogenesis 30
2.2.1 Commitment to Lymphatic Endothelium Cell Identity 30
2.2.2 Lymphatic Endothelial Cell Proliferation and Migration 37
2.2.3 Lymphovascular Separation 39
2.2.4 Remodeling and Maturation of Primitive Lymphatic Vessels 40
2.2.5 Interactions Between LECs and Surrounding Extracellular Matrix 41
2.2.6 Postnatal Maturation of Lymphatic Vessels 43
2.3 Lymphangiogenesis in Human Diseases 43
2.3.1 Inflammatory Lymphangiogenesis 43
2.3.2 Lymphangiogenesis in Cancer 46
2.3.3 Lymphangiogenesis in Hypertension 49
2.4 Therapeutic Modulation of Lymphangiogenesis 49
2.4.1 Pro- and Anti-lymphangiogenic Therapy in Inflammatory Disorders 51
2.4.2 Anti-lymphangiogenic Therapy in Graft Rejection 52
2.4.3 Pro-lymphangiogenic Therapy in Secondary Lymphedema 52
2.4.4 Anti-lymphangiogenic Therapy in Cancer 53
References 54

A. Chachaj
Department of Internal Medicine, Wrocław Medical University, Pasteura 1 street, 50-367 Wrocław, Poland

A. Szuba (✉)
Department of Internal Medicine, 4th Military Hospital in Wrocław, Weigla 5 street, 50-981 Wrocław, Poland

Wrocław Medical University, Pasteura 1 street, 50-367 Wrocław, Poland
e-mail: szubaa@yahoo.com

J. Dulak et al. (eds.), *Angiogenesis and Vascularisation*,
DOI 10.1007/978-3-7091-1428-5_2, © Springer-Verlag Wien 2013

Abstract

The past two decades have significantly improved our understanding of the mechanisms of lymphangiogenesis. Identification of lymphatic endothelial specific immunohistochemical markers and development of new experimental animal models have been instrumental in the identification of a number of molecular players regulating growth and remodeling of the lymphatic vasculature during embryonic development. Studies with different models of genetically deficient mice identified also a spectrum of primary lymphedema phenotypes. Recent findings have highlighted the role of lymphangiogenesis in various pathological conditions. It was clearly demonstrated that lymph vessels are active participants in acute and chronic inflammation, organ transplant rejection, metastatic tumor dissemination, and hypertension. Therefore, the specific targeting of the lymphatic vasculature could be a promising approach for pro- or anti-lymphangiogenic treatment in inflammatory disorders, graft rejection, lymphedema, and cancer.

Keywords

Lymphatic endothelial cells • VEGF-A • VEGF-C • VEGF-D • VEGFR-3 • Inflammation • Tumor • Metastasis

2.1 Developmental and Physiological Lymphangiogenesis

Lymphangiogenesis means the growth and formation of new lymphatic vessels. It is an important biological process which is essential in both, physiological and in pathological processes. In adults, physiological lymphangiogenesis occurs only in corpus luteum formation during luteal phase of the menstrual cycle and in the endometrium during pregnancy [1, 2]. Except for these situations, under normal physiological conditions and similarly to blood system, lymphatic vessels remain in inactive state. Pathological lymphangiogenesis, defined as the abnormal formation of new lymphatic vessels from preexisting ones, is known to occur upon inflammatory and tumor-related stimulations. It occurs in processes of wound healing, tissue and organ regeneration, in acute and chronic inflammation, autoimmunity, allograft rejection, and tumor growth and tumor metastasis [1, 3, 4].

During embryogenesis lymphatic vascular system starts to develop significantly later than the blood vascular system. It arises when the embryonic cardiovascular system is already established and functional. In mice, blood vessels start to develop at embryonic day 7.5 (E 7.5) [5], whereas molecular markers of lymphatics can be seen in the jugular region at approximately E 9.5 [6]. In humans embryos, lymphatic vessels development starts at about gestation week 6–7, and this is 3–4 weeks after the development of the first blood vessels [7]. At that time, distinct subpopulations of endothelial cells in the lateral parts of the anterior cardinal primitive veins commit to the lymphatic lineage and then sprout laterally to form the first morphological structures of developing lymphatics—primary lymph sacs [1, 6]. Research on human embryos revealed that there are six sacs in humans and eight sacs in mice [8, 9]. In human embryos, the lymph sacs consist of paired jugulo-axillary lymph sacs, located in the area of the junction of the subclavian vein and the internal jugular vein, paired lumbo-iliac lymph sacs, which give the origin of the lymph vessels of the inferior part of the body, the single retroperitoneal lymph sac, and the single cisterna chyli [10, 11].

The peripheral lymphatic vessels grow from the primary lymph sacs by endothelial budding and then by sprouting and forming capillaries surrounding tissues and organs, except for the central nervous system, bone marrow, and retina, which remain free of lymphatics [12, 13]. The lymphatic vasculature growth resembles that of blood vascular system development—new lymphatic vessels integrate into primitive lymphatic capillary network, which is followed by remodeling and maturation to form an established lymphatic system.

Lymphatic vasculature is formed by lymphatic endothelial cells (LECs), which are highly related to blood endothelial cells (BECs). The origin of LECs during embryogenesis has been controversial and the object of debate for the last century. Recent experiments using genetically engineered mouse models and in vivo fluorescent, dynamic imaging of embryonic lymphangiogenesis in zebrafish strongly confirmed the venous origin of mammalian lymphatic vasculature [6, 14–17]. This is in accordance to "centrifugal" theory, first proposed by Florence Sabin (1902), the American anatomist, who postulated that lymph sacs are venous derivatives [12]. The alternative, "centripetal" theory, first postulated by Huntington and McClure (1910), proposes that primary lymphatic endothelial cells develop directly from mesenchymal progenitor cells, which are called lymphangioblasts [18].

Even though the venous origin of mammalian lymphatic vasculature is now established, other mechanisms of LECs differentiation are still possible. It was showed that lymphatic vascular system of birds has dual origin—from embryonic veins and mesenchymal lymphoblasts. The deep parts of the jugular lymph sacs are derived from the jugular segments of the cardinal veins and the superficial/dermal lymphatics are derived from local lymphangioblasts. The two compartments fuse then to form the whole lymphatic system [19]. Lymphangioblasts have also been revealed in murine embryos in which scattered mesenchymal cells expressed both lymphoendothelial (e.g., LYVE-1—lymphatic vessel endothelial hyaluronan receptor 1, PROX1—Prospero homeobox protein 1) and leukocyte (e.g., CD45) markers. These cells seem to form peripheral parts of the lymphatic vessels during development [20, 21]. At present, it is not known whether similar resident mesenchymal precursors contribute to the lymphatic vasculature during normal development in humans. However, it was demonstrated that circulating bone marrow-derived endothelial precursor cells may be the source of LECs during pathological lymphangiogenesis. These cells express both myeloid progenitor markers (e.g., CD68, CD11b) and LECs markers (e.g., LYVE-1, PROX1, podoplanin) and are capable of undergoing lymphatic differentiation upon inflammatory stimulation. Although LECs can be derived from several progenitor types, CD11b+ monocytes have been identified as the main source [22]. Transdifferentiation of macrophages into LECs was observed in vivo [23] and has been described in human transplanted kidney [24], in a mouse corneal injury model [25], and in peritumoral lymphatic vessels of fibrosarcoma in irradiated mice [24]. Nevertheless, incorporation of bone marrow-derived progenitor cells into the endothelium of newly formed lymphatic vessels seems to be at the level of few percent of all LECs [26]. Growing body of evidence suggest that they rather support growth of lymphatic vessels via locally secreting pro-angiogenic activities than by transforming into LECs [27].

2.2 Molecular Mechanisms of Lymphangiogenesis

Our knowledge of the molecular mechanisms of lymphangiogenesis has improved significantly in recent years, mostly thanks to progress in the identification of a number of regulators and markers specific to the crucial stages of lymphatic vascular development (Table 2.1).

Studies with various models of genetically deficient mice identified also a spectrum of lymphedema phenotypes, ranging from edematous embryos with chylous ascites and severe vascular defects associated with perinatal lethality to more subtle manifestations of lymphatic function defects [28].

The crucial steps of mammalian developmental lymphangiogenesis are presented in Fig. 2.1 and are described below.

2.2.1 Commitment to Lymphatic Endothelium Cell Identity

Shortly after the separation of arteries and veins, at E 9.0 in mice and gestation week 6 in humans, distinct population of VEGFR-3+ cells of the anterior cardinal vein start to lose blood endothelial characteristics and acquire LECs identity. The first indicator of lymphatic endothelial competence among subset of VEGFR-3+ endothelial cells in the cardinal vein is an integral membrane glycoprotein—LYVE-1 [17]. Physiological function of LYVE-1, especially during embryogenesis, has remained unclear. It has been suggested that LYVE-1 is involved in both cell adhesion/transmigration and as a scavenger for hyaluronan turnover [29]. However, LYVE-1 has been found not to be obligatory for normal lymphatic development and function. *Lyve-1*-knockout mice develop normal lymphatic system [30].

In contrast to LYVE-1, homeobox transcription factor PROX1 is essential in lymphangiogenesis [6]. It regulates expression of the lymphatic phenotype genes of *Lyve-1* and *podoplanin* in VEGFR-3+/LYVE-1+ cells. PROX1 expression in the VEGFR-3+/LYVE-1+ cells of the cardinal vein is followed by the budding and sprouting migration in a polarized manner, and finally formation of primary lymphatic sacs [6, 17]. *Prox1*-knockout mice embryos fail to develop primary lymphatic sacs and consequently lymphatic vessels. They develop edema and die before birth at E13.5 [31].

The regulation of *Prox1* in differentiating LECs is currently poorly understood. Recent experiments on genetically deficient mice demonstrated that *Prox1* gene expression is directly induced by the transcription factor SOX18 (sex determining region Y, SRY-box 18) and by orphan member of the nuclear receptor superfamily, COUP-TFII (Chicken ovalbumin upstream promoter transcription factor II)

Table 2.1 Key molecular factors during developmental lymphangiogenesis

		Defects in lymphatic vascular system in genetic deficiency	
Factors	Function	Animal model	Human syndrome
Transcription factors			
SOX18	– Direct induction of expression of PROX1	(–/–) Mice embryos develop edema and die before birth due to lack of lymphatic vasculature	Recessive and dominant forms of hypotrichosis–lymphedema–telangiectasia
COUP-TFII	– Interacts with PROX1 to maintain LEC phenotype	(–/–) at an early embryonic stage results in failed formation of pre-LECs; (–/–) at a later developmental stage results in loss of LEC identity, gain of blood ECs fate, and impaired lymphatic vessel sprouting	
PROX1	– Master regulator that determines LECs identity – Regulator of *LYVE-1* and *podoplanin* gene expression	(–/–) Mice embryos develop edema and die before birth due to lack of lymphatic vasculature	
FOXC2	– Major regulator of lymphatic capillaries versus collecting lymphatic vessels phenotype specification	(–/–) Mice show loss of valves in the collecting vessels and impaired patterning of capillaries	Lymphedema-distichiasis (LD) syndrome
Receptors/transmembrane proteins			
LYVE-1	– Mediates cell adhesion and transmigration – Scavenger for hyaluronan turnover	(–/–) Mice develop normal lymphatic system	
VEGFR-3	– Mediates a critical pathway for lymphatic vessel growth	(–/–) Mice die during early embryonic development, before emergence of lymphatic vessels, due to	Heterozygous missense point mutation have been found in several families with Milroy's disease

(continued)

Table 2.1 (continued)

Factors	Function	Defects in lymphatic vascular system in genetic deficiency: Animal model	Human syndrome
		defective remodeling and maturation of large blood vessels (−/+) Heterozygous missense point mutation (Chy mutant mice) results in hypoplasia of cutaneous lymphatic capillaries and lymphedema of posterior limbs	
Nrp2	– Co-receptor of VEGFR-3	(−/−) Mice develop transient hypoplasia of lymphatic capillaries; surviving *Nrp2*-deficient mice regenerate lymphatic capillaries starting at 7th day after birth	
Podoplanin	– Transmembrane glycoprotein – Essential in lymphovascular separation	(−/−) Mice develop abnormal lympho-venous connections, dilated and dysfunctional lymphatic vessels, lymphedema, and die perinatally due to respiratory failure	
Ephβ2	– Role in remodeling and maturation of lymphatic vasculature	(−/−) Mice develop major lymphatic defects, including lack of hierarchically organized network, ectopic SMCs coverage in	

(continued)

Table 2.1 (continued)

Factors	Function	Defects in lymphatic vascular system in genetic deficiency	
		Animal model	Human syndrome
		lymphatic capillaries, lack of luminal valve formation and chylothorax; they die 3 weeks after birth	
Integrin α9β1	– Adhesion receptor – Directly binds VEGF-C and -D and promotes LECs adhesion and migration	(−/−) Mice embryos display dysplastic lymphatic valves and die 6–12 days after birth because of bilateral chylothorax	Missense mutation is associated with congenital chylothorax in fetuses
Growth factors			
VEGF-C	– Critical for the development of the lymphatic vasculature – Major chemotactic and survival factor for LECs – Essential for the formation of lymph sacs from embryonic veins	(−/−) Mice embryos develop edema and die before birth due to complete absence of lymphatic vasculature (−/+) Newborn mice have hypoplasia of the lymphatic vessels and develop severe lymphedema in adulthood	
VEGF-D	– Promotes migration and proliferation of LECs	(−/−) Newborn mice develop slight lymphatic dysfunction as a decrease in the abundance of lymphatic vessels in the lungs	
AM	– Vasodilator and diuretic peptide – Necessary in early stages of lymphatic vascular development	(−/−) Mice embryos display decreased LECs proliferation, hypoplastic lymphatic sacs, interstitial lymphedema, and die in mid-gestation	

(continued)

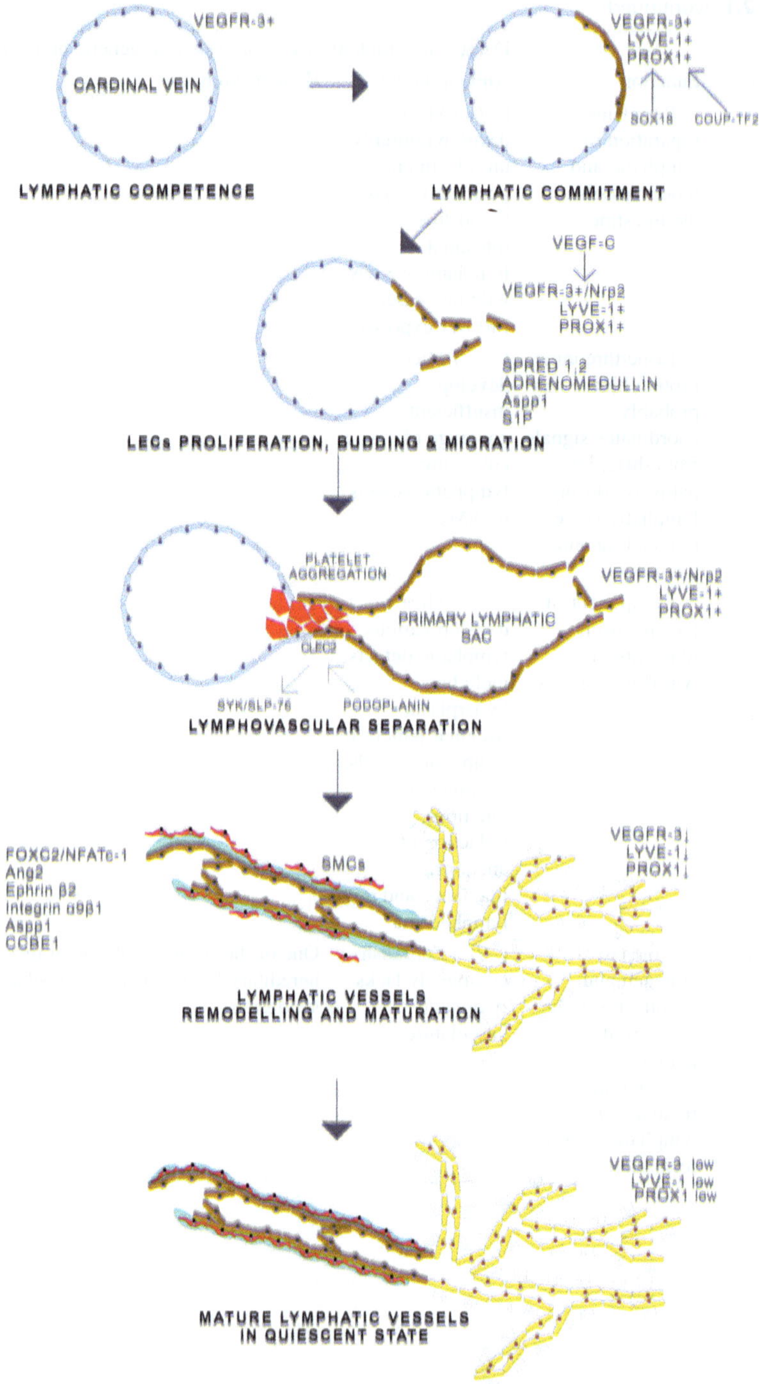

Fig. 2.1 Embryonic development of the mammalian lymphatic vasculature

[32, 33]. *Sox18*-knockout mice, similarly to *Prox1*-knockout mice, develop edema and die before birth [32]. In humans, mutations in the transcription factor SOX18 were recently identified in recessive and dominant forms of hypotrichosis–lymphedema–telangiectasia [34]. *Coup-tf2* deficiency at an early embryonic stage results in failed formation of pre-lymphatic ECs (pre-LECs). *Coup-tf2* deficiency at a later developmental stage results in loss of LEC identity, gain of blood ECs fate, and impaired lymphatic vessel sprouting [35]. Nevertheless, the signals leading to SOX18 and COUP-TFII expression in VEGFR-3+/LYVE-1+ cells remain unknown.

2.2.2 Lymphatic Endothelial Cell Proliferation and Migration

During early embryonic development of vascular endothelial growth factor receptor 3 (VEGFR-3), a cell surface tyrosine kinase receptor (also known as Flt4) is expressed at similar levels in BECs and LECs. As embryonic development progresses, its expression is decreasing in BECs and becomes restricted to the endothelium of lymphatic vessels [36, 37].

VEGFR-3 mediates a critical pathway for lymphatic vessel growth and maintenance. Mutant mice deficient for *Vegfr-3* die during early embryonic development (at E 9.5), before emergence of lymphatic vessels, due to defective remodeling and maturation of large blood vessels [37]. Heterozygous missense mutations in *Vegfr-3* gene that inactivate tyrosine kinase in the encoded protein in mice (Chy mutant mice) results in hypoplasia of cutaneous lymphatic capillaries and lymphedema of posterior limbs. Chy mutant mice are used as a useful model for studies of hereditary lymphedema and its therapy [38]. In humans, similar heterozygous missense point mutations in the gene encoding VEGFR-3 were found in several families with Milroy's disease, a rare autosomal dominant lymphedema, characterized by hypoplastic lymphatic capillaries in the skin [39].

VEGFR-3 is exclusively activated by two ligands—vascular endothelial growth factor-C (VEGF-C) and vascular endothelial growth factor-D (VEGF-D). They are homodimeric glycoproteins closely related in structure and secreted during embryogenesis by mesenchymal cells surrounding cardinal veins, vascular smooth muscle cells (SMCs), and BECs [3, 4, 40]. In humans, both VEGF-C and VEGF-D bind to not only VEGFR-3 but also to VEGFR-2 (also known as Flk-1/KDR). VEGFR-2 is a blood vessel receptor [41], implicated in signaling for angiogenesis [42]. Therefore, in humans, VEGF-C and VEGF-D induce not only lymphangiogenesis but also blood vessel growth, the latter through VEGFR-2 activation in BECs [43–45]. In mice, VEGF-C activates both receptors [46], whereas VEGF-D activates only VEGFR-3 [41].

VEGF-C, like VEGFR-3, is critical for the development of the lymphatic vasculature. Sprouting and growing lymphatic vessels contain specialized tip cells at their leading edge, which sense and respond to chemoattractants. VEGF-C is a major chemotactic and survival factor for LECs [16, 46]. It is essential for the formation of lymph sacs from embryonic veins. Homozygous *Vegfc*(−/−) mice die before birth due to fluid accumulation in tissues as a result of a complete absence of lymphatic vasculature [16]. Heterozygous *Vegfc*(+/−) newborn mice have hypoplasia of the lymphatic vessels and develop severe lymphedema in adulthood [16].

VEGF-D promotes migration and proliferation of LECs like VEGF-C [47]. However, surprisingly, mice with targeted inactivation of *Vegf-d* develop only a slight lymphatic dysfunction as a decrease in the abundance of lymphatic vessels in the lungs, suggesting that VEGF-D is dispensable for lymphangiogenesis [48].

An integral part of VEGF-C/D-VEGFR-3 signaling pathway is the interaction of VEGF-C and VEGF-D with neuropilin-2 (Nrp2). Neuropilin-1 (Nrp1) is expressed by arterial BECs, while Nrp2 is expressed by venous BECs and by LECs. Nrps do not have their own enzymatic signaling activity, but they act as co-receptors of VEGF receptors. Both, VEGF-C and VEGF-D, bind to Nrp2 and this way increase the affinity of LECs toward VEGF-C/D [49]. Nrp2 is associated with VEGFR-3 only in visceral and deep lymphatics, but not in cutaneous ones. Nrp2 binds VEGF-C and enables lymphangiogenesis in visceral organs in VEGFR-3 mutant mouse (Chy mouse). Similar mechanism may explain normal development of visceral lymphatics and dysplasia of cutaneous lymphatics in patients with Milroy's disease [38]. *Nrp2*-deficient mice develop hypoplasia of lymphatic capillaries, but the development of collecting vessels, such as the thoracic duct, is not affected. These defects are transient, and surviving *Nrp2*-deficient mice regenerate lymphatic capillaries starting at 7th day after birth [50, 51]. It was also shown that *Nrp1*-deficient mice survive only until around E13.5, whereas *Nrp1/Nrp2* double-deficient mice die around E8.5 [52]. Therefore, it was suggested that *Nrp1* compensate for the loss of *Nrp2*, which may explain the mild consequences observed in *Nrp2*-deficient mice [53].

VEGF-C-VEGFR-3 signaling pathway was shown to be negatively regulated by the action of Spred (Sprouty-related) proteins, which suppress the phosphorylation of growth factor-induced MAP kinase signaling. *Spred1/2* double-knockout mice develop subcutaneous hemorrhages, edema, blood-filled dilated lymphatic vessels and die by E15.5 [54]. This phenotype closely resembles the one observed in *Syk*(−/−) or *slp-76*(−/−) mice with defects in the separation of lymphatic vessels from blood vessels (see below) [55].

Adrenomedullin (AM) signaling is another transduction pathway necessary in early stages of lymphatic vascular development. AM is a multifunctional vasodilator and diuretic peptide, which was shown to be under the control of the PROX1. Genetic blockade of AM signaling in mice results in decreased LECs proliferation, occurrence of hypoplastic lymphatic sacs, development of interstitial lymphedema, and death in mid-gestation [56].

Defects in lymphatic vessel development have been also reported in apoptosis-stimulating protein of *p53* (*Aspp1*)-knockout mice [57]. *Aspp1*(−/−) mice embryos showed subcutaneous edema and disorganized lymphatic vasculature. However, *Aspp1*(−/−) adult mice was shown to have resolved impaired function of cutaneous lymphatic drainage seen during embryogenesis, but still had abnormal patterns in collecting lymphatic vessels [57]. The lymphatic vascular function of Aspp1 was demonstrated to be independent of p53 and may be linked instead to C-terminal Src kinase (Csk) [57, 58].

Additionally, sphingosine-1-phosphate (S1P), known to be implicated in a variety of biologic processes, such as inflammatory responses and angiogenesis, recently was also reported to be the first identified bioactive lipid which plays a role during lymphangiogenesis. S1P induces migration and capillary-like tube formation, but not proliferation of human LECs in vitro, and promotes the outgrowth of new lymphatic vessels in vivo. S1P promotes lymphangiogenesis by stimulating S1P/G_i/phospholipase C/Ca^{2+} signaling pathways [59].

Finally, there is a number of other growth factors which also were revealed to regulate early stages of lymphangiogenesis, including transforming growth factor-β 1 (TGF-β1) [60], platelet-derived growth factor - β (PDGF-β) [61], fibroblast growth factor 1 (FGF-1), FGF-2 [44, 62], insulin-like growth factor 1 (IGF-1), IGF-2 [63], hepatocyte growth factor (HGF) [64], and growth hormone (GH) [65]. However, because global effects of these growth factors gene targeting result in other early-onset phenotypes, it is not known whether they directly contribute to the development of the lymphatic vascular system [1, 4].

2.2.3 Lymphovascular Separation

There are only two sites where lymph mixes with the blood in the subclavian veins. This suggests the existence of molecular mechanisms underlying the process of lymphovascular separation during embryonic development.

One such mechanism involve the function of podoplanin, transmembrane mucin-like protein, which is expressed at E11.5 in the cardinal vein and later in PROX1-positive LECs [66, 67]. Targeted inactivation of *podoplanin* results in abnormal lung development, abnormal lympho-venous connections, dilated and dysfunctional lymphatic vessels, lymphedema, and perinatal lethality due to respiratory failure [67]. The defect in lympho-venous separation in these mice is a result of deficiency of platelet activation by podoplanin. Platelet aggregation is required for constriction of the opening between the cardinal vein and the lymphatic sacs. It is induced by binding of podoplanin to C-type lectin receptor 2 (CLEC-2) expressed on the surface of platelets. CLEC-2 activation by podoplanin leads to activation of hematopoietic signaling proteins SYK (spleen tyrosine kinase) via its substrate adaptor protein SLP-76 (Src-homology 2 domain-containing leukocyte protein of 76 kDa) in the platelets [68]. Mice with homozygous mutations in either *Syk* or *slp-76* develop abnormal arterio-venous shunts, manifested as blood filled lymphatic vessels with mosaic expression of LYVE-1 [55, 69].

Fasting-induced adipose factor (Fiaf), also known as angiopoietin-like protein 4 (Angptl4), is the other molecular player that regulates separation of lymphatic and blood vessels, but only in the intestine. *Fiaf*-knockout mice develop normally until birth but postnatally have blood-filled intestinal lymphatic vessels with decreased PROX1 expression [70].

2.2.4 Remodeling and Maturation of Primitive Lymphatic Vessels

Remodeling and maturation of lymphatic vessels begins around E14.5 and lasts until birth. These processes are less understood than earlier events of lymphangiogenesis. They include the transformation of primitive lymphatic plexus into a hierarchical network of lymph vessels, composed of collecting lymphatic vessels and lymphatic capillaries.

Collecting lymphatic vessels contain intraluminal valves, basement membrane, and are covered with SMCs. Lymphatic capillaries consist of basement membrane and pericytes. Remodeling of collecting lymphatic vessels involves formation of intraluminal valves and recruitment of SMCs. New capillaries arise by sprouting from preexisting vasculature.

During these processes interactions between LECs and surrounding extracellular matrix are essential. Maturation steps are accompanied by reduced expression of LECs markers that were implicated in actively growing vessels, including VEGFR-3, LYVE-1, and PROX1 [4, 71, 72].

FOXC2, the forkhead transcription factor, is the major molecular regulator responsible for the specification of lymphatic capillaries versus collecting lymphatic vessels phenotype. Early steps of developmental lymphangiogenesis proceed normally in the absence of FOXC2. Its expression is upregulated in all developing lymphatic vessels between E14.5 and E15.5, and in adults the highest expression is observed in the endothelial cells of lymphatic valves [73, 74]. The nuclear factor of activated T cells (NFATc1), which expression is controlled by VEGF-C-VEGFR-3 signaling, cooperates with FoxC2 to generate the genetic program involved in the maturation process [75]. Mutations in the human *FoxC2* gene are the cause for the hereditary lymphedema-distichiasis syndrome (LD), an autosomal dominant disease characterized by late onset leg lymphedema and metaplasia of meibomian glands [76]. In contrast to Milroy's disease, lymphatic vasculature in LD is normal or hyperplastic, but lymphatic vessels are insufficient, most probably due to lack of lymphatic valves abnormal patterning and the presence of ectopic SMCs [73]. Inactivation of *FoxC2* in mice has similar consequences like in LD syndrome—collecting lymphatic vessels lack valves, the lymphatic capillaries acquire an excessive sheath with basal lamina components and SMCs, and they begin to express blood vascular endothelial cell markers such as endoglin and PDGF-β [73].

Angiopoietins and Tie receptors are also essential during remodeling and maturation of lymphatic vessels. There are three members of the angiopoietin family: Ang1, Ang2, and Ang3 (in mouse)/4 (in human). They bind to the Tie1/Tie2 endothelial-specific receptors tyrosine kinase, which are expressed at low levels in LECs and regulate interactions between endothelial cells and mural cells [77]. Ang1 can activate both Tie1 and Tie2 [78, 79], Ang2 is suggested to be a context-dependent agonist or antagonist of Tie2 signaling [80], and Ang3 (from mouse) and Ang4 (from human) are interspecies orthologs that both act as Tie2 agonists [81]. Ang1 and Ang2 were shown to be indispensable in remodeling and maturation of blood vessels and normal development of heart, whereas in lymphatic vessel development activation of Ang2 seems to be crucial [80, 82]. *Ang2*-deficient mice have defective postnatal blood and lymphatic vessel structures, such as incomplete regression of hyaloid artery, and distorted leaky lymphatic vessels—irregularly patterned hypoplastic lymphatic capillaries and collecting vessels lacking proper smooth muscle cell coverage. They present subcutaneous edema and chylous ascites shortly after birth [80]. Ang2 also seems to prevent premature coverage of lymphatic vessels with SMCs [80]. FoxC2 and NFATc1, by repressing Ang2 expression, initiate the recruitment of pericytes and SMCs to the maturing lymphatic vessels [75]. Lack of Ang2 can be largely corrected by replacing it with Ang1. However, abnormal persistence of the hyaloid artery cannot be rescued, indicating that although Ang1 and Ang2 have redundant roles in the development of the lymphatics, they do not have that in the development of blood vessels [80].

EphrinB2, one of the transmembrane ligands of the Eph group, and its receptor, Ephβ4, regulate lymphatic endothelial cell sprouting during the remodeling process [83]. In mice, genetic deficiency in the cytoplasmic PDZ-interaction domain of *Ephrinβ2* results in lack of hierarchically organized lymphatic vessel network consisting of lymphatic capillaries and collecting vessels and lead to postnatal lethality 3 weeks after birth [83].

The serine/threonine protein kinase AKT1 is a major signal transducer and is downstream of various receptor protein kinases involved in the regulation of lymphatic development, including VEGFR-3, TIE-2, PDGFR-β, and Ephβ4. It is therefore plausible that AKT1 may coordinate signals from the different pathways during lymphatic vessel remodeling and maturation. *Akt1*-deficient mice showed insufficient coverage of collecting lymphatic vessels by SMCs [84].

2.2.5 Interactions Between LECs and Surrounding Extracellular Matrix

Recent studies have highlighted the importance of the extracellular matrix (ECM) in the generation of new lymphatic vessels. However, the molecular mechanisms underlying interactions between LECs and surrounding ECM remain largely unknown.

Unlike BECs, LECs do not adhere to continuous basement membrane and form an intimate association with adjacent interstitial tissue. Instead of adhering to basal lamina, capillary LECs are attached to anchoring filaments composed of fibrillin and emilin-1, which anchor the LECs cytoskeleton with the fibrous components of ECM (i.e., collagen, fibronectin, tenascin C, vitronectin, laminin) through the transmembrane integrin and FAK molecule [85, 86]. Anchoring filaments are the characteristic feature of initial lymphatic vessels. Forces applied through these anchoring filaments enhance lymphatic permeability and promote tissue drainage [85].

ECM provides important growth-promoting signals for lymphatic vessels. In addition, proteins of extracellular matrix, like collagen and fibronectin, were shown to enhance tyrosine phosphorylation of VEGFR-3 through activation of integrin $\alpha 9\beta 1$ [85].

LECs express multiple integrins other than $\alpha 9\beta 1$, including $\alpha 1\beta 1$, $\alpha 2\beta 1$, $\alpha 3\beta 1$, $\alpha 4\beta 1$, $\alpha 5\beta 1$, $\alpha 6\beta 1$, and $\alpha v\beta 3$, but their importance in developmental lymphangiogenesis remain unclear, since majority of integrin-deficient mice show no apparent developmental defects in lymphangiogenesis or in lymphatic function [85].

So far, only one of the integrins expressed by LECs, α9β1, was shown to be essential in embryogenic lymphangiogenesis. α9β1 directly binds VEGF-C and VEGF-D and promotes LECs adhesion and migration [87]. *Integrin* $\alpha 9\beta 1$-deficient mice die 6–12 days after birth because of bilateral chylothorax. This mouse embryos display dysplastic lymphatic valves, characterized by disorganized fibronectin matrix and short cusps, which results in retrograde lymphatic flow [88]. In humans, missense mutation in the *α9 integrin* gene is associated with congenital chylothorax in fetuses [89]. Among several extracellular matrix ligands for $\alpha 9\beta 1$ (fibronectin, tenascin-C, fibronectin, osteopontin), fibronectin seems to be the most important. Mice deficient for the EIIIA domain of fibronectin exhibit similar defects in lymphatic valves like those deficient for $\alpha 9\beta 1$ [88].

Emilin-1, an ECM glycoprotein associated with elastic fibers, is a component of the anchoring filaments in lymphatic vessels [90]. *Emilin1*-deficient mice have hyperplastic and disorganized lymphatic vessels with a reduced number of anchoring filaments and dysfunctional junctions, and they show impaired lymphatic drainage function resulted in mild lymphedema [90].

Recent studies demonstrated also important function of extracellular collagen in the regulation of developmental lymphangiogenesis. In the in vitro study LECs incorporated into collagen type I scaffolds exhibited a significantly higher survival rate than BECs [91]. In the experiment using zebrafish and mouse models, CCBE1 (collagen and calcium-binding EGF domain-containing protein 1) was identified as an essential regulator of lymphatic vessel formation [92]. CCBE1 binds to collagen and vitronectin in ECM. Lack of CCBE1 expression prevents the budding of new lymphatic endothelial cells from the cardinal vein [92]. *Ccbe1*-targeted zebrafish completely lacks lymphatic vasculature [92]. In humans, mutations in *CCBE1* were found to cause one of a form of autosomal recessive hereditary lymphedema—Hennekam syndrome, in which patients have lymphedema, lymphangiectasias, and other cardiovascular anomalies [93]. However, whether CCBE1 acts as only a guiding molecule for growing lymphatics or whether it also promotes LECs proliferation warrants further studies.

Apart from the influence of various growth factors and ECM elements on lymphangiogenesis, it has become increasingly clear that the biophysical microenvironment may also influence the growth of lymph vessels. Changes in blood flow, which are the mechanical stimuli in the form of shear stress [94] were shown to be sufficient to induce blood vessel growth and remodeling of ECM [95]. Recent in vitro and in vivo studies point either to a role of a slow flow of interstitial fluid through the interstitium as an important morphoregulator and lymphangiogenic factor [96]. Interstitial flow was found to regulate the generation of VEGF-C in the tissue [97] and to induce lymphatic capillary formation [98].

2.2.6 Postnatal Maturation of Lymphatic Vessels

Recent data reveal that postnatal maturation of lymphatic vessels differs from embryonic one. VEGF-C-VEGFR-3 signaling pathway is absolutely required for the maintenance of lymphatic vasculature only during the first 2 weeks after birth. After that time postnatal lymphatic vascular mechanism of lymphatic vessels maturation becomes independent of VEGF-C, VEGF-D, and VEGFR-3 [99]. However, although the VEGF-C-VEGFR-3 pathway is not needed for the growth of the lymphatic vessels after 2 weeks after birth, it is functional, because lymphangiogenesis can be stimulated by VEGF-C [100].

Possible postnatal factors responsible for the VEGF-C/VEGF-D-VEGFR-3-independent pathway of maturation of lymphatic vessels include VEGF-A, HGF, angiopoietins, and integrins [99].

Furthermore, it was shown that whereas transgenic overexpression of VEGF-C induces lymphangiogenesis during embryogenesis, after birth lymphatic vessel growth predominantly is stimulated by VEGF-D. That indicates additional, still unrecognized, maturation signals in postnatal lymphatic maturation [99].

2.3 Lymphangiogenesis in Human Diseases

2.3.1 Inflammatory Lymphangiogenesis

Inflammation is the major cause of lymphangiogenesis in adults. There is increasing evidence that lymphatic vessels have an active role in acute and chronic inflammation and contribute to the pathophysiology of these conditions.

Increased lymphatic vessel density (LVD) has been documented in both clinical studies and experimental models of acute and chronic inflammation, including chronic airway infection [101], inflammatory bowel disease [102], rheumatoid arthritis [103], psoriasis [104, 105], corneal injury [25], UVB skin inflammation [106], skin irradiation [107], and rejecting renal transplants [108].

In acute inflammation, like in wound healing, new lymphatic vessels appear in the area of inflammation concurrently with blood vessels but regress earlier [109, 110]. In wound healing of the pig's skin lymphatic vessels were observed from day 5 after injury. Only few lymphatic vessels persisted on day 9, and none was found on day 14 [110]. In acute inflammation, unlike in chronic one, lymphatic vessels rather enlarge than sprout [109]. In chronic inflammation, newly formed lymphatic vessels persist for at least 12 weeks after recovery [101, 111].

Inflammatory lymphangiogenesis facilitates the absorption of tissue edema from inflammatory site [101, 112]. Inhibition of VEGF-C/VEGF-D–VEGFR-3 signaling pathway, by blocking lymphangiogenesis, increases edema and inflammation, e.g., exacerbates pulmonary edema in a mouse model of *Mycoplasma pulmonis* infection [101] or leads to prolonged edema and inflammation after UVB irradiation [113]. Lymphatic vessels serve also as the principal conduit for microbial and self-antigens, antigen-presenting cells, and leukocytes from inflamed tissues to the lymph nodes (LNs) and other secondary lymphoid organs. They also contribute to inflammation resolution. Thus, lymphangiogenesis enhances immune responses during inflammation and help to clear the sites of resolving inflammation.

Lymphatic vessels may actively participate in the inflammatory process locally by secreting number of chemokines that enhance trafficking of immune cells to LNs, including CCL21 (also known as secondary lymphoid chemokine, SLC), CXCL12, and CCL27. CCL21 is a ligand for CCR7 receptor, expressed on mature dendritic cells, CCR7+ T, and B cells [114]. CCL21 expression in LECs was shown to be upregulated by VEGF-C or enhanced fluid flow [115]. CXCL12 is a chemokine, the expression of which is upregulated by hypoxia and which acts via its CXCR4 receptor located on dendritic cells [116]. CCL27 is a ligand for CCR10 receptor, located on CCR10-positive T cells [117].

LECs secret also sphingosine-1-phosphate (S1P), a bioactive lipid synthesized by sphingosine kinases, which guide migration of lymphocytes expressing S1P1 receptor [118]. In addition, LECs can express some specific adhesion molecules mediating leukocyte adhesion to the lymphatics, including ICAM-1 and VCAM-1 [119], E-selectin (also named CD62E or ELAM-1, endothelial leukocyte adhesion molecule-1) [120], CLEVER-1 (Common Lymphatic Endothelial and Vascular Receptor-1) [121], Mannose receptor 1 [122], podoplanin [108], and LYVE-1 [123].

The most important signaling pathway for inflammatory lymphangiogenesis, like in the developmental one, is VEGF-C–VEGFR-3 axis. VEGF-D has a similar but less potent effect than that of VEGF-C [109]. VEGF-C and VEGF-D are elevated during inflammation and are produced by a variety of cells residing at inflamed sites, including macrophages, dendritic cells, neutrophils, mast cells, and fibroblasts [1].

VEGF-A is also elevated at the sites of various inflammatory conditions. It is the major driver of angiogenesis, blood vessel remodeling, and hyperpermeability in inflamed tissue, leading to edema formation, but also plays an important role in inflammatory lymphangiogenesis [124]. VEGF-A mediates process of lymphangiogenesis directly, by binding to its receptor, VEGFR-2. VEGFR-2 is predominantly located on vascular endothelium but has also been demonstrated in LECs. Binding to this receptor has been shown to promote proliferation and migration of VEGFR-2-positive LECs in vitro [125], but only hypertrophy and enlargement of lymphatic vessels without inducing sprouting of LECs in vivo [126]. Indirect VEGF-A effect on lymphangiogenesis include activation of macrophages that produce high levels of VEGF-C/-D [127] and increasing of VEGFR-3 expression in LECs [128].

Expression of VEGF-A is strongly upregulated by activation of NF-κB (nuclear factor kappa-light-chain-enhancer of activated B cells) during inflammation [124]. NF-κB is constitutively expressed in LECs. It also upregulates expression of PROX-1 and VEGFR-3 in them [129] and stimulates secretion of several chemoattractants, such as CCL2, CCL5, and CX3CL1 [130]. It results in greater sensitivity of lymphatic vessels to VEGF-C and VEGF-D produced by immune cells [130] and in increased mobilization of dendritic cells and macrophages at inflammatory sites [112].

Other stimulators of inflammatory lymphangiogenesis at the sites of tissue inflammation are several proinflammatory cytokines and growth factors produced mainly by immune cells, including TNF-α [1], IL-3 [130], IL-7 [131], FGF-2 [44], PDGF-β [61], IGF-1, IGF-2 [63], HGF [132], GH [65], Fiaf [70], and S1P [59]. Some of these factors promote lymphangogenesis directly, independently of VEGR-3 signaling, e.g. PDGF-β [61], IGF-1, IGF-2 [63], and GH [65], while others—indirectly, i.e., via upregulation of VEGF-C.

Inflammatory lymphangiogenesis can be observed not only in inflamed peripheral tissue (extranodal lymphangiogenesis) but also in its draining LNs (intranodal lymphangiogensis) and even in uninvolved peripheral tissue that drains to the same lymph node [126]. Moreover, chronic inflammation in autoimmunity, graft rejection, or chronic infection induces accumulation of lymphoid cells that resemble LNs and are termed "tertiary lymphoid organs." They occur, e.g., in thyroiditis, sialitis, rheumatoid arthritis, and chronic kidney graft rejection, and are the sites of locally perpetuating autoimmune reactions that support autoantigenic epitope spreading [133]. Lymphangiogenesis within these "tertiary lymphoid organs" is also known to take place [28].

Pathophysiology of lymphangiogenesis at these different sites largely overlaps. Nevertheless, because of the structure and function of LNs, intranodal lymphangiogenesis mechanisms appear to be more complicated than in inflamed peripheral tissue. LNs collect lymph flowing from multiple areas, heavily populated with various immune cells and, consequently, are the site of tense microenvironment where vigorous immune reactions take place. Like in extranodal lymphangiogenesis, during inflammation in LNs the main signaling pathways are VEGF-C/VEGF-D-VEGFR-3, and VEGF-A-VEGFR-2. The main source of these cytokines are follicular B cells [126, 134], CD11b macrophages (migrated from the peripheral tissue) [112, 135], and fibroblast-type reticular stromal cells [136]. Several key TNF family members, lymphotoxin λ and LIGHT (homologous to lymphotoxins), have also been shown to play significant roles in intranodal lymphangiogenesis [28, 137, 138]. In addition, intranodal lymphangiogenesis can be regulated not only by the local intranodal signals but also by remote inflammatory factors transported via lymphatics from inflamed tissue into the draining LNs [126].

The most unique features of intranodal lymphangiogenesis are the strong negative regulatory mechanisms. VEGF-A, which initially promotes lymphangiogenesis and angiogenesis within LNs, was demonstrated to suppress certain aspects of the ensuing immune responses at the later phase of lymphangiogenesis within LNs [134]. T cells, which constitutionally reside in large quantities inside LNs, appear to antagonize B cell-driven lymphangiogenesis in the lymph node via interferon-γ signals [139]. TGF-β inhibits lymphangiogenesis, either in cultured human LECs, as well as in inflammatory settings and tumors [60, 140, 141]. This negative regulation indicates that lymphangogenesis in LNs is determined by the balance between pro-lymphangiogenic and anti-lymphangiogenic molecules, produced locally in lymph node and drained from the peripheral tissue [112, 142]. That explains, e.g., that LPS injection promotes intranodal lymphangiogenesis, whereas concanavalin-A, as T cell mitogen, does not, although both agents induce potent inflammatory reactions in the skin [142]. Negative regulation of lymphangogenesis within LNs also comprises mechanisms for regression of inflammation-induced lymphangiogenesis during resolution of inflammation [143]. Indeed, when inflammation in lymph node resolves, newly formed intranodal lymphatic vessels readily regress [142, 144]. It is in contrast to newly formed lymphatic vessels in peripheral tissues, which may persist after the resolution of chronic inflammation for at least 12 weeks [101, 111]. This relative stability of newly formed lymphatic vessels in chronic inflammation within peripheral tissues also clearly differs from the corresponding changes in newly formed blood vessels which rapidly regress after inflammation is resolved [101]. The mechanism and biological significance of this difference is unknown. Possibly, it may reflect long-term microenvironment remodeling of peripheral tissue that prepares more efficiently the immune system for potentially repeatable inflammatory episodes [143].

2.3.2 Lymphangiogenesis in Cancer

Lymphatic metastases represent the first site of tumor dissemination in many solid tumors. Haematogenous metastasis can occur without sentinel lymph node (SLN) metastasis, but likely in a minority of cases [75, 145]. Status of LNs metastasis plays a major role in cancer staging, prognosis, and applied therapy in most human malignancies.

Historically, lymphatic vessels were regarded as a passive contributor in tumor pathophysiology that simply provides channels for tumor cells to transit to LNs. Recent studies have highlighted a much more complex, active role of the lymphatic vasculature in tumor spread. Tumor-associated lymphatics are not only routes for cancer cells, but they directly facilitate tumor spreading and also comprise an active link between tumor and immunological system [146]. Lymphangiogenesis and tumor metastatic process are mediated by a complex series of interactions between tumor cells, inflammatory cells, endothelial cells, extracellular matrix components and multiple growth factors, chemokines, cytokines, and stromal molecules secreted by them [146].

Studies of several types of cancers overwhelmingly show that lymphovascular invasion of the tumor and lymphatic vessel density (LVD) detected by immunohistochemistry correlate with LNs and distant metastasis and with poor prognosis [147].

The most important signaling pathway in tumor-induced lymphangiogenesis is, like in developmental lymphangiogenesis, VEGF-C–VEGFR-3 axis. VEGF-D is also overexpressed in tumors and its acting resembles that of VEGF-C, but it has been much less studied than VEGF-C [148]. The main sources of VEGF-C are tumor cells and tumor macrophages [149]; however, it may be also produced by many other inflammatory cells that infiltrate tumor site [1].

Studies in mouse tumor models showed that stimulation by VEGF-C results in lymphatic vessel growth and sprouting, both within the tumor and peripherally around the tumor [150]. VEGF-C also promotes formation of intercellular gaps, which facilitate tumor cell entry into the lymphatic vessels, and circumferential dilation of collecting vessels, leading to increased flow rates of lymph and enhanced passage of clusters of tumor cells to the sentinel LNs [1, 151]. Forced expression of VEGF-C or VEGF-D in tumor results in lymphangiogenesis and increased tumor dissemination to regional LNs, even in tumors that normally do not metastasize [3, 150]. Conversely, inhibiting the VEGF-C/VEGF-D–VEGFR-3 pathway results in decreasing by approximately 60–70 % of LNs metastasis in a variety of experimental tumor models, including melanoma, fibrosarcoma, breast cancer, lung cancer, and gastric cancer [152]. An increasing number of studies in human cancers, including prostate, gastric, thyroid, colorectal, esophageal, and lung carcinoma also have shown a strong correlation between the expression of VEGF-C (or VEGF-D), tumor lymphangiogenesis, lymph node and distant organ metastasis, and poor prognosis of survival [1, 3, 151, 153].

VEGF-A is also expressed in tumors, and its overexpression has also been shown to strongly promote lymphangiogenesis within primary tumor and sentinel LNs, and to endorse lymph node metastasis [154].

Lymphangiogenesis within tumor begins at least 1 week after the beginning of angiogenesis [151]. That indicates that tumor lymphangiogenesis is regulated not only by lymphangiogenic growth signals but also by tumor-specific microenvironmental factors. They may include interstitial pressure due to leaky tumor blood vessels and mechanical pressure due to the proliferating tumor cells [155]. Tumor-associated lymphatics are formed mainly by sprouting and splitting from preexisting lymphatic vessels in surrounding tissues [156]. There is also evidence that bone marrow-derived cells have ability to differentiate into LECs [25]; however, it was demonstrated that this contribution of LECs within tumor-associated lymphatics amounts to approximately 3 % [26].

Although the main molecular pathways of lymphangiogenesis during development and in cancer are similar, there are several characteristic features of tumor-associated lymphangiogenesis. During embryogenesis, prolymphangiogenic factors, including VEGF-C, are secreted by BECs and SMCs to induce lymphatic vessel development in a highly organized manner. Tumor site is the area of acting of multiple cytokines, growth factors, proteinases that affect lymphangiogenesis, and are produced not only by BECs and SMCs, but also by tumor cells, tumor infiltrating immune cells, and tumor stromal cells [146, 155]. These lymphangiogenic factors are not secreted in the specific, accurate order, like during embryogenesis, but totally in a chaotic manner. That causes that the tumor-associated lymphatic vessels are disorganised and without hierarchical vascular arrangement. Lymphatics within tumor are also patchy and not homogenously distributed within tumors. They rather grow in necrotic areas or peripheral regions of tumor mass. Therefore, it is still controversial whether intratumoral lymphatics are functional and vital for lymphatic metastasis [147, 155]. Especially, mechanistic studies have elucidated that intratumoral lymphatic vessels may be poorly functional due to high intratumoral pressure [152]. Conversely, lymphatic vessels in the tumor periphery are functional and can drain colloids from the tumor site [1, 151, 152].

To spread from the primary tumor and establish metastases in LNs, tumor cells have to disconnect from the tumor mass, enter lymphatic vessels by interaction with LECs, survive during the passage in lymphatic vessels, and escape from immune surveillance to form metastatic foci [139].

Immune cells that infiltrate tumor stroma facilitate tumor progression via the remodeling of intratumoral ECM and induction of intratumoral angiogenesis and lymphangiogenesis [157]. Macrophages are the major components of infiltrating inflammatory cells in tumor mass [158]. They are attracted by VEGF-C secreted by tumor cells and then, in turn, secrete prolymphangiogenic factors which stimulate the process of tumor lymphangiogenesis [157]. Other immune cells are also recruited to most tumors and contribute to lymphangiogenesis, e.g., tumor-associated LECs recruit leukocytes via secreting various chemokines. Infiltrating leukocytes modulate tumor-associated vascular growth and tumor progression, e.g., by secreting angiogenic and lymphangiogenic factors [155]. Tumor-associated fibroblasts produce high quantities of hyaluronan that promotes tumor lymphangiogenesis, presumably by stimulating malignant cells to secrete specific lymphangiogenic factors [159].

Lymphatic vessels themselves also may facilitate tumor cell recruitment into lymphatic vessels. LECs activated by factors produced by tumors, such as VEGF-C, produce chemokines that can guide tumor cells toward the lymphatic vessels, utilizing similar mechanisms as in inflammation. For example, CCL21 and CXCL12, produced by LECs, appear to attract some tumor cells that express their receptors, accordingly CCR7 and CXCR4 [115, 160, 161]. These chemokines and their receptors play a critical role in mediating lymphatic cancer metastasis. CCL21 secreted by LECs, but also by tumor cells, has also been suggested to participate in the modulation of tumor microenvironment to be immunotolerant [160]. Moreover, alterations of the lymphatic endothelial junctions and expression of cell adhesion molecules by LECs, like CLEVER-1, may also play a role in tumor-induced metastasis via lymphatics [162].

In comparison with the low survival rate of tumor cells in blood circulation, it seems that tumor cells are ready to survive and proliferate in LNs and even in lymphatic vessels when the flow is blocked [151, 163]. It was shown that tumor cells become highly invasive after they spread further from LNs [164]. That indicates that tumor cells could acquire some capacity of leukocytes for vascular invasion [165].

Recent studies have demonstrated that lymphangiogenesis occurs not only at the primary site but also within draining LNs, even before tumor cells colonize the nodes [166, 167]. This "seed-and-soil" theory suggests that "the seed" (cancer) can remotely modify "the soil" (lymph node) to prepare it for its later arrival. The expansion of lymphatic network in the sentinel LNs is mediated mainly by remote secretion of VEGF-C and VEGF-A at the primary tumor site, from which they are transported via lymphatic vessels [167]. The induction of lymphangiogenesis within LNs strongly promotes cancer metastasis to distant LNs and organs. Arrival of metastatic cells further enhances lymphangiogenesis in lymph nodes [167]. Increased levels of lymphangiogenic growth factors are found in blood of cancer patients and they correlate with reduced survival. It indicates that tumor-derived prolymphangiogenic factors can act not only within sentinel LNs but more generally [168]. VEGF-A, for example, was shown to mobilize bone marrow precursor cells at concentrations typically found in the blood of cancer patients [169]. Thus, there might be the mechanisms by which tumor cells prepare distant organs for their metastases; however, this remains to be further investigated [168].

The scheme of lymphangiogenesis in cancer is presented in Fig. 2.2.

2.3.3 Lymphangiogenesis in Hypertension

Recent experimental studies on rats (DSS—Dahl's salt sensitive rats) have demonstrated that lymphatic vessels may play a crucial role in the pathogenesis of hypertension.

A high salt diet in DSS rats induces hypertension and sodium accumulation within skin glycosaminoglycans. The local hypertonicity in the skin interstitium promotes macrophages infiltration [170, 171]. They express tonicity-responsive enhancer binding protein (TonEBP), a transcription factor of vascular endothelial growth factor-C (VEGF-C) [171]. VEGF-C, secreted by macrophages, stimulates not only lymphangiogenesis but also induces endothelial nitric oxide synthase (eNOS) expression. Lymphangiogenesis in the skin of rats on high salt diet was displayed by both increased number of lymph capillaries and also by its increased diameters, reflecting hyperplasia of the preexisting lymph capillary network in response to dietary salt loading [171]. Macrophages infiltration into the skin in response to osmotic stress induces the TonEBP/VEGF-C regulatory axis without concomitant TNFλ expression. It proves that skin lymphangiogenesis in response to osmotic stress differs from inflammatory lymphangiogenesis [172].

Macrophages-driven lymphangiogenesis and increased nitric oxide (NO) production are like buffering system to maintain blood pressure in the states of high salt load. New, hyperplastic, and dilated lymph vessels are enabled to efficiently drain increased interstitial fluid. Inhibition of these buffering mechanisms in the skin of rats on high salt diet (by macrophages depletion or inhibition of VEGF-C signaling) aggravates salt-sensitive hypertension. Further researches should elucidate if these mechanisms can lead to additional regulation of blood pressure in humans.

2.4 Therapeutic Modulation of Lymphangiogenesis

Pro- and anti-lymphangiogenic therapies are still in experimental phase. Studies on therapeutic modulation of lymphangiogenesis are currently conducted on animal models (Table 2.2). However, as more insights on the molecular pathways of lymphatic differentiation emerge, novel potential targeted therapies might be established.

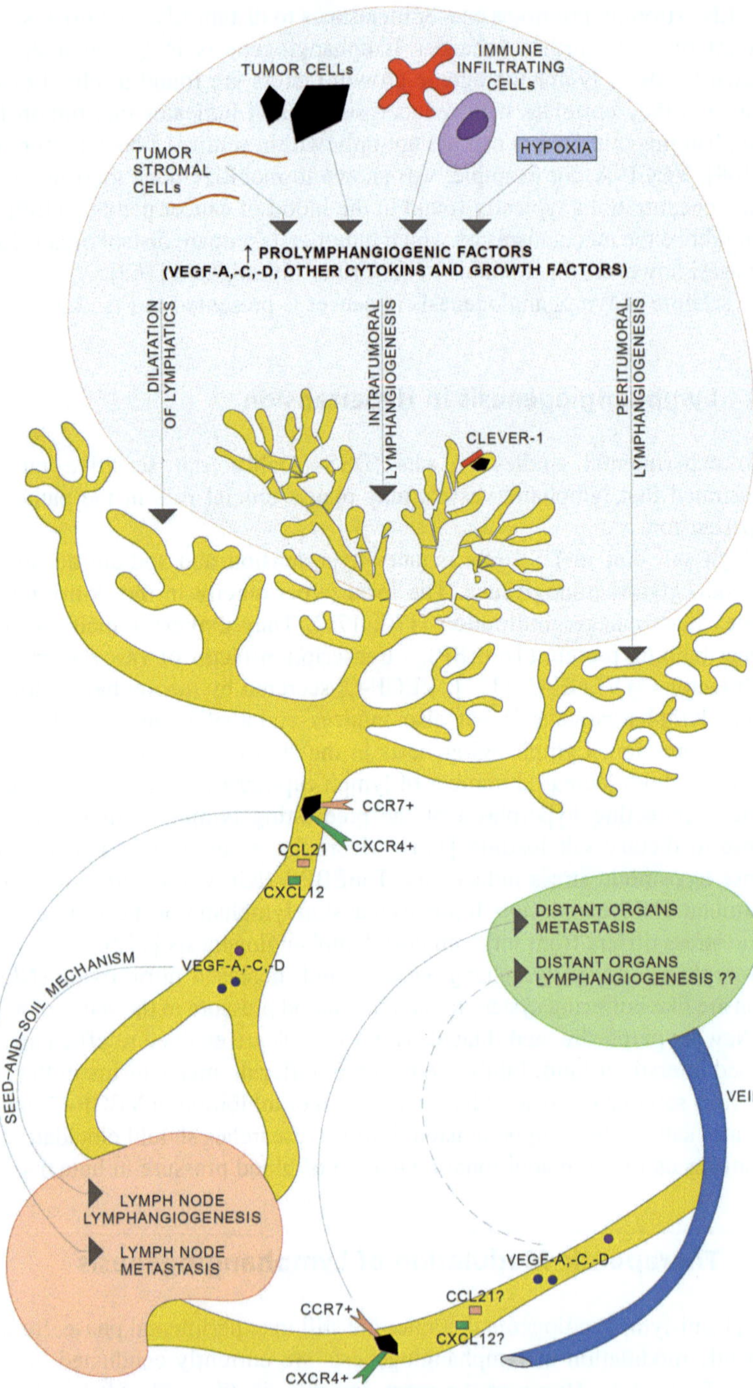

Fig. 2.2 Lymphangiogenesis in cancer

Table 2.2 Therapeutic modulation of lymphangiogenesis established in different diseases on animal models

Pro-lymphangiogenic therapies	Anti-lymphangiogenic therapies
Therapy in inflammation	
VEGF-C, VEGF-D	Anti-VEGF-2 Ab
– In acute UVB-induced inflammation – In chronic skin inflammation	– In psoriasis – In rheumatoid arthritis – In chronic skin inflammation
Therapy in graft rejection	
	Anti-VEGFR-3 Ab Inhibitor of Ang2 (?)
Therapy in secondary lymphedema	
VEGF-C, VEGF-D	
Therapy in cancer	
	1. Inhibitors of VEGF-C—VEGFR-3 axis (anti-VEGR-3 Ab, soluble VEGFR-3, anti Nrp2, indoline (MAZ51), enediynyl peptide inhibitor celecoxib) 2. Hyperforin, aristoforin 3. Antagonists of integrin $\alpha4\beta1$ 4. Inhibitors of VEGF-A pathway

2.4.1 Pro- and Anti-lymphangiogenic Therapy in Inflammatory Disorders

Anti-lymphangiogenic therapies using monoclonal antibodies against VEGFR-2 (anti-VEGFR-2 Ab) inhibit angiogenesis and lymphangiogenesis in animal models of psoriasis [173, 174] and chronic skin inflammation [175] and have produced promising results in preclinical studies. Similarly, treatment with anti-VEGFR-2 Ab in mouse model of rheumatoid arthritis impeded both lymphangiogenesis and joint inflammation [176].

Conversely, therapy with monoclonal antibodies against VEGFR-3 (anti-VEGFR-3 Ab) in the same model of rheumatoid arthritis resulted in the intensification of inflammation, despite inhibition of lymphangiogenesis [176]. Anti-VEGFR-3 Ab was also used in mouse model of chronic skin inflammation and the results were similar. This therapy prolonged inflammatory edema formation regardless of inhibition of lymphangiogenesis. In contrast, increased production of VEGF-C or VEGF-D reduced severity of both chronic and acute skin inflammation and diminished dermal edema, by enhanced lymphatic drainage from the area of inflammation [109, 175].

Pro-lymphangiogenic therapy was also successfully applied in mouse model of acute UVB-induced inflammation of the skin. UVB irradiation of the skin causes enlargement of lymphatic vessels with impairment of their clearance function and inflammatory macrophage infiltration in the cutis. These symptoms were associated with downregulation of VEGF-C expression. Activation of VEGF-C-VEGFR-3 signaling pathway via injection of VEGF-C reversed the effects caused by UVB irradiation and might represent a strategy to treat or prevent cutaneous photodamage [106].

Although pro-lymphangiogenic therapies seem to be beneficial in acute and chronic inflammatory disorders, they also have their risks. Endorsement of lymphatic drainage theoretically may result in increased systemic exposure to unfiltered pathogens and inflammatory mediators. In addition, stimulation of lymphatic vessel growth in cancer patients may enhance metastatic spread and modulation of the immune system.

2.4.2 Anti-lymphangiogenic Therapy in Graft Rejection

Another major application of anti-lymphangiogenic treatments has been attempted to control allogenic immune response in organ transplantation.

Increased lymphangiogenesis after organ transplantation enhances delivery of antigen-presenting cells to draining LNs and provokes unwanted immune responses [177, 178].

Preventing graft rejection after transplantation were investigated by application of anti-VEGFR-3 Ab in animal models of transplanted cornea [179] and pancreas islet cells [177]. Suppression of lymphangiogenesis in these experimentally designed transplantations improves survival of cornea and pancreatic islet cells [177]. It has been suggested that short-term selective inhibition of lymphangiogenesis before transplantation might help prevent early postoperative access of inflammatory cells to the lymphatic system and in this manner improve overall graft survival [178]. In addition, the novel anti-lymphangiogenic candidate—inhibitor of Ang-2—may prevent graft rejection [180].

2.4.3 Pro-lymphangiogenic Therapy in Secondary Lymphedema

VEGF-C gene transfer via adenoviruses, adeno-associated viruses, or naked plasmids, as well as the application of recombinant VEGF-C protein represent novel promising therapies for lymphedema. Accordingly, VEGF-C gene therapies were shown to stimulate formation of new lymphatic capillaries and alleviated edema in preclinical animal models of primary or secondary lymphedema [1, 181].

Pro-lymphagiogenic therapy was also used as a part of experimental LNs transplantation in mouse model of axillary lymphadenectomy. It was demonstrated that collecting lymphatic vessels can be regenerated and fused to LNs transplants after LNs removal. Treatment with adenovirally delivered VEGF-C or VEGF-D induced robust growth of the lymphatic capillaries, which matured and became functional. VEGF-C therapy improved also the success of LNs survival. Without application of VEGF-C the transplantation would lead to atrophy of LNs structure. LNs transplantation associated with VEGF-C therapy provides thus a working model for treating individuals with secondary lymphedema due to cancer [182].

2.4.4 Anti-lymphangiogenic Therapy in Cancer

Blocking tumor-induced lymphangiogenesis has emerged as an attractive therapeutic goal in cancer treatment. Although over the past two decades many key factors have been identified as important regulators for tumor lymphangiogenesis, the major focus in anti-lymphangiogenic therapy has been targeting VEGF-C/D–VEGFR-3 signaling pathway [183]. Many experimental mouse models of tumor lymphangiogenesis have been employed to test different approaches for repressing this signaling pathway. Preclinical and clinical trials on animal models of many types of cancer using antibodies to neutralize lymphatic growth factors, anti-VEGFR-3 Ab or soluble VEGFR-3, demonstrated reduction of tumor lymphangiogenesis and, to varying degrees, metastasis to LNs and distant organs [151, 184–187]. Experiments using indirect approaches to inhibit VEGF-C–VEGFR-3 signaling axis, e.g., with antibodies against the Nrp-2 [188], with enediynyl peptide inhibitor, that has been developed to block the furin-mediated processing of pro-VEGF-C to mature VEGF-C [189], with indoline (MAZ51), that is able to inhibit activation of VEGFR-3 in response to ligands [190, 191], or with celecoxib (COX-2 inhibitor), that causes downregulation of VEGF-C expression [192], showed similar results.

Beyond inhibiting VEGF-C/D—VEGFR-3 signaling pathway, a variety of other lymphangiogenesis inhibitors have been defined and developed. Among them is hyperforin, and his synthetic form—aristoforin. Both of them have potent antitumor, anti-angiogenesis, and anti-lymphangiogenesis properties [193]. At low concentrations of both substances, LECs enter cell-cycle arrest, whereas higher concentrations induce apoptosis of LECs [193]. Antagonists of integrin $\alpha4\beta1$ were also demonstrated to significantly inhibit lymphangiogenesis in wound healing and pathologic lymphangiogenesis in skin tumor models [194].

Although these anti-lymphangiogenic therapies in cancer have encouraging results tested in animal models, they may cause possible significant side effects in humans [168]. Fluid homeostasis requires an intact lymphatic system and inhibition of lymphangiogenesis may increase the risk of lymphedema. Theoretically, therapies that inhibit VEGF-C/D—VEGFR-3 signaling should be safe for preexisting lymphatic vessels in the adults, because VEGF-C/D—VEGFR-3 signaling is not required for the maintenance of preexisting lymphatic vasculature [186]. Indeed, inhibition of the VEGFR-3 pathway in adult mice did not affect normal lymphatic vessels [99]. Nevertheless, the risk of lymphedema during such therapies have to be taken into account. In addition, inhibition of lymphangiogenesis may interfere with physiological processes such as wound healing or tissue regeneration, and preexisting vessels that will be not affected by targeted therapy may serve as potential routes for cancer cell dissemination [168, 195].

Tyrosine kinase inhibitors of the VEGF-A pathway, as inhibitors of angiogenesis and lymphangiogenesis, are being evaluated in clinical trials in advanced colorectal cancer, renal cell, breast, and non-small-cell lung cancer. Although they have some anti-lymphangiogenic properties, they are focused on anti-angiogenic and anti-tumor growth effects, and not anti-lymphangiogenic ones [183]. It was demonstrated that these inhibitors are well tolerated, display low toxicity, and have beneficial effects on overall and regression-free survival. However, there are also observations that they inhibit only tumor growth and angiogenesis but not LNs metastasis. Therefore, applying a combination of inhibitors of both, VEGF-A and VEGF-C, possibly may be more effective in inhibition of tumor progression [196].

References

1. Alitalo K, Tammela T, Petrova TV (2005) Lymphangiogenesis in development and human disease. Nature 438:946–953
2. Xu F, Stouffer RL (2009) Existence of the lymphatic system in the primate corpus luteum. Lymphat Res Biol 7:159–168
3. Achen MG, McColl BK, Stacker SA (2005) Focus on lymphangiogenesis in tumor metastasis. Cancer Cell 7:121–127
4. Karpanen T, Alitalo K (2008) Molecular biology and pathology of lymphangiogenesis. Annu Rev Pathol 3:367–397
5. Breier G, Breviario F, Caveda L, Berthier R, Schnurch H, Gotsch U, Vestweber D, Risau W, Dejana E (1996) Molecular cloning and expression of murine vascular endothelial-cadherin in early stage development of cardiovascular system. Blood 87:630–641
6. Wigle JT, Oliver G (1999) Prox1 function is required for the development of the murine lymphatic system. Cell 98:769–778
7. van der Putte SC (1975) The development of the lymphatic system in man. Adv Anat Embryol Cell Biol 51:3–60
8. Lewis FT (1909) On the cervical veins and lymphatics in four human embryos. With an interpretation of anomalies of the subclavian and jugular veins in the adult. Am J Anat 9:33–43
9. Sabin FR (1909) The lymphatic system in human embryos, with consideration of the system as a whole. Am J Anat 9:43–91
10. Faul JL, Berry GJ, Colby TV, Ruoss SJ, Walter MB, Rosen GD, Raffin TA (2000) Thoracic lymphangiomas, lymphangiectasis, lymphangiomatosis, and lymphatic dysplasia syndrome. Am J Respir Crit Care Med 161:1037–1046
11. Wiegand S, Eivazi B, Barth PJ, von Rautenfeld DB, Folz BJ, Mandic R, Werner JA (2008) Pathogenesis of lymphangiomas. Virchows Arch 453:1–8
12. Sabin F (1902) On the origin of the lymphatics system from the veins and the development of the lymph hearts and thorarcic duct in the pig. Am J Anat 1:367–391
13. Sabin F (1904) On the development of the superficial lymphatics in the skin of the pig. Am J Anat 3:183–195
14. Srinivasan RS, Dillard ME, Lagutin OV, Lin FJ, Tsai S, Tsai MJ, Samokhvalov IM, Oliver G (2007) Lineage tracing demonstrates the venous origin of the mammalian lymphatic vasculature. Genes Dev 21:2422–2432
15. Yaniv K, Isogai S, Castranova D, Dye L, Hitomi J, Weinstein BM (2006) Live imaging of lymphatic development in the zebrafish. Nat Med 12:711–716
16. Karkkainen MJ, Haiko P, Sainio K, Partanen J, Taipale J, Petrova TV, Jeltsch M, Jackson DG, Talikka M, Rauvala H, Betsholtz C, Alitalo K (2004) Vascular endothelial growth factor C is required for sprouting of the first lymphatic vessels from embryonic veins. Nat Immunol 5:74–80
17. Oliver G (2004) Lymphatic vasculature development. Nat Rev Immunol 4:35–45
18. Huntington GC, Mc Clure CFW (1910) The anatomy and development of the jugular lymph sacs in the domestic cat (Felis domestica). Am J Anat 10:177–311

19. Wilting J, Aref Y, Huang R, Tomarev SI, Schweigerer L, Christ B, Valasek P, Papoutsi M (2006) Dual origin of avian lymphatics. Dev Biol 292:165–173
20. Buttler K, Kreysing A, von Kaisenberg CS, Schweigerer L, Gale N, Papoutsi M, Wilting J (2006) Mesenchymal cells with leukocyte and lymphendothelial characteristics in murine embryos. Dev Dyn 235:1554–1562
21. Buttler K, Ezaki T, Wilting J (2008) Proliferating mesodermal cells in murine embryos exhibiting macrophage and lymphendothelial characteristics. BMC Dev Biol 8:43
22. Ran S, Montgomery KE (2012) Macrophage-mediated lymphangiogenesis: the emerging role of macrophages as lymphatic endothelial progenitors. Cancers (Basel) 4:618–657
23. Asahara T, Kawamoto A (2004) Endothelial progenitor cells for postnatal vasculogenesis. Am J Physiol Cell Physiol 287:C572–579
24. Religa P, Cao R, Bjorndahl M, Zhou Z, Zhu Z, Cao Y (2005) Presence of bone marrow-derived circulating progenitor endothelial cells in the newly formed lymphatic vessels. Blood 106:4184–4190
25. Maruyama K, Ii M, Cursiefen C, Jackson DG, Keino H, Tomita M, Van Rooijen N, Takenaka H, D'Amore PA, Stein-Streilein J, Losordo DW, Streilein JW (2005) Inflammation-induced lymphangiogenesis in the cornea arises from CD11b-positive macrophages. J Clin Invest 115:2363–2372
26. Zumsteg A, Baeriswyl V, Imaizumi N, Schwendener R, Ruegg C, Christofori G (2009) Myeloid cells contribute to tumor lymphangiogenesis. PLoS One 4:e7067
27. Grunewald M, Avraham I, Dor Y, Bachar-Lustig E, Itin A, Jung S, Chimenti S, Landsman L, Abramovitch R, Keshet E (2006) VEGF-induced adult neovascularization: recruitment, retention, and role of accessory cells. Cell 124:175–189
28. Mounzer RH, Svendsen OS, Baluk P, Bergman CM, Padera TP, Wiig H, Jain RK, McDonald DM, Ruddle NH (2010) Lymphotoxin-alpha contributes to lymphangiogenesis. Blood 116:2173–2182
29. Fraser JR, Kimpton WG, Laurent TC, Cahill RN, Vakakis N (1988) Uptake and degradation of hyaluronan in lymphatic tissue. Biochem J 256:153–158
30. Gale NW, Prevo R, Espinosa J, Ferguson DJ, Dominguez MG, Yancopoulos GD, Thurston G, Jackson DG (2007) Normal lymphatic development and function in mice deficient for the lymphatic hyaluronan receptor LYVE-1. Mol Cell Biol 27:595–604
31. Hong YK, Harvey N, Noh YH, Schacht V, Hirakawa S, Detmar M, Oliver G (2002) Prox1 is a master control gene in the program specifying lymphatic endothelial cell fate. Dev Dyn 225:351–357
32. Francois M, Caprini A, Hosking B, Orsenigo F, Wilhelm D, Browne C, Paavonen K, Karnezis T, Shayan R, Downes M, Davidson T, Tutt D, Cheah KS, Stacker SA, Muscat GE, Achen MG, Dejana E, Koopman P (2008) Sox18 induces development of the lymphatic vasculature in mice. Nature 456:643–647
33. Srinivasan RS, Geng X, Yang Y, Wang Y, Mukatira S, Studer M, Porto MP, Lagutin O, Oliver G (2010) The nuclear hormone receptor Coup-TFII is required for the initiation and early maintenance of Prox1 expression in lymphatic endothelial cells. Genes Dev 24:696–707
34. Irrthum A, Devriendt K, Chitayat D, Matthijs G, Glade C, Steijlen PM, Fryns JP, Van Steensel MA, Vikkula M (2003) Mutations in the transcription factor gene SOX18 underlie recessive and dominant forms of hypotrichosis-lymphedema-telangiectasia. Am J Hum Genet 72:1470–1478
35. Lin FJ, Chen X, Qin J, Hong YK, Tsai MJ, Tsai SY (2010) Direct transcriptional regulation of neuropilin-2 by COUP-TFII modulates multiple steps in murine lymphatic vessel development. J Clin Invest 120:1694–1707
36. Kaipainen A, Korhonen J, Mustonen T, van Hinsbergh VW, Fang GH, Dumont D, Breitman M, Alitalo K (1995) Expression of the fms-like tyrosine kinase 4 gene becomes restricted to lymphatic endothelium during development. Proc Natl Acad Sci USA 92:3566–3570

37. Dumont DJ, Jussila L, Taipale J, Lymboussaki A, Mustonen T, Pajusola K, Breitman M, Alitalo K (1998) Cardiovascular failure in mouse embryos deficient in VEGF receptor-3. Science 282:946–949
38. Karkkainen MJ, Saaristo A, Jussila L, Karila KA, Lawrence EC, Pajusola K, Bueler H, Eichmann A, Kauppinen R, Kettunen MI, Yla-Herttuala S, Finegold DN, Ferrell RE, Alitalo K (2001) A model for gene therapy of human hereditary lymphedema. Proc Natl Acad Sci USA 98:12677–12682
39. Karkkainen MJ, Ferrell RE, Lawrence EC, Kimak MA, Levinson KL, McTigue MA, Alitalo K, Finegold DN (2000) Missense mutations interfere with VEGFR-3 signalling in primary lymphoedema. Nat Genet 25:153–159
40. Achen MG, Jeltsch M, Kukk E, Makinen T, Vitali A, Wilks AF, Alitalo K, Stacker SA (1998) Vascular endothelial growth factor D (VEGF-D) is a ligand for the tyrosine kinases VEGF receptor 2 (Flk1) and VEGF receptor 3 (Flt4). Proc Natl Acad Sci USA 95:548–553
41. Baldwin ME, Catimel B, Nice EC, Roufail S, Hall NE, Stenvers KL, Karkkainen MJ, Alitalo K, Stacker SA, Achen MG (2001) The specificity of receptor binding by vascular endothelial growth factor-d is different in mouse and man. J Biol Chem 276:19166–19171
42. Millauer B, Wizigmann-Voos S, Schnurch H, Martinez R, Moller NP, Risau W, Ullrich A (1993) High affinity VEGF binding and developmental expression suggest Flk-1 as a major regulator of vasculogenesis and angiogenesis. Cell 72:835–846
43. Cao R, Eriksson A, Kubo H, Alitalo K, Cao Y, Thyberg J (2004) Comparative evaluation of FGF-2-, VEGF-A-, and VEGF-C-induced angiogenesis, lymphangiogenesis, vascular fenestrations, and permeability. Circ Res 94:664–670
44. Kubo H, Cao R, Brakenhielm E, Makinen T, Cao Y, Alitalo K (2002) Blockade of vascular endothelial growth factor receptor-3 signaling inhibits fibroblast growth factor-2-induced lymphangiogenesis in mouse cornea. Proc Natl Acad Sci USA 99:8868–8873
45. Saaristo A, Veikkola T, Enholm B, Hytonen M, Arola J, Pajusola K, Turunen P, Jeltsch M, Karkkainen MJ, Kerjaschki D, Bueler H, Yla-Herttuala S, Alitalo K (2002) Adenoviral VEGF-C overexpression induces blood vessel enlargement, tortuosity, and leakiness but no sprouting angiogenesis in the skin or mucous membranes. FASEB J 16:1041–1049
46. Kukk E, Lymboussaki A, Taira S, Kaipainen A, Jeltsch M, Joukov V, Alitalo K (1996) VEGF-C receptor binding and pattern of expression with VEGFR-3 suggests a role in lymphatic vascular development. Development 122:3829–3837
47. Byzova TV, Goldman CK, Jankau J, Chen J, Cabrera G, Achen MG, Stacker SA, Carnevale KA, Siemionow M, Deitcher SR, DiCorleto PE (2002) Adenovirus encoding vascular endothelial growth factor-D induces tissue-specific vascular patterns in vivo. Blood 99:4434–4442
48. Baldwin ME, Halford MM, Roufail S, Williams RA, Hibbs ML, Grail D, Kubo H, Stacker SA, Achen MG (2005) Vascular endothelial growth factor D is dispensable for development of the lymphatic system. Mol Cell Biol 25:2441–2449
49. Karpanen T, Heckman CA, Keskitalo S, Jeltsch M, Ollila H, Neufeld G, Tamagnone L, Alitalo K (2006) Functional interaction of VEGF-C and VEGF-D with neuropilin receptors. FASEB J 20:1462–1472
50. Yuan L, Moyon D, Pardanaud L, Breant C, Karkkainen MJ, Alitalo K, Eichmann A (2002) Abnormal lymphatic vessel development in neuropilin 2 mutant mice. Development 129:4797–4806
51. Xu Y, Yuan L, Mak J, Pardanaud L, Caunt M, Kasman I, Larrivee B, Del Toro R, Suchting S, Medvinsky A, Silva J, Yang J, Thomas JL, Koch AW, Alitalo K, Eichmann A, Bagri A (2010) Neuropilin-2 mediates VEGF-C-induced lymphatic sprouting together with VEGFR3. J Cell Biol 188:115–130
52. Takashima S, Kitakaze M, Asakura M, Asanuma H, Sanada S, Tashiro F, Niwa H, Miyazaki Ji J, Hirota S, Kitamura Y, Kitsukawa T, Fujisawa H, Klagsbrun M, Hori M (2002) Targeting of both mouse neuropilin-1 and neuropilin-2 genes severely impairs developmental yolk sac and embryonic angiogenesis. Proc Natl Acad Sci USA 99:3657–3662

53. Oliver G, Srinivasan RS (2008) Lymphatic vasculature development: current concepts. Ann N Y Acad Sci 1131:75–81
54. Taniguchi K, Kohno R, Ayada T, Kato R, Ichiyama K, Morisada T, Oike Y, Yonemitsu Y, Maehara Y, Yoshimura A (2007) Spreds are essential for embryonic lymphangiogenesis by regulating vascular endothelial growth factor receptor 3 signaling. Mol Cell Biol 27:4541–4550
55. Abtahian F, Guerriero A, Sebzda E, Lu MM, Zhou R, Mocsai A, Myers EE, Huang B, Jackson DG, Ferrari VA, Tybulewicz V, Lowell CA, Lepore JJ, Koretzky GA, Kahn ML (2003) Regulation of blood and lymphatic vascular separation by signaling proteins SLP-76 and Syk. Science 299:247–251
56. Fritz-Six KL, Dunworth WP, Li M, Caron KM (2008) Adrenomedullin signaling is necessary for murine lymphatic vascular development. J Clin Invest 118:40–50
57. Hirashima M, Sano K, Morisada T, Murakami K, Rossant J, Suda T (2008) Lymphatic vessel assembly is impaired in Aspp1-deficient mouse embryos. Dev Biol 316:149–159
58. Langton PF, Colombani J, Aerne BL, Tapon N (2007) Drosophila ASPP regulates C-terminal Src kinase activity. Dev Cell 13:773–782
59. Yoon CM, Hong BS, Moon HG, Lim S, Suh PG, Kim YK, Chae CB, Gho YS (2008) Sphingosine-1-phosphate promotes lymphangiogenesis by stimulating S1P1/Gi/PLC/Ca2+ signaling pathways. Blood 112:1129–1138
60. Clavin NW, Avraham T, Fernandez J, Daluvoy SV, Soares MA, Chaudhry A, Mehrara BJ (2008) TGF-beta1 is a negative regulator of lymphatic regeneration during wound repair. Am J Physiol Heart Circ Physiol 295:H2113–2127
61. Cao R, Bjorndahl MA, Religa P, Clasper S, Garvin S, Galter D, Meister B, Ikomi F, Tritsaris K, Dissing S, Ohhashi T, Jackson DG, Cao Y (2004) PDGF-BB induces intratumoral lymphangiogenesis and promotes lymphatic metastasis. Cancer Cell 6:333–345
62. Chang LK, Garcia-Cardena G, Farnebo F, Fannon M, Chen EJ, Butterfield C, Moses MA, Mulligan RC, Folkman J, Kaipainen A (2004) Dose-dependent response of FGF-2 for lymphangiogenesis. Proc Natl Acad Sci USA 101:11658–11663
63. Bjorndahl M, Cao R, Nissen LJ, Clasper S, Johnson LA, Xue Y, Zhou Z, Jackson D, Hansen AJ, Cao Y (2005) Insulin-like growth factors 1 and 2 induce lymphangiogenesis in vivo. Proc Natl Acad Sci USA 102:15593–15598
64. Kajiya K, Hirakawa S, Ma B, Drinnenberg I, Detmar M (2005) Hepatocyte growth factor promotes lymphatic vessel formation and function. EMBO J 24:2885–2895
65. Banziger-Tobler NE, Halin C, Kajiya K, Detmar M (2008) Growth hormone promotes lymphangiogenesis. Am J Pathol 173:586–597
66. Breiteneder-Geleff S, Soleiman A, Kowalski H, Horvat R, Amann G, Kriehuber E, Diem K, Weninger W, Tschachler E, Alitalo K, Kerjaschki D (1999) Angiosarcomas express mixed endothelial phenotypes of blood and lymphatic capillaries: podoplanin as a specific marker for lymphatic endothelium. Am J Pathol 154:385–394
67. Schacht V, Ramirez MI, Hong YK, Hirakawa S, Feng D, Harvey N, Williams M, Dvorak AM, Dvorak HF, Oliver G, Detmar M (2003) T1alpha/podoplanin deficiency disrupts normal lymphatic vasculature formation and causes lymphedema. EMBO J 22:3546–3556
68. Suzuki-Inoue K, Kato Y, Inoue O, Kaneko MK, Mishima K, Yatomi Y, Yamazaki Y, Narimatsu H, Ozaki Y (2007) Involvement of the snake toxin receptor CLEC-2, in podoplanin-mediated platelet activation, by cancer cells. J Biol Chem 282:25993–26001
69. Sebzda E, Hibbard C, Sweeney S, Abtahian F, Bezman N, Clemens G, Maltzman JS, Cheng L, Liu F, Turner M, Tybulewicz V, Koretzky GA, Kahn ML (2006) Syk and Slp-76 mutant mice reveal a cell-autonomous hematopoietic cell contribution to vascular development. Dev Cell 11:349–361
70. Backhed F, Crawford PA, O'Donnell D, Gordon JI (2007) Postnatal lymphatic partitioning from the blood vasculature in the small intestine requires fasting-induced adipose factor. Proc Natl Acad Sci USA 104:606–611

71. Maby-El Hajjami H, Petrova TV (2008) Developmental and pathological lymphangiogenesis: from models to human disease. Histochem Cell Biol 130:1063–1078
72. Albrecht I, Christofori G (2011) Molecular mechanisms of lymphangiogenesis in development and cancer. Int J Dev Biol 55:483–494
73. Petrova TV, Karpanen T, Norrmen C, Mellor R, Tamakoshi T, Finegold D, Ferrell R, Kerjaschki D, Mortimer P, Yla-Herttuala S, Miura N, Alitalo K (2004) Defective valves and abnormal mural cell recruitment underlie lymphatic vascular failure in lymphedema distichiasis. Nat Med 10:974–981
74. Dagenais SL, Hartsough RL, Erickson RP, Witte MH, Butler MG, Glover TW (2004) Foxc2 is expressed in developing lymphatic vessels and other tissues associated with lymphedema-distichiasis syndrome. Gene Expr Patterns 4:611–619
75. Norrmen C, Ivanov KI, Cheng J, Zangger N, Delorenzi M, Jaquet M, Miura N, Puolakkainen P, Horsley V, Hu J, Augustin HG, Yla-Herttuala S, Alitalo K, Petrova TV (2009) FOXC2 controls formation and maturation of lymphatic collecting vessels through cooperation with NFATc1. J Cell Biol 185:439–457
76. Fang J, Dagenais SL, Erickson RP, Arlt MF, Glynn MW, Gorski JL, Seaver LH, Glover TW (2000) Mutations in FOXC2 (MFH-1), a forkhead family transcription factor, are responsible for the hereditary lymphedema-distichiasis syndrome. Am J Hum Genet 67:1382–1388
77. Augustin HG, Koh GY, Thurston G, Alitalo K (2009) Control of vascular morphogenesis and homeostasis through the angiopoietin-Tie system. Nat Rev Mol Cell Biol 10:165–177
78. Saharinen P, Kerkela K, Ekman N, Marron M, Brindle N, Lee GM, Augustin H, Koh GY, Alitalo K (2005) Multiple angiopoietin recombinant proteins activate the Tie1 receptor tyrosine kinase and promote its interaction with Tie2. J Cell Biol 169:239–243
79. Davis S, Aldrich TH, Jones PF, Acheson A, Compton DL, Jain V, Ryan TE, Bruno J, Radziejewski C, Maisonpierre PC, Yancopoulos GD (1996) Isolation of angiopoietin-1, a ligand for the TIE2 receptor, by secretion-trap expression cloning. Cell 87:1161–1169
80. Gale NW, Thurston G, Hackett SF, Renard R, Wang Q, McClain J, Martin C, Witte C, Witte MH, Jackson D, Suri C, Campochiaro PA, Wiegand SJ, Yancopoulos GD (2002) Angiopoietin-2 is required for postnatal angiogenesis and lymphatic patterning, and only the latter role is rescued by Angiopoietin-1. Dev Cell 3:411–423
81. Lee HJ, Cho CH, Hwang SJ, Choi HH, Kim KT, Ahn SY, Kim JH, Oh JL, Lee GM, Koh GY (2004) Biological characterization of angiopoietin-3 and angiopoietin-4. FASEB J 18:1200–1208
82. Vikkula M, Boon LM, Carraway KL 3rd, Calvert JT, Diamonti AJ, Goumnerov B, Pasyk KA, Marchuk DA, Warman ML, Cantley LC, Mulliken JB, Olsen BR (1996) Vascular dysmorphogenesis caused by an activating mutation in the receptor tyrosine kinase TIE2. Cell 87:1181–1190
83. Makinen T, Adams RH, Bailey J, Lu Q, Ziemiecki A, Alitalo K, Klein R, Wilkinson GA (2005) PDZ interaction site in ephrinB2 is required for the remodeling of lymphatic vasculature. Genes Dev 19:397–410
84. Zhou F, Chang Z, Zhang L, Hong YK, Shen B, Wang B, Zhang F, Lu G, Tvorogov D, Alitalo K, Hemmings BA, Yang Z, He Y (2010) Akt/Protein kinase B is required for lymphatic network formation, remodeling, and valve development. Am J Pathol 177:2124–2133
85. Chen J, Alexander JS, Orr AW (2012) Integrins and their extracellular matrix ligands in lymphangiogenesis and lymph node metastasis. Int J Cell Biol 2012:853703
86. Gerli R, Solito R, Weber E, Agliano M (2000) Specific adhesion molecules bind anchoring filaments and endothelial cells in human skin initial lymphatics. Lymphology 33:148–157
87. Vlahakis NE, Young BA, Atakilit A, Sheppard D (2005) The lymphangiogenic vascular endothelial growth factors VEGF-C and -D are ligands for the integrin alpha9beta1. J Biol Chem 280:4544–4552
88. Bazigou E, Xie S, Chen C, Weston A, Miura N, Sorokin L, Adams R, Muro AF, Sheppard D, Makinen T (2009) Integrin-alpha9 is required for fibronectin matrix assembly during lymphatic valve morphogenesis. Dev Cell 17:175–186

89. Ma GC, Liu CS, Chang SP, Yeh KT, Ke YY, Chen TH, Wang BB, Kuo SJ, Shih JC, Chen M (2008) A recurrent ITGA9 missense mutation in human fetuses with severe chylothorax: possible correlation with poor response to fetal therapy. Prenat Diagn 28:1057–1063
90. Danussi C, Spessotto P, Petrucco A, Wassermann B, Sabatelli P, Montesi M, Doliana R, Bressan GM, Colombatti A (2008) Emilin1 deficiency causes structural and functional defects of lymphatic vasculature. Mol Cell Biol 28:4026–4039
91. Podgrabinska S, Braun P, Velasco P, Kloos B, Pepper MS, Skobe M (2002) Molecular characterization of lymphatic endothelial cells. Proc Natl Acad Sci USA 99:16069–16074
92. Hogan BM, Bos FL, Bussmann J, Witte M, Chi NC, Duckers HJ, Schulte-Merker S (2009) Ccbe1 is required for embryonic lymphangiogenesis and venous sprouting. Nat Genet 41:396–398
93. Alders M, Hogan BM, Gjini E, Salehi F, Al-Gazali L, Hennekam EA, Holmberg EE, Mannens MM, Mulder MF, Offerhaus GJ, Prescott TE, Schroor EJ, Verheij JB, Witte M, Zwijnenburg PJ, Vikkula M, Schulte-Merker S, Hennekam RC (2009) Mutations in CCBE1 cause generalized lymph vessel dysplasia in humans. Nat Genet 41:1272–1274
94. Hahn C, Schwartz MA (2009) Mechanotransduction in vascular physiology and atherogenesis. Nat Rev Mol Cell Biol 10:53–62
95. Chiquet M, Gelman L, Lutz R, Maier S (2009) From mechanotransduction to extracellular matrix gene expression in fibroblasts. Biochim Biophys Acta 1793:911–920
96. Rutkowski JM, Swartz MA (2007) A driving force for change: interstitial flow as a morphoregulator. Trends Cell Biol 17:44–50
97. Goldman J, Conley KA, Raehl A, Bondy DM, Pytowski B, Swartz MA, Rutkowski JM, Jaroch DB, Ongstad EL (2007) Regulation of lymphatic capillary regeneration by interstitial flow in skin. Am J Physiol Heart Circ Physiol 292:H2176–2183
98. Ng CP, Helm CL, Swartz MA (2004) Interstitial flow differentially stimulates blood and lymphatic endothelial cell morphogenesis in vitro. Microvasc Res 68:258–264
99. Karpanen T, Wirzenius M, Makinen T, Veikkola T, Haisma HJ, Achen MG, Stacker SA, Pytowski B, Yla-Herttuala S, Alitalo K (2006) Lymphangiogenic growth factor responsiveness is modulated by postnatal lymphatic vessel maturation. Am J Pathol 169:708–718
100. Enholm B, Karpanen T, Jeltsch M, Kubo H, Stenback F, Prevo R, Jackson DG, Yla-Herttuala S, Alitalo K (2001) Adenoviral expression of vascular endothelial growth factor-C induces lymphangiogenesis in the skin. Circ Res 88:623–629
101. Baluk P, Tammela T, Ator E, Lyubynska N, Achen MG, Hicklin DJ, Jeltsch M, Petrova TV, Pytowski B, Stacker SA, Yla-Herttuala S, Jackson DG, Alitalo K, McDonald DM (2005) Pathogenesis of persistent lymphatic vessel hyperplasia in chronic airway inflammation. J Clin Invest 115:247–257
102. Kaiserling E, Krober S, Geleff S (2003) Lymphatic vessels in the colonic mucosa in ulcerative colitis. Lymphology 36:52–61
103. Zhang Q, Lu Y, Proulx ST, Guo R, Yao Z, Schwarz EM, Boyce BF, Xing L (2007) Increased lymphangiogenesis in joints of mice with inflammatory arthritis. Arthritis Res Ther 9:R118
104. Kunstfeld R, Hirakawa S, Hong YK, Schacht V, Lange-Asschenfeldt B, Velasco P, Lin C, Fiebiger E, Wei X, Wu Y, Hicklin D, Bohlen P, Detmar M (2004) Induction of cutaneous delayed-type hypersensitivity reactions in VEGF-A transgenic mice results in chronic skin inflammation associated with persistent lymphatic hyperplasia. Blood 104:1048–1057
105. Henno A, Blacher S, Lambert C, Colige A, Seidel L, Noel A, Lapiere C, de la Brassinne M, Nusgens BV (2009) Altered expression of angiogenesis and lymphangiogenesis markers in the uninvolved skin of plaque-type psoriasis. Br J Dermatol 160:581–590
106. Kajiya K, Sawane M, Huggenberger R, Detmar M (2009) Activation of the VEGFR-3 pathway by VEGF-C attenuates UVB-induced edema formation and skin inflammation by promoting lymphangiogenesis. J Invest Dermatol 129:1292–1298

142. Kataru RP, Kim H, Jang C, Choi DK, Koh BI, Kim M, Gollamudi S, Kim YK, Lee SH, Koh GY (2011) T lymphocytes negatively regulate lymph node lymphatic vessel formation. Immunity 34:96–107
143. Kim H, Kataru RP, Koh GY (2012) Regulation and implications of inflammatory lymphangiogenesis. Trends Immunol 33:350–356
144. Mumprecht V, Roudnicky F, Detmar M (2012) Inflammation-induced lymph node lymphangiogenesis is reversible. Am J Pathol 180:874–879
145. Fisher B, Fisher ER (1966) The interrelationship of hematogenous and lymphatic tumor cell dissemination. Surg Gynecol Obstet 122:791–798
146. Li T, Yang J, Zhou Q, He Y (2012) Molecular regulation of lymphangiogenesis in development and tumor microenvironment. Cancer Microenviron 5:249–260
147. Van der Auwera I, Cao Y, Tille JC, Pepper MS, Jackson DG, Fox SB, Harris AL, Dirix LY, Vermeulen PB (2006) First international consensus on the methodology of lymphangiogenesis quantification in solid human tumours. Br J Cancer 95:1611–1625
148. Stacker SA, Caesar C, Baldwin ME, Thornton GE, Williams RA, Prevo R, Jackson DG, Nishikawa S, Kubo H, Achen MG (2001) VEGF-D promotes the metastatic spread of tumor cells via the lymphatics. Nat Med 7:186–191
149. Schoppmann SF, Fenzl A, Nagy K, Unger S, Bayer G, Geleff S, Gnant M, Horvat R, Jakesz R, Birner P (2006) VEGF-C expressing tumor-associated macrophages in lymph node positive breast cancer: impact on lymphangiogenesis and survival. Surgery 139:839–846
150. Mandriota SJ, Jussila L, Jeltsch M, Compagni A, Baetens D, Prevo R, Banerji S, Huarte J, Montesano R, Jackson DG, Orci L, Alitalo K, Christofori G, Pepper MS (2001) Vascular endothelial growth factor-C-mediated lymphangiogenesis promotes tumour metastasis. EMBO J 20:672–682
151. He Y, Rajantie I, Pajusola K, Jeltsch M, Holopainen T, Yla-Herttuala S, Harding T, Jooss K, Takahashi T, Alitalo K (2005) Vascular endothelial cell growth factor receptor 3-mediated activation of lymphatic endothelium is crucial for tumor cell entry and spread via lymphatic vessels. Cancer Res 65:4739–4746
152. Padera TP, Kadambi A, di Tomaso E, Carreira CM, Brown EB, Boucher Y, Choi NC, Mathisen D, Wain J, Mark EJ, Munn LL, Jain RK (2002) Lymphatic metastasis in the absence of functional intratumor lymphatics. Science 296:1883–1886
153. McAllaster JD, Cohen MS (2011) Role of the lymphatics in cancer metastasis and chemotherapy applications. Adv Drug Deliv Rev 63:867–875
154. Hirakawa S, Kodama S, Kunstfeld R, Kajiya K, Brown LF, Detmar M (2005) VEGF-A induces tumor and sentinel lymph node lymphangiogenesis and promotes lymphatic metastasis. J Exp Med 201:1089–1099
155. Alitalo K (2011) The lymphatic vasculature in disease. Nat Med 17:1371–1380
156. He Y, Rajantie I, Ilmonen M, Makinen T, Karkkainen MJ, Haiko P, Salven P, Alitalo K (2004) Preexisting lymphatic endothelium but not endothelial progenitor cells are essential for tumor lymphangiogenesis and lymphatic metastasis. Cancer Res 64:3737–3740
157. Condeelis J, Pollard JW (2006) Macrophages: obligate partners for tumor cell migration, invasion, and metastasis. Cell 124:263–266
158. Pollard JW (2004) Tumour-educated macrophages promote tumour progression and metastasis. Nat Rev Cancer 4:71–78
159. Koyama H, Kobayashi N, Harada M, Takeoka M, Kawai Y, Sano K, Fujimori M, Amano J, Ohhashi T, Kannagi R, Kimata K, Taniguchi S, Itano N (2008) Significance of tumor-associated stroma in promotion of intratumoral lymphangiogenesis: pivotal role of a hyaluronan-rich tumor microenvironment. Am J Pathol 172:179–193
160. Shields JD, Kourtis IC, Tomei AA, Roberts JM, Swartz MA (2010) Induction of lymphoidlike stroma and immune escape by tumors that express the chemokine CCL21. Science 328:749–752

161. Muller A, Homey B, Soto H, Ge N, Catron D, Buchanan ME, McClanahan T, Murphy E, Yuan W, Wagner SN, Barrera JL, Mohar A, Verastegui E, Zlotnik A (2001) Involvement of chemokine receptors in breast cancer metastasis. Nature 410:50–56
162. Irjala H, Alanen K, Grenman R, Heikkila P, Joensuu H, Jalkanen S (2003) Mannose receptor (MR) and common lymphatic endothelial and vascular endothelial receptor (CLEVER)-1 direct the binding of cancer cells to the lymph vessel endothelium. Cancer Res 63:4671–4676
163. Karpanen T, Egeblad M, Karkkainen MJ, Kubo H, Yla-Herttuala S, Jaattela M, Alitalo K (2001) Vascular endothelial growth factor C promotes tumor lymphangiogenesis and intralymphatic tumor growth. Cancer Res 61:1786–1790
164. Kozaki K, Miyaishi O, Tsukamoto T, Tatematsu Y, Hida T, Takahashi T (2000) Establishment and characterization of a human lung cancer cell line NCI-H460-LNM35 with consistent lymphogenous metastasis via both subcutaneous and orthotopic propagation. Cancer Res 60:2535–2540
165. Kraizer Y, Mawasi N, Seagal J, Paizi M, Assy N, Spira G (2001) Vascular endothelial growth factor and angiopoietin in liver regeneration. Biochem Biophys Res Commun 287:209–215
166. Qian CN, Berghuis B, Tsarfaty G, Bruch M, Kort EJ, Ditlev J, Tsarfaty I, Hudson E, Jackson DG, Petillo D, Chen J, Resau JH, Teh BT (2006) Preparing the "soil": the primary tumor induces vasculature reorganization in the sentinel lymph node before the arrival of metastatic cancer cells. Cancer Res 66:10365–10376
167. Hirakawa S, Brown LF, Kodama S, Paavonen K, Alitalo K, Detmar M (2007) VEGF-C-induced lymphangiogenesis in sentinel lymph nodes promotes tumor metastasis to distant sites. Blood 109:1010–1017
168. Witte MH, Dellinger MT, McDonald DM, Nathanson SD, Boccardo FM, Campisi CC, Sleeman JP, Gershenwald JE (2011) Lymphangiogenesis and hemangiogenesis: potential targets for therapy. J Surg Oncol 103:489–500
169. Spring H, Schuler T, Arnold B, Hammerling GJ, Ganss R (2005) Chemokines direct endothelial progenitors into tumor neovessels. Proc Natl Acad Sci USA 102:18111–18116
170. Titze J, Machnik A (2010) Sodium sensing in the interstitium and relationship to hypertension. Curr Opin Nephrol Hypertens 19:385–392
171. Machnik A, Neuhofer W, Jantsch J, Dahlmann A, Tammela T, Machura K, Park JK, Beck FX, Muller DN, Derer W, Goss J, Ziomber A, Dietsch P, Wagner H, van Rooijen N, Kurtz A, Hilgers KF, Alitalo K, Eckardt KU, Luft FC, Kerjaschki D, Titze J (2009) Macrophages regulate salt-dependent volume and blood pressure by a vascular endothelial growth factor-C-dependent buffering mechanism. Nat Med 15:545–552
172. Machnik A, Dahlmann A, Kopp C, Goss J, Wagner H, van Rooijen N, Eckardt KU, Muller DN, Park JK, Luft FC, Kerjaschki D, Titze J (2010) Mononuclear phagocyte system depletion blocks interstitial tonicity-responsive enhancer binding protein/vascular endothelial growth factor C expression and induces salt-sensitive hypertension in rats. Hypertension 55:755–761
173. Halin C, Fahrngruber H, Meingassner JG, Bold G, Littlewood-Evans A, Stuetz A, Detmar M (2008) Inhibition of chronic and acute skin inflammation by treatment with a vascular endothelial growth factor receptor tyrosine kinase inhibitor. Am J Pathol 173:265–277
174. Schonthaler HB, Huggenberger R, Wculek SK, Detmar M, Wagner EF (2009) Systemic anti-VEGF treatment strongly reduces skin inflammation in a mouse model of psoriasis. Proc Natl Acad Sci USA 106:21264–21269
175. Huggenberger R, Ullmann S, Proulx ST, Pytowski B, Alitalo K, Detmar M (2010) Stimulation of lymphangiogenesis via VEGFR-3 inhibits chronic skin inflammation. J Exp Med 207:2255–2269
176. Guo R, Zhou Q, Proulx ST, Wood R, Ji RC, Ritchlin CT, Pytowski B, Zhu Z, Wang YJ, Schwarz EM, Xing L (2009) Inhibition of lymphangiogenesis and lymphatic drainage via vascular endothelial growth factor receptor 3 blockade increases the severity of inflammation in a mouse model of chronic inflammatory arthritis. Arthritis Rheum 60:2666–2676
177. Yin N, Zhang N, Xu J, Shi Q, Ding Y, Bromberg JS (2011) Targeting lymphangiogenesis after islet transplantation prolongs islet allograft survival. Transplantation 92:25–30

178. Dietrich T, Bock F, Yuen D, Hos D, Bachmann BO, Zahn G, Wiegand S, Chen L, Cursiefen C (2010) Cutting edge: lymphatic vessels, not blood vessels, primarily mediate immune rejections after transplantation. J Immunol 184:535–539
179. Albuquerque RJ, Hayashi T, Cho WG, Kleinman ME, Dridi S, Takeda A, Baffi JZ, Yamada K, Kaneko H, Green MG, Chappell J, Wilting J, Weich HA, Yamagami S, Amano S, Mizuki N, Alexander JS, Peterson ML, Brekken RA, Hirashima M, Capoor S, Usui T, Ambati BK, Ambati J (2009) Alternatively spliced vascular endothelial growth factor receptor-2 is an essential endogenous inhibitor of lymphatic vessel growth. Nat Med 15:1023–1030
180. Yan ZX, Jiang ZH, Liu NF (2012) Angiopoietin-2 promotes inflammatory lymphangiogenesis and its effect can be blocked by the specific inhibitor L1-10. Am J Physiol Heart Circ Physiol 302:H215–223
181. Szuba A, Skobe M, Karkkainen MJ, Shin WS, Beynet DP, Rockson NB, Dakhil N, Spilman S, Goris ML, Strauss HW, Quertermous T, Alitalo K, Rockson SG (2002) Therapeutic lymphangiogenesis with human recombinant VEGF-C. FASEB J 16:1985–1987
182. Tammela T, Saaristo A, Holopainen T, Lyytikka J, Kotronen A, Pitkonen M, Abo-Ramadan U, Yla-Herttuala S, Petrova TV, Alitalo K (2007) Therapeutic differentiation and maturation of lymphatic vessels after lymph node dissection and transplantation. Nat Med 13:1458–1466
183. Duong T, Koopman P, Francois M (2012) Tumor lymphangiogenesis as a potential therapeutic target. J Oncol 2012:204946
184. Roberts N, Kloos B, Cassella M, Podgrabinska S, Persaud K, Wu Y, Pytowski B, Skobe M (2006) Inhibition of VEGFR-3 activation with the antagonistic antibody more potently suppresses lymph node and distant metastases than inactivation of VEGFR-2. Cancer Res 66:2650–2657
185. Wong SY, Haack H, Crowley D, Barry M, Bronson RT, Hynes RO (2005) Tumor-secreted vascular endothelial growth factor-C is necessary for prostate cancer lymphangiogenesis, but lymphangiogenesis is unnecessary for lymph node metastasis. Cancer Res 65:9789–9798
186. Pytowski B, Goldman J, Persaud K, Wu Y, Witte L, Hicklin DJ, Skobe M, Boardman KC, Swartz MA (2005) Complete and specific inhibition of adult lymphatic regeneration by a novel VEGFR-3 neutralizing antibody. J Natl Cancer Inst 97:14–21
187. Lin J, Lalani AS, Harding TC, Gonzalez M, Wu WW, Luan B, Tu GH, Koprivnikar K, VanRoey MJ, He Y, Alitalo K, Jooss K (2005) Inhibition of lymphogenous metastasis using adeno-associated virus-mediated gene transfer of a soluble VEGFR-3 decoy receptor. Cancer Res 65:6901–6909
188. Caunt M, Mak J, Liang WC, Stawicki S, Pan Q, Tong RK, Kowalski J, Ho C, Reslan HB, Ross J, Berry L, Kasman I, Zlot C, Cheng Z, Le Couter J, Filvaroff EH, Plowman G, Peale F, French D, Carano R, Koch AW, Wu Y, Watts RJ, Tessier-Lavigne M, Bagri A (2008) Blocking neuropilin-2 function inhibits tumor cell metastasis. Cancer Cell 13:331–342
189. Basak A, Khatib AM, Mohottalage D, Basak S, Kolajova M, Bag SS (2009) A novel enediynyl peptide inhibitor of furin that blocks processing of proPDGF-A, B and proVEGF-C. PLoS One 4:e7700
190. Kirkin V, Mazitschek R, Krishnan J, Steffen A, Waltenberger J, Pepper MS, Giannis A, Sleeman JP (2001) Characterization of indolinones which preferentially inhibit VEGF-C- and VEGF-D-induced activation of VEGFR-3 rather than VEGFR-2. Eur J Biochem 268:5530–5540
191. Kirkin V, Thiele W, Baumann P, Mazitschek R, Rohde K, Fellbrich G, Weich H, Waltenberger J, Giannis A, Sleeman JP (2004) MAZ51, an indolinone that inhibits endothelial cell and tumor cell growth in vitro, suppresses tumor growth in vivo. Int J Cancer 112:986–993
192. Liu H, Yang Y, Xiao J, Lv Y, Liu Y, Yang H, Zhao L (2009) Inhibition of cyclooxygenase-2 suppresses lymph node metastasis via VEGF-C. Anat Rec (Hoboken) 292:1577–1583
193. Rothley M, Schmid A, Thiele W, Schacht V, Plaumann D, Gartner M, Yektaoglu A, Bruyere F, Noel A, Giannis A, Sleeman JP (2009) Hyperforin and aristoforin inhibit lymphatic

endothelial cell proliferation in vitro and suppress tumor-induced lymphangiogenesis in vivo. Int J Cancer 125:34–42
194. Garmy-Susini B, Avraamides CJ, Schmid MC, Foubert P, Ellies LG, Barnes L, Feral C, Papayannopoulou T, Lowy A, Blair SL, Cheresh D, Ginsberg M, Varner JA (2010) Integrin alpha4beta1 signaling is required for lymphangiogenesis and tumor metastasis. Cancer Res 70:3042–3051
195. Thiele W, Sleeman JP (2006) Tumor-induced lymphangiogenesis: a target for cancer therapy? J Biotechnol 124:224–241
196. Schomber T, Zumsteg A, Strittmatter K, Crnic I, Antoniadis H, Littlewood-Evans A, Wood J, Christofori G (2009) Differential effects of the vascular endothelial growth factor receptor inhibitor PTK787/ZK222584 on tumor angiogenesis and tumor lymphangiogenesis. Mol Cancer Ther 8:55–63

Vascular-Resident Endothelial Side Population Cells

3

Hisamichi Naito and Nobuyuki Takakura

Contents

3.1 Tissue-Resident Adult Stem/Progenitor Cells and Side Population Cells 69
3.2 Stem/Progenitor Cells in the Peripheral Vascular System 71
 3.2.1 Vascular Wall Stem and Progenitor Cells (Fig. 3.1) 71
 3.2.2 Endothelial Progenitor Cells 73
3.3 Isolation and Characterization of Endothelial Side Population Cells 73
 3.3.1 Isolation of Endothelial Side Population Cells 73
 3.3.2 EC–SP Cells Do Not Originate from Bone Marrow Cells 75
 3.3.3 Proliferation and Colony Formation of EC–SP Cells 76
3.4 Therapeutic Potential of Endothelial Side Population Cells and Perspectives 76
References 80

Abstract

Postnatal neovascular formation was originally thought to be mediated by angiogenesis, which is defined as the formation of new blood vessels from pre-existing endothelial cells (ECs). However, over the last decade, it has been proposed that vasculogenesis, that is, de novo blood vessel generation from endothelial progenitor cells (EPCs) derived from bone marrow, may persist into adult life. Nonetheless, it is still a matter of debate as to what extent the EPCs contribute to new vessel formation in the adult.

H. Naito (✉)
Department of Signal Transduction, Research Institute for Microbial Diseases, Osaka University, 3-1 Yamada-oka, Suita, Osaka 565-0871, Japan
e-mail: naitohi@biken.osaka-u.ac.jp

N. Takakura
Department of Signal Transduction, Research Institute for Microbial Diseases, Osaka University, 3-1 Yamada-oka, Suita, Osaka 565-0871, Japan

JST, CREST, K's Gobancho, 7, Gobancho, Chiyoda-ku, Tokyo 102-0076, Japan
e-mail: ntakaku@biken.osaka-u.ac.jp

J. Dulak et al. (eds.), *Angiogenesis and Vascularisation*,
DOI 10.1007/978-3-7091-1428-5_3, © Springer-Verlag Wien 2013

In the peripheral blood vessels, the presence of different stem and progenitor cell types residing in the media and adventitia of the vascular wall has been suggested. These stem/progenitor cells were reported to have the ability to differentiate into ECs in culture and form capillary-like microvessels in ex vivo assays. However, the precise roles of these cells during angiogenic growth are not well defined. We recently isolated a novel endothelial stem/progenitor-like cells from the intima of adult murine blood vessels using the Hoechst method in which stem cell populations are identified as side populations. This vascular endothelial side population cell possesses colony-forming ability, generates large numbers of ECs, and when transplanted into ischemic lesions, these cells contribute to the newly formed long-term surviving blood vessels and restore blood flow completely. Our discovery of vascular endothelial side population cells that have features of endothelial stem/progenitor cells may lead to the identification of new targets for vascular regeneration therapy as well as vascular disrupting therapy.

Keywords
Endothelium • Side population • FACS analysis • Heterogeneity • Stemness • Regeneration

Abbreviations

BM	Bone marrow
ECs	Endothelial cells
EPCs	Endothelial progenitor cells
FACS	Fluorescence-activated cell sorting
HAECs	Human aortic endothelial cells
HC	Hematopoietic cell
HSC	Hematopoietic stem cell
HUVECs	Human umbilical vein endothelial cells
KSL	c-Kit$^+$ Sca-1$^+$, and Lineage$^-$
MP	Main population
SMCs	Smooth muscle cells
SP	Side population

3.1 Tissue-Resident Adult Stem/Progenitor Cells and Side Population Cells

Tissue stem cells are undifferentiated cells that have the ability to perpetuate themselves through self-renewal and to generate committed progenitor cells that differentiate into the tissue-specific cell lineages [1, 2]. Identification of these rare populations of adult stem cells in most tissues or organs has emerged as an attractive source of multipotent stem/progenitor cells for regenerative medicine. The tissues and organs harboring a small number of these specific adult stem/progenitor cells include bone marrow (BM), lung, liver, pancreas, heart, brain, intestinal epithelium, skin, adipose tissues, skeletal muscles, limbus, retina, breast, vascular walls, ovaries, uterus, prostate and testis. In general, these adult stem/progenitor cells are localized within a restricted tissue microenvironment known as a niche, consisting of neighboring cells that tightly regulate their functions through direct interactions and release of specific soluble factors [2, 3].

Adult stem cells have most commonly been isolated by fluorescence-activated cell sorting (FACS) with antibodies directed against specific stem cell markers. Among mammalian tissues, the hematopoietic system is the most advanced in terms of our knowledge of stem cell markers. The surface marker schemes for hematopoietic stem cells (HSCs) are primarily based on using c-Kit (K), Sca-1 (S), and Lineage markers (L), resulting in the so-called KSL population (c-Kit$^+$, Sca-1$^+$, and Lineage$^-$). KSL cells represent about 0.1 % of all BM cells and are enriched for progenitors, but bona fide stem cells comprise only about 10 % of the KSL population [4]. All long-term HSC activity in adult mouse BM is believed to be contained within a population marked by the composite phenotype of KSL, Thy-1.1^{lo}, Slamf1$^+$, Tie2$^+$, Flk2$^-$, and CD34$^-$ [5–10].

The other strategy to identify HSCs is based on their marked ability to efflux fluorescent dyes. Previous studies using Hoechst 33342 for enrichment of HSC exploited the fluorescence only at the blue emission wavelength [11]. With the use of two emission wavelengths, Goodell et al. reported a small and distinct cell population that has HSC properties [12]. They showed that this population correlates closely with cells expressing other cell surface markers previously identified on HSC. These cells contain the vast majority of HSC activity in murine BM and were shown by competitive repopulation experiments to be enriched at least 1,000-fold for in vivo reconstitution activity. These investigators also demonstrated that the exclusion of Hoechst 33342 by SP cells is an active process involving multidrug resistance transporter 1 (MDR1), a member of the ATP-binding cassette (ABC) family of transporter proteins. To confirm this notion, they applied the MDR1 inhibitor verapamil and showed that Hoechst 33342 exclusion by SP cells is blocked in its presence. However, verapamil is not specific only for the MDR1 transporter. Furthermore, an antibody to MDR1 does not identify SP cells exclusively, because MDR1 is expressed by 60–65 % of BM cells [13]. Thus, MDR1 cannot be taken as a single marker to identify and isolate SP cells, and additional transporters are likely to be involved in defining the SP phenotype. Consistent with this view, although both MDR1 and ABCG2 are expressed in BM SP cells, it is now believed that ABCG2 is the sole transporter accounting for dye efflux by SP cells isolated from BM [14].

The main advantage of this SP purification approach is its potential application to stem cells of other tissues for which no stem cell-specific surface markers are known. Since this method was reported, stem and progenitor populations in different tissue have been identified including in the heart, mammary gland, liver, lung, testis, epidermis, skeletal muscle, and others [15–21]. Moreover, identification of cancer stem-like cells has further raised interest in the Hoechst–SP method, and SP cells have been identified in a number of cancers [22, 23].

On the other hand, some reports suggest that Hoechst dye efflux is not a common property of all stem cells. Indeed, several differentiated tissues including intestine, kidney, and ECs of the blood–brain barrier constitutively express ABC transporters [24–26]. Moreover, the main disadvantage of the Hoechst–SP method is that ABC transporter-mediated dye efflux is an active metabolic process that is easily affected by the staining technique employed [27]. This high sensitivity to staining conditions led to discrepancies in results between laboratories, which in turn led to questioning the reliability of the Hoechst–SP method itself. Also, the FACS equipment, especially the UV laser, needed for SP analysis is not available to all investigators, which is a general limitation of SP analysis. However, if these difficulties can be overcome, the Hoechst–SP method can be applied to identify stem and progenitor cells of less-studied tissues for which there are as yet no specific surface markers. Among the latter, heart-resident progenitor cells may be one of the most well-characterized cell populations in terms of Hoechst–SP analysis. Hierlihy et al. presented the first evidence for the existence of an SP cell population in the adult heart [28]. They showed the stem cell-like activity of cardiac SP cells in a methylcellulose stem cell culture system, where approximately one colony per 50,000 cells plated was seen. These cells were capable of differentiating into the cardiomyocyte lineage in coculture experiments with primary cardiomyocytes. Cardiac SP cells represent 0.03–2.0 % of total mononuclear cardiac cells and generally express Sca1 but not hematopoietic markers, such as CD45 [21, 29–31]. Although the dye efflux property has been found to be a valid feature for isolating potential stem/progenitor cells, the SP populations are often heterogeneous. In several studies, combining the SP assay with additional phenotyping significantly increased the purity of the stem cells. Pfister et al. identified specific subpopulations within cardiac SP cells, including $Sca1^{+}CD31^{-}$ SP cells, which were found to be enriched for cardiomyogenic potential [31]. This group also showed that following myocardial injury, cardiac SP populations are acutely depleted and are reconstituted to normal levels by both self-proliferation and selective homing of BM cells [32]. Furthermore, Oyama et al. demonstrated the functional capacity of exogenous cardiac SP cells in vivo. They showed that when delivered intravenously, neonatal cardiac SP cells were able to migrate to areas of myocardial injury in a rat model of cryo-injury. Using immunohistochemical staining, they also showed that implanted cells were able to differentiate into cardiomyocytes, endothelial cells, smooth muscle cells, and fibroblasts [21]. However, the ability of

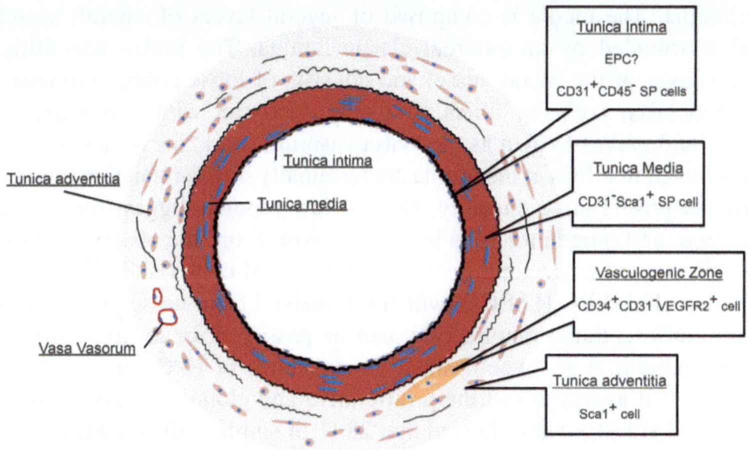

Fig. 3.1 Stem and progenitor cells in the vascular wall. Resident stem/progenitor cells have been identified in the vascular wall in large arteries, i.e., endothelial progenitor cells (EPCs) and $CD31^+CD45^-$ side population (SP) cells in the tunica intima, $CD31^+Sca1^+$ SP cells in the tunica media, $CD34^+CD31^-VEGFR2^+$ immature cells in the vasculogenic zone, and $Sca1^+$ immature cells in the tunica adventitia

endogenous resident cardiac SP cells to differentiate into cardiac myocytes or to contribute to functional repair of damaged heart has not yet been evaluated. Furthermore, the localization of cardiac SP cells and their niche are both unknown.

3.2 Stem/Progenitor Cells in the Peripheral Vascular System

The blood vessels in the body had long been considered to merely function as a transport compartment for the blood. However, it is now appreciated that the vasculature is one of the main tissues in the body with an area extending to more than 900 m^2 [33]. Regarding the vasculature as a large tissue, it would not be unexpected that blood vessels should harbor resident stem cells and progenitor cells. Indeed, over the past decade, several types of stem and progenitor-like cells have been proposed in the peripheral vasculature [34, 35].

3.2.1 Vascular Wall Stem and Progenitor Cells (Fig. 3.1)

The normal blood vessel is composed of three layers with different characteristics: the tunica intima, tunica media, and tunica adventitia. The tunica intima is the innermost layer containing a layer of ECs supported by an internal elastic lamina. The media is composed of several layers of smooth muscle cells (SMCs) surrounded by an external elastic lamina. The tunica adventitia is the outermost layer of the blood vessel and consists of loose connective tissue made up of fibroblasts, collagen fibers, and adipose tissue with a mixture of small capillaries and nerves known as the "vasa vasorum" and "nervi vasorum."

It is believed that ECs at the intima are terminally differentiated mature cells that still possess proliferative capacity. Only a few reports suggest the existence of resident stem and progenitor cells in the EC layer. From the observation that early passage (3 and 4) human umbilical vein endothelial cells (HUVECs) and human aortic endothelial cells (HAECs), which are derived from vessel walls and considered to be differentiated mature ECs, can be passaged for at least 40 population doublings, Ingram et al. hypothesized the existence of EPCs in vessel walls. By using single cell assays to test the proliferative and clonogenic potential of single HUVECs and HAECs, they showed that 28 % of single cells in cultured HUVECs and 27 % of HAECs formed EC colonies containing more than 2,000 cells. Based on these results, they suggested that a complete hierarchy of EPCs may exist in the vessels [36]. More recently, we isolated a novel endothelial stem/progenitor-like cell from the intima of adult murine blood vessels [37].

Sainz et al. isolated progenitor cells from the tunica media of adult murine aorta [38]. They digested aorta enzymatically and mechanically and found side population cells by FACS analysis in the cell fraction. These side population cells were characterized as $CD31^-$ $Sca1^+$ and could differentiate into ECs and SMCs in vitro. However, as the authors declared that no living ECs from the intima were found in the digestions of the intima/media fraction, whether SP cells were present in the EC fraction could not be assessed.

For a long time, the tunica adventitia has been considered as an inactive component only supporting the blood vessel structurally. However, it is now well accepted that the adventitial cells can be activated in response to a variety of stimuli such as injury or atherosclerotic lesion formation [39]. Recently, the presence of stem and progenitor cells in the adventitia that have differentiation potential to ECs and/or SMCs has been reported by several groups. Hu et al. isolated $Sca1^+$ progenitor cells from the adventitia in aortic roots of adult apolipoprotein E-deficient mice that are used as atherosclerosis models and showed that these cells were able to differentiate into SMCs in vitro and in vivo [40]. At the region between the media and adventitia, Zengin et al. identified $CD34^+CD31^-VEGFR2^+$ vascular wall progenitor cells that can form capillary sprouts and designated this region the "Vasculogenic Zone" of blood vessels [41]. Moreover, Passman et al. described a novel domain of sonic hedgehog (Shh) signaling that is restricted to the adventitia of large and medium-sized vessels and which co-localizes with the "Vasculogenic Zone" [42]. Furthermore, Pasquinelli et al. identified resident angiogenic mesenchymal stromal cells that can differentiate into ECs in vitro at the region between the media and adventitia [43]. However, the contribution of these stem and progenitor cells to neovascular ECs in vivo has not been elucidated yet.

3.2.2 Endothelial Progenitor Cells

BM-derived EPCs recruited to the site of new vessel formation contribute to new blood vessel formation by differentiating into ECs and can thus be considered as EC progenitor cells. Following the proposal that EPCs are involved in an important mechanism for new vessel formation via postnatal vasculogenesis [44], several groups have demonstrated their functional contribution during many pathological states including limb ischemia, myocardial infarction, cerebral infarction, diabetic retinopathy, and tumor neovascularization [45]. Notably, it has been proposed that BM-derived ECs could constitute more than 50 % of all ECs in tumor neovessels [46]. However, there is no simple definition of EPCs and various methods to isolate them have been reported. In most studies, EPCs are defined as cells expressing CD34, CD133, or VEGFR2 (KDR) in the peripheral blood or BM. Currently, there are several ongoing clinical trials based on this concept. On the other hand, some reports question whether EPCs indeed contribute to ECs in the newly formed blood vessels. De Palma et al. reported the absence of GFP-positive ECs in mouse tumor xenograft vasculature after transplantation of BM progenitor cells expressing GFP [47]. They further demonstrated that monocytes expressing the Tie2 receptor (Tie2-expressing monocytes) are proangiogenic cells that are selectively recruited to tumors and promote angiogenesis in a paracrine manner [48]. Gothert et al. suggested that BM cells do not contribute to tumor endothelium and demonstrated a lineage relationship between pre-existing endothelium and newly generated tumor endothelial cells by using endothelial-specific tamoxifen-inducible reporter mice, namely endothelial-SCL-Cre-ERT [49]. Also, several groups showed that BM-derived cells are not incorporated into new blood vessels in a hind limb ischemia model, tumor xenograft model, and APCmin tumor model [50–52]. Moreover, Okuno et al. showed that proangiogenic effects of macrophages, but not endothelial differentiation, are the major contribution of BM-derived cells to wound healing [53]. More recently, Rinkevich et al. demonstrated that circulating cells do not contribute to angiogenesis by genetic fate mapping and clonal analysis of individual cells [54]. These findings suggest that in the adult organism, BM-derived EPCs do not contribute to vascular growth by incorporating into blood vessels but may function as supporting cells. Therefore, the concept that ECs residing at peripheral blood vessels have proliferative capacity and thus that ECs of the pre-existing vasculature play a central role in supplying neovascular ECs during angiogenesis needs to be reconsidered.

3.3 Isolation and Characterization of Endothelial Side Population Cells

3.3.1 Isolation of Endothelial Side Population Cells

The Hoechst–SP method has been applied by Sainz et al. to analyze cells from the aorta, as described above [38]. However, because of the difficulty in isolating viable ECs, these investigators were not able to test for the possible existence of

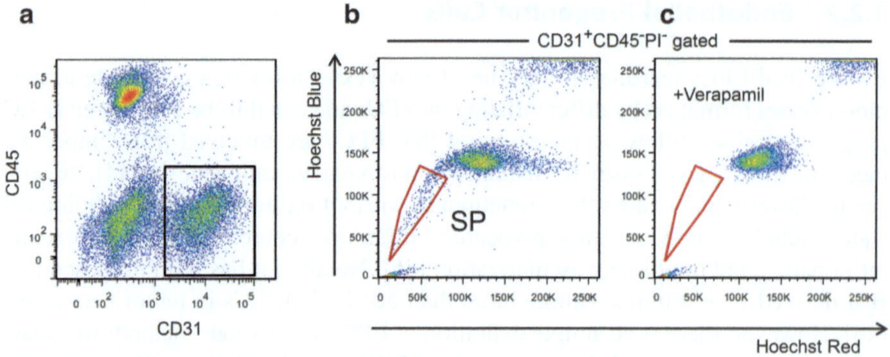

Fig. 3.2 Identification of endothelial side population cells. (**a**) Flow cytometric analysis of cells from hind limb muscle stained with anti-CD31 and anti-CD45 antibodies. $CD31^+CD45^-$ ECs are shown in the *black gate*. (**b**) Hoechst 33342 staining of $CD31^+CD45^-$ ECs shown in (**a**). Approximately 1 % of the ECs are in the SP gate (*red*). (**c**) Flow cytometric analysis following treatment with the ATP-binding cassette (ABC) transporter inhibitor verapamil, demonstrating inhibition of Hoechst efflux from SP cells

endothelial SP cells. The isolation of living ECs has often proven a difficult task and different techniques have been employed to purify primary ECs. In early reports, tissues were minced and digested enzymatically, after which they were cultured in dishes and ECs subsequently selected by complex purification techniques [55]. A more successful protocol has been described by several groups using magnetic bead sorting or FACS [56, 57]. However, these techniques often require culturing of cells and there are few protocols for separating large numbers of purified primary ECs [58]. When cells are isolated from their original environment, molecular changes will occur, which might affect the characteristics of the primary EC. In order to avoid these changes to primary ECs, a method that can isolate a large number of primary ECs by FACS was applied to test whether endothelial SP cells were present in ECs from hind limbs of adult mice [37].

Because previous studies of tissue-resident SP cells had analyzed total cell suspensions from tissues or organs, they were very heterogeneous and contained several types of cell including hematopoietic progenitors [59]. This led to difficulties in isolating lineage-specific stem/progenitor cells. Therefore, to test the possibility that SP cells are present within the EC fraction, we first gated cells using the EC marker CD31 and the hematopoietic cell (HC) marker CD45. Among cells positive for CD31, but negative for CD45 ($CD31^+$ $CD45^-$ ECs), about 1 % were in the SP gate, confirmed by their disappearance in the presence of the drug efflux pump inhibitor, verapamil. This small endothelial SP (EC–SP) population was distinct from the main population (MP) of ECs (Fig. 3.2). In addition, these EC–SP cells were also identified by using transgenic mice with a GFP reporter under the transcriptional control of the *VE-cadherin* promoter, rendering ECs recognizable as GFP positive. This clearly demonstrated that EC–SP cells are true ECs and not contaminations of HCs or mesenchymal cells. Moreover surface marker phenotyping revealed that these EC–SP cells were positive for VE-cadherin, Flk-1, and Sca-1, but negative for the hematopoietic lineage markers CD45, B220, CD4, CD8, Ter119, Gr-1, or CD11b or the pericyte marker PDGFR-beta, which is identical to EC–MP cells. However, in contrast to EC–MP cells, EC–SP cells strongly expressed CD133, which is a stem/progenitor cell marker in several tissues. They also expressed little CD34, similar to CD34-negative long-term repopulating HSCs. Regarding ABC transporters, the expression of which correlates with SP phenotype, ABCB1a, ABCG2, ABCB2, and ABCA5 were higher in EC–SP than EC–MP cells. However, the difference in the level of expression of each of these transporter genes was not great, but it remains possible that several together may mediate effects in a synergistic manner and could determine the functionality of the EC–SP phenotype [60]. The quiescence of stem cells is a critical biological phenomenon for protecting the stem cell compartment from senescence [61]. As with other tissue-resident stem cells, about 95 % of the EC–SP cells were in G0 (dormant), as shown by cell cycle analysis with simultaneous DNA/RNA staining with Hoechst 33342 and Pyronin Y.

3.3.2 EC–SP Cells Do Not Originate from Bone Marrow Cells

Hoechst analysis using VE cadherin-promoter GFP reporter mice showed that EC–SP cells are phenotypically true ECs. Furthermore, to confirm that EC–SP cells are not identical to EPCs, BM transplantation from GFP mice into irradiated wild-type mice was performed and the absence of GFP-positive EC–SP cells documented. The contribution of EPCs to the vasculature is a matter of debate; however; the hind limb ischemia model using these BM chimeric mice suggested only a minor contribution of EPC to peripheral ECs, as assessed by histology. In addition, in our hands GFP-positive ECs were rarely observed by FACS analysis. Among a few GFP-positive ECs in this model, EC–SP cells could not be detected [37]. Moreover, this was also confirmed in a BM transplantation model using neonates, in which their BM cells were replaced by the injection of BM cells from GFP mice into the liver of wild-type neonates within 12 h of birth [62]. This model allows us to ask whether EPCs derived from BM undergo an EC transition at the growing stage and become ECs and/or EC–SP cells. Again, FACS analysis of adult mice that had undergone BM transplantation shortly after birth showed no contribution of GFP-positive BM-derived cells to the $CD31^{+}CD45^{-}$ ECs in the periphery [37]. These two BM transplantation models clearly show that at least for the ECs of lower limb muscle, ECs and EC–SP cells do not originate from BM cells and are distinct from EPCs. The origin of these EC–SP cells is currently unknown. The lectin perfusion assay revealed that more than 90 % of the EC–SP cells are lectin positive, indicating that EC–SP cells reside in the intra-luminal cavity of the blood vessels [37]. It is possible that EC–SP cells originate from angioblasts or undifferentiated embryonic/neonatal-type ECs that reside in the inner surface of the blood vessel from the early stage of development.

Another possibility is that EC–SP cells may emerge at different sites of the body and migrate to the inner surface of blood vessels and reside there as $CD31^+$ VE cadherin$^+$ cells undistinguishable from other ECs.

3.3.3 Proliferation and Colony Formation of EC–SP Cells

A common hallmark of stem cells and progenitor cells is their ability to proliferate and generate differentiated progeny [63]. In hematopoiesis, the most proliferative progenitor that can be cultured in the absence of a stromal cell monolayer is termed the high proliferative potential colony-forming cell (HPP-CFC) [64]. For the ECs, Ingram et al. suggested the existence of high proliferative potential endothelial colony-forming cell (HPP-ECFC) in peripheral and umbilical cord blood EPCs, HUVECs, and HAECs [36]. However, in the case of primary ECs, the existence of HPP-ECFC was not demonstrated, perhaps because of the difficulty of isolating and culturing EC, as described above. To address the question of whether EC–SP cells possess colony-forming potential, culture systems supporting primary ECs without feeder cells are required. However, culturing single EC without contact with neighboring ECs is very difficult. Therefore, we used OP9 feeder cells established from calvaria of newborn osteopetrosis (op/op) mutant mice that possess a mutant csf-1 gene encoding macrophage colony-stimulating factor, which are used to support growth of EC and HCs [65]. To assess the proliferative potential of the EC–SP cells, 10^3 EC–SP cells, EC–MP cells, or $CD31^+CD45^-$ ECs were sorted and cultured on OP9 feeder cells. After 10 days, EC–SP cells generated more than tenfold the number of cord-like vascular colonies, as well as five- to tenfold more ECs than EC–MP cells or $CD31^+CD45^-$ ECs. Time-lapse analysis of a single EC–SP cell confirmed clonal expansion of cord-like vascular colonies from single cells, clearly showing the existence of high proliferative potential colony-forming ECs (Fig. 3.3). Moreover, long-term culture-initiating cell (LTC-IC) assays revealed that more ECs having higher proliferative potential were generated from EC–SP cells than could be generated by EC–MP cells. The frequency of cells with the capacity to form colonies was assessed by limiting dilution assays. In the EC–SP and EC–MP cells, 1 in 6.6 and 1 in 66 of the cells possesses colony-forming ability, respectively [37]. These results clearly show that EC–SP cells are colony-forming stem/progenitor-like EC in the peripheral vasculature of adult mice.

3.4 Therapeutic Potential of Endothelial Side Population Cells and Perspectives

The potential of stem cell therapies and regenerative medicine is seductive, offering the possibility of repairing or replacing the tissue damaged by disease or injury. For vascular regenerative medicine, two modalities are practiced for induction of new vessel formation: administration of angiogenic growth factors and cell

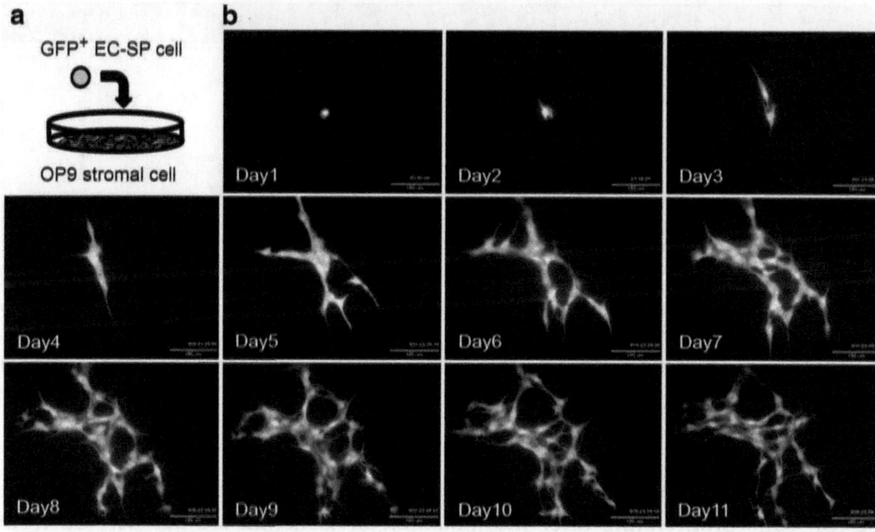

Fig. 3.3 Single EC–SP cells form EC colonies. (**a**) Schematic representation of the co-culture system of EC–SP cells from the EGFP mouse with OP9 stromal cells. (**b**) Time-lapse analysis of EC–SP cells. A single EC–SP cell starts to move and divide on culture days 3–4. ECs continuously divide and form large colonies

transplantation. The angiogenic growth factors employed in preclinical and clinical studies have been administered as recombinant proteins or by gene transfer and include vascular endothelial growth factor (VEGF), fibroblast growth factor (FGF), and hepatocyte growth factor (HGF). Thus far, clinical trials using angiogenic growth factors delivered in this way for the treatment of myocardial or peripheral ischemia have not shown convincing efficacy [66–68]. Transplantation of peripheral blood or BM-derived mononuclear cells has shown a certain degree of efficacy in therapeutic angiogenesis [69, 70]. However, it has become apparent that beneficial effects of these cell transplantation therapies were mainly achieved by their direct paracrine action on stromal cells and ECs, rather than through their physical engraftment into host vasculature [71]. In addition, conditioned medium from CD34-positive cells from human umbilical cord was shown to be therapeutically equivalent to the cells themselves, at least in a wound healing model [72]. Taken together, these data show that it is still an arduous task to regenerate whole blood vessels. In this regard, EC–SP cells that possess colony-forming stem/progenitor-like properties may be strong candidates for stem and progenitor cell transplantation therapy.

The hind limb ischemia model revealed that induction of ischemia triggered EC–SP cells to leave the dormant state, enter the cell cycle, and begin to proliferate. The percentage and absolute number of EC–SP cells increased 1 day after induction of ischemia, peaked after 3 days at about 4 % of total ECs, and gradually declined to the steady-state level after 2 weeks [37]. The colony-forming ability of the increased numbers of EC–SP cells was comparable to that in the steady state.

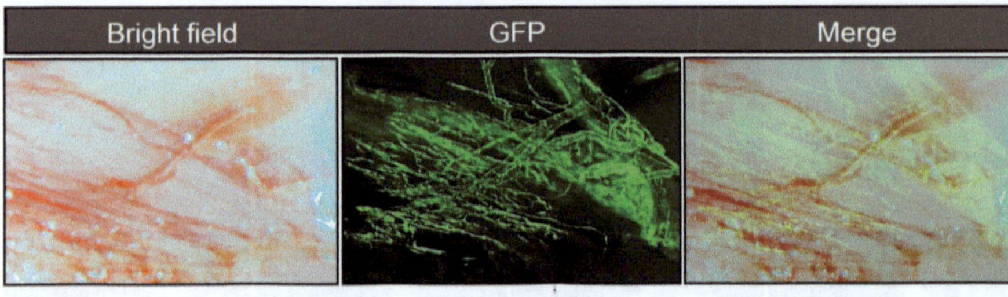

Fig. 3.4 EC–SP cells generate functional blood vessels. GFP^+EC–SP cells (5×10^3) were transplanted into the ischemic limb of wild-type mice. Fluorescent stereomicroscopic images of hind limb muscle of EC–SP-transplanted mice are shown. Many GFP^+ blood vessels filled with blood are generated. Note that EC–SP cells contribute not only to small capillaries but also larger arterioles and venules

Moreover, EC–SP analysis of the BM transplantation model revealed that the EC–SP cells proliferating after the induction of ischemia are not derived from BM cells. Transplantation of EC–SP cells in the hind limb ischemia model completely restored the blood flow by a mechanism involving their physical engraftment into the host vasculature in the form of ECs partially covered by host smooth muscle cells. Moreover, the blood vessels constituted by EC–SP cells persisted for at least 6 months thereafter, implicating a long-term contribution of EC–SP to ECs as a component of functional blood vessels (Fig. 3.4). This is quite different from the situation with BM-derived progenitor cells because endothelial-like cells derived from BM cells have disappeared 3 months after injection in the hind limb ischemia model [73]. A stem cell system is required for the tissue maintenance [74], and this must also apply to the vascular system. Furthermore, long-term contribution of the cells constituting tissues and maintaining tissue homeostasis is a required stem cell property. Therefore, the long-term contribution of EC–SP cells as ECs must be defined as a stem cell property.

Antiangiogenic approaches aimed at blocking vessel growth in cancer and eye disease led to the approval of therapeutics targeting VEGF [75]. However, only a fraction of cancer patients show benefit, because tumors acquire resistance mechanisms or become refractory to VEGF inhibitors [76]. In the tumor vasculature, although their function remains to be elucidated, EC–SP cells are present at higher proportions than in the normal vasculature. We confirmed that they possess high colony-forming ability in vitro [37]. The development of resistance to drugs is often associated with multidrug efflux pumps; it is possible that EC–SP cells that have high expression of drug pumps may be a cause of drug resistance targeting ECs.

Further analysis is required to clearly show that EC–SP cells can constitute the endothelial stem cell systems in the peripheral vasculature. To establish this, tissue localization of stem cells may help to identify niche cells supporting the latter. So far, no markers identifying EC–SP cells have been discovered. Therefore, localization of these cells in vivo cannot be elucidated. In vivo lineage tracing studies with specific markers for EC–SP cells would open new perspectives on understanding mechanisms of angiogenesis.

Self-renewal is the process by which stem cells produce at least one stem cell by symmetric or asymmetric cell division to perpetuate the stem cell pool [2]. Serial transplantation assays are required to test the ability of EC–SP cells to undergo self-renewal. Furthermore, serial transplantation of EC–SP cells and EC–MP cells could reveal a hierarchy system of ECs. Moreover, it would be interesting to know whether EC–SP cells in different organs produce ECs having different characteristics, such as morphological heterogeneity referred to as continuous, fenestrated, and discontinuous.

Recent studies provide a new mechanistic concept of functional specialization of ECs during different phases of angiogenesis. After receiving pro-angiogenic signals, EC become motile and invasive and protrude filopodia. These spearhead new sprouts called tip cells, which probe for environmental guidance cues. In contrast, stalk cells following the tip cell establish a lumen and proliferate to support sprout elongation [77]. During the transition from active sprouting to quiescent ECs, tip and stalk cells adopt a "phalanx cell" phenotype (resembling a phalanx formation of ancient Greek soldiers) characterized as lumenized, nonproliferating, and immobile, which promotes vessel integrity and stabilizes the vasculature [78, 79]. Mechanisms for selection of the tip cell phenotype were recently proposed to be based on initial stochastic differences in soluble (s) Flt-1 expression by individual ECs of a developing vessel [80]. This results in localized VEGF inactivation and thus more active angiogenic stimuli in the area of the migrating sprout. In our in vivo regeneration assay using EC–SP cells, we observed new vessels sprouting from newly developed blood vessels generated by EC–SP cells. This strongly suggests that EC–SP cells have the potential to generate both tip cell and stalk cell phenotypes in the active sprouting state and a phalanx cell phenotype in the quiescent state. In the latter, the level of sFlt1 mRNA is not high, as it is in EC–SP cells; however, it is possible that tip cells generated from EC–SP cells start to express high levels of sFlt1. Little is known about the mechanisms of EC–SP cell activation under ischemic conditions. Identifying the signals that are responsible for EC–SP activation would help clarify the relationship between these cells and EC–SP cells.

In consideration of the evidence that EC–SP cells do exist, we hypothesize that the ECs in peripheral blood vessels are not homogeneous and that certain of them retain their original hierarchical stem/progenitor characteristics. We propose a "heterogeneous angiogenesis model" in which resident stem/progenitor ECs play a central role in producing a number of ECs (Fig. 3.5).

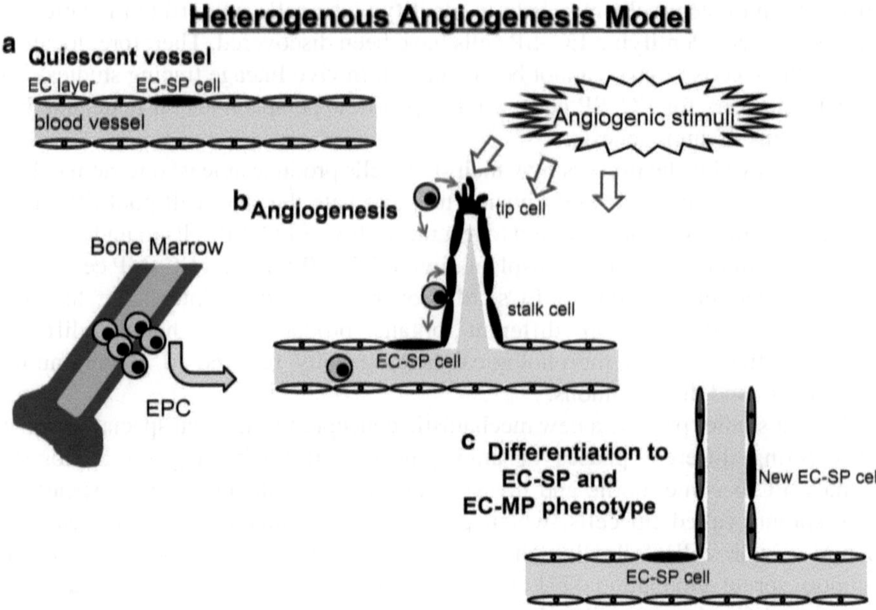

Fig. 3.5 Proposed model for EC–SP mediated angiogenesis. (**a**) Quiescent blood vessels harbor EC–SP cells at their inner surface (shown in *black*). (**b**) Angiogenic stimuli activate EC–SP cells to divide and produce large numbers of progeny (all *black cells* are derived from EC–SP cells). Bone marrow-derived EPCs are recruited to the site of active neovascularization where they secrete cytokines and activate surrounding cells. (**c**) As newly formed blood vessel become stable, EC–SP cell-derived ECs differentiate to the EC–MP phenotype (*gray cells*). Some of the newly produced EC may remain as EC–SP cells

Acknowledgments This work was partially supported by a Grant-in-Aid for Scientific Research from the Ministry of Education, Culture, Sports, Science and Technology and Japan Society for the Promotion of Science of Japan.

References

1. Reya T, Morrison SJ, Clarke MF, Weissman IL (2001) Stem cells, cancer, and cancer stem cells. Nature 414(6859):105–111
2. Morrison SJ, Spradling AC (2008) Stem cells and niches: mechanisms that promote stem cell maintenance throughout life. Cell 132(4):598–611
3. Mimeault M, Batra SK (2008) Recent progress on tissue-resident adult stem cell biology and their therapeutic implications. Stem Cell Rev 4(1):27–49
4. Lin KK, Goodell MA (2011) Detection of hematopoietic stem cells by flow cytometry. Methods Cell Biol 103:21–30
5. Okada S, Nakauchi H, Nagayoshi K, Nishikawa S, Miura Y, Suda T (1992) In vivo and in vitro stem cell function of c-kit- and Sca-1-positive murine hematopoietic cells. Blood 80 (12):3044–3050
6. Spangrude GJ, Heimfeld S, Weissman IL (1988) Purification and characterization of mouse hematopoietic stem cells. Science 241(4861):58–62

7. Kiel MJ, Yilmaz OH, Iwashita T, Yilmaz OH, Terhorst C, Morrison SJ (2005) SLAM family receptors distinguish hematopoietic stem and progenitor cells and reveal endothelial niches for stem cells. Cell 121(7):1109–1121
8. Takakura N, Watanabe T, Suenobu S, Yamada Y, Noda T, Ito Y, Satake M, Suda T (2000) A role for hematopoietic stem cells in promoting angiogenesis. Cell 102(2):199–209
9. Christensen JL, Weissman IL (2001) Flk-2 is a marker in hematopoietic stem cell differentiation: a simple method to isolate long-term stem cells. Proc Natl Acad Sci U S A 98 (25):14541–14546
10. Osawa M, Hanada K, Hamada H, Nakauchi H (1996) Long-term lymphohematopoietic reconstitution by a single CD34-low/negative hematopoietic stem cell. Science 273 (5272):242–245
11. Pallavicini MG, Summers LJ, Dean PN, Gray JW (1985) Enrichment of murine hemopoietic clonogenic cells by multivariate analyses and sorting. Exp Hematol 13(11):1173–1181
12. Goodell MA, Brose K, Paradis G, Conner AS, Mulligan RC (1996) Isolation and functional properties of murine hematopoietic stem cells that are replicating in vivo. J Exp Med 183 (4):1797–1806
13. Chaudhary PM, Roninson IB (1991) Expression and activity of P-glycoprotein, a multidrug efflux pump, in human hematopoietic stem cells. Cell 66(1):85–94
14. Zhou S, Schuetz JD, Bunting KD, Colapietro AM, Sampath J, Morris JJ, Lagutina I, Grosveld GC, Osawa M, Nakauchi H, Sorrentino BP (2001) The ABC transporter Bcrpl/ABCG2 is expressed in a wide variety of stem cells and is a molecular determinant of the side-population phenotype. Nat Med 7(9):1028–1034
15. Asakura A, Seale P, Girgis-Gabardo A, Rudnicki MA (2002) Myogenic specification of side population cells in skeletal muscle. J Cell Biol 159(1):123–134
16. Welm BE, Tepera SB, Venezia T, Graubert TA, Rosen JM, Goodell MA (2002) Sca-1(pos) cells in the mouse mammary gland represent an enriched progenitor cell population. Dev Biol 245(1):42–56
17. Shimano K, Satake M, Okaya A, Kitanaka J, Kitanaka N, Takemura M, Sakagami M, Terada N, Tsujimura T (2003) Hepatic oval cells have the side population phenotype defined by expression of ATP-binding cassette transporter ABCG2/BCRP1. Am J Pathol 163(1):3–9
18. Summer R, Kotton DN, Sun X, Ma B, Fitzsimmons K, Fine A (2003) Side population cells and Bcrpl expression in lung. Am J Physiol Lung Cell Mol Physiol 285(1):L97–L104
19. Lassalle B, Bastos H, Louis JP, Riou L, Testart J, Dutrillaux B, Fouchet P, Allemand I (2004) 'Side Population' cells in adult mouse testis express Bcrpl gene and are enriched in spermatogonia and germinal stem cells. Development 131(2):479–487
20. Redvers RP, Li A, Kaur P (2006) Side population in adult murine epidermis exhibits phenotypic and functional characteristics of keratinocyte stem cells. Proc Natl Acad Sci U S A 103 (35):13168–13173
21. Oyama T, Nagai T, Wada H, Naito AT, Matsuura K, Iwanaga K, Takahashi T, Goto M, Mikami Y, Yasuda N, Akazawa H, Uezumi A, Takeda S, Komuro I (2007) Cardiac side population cells have a potential to migrate and differentiate into cardiomyocytes in vitro and in vivo. J Cell Biol 176(3):329–341
22. Kondo T, Setoguchi T, Taga T (2004) Persistence of a small subpopulation of cancer stem-like cells in the C6 glioma cell line. Proc Natl Acad Sci U S A 101(3):781–786
23. Hirschmann-Jax C, Foster AE, Wulf GG, Nuchtern JG, Jax TW, Gobel U, Goodell MA, Brenner MK (2004) A distinct "side population" of cells with high drug efflux capacity in human tumor cells. Proc Natl Acad Sci U S A 101(39):14228–14233
24. Croop JM, Raymond M, Haber D, Devault A, Arceci RJ, Gros P, Housman DE (1989) The three mouse multidrug resistance (mdr) genes are expressed in a tissue-specific manner in normal mouse tissues. Mol Cell Biol 9(3):1346–1350
25. Huls M, van den Heuvel JJ, Dijkman HB, Russel FG, Masereeuw R (2006) ABC transporter expression profiling after ischemic reperfusion injury in mouse kidney. Kidney Int 69 (12):2186–2193

26. Miller DS (2010) Regulation of P-glycoprotein and other ABC drug transporters at the blood–brain barrier. Trends Pharmacol Sci 31(6):246–254
27. Golebiewska A, Brons NH, Bjerkvig R, Niclou SP (2011) Critical appraisal of the side population assay in stem cell and cancer stem cell research. Cell Stem Cell 8(2):136–147
28. Hierlihy AM, Seale P, Lobe CG, Rudnicki MA, Megeney LA (2002) The post-natal heart contains a myocardial stem cell population. FEBS Lett 530(1–3):239–243
29. Oh H, Bradfute SB, Gallardo TD, Nakamura T, Gaussin V, Mishina Y, Pocius J, Michael LH, Behringer RR, Garry DJ, Entman ML, Schneider MD (2003) Cardiac progenitor cells from adult myocardium: homing, differentiation, and fusion after infarction. Proc Natl Acad Sci U S A 100(21):12313–12318
30. Martin CM, Meeson AP, Robertson SM, Hawke TJ, Richardson JA, Bates S, Goetsch SC, Gallardo TD, Garry DJ (2004) Persistent expression of the ATP-binding cassette transporter, Abcg2, identifies cardiac SP cells in the developing and adult heart. Dev Biol 265(1):262–275
31. Pfister O, Mouquet F, Jain M, Summer R, Helmes M, Fine A, Colucci WS, Liao R (2005) CD31- but Not CD31+ cardiac side population cells exhibit functional cardiomyogenic differentiation. Circ Res 97(1):52–61
32. Mouquet F, Pfister O, Jain M, Oikonomopoulos A, Ngoy S, Summer R, Fine A, Liao R (2005) Restoration of cardiac progenitor cells after myocardial infarction by self-proliferation and selective homing of bone marrow-derived stem cells. Circ Res 97(11):1090–1092
33. Griffioen AW, Molema G (2000) Angiogenesis: potentials for pharmacologic intervention in the treatment of cancer, cardiovascular diseases, and chronic inflammation. Pharmacol Rev 52 (2):237–268
34. Torsney E, Xu Q (2011) Resident vascular progenitor cells. J Mol Cell Cardiol 50(2):304–311
35. Majesky MW, Dong XR, Regan JN, Hoglund VJ (2011) Vascular smooth muscle progenitor cells: building and repairing blood vessels. Circ Res 108(3):365–377
36. Ingram DA, Mead LE, Moore DB, Woodard W, Fenoglio A, Yoder MC (2005) Vessel wall-derived endothelial cells rapidly proliferate because they contain a complete hierarchy of endothelial progenitor cells. Blood 105(7):2783–2786
37. Naito H, Kidoya H, Sakimoto S, Wakabayashi T, Takakura N (2012) Identification and characterization of a resident vascular stem/progenitor cell population in preexisting blood vessels. EMBO J 31(4):842–855
38. Sainz J, Al Haj Zen A, Caligiuri G, Demerens C, Urbain D, Lemitre M, Lafont A (2006) Isolation of "side population" progenitor cells from healthy arteries of adult mice. Arterioscler Thromb Vasc Biol 26(2):281–286
39. Scott NA, Cipolla GD, Ross CE, Dunn B, Martin FH, Simonet L, Wilcox JN (1996) Identification of a potential role for the adventitia in vascular lesion formation after balloon overstretch injury of porcine coronary arteries. Circulation 93(12):2178–2187
40. Hu Y, Zhang Z, Torsney E, Afzal AR, Davison F, Metzler B, Xu Q (2004) Abundant progenitor cells in the adventitia contribute to atherosclerosis of vein grafts in ApoE-deficient mice. J Clin Invest 113(9):1258–1265
41. Zengin E, Chalajour F, Gehling UM, Ito WD, Treede H, Lauke H, Weil J, Reichenspurner H, Kilic N, Ergun S (2006) Vascular wall resident progenitor cells: a source for postnatal vasculogenesis. Development 133(8):1543–1551
42. Passman JN, Dong XR, Wu SP, Maguire CT, Hogan KA, Bautch VL, Majesky MW (2008) A sonic hedgehog signaling domain in the arterial adventitia supports resident Sca1+ smooth muscle progenitor cells. Proc Natl Acad Sci U S A 105(27):9349–9354
43. Pasquinelli G, Tazzari PL, Vaselli C, Foroni L, Buzzi M, Storci G, Alviano F, Ricci F, Bonafe M, Orrico C, Bagnara GP, Stella A, Conte R (2007) Thoracic aortas from multiorgan donors are suitable for obtaining resident angiogenic mesenchymal stromal cells. Stem Cells 25 (7):1627–1634
44. Asahara T, Murohara T, Sullivan A, Silver M, van der Zee R, Li T, Witzenbichler B, Schatteman G, Isner JM (1997) Isolation of putative progenitor endothelial cells for angiogenesis. Science 275(5302):964–967

45. Asahara T, Kawamoto A, Masuda H (2011) Concise review: circulating endothelial progenitor cells for vascular medicine. Stem Cells 29(11):1650–1655
46. Lyden D, Hattori K, Dias S, Costa C, Blaikie P, Butros L, Chadburn A, Heissig B, Marks W, Witte L, Wu Y, Hicklin D, Zhu Z, Hackett NR, Crystal RG, Moore MA, Hajjar KA, Manova K, Benezra R, Rafii S (2001) Impaired recruitment of bone-marrow-derived endothelial and hematopoietic precursor cells blocks tumor angiogenesis and growth. Nat Med 7 (11):1194–1201
47. De Palma M, Venneri MA, Roca C, Naldini L (2003) Targeting exogenous genes to tumor angiogenesis by transplantation of genetically modified hematopoietic stem cells. Nat Med 9 (6):789–795
48. De Palma M, Venneri MA, Galli R, Sergi Sergi L, Politi LS, Sampaolesi M, Naldini L (2005) Tie2 identifies a hematopoietic lineage of proangiogenic monocytes required for tumor vessel formation and a mesenchymal population of pericyte progenitors. Cancer Cell 8(3):211–226
49. Gothert JR, Gustin SE, van Eekelen JA, Schmidt U, Hall MA, Jane SM, Green AR, Gottgens B, Izon DJ, Begley CG (2004) Genetically tagging endothelial cells in vivo: bone marrow-derived cells do not contribute to tumor endothelium. Blood 104(6):1769–1777
50. Ziegelhoeffer T, Fernandez B, Kostin S, Heil M, Voswinckel R, Helisch A, Schaper W (2004) Bone marrow-derived cells do not incorporate into the adult growing vasculature. Circ Res 94 (2):230–238
51. Shinde Patil VR, Friedrich EB, Wolley AE, Gerszten RE, Allport JR, Weissleder R (2005) Bone marrow-derived lin(−)c-kit(+)Sca-1+ stem cells do not contribute to vasculogenesis in Lewis lung carcinoma. Neoplasia 7(3):234–240
52. Purhonen S, Palm J, Rossi D, Kaskenpaa N, Rajantie I, Yla-Herttuala S, Alitalo K, Weissman IL, Salven P (2008) Bone marrow-derived circulating endothelial precursors do not contribute to vascular endothelium and are not needed for tumor growth. Proc Natl Acad Sci U S A 105 (18):6620–6625
53. Okuno Y, Nakamura-Ishizu A, Kishi K, Suda T, Kubota Y (2011) Bone marrow-derived cells serve as proangiogenic macrophages but not endothelial cells in wound healing. Blood 117 (19):5264–5272
54. Rinkevich Y, Lindau P, Ueno H, Longaker MT, Weissman IL (2011) Germ-layer and lineage-restricted stem/progenitors regenerate the mouse digit tip. Nature 476(7361):409–413
55. Folkman J, Haudenschild CC, Zetter BR (1979) Long-term culture of capillary endothelial cells. Proc Natl Acad Sci U S A 76(10):5217–5221
56. Wu Z, Hofman FM, Zlokovic BV (2003) A simple method for isolation and characterization of mouse brain microvascular endothelial cells. J Neurosci Methods 130(1):53–63
57. Ewing P, Wilke A, Brockhoff G, Andreesen R, Eissner G, Holler E, Gerbitz A (2003) Isolation and transplantation of allogeneic pulmonary endothelium derived from GFP transgenic mice. J Immunol Methods 283(1–2):307–315
58. van Beijnum JR, Rousch M, Castermans K, van der Linden E, Griffioen AW (2008) Isolation of endothelial cells from fresh tissues. Nat Protoc 3(6):1085–1091
59. Asakura A, Rudnicki MA (2002) Side population cells from diverse adult tissues are capable of in vitro hematopoietic differentiation. Exp Hematol 30(11):1339–1345
60. Jonker JW, Freeman J, Bolscher E, Musters S, Alvi AJ, Titley I, Schinkel AH, Dale TC (2005) Contribution of the ABC transporters Bcrp1 and Mdr1a/1b to the side population phenotype in mammary gland and bone marrow of mice. Stem Cells 23(8):1059–1065
61. Pietras EM, Warr MR, Passegue E (2011) Cell cycle regulation in hematopoietic stem cells. J Cell Biol 195(5):709–720
62. Bruscia EM, Ziegler EC, Price JE, Weiner S, Egan ME, Krause DS (2006) Engraftment of donor-derived epithelial cells in multiple organs following bone marrow transplantation into newborn mice. Stem Cells 24(10):2299–2308
63. Simons BD, Clevers H (2011) Strategies for homeostatic stem cell self-renewal in adult tissues. Cell 145(6):851–862

64. McNiece IK, Stewart FM, Deacon DM, Temeles DS, Zsebo KM, Clark SC, Quesenberry PJ (1989) Detection of a human CFC with a high proliferative potential. Blood 74(2):609–612
65. Takakura N, Huang XL, Naruse T, Hamaguchi I, Dumont DJ, Yancopoulos GD, Suda T (1998) Critical role of the TIE2 endothelial cell receptor in the development of definitive hematopoiesis. Immunity 9(5):677–686
66. Rajagopalan S, Mohler ER III, Lederman RJ, Mendelsohn FO, Saucedo JF, Goldman CK, Blebea J, Macko J, Kessler PD, Rasmussen HS, Annex BH (2003) Regional angiogenesis with vascular endothelial growth factor in peripheral arterial disease: a phase II randomized, double-blind, controlled study of adenoviral delivery of vascular endothelial growth factor 121 in patients with disabling intermittent claudication. Circulation 108(16):1933–1938
67. Nikol S, Baumgartner I, Van Belle E, Diehm C, Visona A, Capogrossi MC, Ferreira-Maldent N, Gallino A, Wyatt MG, Wijesinghe LD, Fusari M, Stephan D, Emmerich J, Pompilio G, Vermassen F, Pham E, Grek V, Coleman M, Meyer F (2008) Therapeutic angiogenesis with intramuscular NV1FGF improves amputation-free survival in patients with critical limb ischemia. Mol Ther 16(5):972–978
68. Powell RJ, Simons M, Mendelsohn FO, Daniel G, Henry TD, Koga M, Morishita R, Annex BH (2008) Results of a double-blind, placebo-controlled study to assess the safety of intramuscular injection of hepatocyte growth factor plasmid to improve limb perfusion in patients with critical limb ischemia. Circulation 118(1):58–65
69. Tateishi-Yuyama E, Matsubara H, Murohara T, Ikeda U, Shintani S, Masaki H, Amano K, Kishimoto Y, Yoshimoto K, Akashi H, Shimada K, Iwasaka T, Imaizumi T (2002) Therapeutic angiogenesis for patients with limb ischaemia by autologous transplantation of bone-marrow cells: a pilot study and a randomised controlled trial. Lancet 360(9331):427–435
70. Minamino T, Toko H, Tateno K, Nagai T, Komuro I (2002) Peripheral-blood or bone-marrow mononuclear cells for therapeutic angiogenesis? Lancet 360(9350):2083–2084
71. Ferrara N, Kerbel RS (2005) Angiogenesis as a therapeutic target. Nature 438(7070):967–974
72. Kim JY, Song SH, Kim KL, Ko JJ, Im JE, Yie SW, Ahn YK, Kim DK, Suh W (2010) Human cord blood-derived endothelial progenitor cells and their conditioned media exhibit therapeutic equivalence for diabetic wound healing. Cell Transplant 19(12):1635–1644
73. Yamada Y, Takakura N (2006) Physiological pathway of differentiation of hematopoietic stem cell population into mural cells. J Exp Med 203(4):1055–1065
74. Biteau B, Hochmuth CE, Jasper H (2011) Maintaining tissue homeostasis: dynamic control of somatic stem cell activity. Cell Stem Cell 9(5):402–411
75. Ferrara N, Hillan KJ, Gerber HP, Novotny W (2004) Discovery and development of bevacizumab, an anti-VEGF antibody for treating cancer. Nat Rev Drug Discov 3(5):391–400
76. Bergers G, Hanahan D (2008) Modes of resistance to anti-angiogenic therapy. Nat Rev Cancer 8(8):592–603
77. Gerhardt H, Golding M, Fruttiger M, Ruhrberg C, Lundkvist A, Abramsson A, Jeltsch M, Mitchell C, Alitalo K, Shima D, Betsholtz C (2003) VEGF guides angiogenic sprouting utilizing endothelial tip cell filopodia. J Cell Biol 161(6):1163–1177
78. Mazzone M, Dettori D, Leite de Oliveira R, Loges S, Schmidt T, Jonckx B, Tian YM, Lanahan AA, Pollard P, Ruiz de Almodovar C, De Smet F, Vinckier S, Aragones J, Debackere K, Luttun A, Wyns S, Jordan B, Pisacane A, Gallez B, Lampugnani MG, Dejana E, Simons M, Ratcliffe P, Maxwell P, Carmeliet P (2009) Heterozygous deficiency of PHD2 restores tumor oxygenation and inhibits metastasis via endothelial normalization. Cell 136(5):839–851
79. Potente M, Gerhardt H, Carmeliet P (2011) Basic and therapeutic aspects of angiogenesis. Cell 146(6):873–887
80. Chappell JC, Taylor SM, Ferrara N, Bautch VL (2009) Local guidance of emerging vessel sprouts requires soluble Flt-1. Dev Cell 17(3):377–386

Multiple Roles of Proteases in Angiogenesis

4

Ester M. Weijers and Victor W.M. van Hinsbergh

Contents

4.1 Introduction 86
4.2 Classes of Proteases Involved in Angiogenesis 87
4.2.1 Serine Proteases 87
4.2.2 Matrix Metalloproteinases 88
4.2.3 Cysteine Proteases 89
4.3 Proteases in VEGF-Induced Sprouting 89
4.4 Migration, Invasion, and Pericellular Proteolysis 91
4.5 Proteolytic Modification of Growth Factors and Receptors 92
4.6 Proteolytic Modification of Matrix Proteins: Matrikines and Fibrin Forms 93
4.7 Perspective 94
References 96

Abstract

Proteases play an important role in various aspects of angiogenesis. First, pericellular proteolysis facilitates matrix degradation as well as migration and invasion of capillary sprouts into its surrounding matrix. (Membrane-type) matrix metalloproteases [(MT-)MMPs], ADAM(TS) metalloproteases, serine proteases, their inhibitors, and occasionally cysteine proteases are important players in pericellular proteolysis. The specific location and short exposure of protease activity can be explained by a tread-milling model with internalization of protease–receptor complexes as was proposed for MT1-MMP/MMP2 and urokinase/urokinase receptor complexes. Second, proteolytic events play a role in regulating VEGF-induced sprouting, such as proteasomal degradation of hypoxia inducible factors (HIFs), Notch-mediated cell signaling, and turnover of VEGF receptors. Third, proteases can activate and modify growth factors and

E.M. Weijers • V.W.M. van Hinsbergh (✉)
Laboratory for Physiology, Institute for Cardiovascular Research, VU University Medical Center, Van der Boechorststraat 7, 1081 BT Amsterdam, The Netherlands
e-mail: v.vanhinsbergh@vumc.nl

J. Dulak et al. (eds.), *Angiogenesis and Vascularisation*,
DOI 10.1007/978-3-7091-1428-5_4, © Springer-Verlag Wien 2013

receptors and thus generate split products with new biological functions and contribute to cell recruitment. Proteases likely also contribute to formation of anastomoses, which is required to restore circulation. Finally, new biological functions are also acquired by generation of split products from and modifications of matrix proteins. They include angiogenesis-inhibiting matrikines, and modified fibrin forms with altered angiogenic properties, which may bear perspective for tissue engineering applications.

Keywords
Matrix metalloproteinases • Cysteine proteases • VEGF • Hypoxia • HIF • Notch signaling

4.1 Introduction

In many physiological and pathological situations, angiogenesis is an important process. Its regulation by a tight balance between pro- and anti-angiogenic factors is fascinating. An important group of proteins that contributes to this balance is that of proteases. The importance of protease activity for cardiovascular development is reflected by prenatal death or major defects during development in animals with deficiency of many different proteases. Deficiencies of various hemostasis regulating proteases cause early developmental problems due to angiogenesis defects rather than to coagulation disturbances [1].

It is not surprising that proteolytic activity is required for detachment of endothelial cells from their basement membrane and for creating space in the interstitial tissue to facilitate the formation of new blood vessels. In addition, proteases are required for the migration of endothelial cells, pericytes, and angiogenesis-modulating monocytes. Movement within tissues involves the interplay between iterating degradation and generation of cell–matrix interactions, which act together with contractile changes in the intracellular cytoskeleton. However, proteases are also involved in many more processes that regulate angiogenesis. On one hand, the action of proteases has intracellular effects, for example, the γ-secretase complex facilitates receptor-mediated cell signaling for specific receptors; caspases regulate apoptosis; and lysosomal enzymes cause degradation and turnover of cellular proteins and receptor–ligand complexes. On the other hand, proteolytic modification of growth factors and in particular matrix proteins outside the cell converts existing molecules into proteins or peptides with new angiogenesis-modulating features.

As several reviews have surveyed a number of these aspects, we refer to more extensive references in previous reviews [2–6]. In this chapter we describe several aspects of proteases in the regulation of angiogenesis. After a short survey of the major groups of proteases that are involved in angiogenesis, we elaborate on the action of proteases in sprouting via vascular endothelial growth factor (VEGF)-signaling, cell migration and invasion, stem/progenitor cell recruitment, and proteolytic modifications of growth factors, receptors, and matrix proteins that modulate angiogenesis.

4.2 Classes of Proteases Involved in Angiogenesis

Proteases are classified based on their mechanism of catalysis; a distinction into six different classes can be made: metalloproteinases and serine-, aspartic-, cysteine-, threonine-, and glutamic proteases [7]. In angiogenesis the most dominant protease classes are serine proteases, metalloproteinases, in particular the matrix metalloproteinase (MMP) family and two related families ADAM (a disintegrin and metalloproteinase domain) and ADAMTS (ADAM with thrombospondin motifs), and cysteine proteases.

4.2.1 Serine Proteases

The class of serine proteases is the largest and most identified class of proteases in the human degradome. This endopeptidase group contains well-known proteases, e.g., chymotrypsin, trypsin, elastase, plasmin, and kallikreins. Generally, the inactive zymogen is activated by removal of a pro-peptide on the N-terminal side of the protease, leading to exposure of its catalytic triad. With respect to vascular remodeling, a well-studied serine protease family is the plasminogen activator—plasmin system, consisting of tissue-type plasminogen activator (tPA), urokinase plasminogen activator (uPA), and their substrate plasminogen. The zymogen plasminogen can be activated by two different plasminogen activators, tPA and uPA, both cleaving Arg^{561}–Val^{562} between the kringle 5 and protease domain. While tPA is mainly responsible for active plasmin generation during fibrinolysis in blood and body fluids, uPA is responsible for plasmin activity at the cell surface during cell migration and invasion. Activation of plasminogen by uPA is facilitated by the uPA receptor (UPAR) on the cell surface, leading to very specifc and localized proteolysis [8, 9]. The proteolytic activity of plasmin spans a wide range of extracellular matrix components, including collagen, fibrin, laminin, fibronectin, and proteoglycans [10]. This range is broadened by the fact that plasmin is involved in the activation of other proteases, as was shown for several matrix metalloproteinases [11].

The activation of plasminogen into active plasmin is not only tightly regulated by tPA and uPA but also by plasminogen activator inhibitor 1 and 2 (PAI-1, PAI-2). Plasmin activity is rapidly quenched by α2-antiplasmin and α2-macroglobulin. Proteolysis via the uPA-plasmin system favors tumor invasion and angiogenesis. Accordingly, high levels of PAI-1 are correlated with a poor prognosis of cancer patients [12]. The endogenous plasmin inhibitor α2-antiplasmin inactivates plasmin very rapidly in the fluid phase, whereas membrane-bound plasmin is rather inaccessible for α2-antiplasmin [13]. VEGF cleavage by plasmin regulated by α2-antiplasmin was shown to alter re-endothelialization after vascular injury [14]. Circulating α2-macroglobulin is an inhibitor of both plasmin and MMP activity.

4.2.2 Matrix Metalloproteinases

The class of matrix metalloproteinases contains over 20 different zinc-dependent endopeptidases. The members of this family can be classified into two groups based on their structural features: secreted MMPs and membrane-type MMPs (MT-MMP), containing an additional transmembrane and intracellular domain (MT1,2,3,5-MMP) or a GPI anchor (MT4,6-MMP). The soluble MMPs occur as a zymogen until its pro-peptide is removed or disrupted; upon activation the MMP catalytic domain with its highly conserved zinc-binding region becomes available. The hemopexin domain determines the substrate specificity of the MMPs. During transport from the Golgi complex to the cell membrane, MT-MMPs are processed intracellularly by furin-like enzymes, by which they become active. Together, the members of the MMP class of proteases can digest all known extracellular matrix macromolecules, affecting a variety of pathological conditions, e.g., tumor growth, atherosclerosis, and heart failure [15–17]. For specific information on MMP structure, biochemistry, and substrate specificities, we refer to comprehensive reviews on these topics [10, 18–21]. MMP-2, MMP-9, and MT1-MMP are in particular recognized as important regulators of angiogenesis [2, 5, 22]. Besides MMP regulation on transcriptional and activation levels, their activities on the cell surface are regulated by complex formation with other proteins and internalization [23]. Trafficking and turnover of MT1-MMP is essential for its proper function in cell signaling and invasion [24, 25]. Moreover, MMP activity is regulated by tissue inhibitors of metalloproteinases (TIMPs) and the inhibitor RECK (Reverse-inducing cysteine-rich protein with Kazal motifs) [26, 27]. TIMPs not only play a role in the inhibition of MMP activity but also can possess other biological functions. TIMP-2—via sphingosine-1-phosphate (S1P) and Rho-kinase—can interfere with the formation of adherence junctions in endothelial cells, leading to reduced vascular sprouting [28]. TIMP-2 can suppress fibroblast growth factor-induced angiogenesis, independent of MMP inhibition [29]. Differential regulation and broad localization of TIMPs is associated with their specific biological functions.

Two large protease families that are related to MMPs are the ADAMs and ADAMTSs. ADAMs have a membrane-spanning domain and can act as ectodomain sheddases of cell surface receptors or cell adhesion molecules. In general, ADAMs influence cell surface remodeling, ectodomain shedding, growth factor availability, and cell–cell and cell–matrix interactions. For example, ADAM-10 and -17 facilitate the cleavage of ectodomains of Notch-1 and Tie-1 upon binding with their ligands Delta-like ligand 4 and angiopoietin-1. In contrast to the latter proteases, ADAMTS are secreted proteins, wherein the thrombospondin motif facilitates its localization in the extracellular matrix (ECM) and occasionally causes inhibition of angiogenesis. Most of the ADAMTSs are anti-angiogenic or tumor suppressors, whereas ADAMTS1, ADAMTS4, and ADAMTS18 were reported to be pro-angiogenic [30]. Where the inhibition of TIMPs on secreted

MMPs is universal, the inhibition appears to be more selective for the ADAM and ADAMTS family members. For example, TIMP-3 can inhibit the proteolytic activity of ADAM-12 and 17, and ADAMTS4 and -5. On the other hand, the TIMPs cannot inhibit the action of ADAM-8 and 9.

4.2.3 Cysteine Proteases

Cysteine cathepsins are a subfamily within the cysteine proteases. Cysteine cathepsins are mainly endopeptidases that are intracellularly located in endolysosomal vesicles and responsible for protein degradation. Similarly as serine proteases and MMPs, the zymogen is activated upon removal of the pro-peptide. Cathepsins are mainly active at a rather acid pH and therefore usually active in lysosomes and involved in protein turnover. A few cathepsins including cathepsin B can be active on the cell surface after binding to other proteins (see [3] for details). Several reports suggest a role of cathepsin B in angiogenesis, which might be indirect. Kostoulas et al. proposed that cathepsin B can degrade TIMP-1 and TIMP-2 and thereby enhance MMP activity [31]. Moreover, cathepsin L is involved in the recruitment of progenitor cells that contribute to neovascularization in ischemic hindlimbs [32]. Cathepsins B and S showed increased expression in a tumor environment including possible secretion and membrane exposure. Potential extracellular roles of cathepsins are cleavage of ECM molecules, e.g., laminin, collagen type IV and tenascin C, cell-adhesion molecules, and activation of other enzymes involved in proteolysis [33]. Cysteine cathepsins were shown to enhance tumor growth, invasion, and angiogenesis [33]. Moreover, these proteases can activate UPAR-bound pro-uPA to active uPA leading to plasmin generation [34].

In the following sections we will review specific functions of proteases with respect to positive and negative effects on various aspects of the regulation of angiogenesis.

4.3 Proteases in VEGF-Induced Sprouting

A key feature of injury-induced angiogenesis is hypoxia and the induction of VEGF (VEGF-A). In this sequence of events, proteases are important in cell signaling both in the leading sprouting endothelial cells (tip cell) as well as in the adjacent endothelial cells. Hypoxia causes the survival of HIF-1α (hypoxia inducible factor 1α), which together with HIF-1β forms the transcription factor HIF-1 that induces many genes including VEGF, a key regulator of angiogenesis. In the presence of oxygen, HIF-1α is rapidly hydroxylated at two proline residues and an asparaginyl residue. The proline hydroxylation causes the binding of the Von Hippel Lindau protein, after which HIF-1α becomes degraded in the proteasomes after poly-ubiquitinylation. Additionally, the hydroxylation of the Asp-residue of HIF-1 by the asparaginyl hydroxylase FIH-1 (factor inhibiting HIF-1) interferes with the binding of p300 protein and thereby inhibits the formation of an active transcription initiation complex. Thus, in well perfused tissues HIF-1α, although continuously synthesized, is unable to induce angiogenesis, due to its continuous rapid proteolytic degradation and its inability to form a suitable transcription activation complex. In the absence of oxygen no proline and asparaginyl hydroxylations occur and HIF-1 strongly induces the synthesis of VEGF and other angiogenesis stimulating factors. Macrophages and bone marrow niche cells can maintain some HIF-1 activity even in normoxic conditions. Interestingly, recent studies point to a role of the non-proteolytic cytoplasmic domain of MT1-MMP, which can bind and suppress FIH-1 in these cells and thereby allows nuclear HIF-1 to induce the expression of new genes [35, 36].

VEGF is normally secreted in large isoforms that bind to heparan sulfates in the tissue and is referred to as matrix-bound VEGF. The matrix-bound VEGF provides spatial information [37] that contributes to the formation of an organized vascular network. If the $VEGF_{165}$ is degraded by MMP-9, VEGF split products are generated that resemble the properties of the short isoform $VEGF_{121}$. This (soluble) $VEGF_{121}$ and the split products (e.g., 113N-terminal amino acids of VEGF, $VEGF_{113}$) cannot bind to heparan sulfates and induce a chaotic pattern of newly formed sprouts, in contrast to the matrix-bound isoforms [38].

The endothelial cells on the tip of a capillary sprout recognize matrix-bound VEGF by VEGF receptor-2 (VEGFR2) that is present on filopodial extrusions of these cells. Initially, one had observed that the presence of VEGFR2 was not altered at the invasive front of the cell during angiogenesis. However, new insights show rapid internalization and degradation of the VEGF-VEGFR2 complex. Thus, the increased availability of VEGFR2 in angiogenic conditions was masked by the enhanced turnover at the invasive front [39]. This increase is unmasked by inhibition of lysosomal proteins or internalization of the VEGF-VEGFR2 complex.

Sprouting is limited because endothelial tip cells regulate the sprout-initiating properties of adjacent endothelial cells. Suppression of sprout formation occurs via the VEGF-mediated induction of the protein DLL-4 (delta like ligand 4). DLL-4 binds to the receptor Notch-1, upon binding two subsequent proteolytic cleavages of Notch occur. First, Notch-1 is cleaved on the cell surface by the metalloproteinase ADAM-17 or ADAM-10. Subsequently, a second proteolytic cleavage occurs at the cytoplasmic side by the aspartate proteinase presenilin in the γ-secretase complex, after which the liberated cytoplasmic tail translocates to the nucleus where it can induce gene transcription. As a consequence, the gene expression profile of the cell is altered, i.e., the expression of VEGFR2, VEGFR3, and neuropilin-1 are reduced, and simultaneously the VEGF-quenching VEGFR1 is increased [40, 41]. The Notch-1-induced expression profile affects the responsiveness of the endothelial cell, by which it becomes a stalk cell. Stalk cells are less involved in sprouting, but rather in cell division and the subsequent formation of a more stable vascular structure. Notch signaling was reported to mediate the VEGF-induced activation of MMP-9 and MT1-MMP and the subsequent activation of MMP-2 [42]. This is an intriguing observation as Notch activation occurs in the stalk cells. It may contribute to further widening of the initial endothelial sprout and further invasion into the tissue. The lumen of newly formed blood vessels forms either by fusion of intracellular vesicles [43] or by

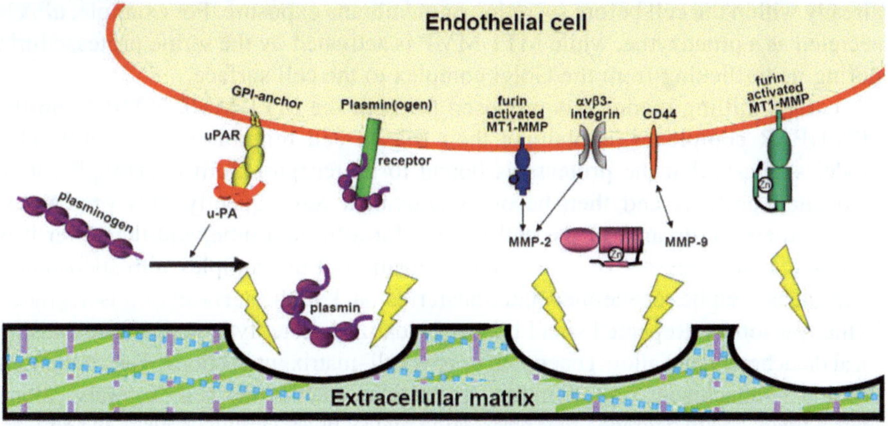

Fig. 4.1 Schematic representation of pericellular proteolytic activity during matrix invasion. Various proteases act on the cell surface, such as uPA bound to the GPI-anchored UPAR, plasmin bound to plasmin(ogen) receptors, MT1-MMPs, and MMP-2 bound to MT1-MMP. *GPI-anchor* glycophosphatidylinositol-anchor

regulated dissociation of part of the intercellular junctions [44]. It is unknown whether proteases are involved in lumen formation per se, but the simultaneous widening of the newly formed vascular structure requires proteases for remodeling of the tissue surrounding the vascular structure. Further vessel stabilization only occurs after formation of a basement membrane and association of pericytes with the newly formed endothelial tubular structure.

4.4 Migration, Invasion, and Pericellular Proteolysis

Migration and invasive growth of cells involved in angiogenesis requires a delicately balanced interplay between detachment and formation of new cell adhesions that enables the cell to pull itself forward through the extracellular matrix. To make this possible, spatially limited proteolytic activities occur on the cell surface, usually at locations of cell–matrix contact. It has become clear that either membrane-bound proteases are involved or multiprotein complexes are built in lipid rafts of the cell membrane which regulate and harbor proteases. The MT-MMPs and several ADAM family members have a transmembane domain or GPI-anchor, which binds them directly to the membrane [10, 30]. uPA and its receptor UPAR form a multiprotein complex with integrins and other proteins in focal adhesions [8]. Other proteases bind indirectly to the surface of the endothelial cells, such as plasmin(ogen) that binds to plasminogen receptors, MMP-2 that binds to MT1-MMP or αvβ3-integrin, and MMP-9 that interacts with CD44 [45]. MMP-2 is activated in complex with MT1-MMP in a process that is facilitated by TIMP-2. Activation of the protease from its pro-form can occur after secretion or directly on the cell surface (Fig. 4.1), but for other proteases the activation occurs already within the cell before secretion or membrane exposure. For example, uPA is secreted as a proenzyme, while MT1-MMP is activated by the serine protease furin during its trafficking from the Golgi complex to the cell surface.

A tread-milling model was proposed for both the MT1-MMP/MMP-2 and the uPA/UPAR complexes, explaining their role in cell migration [8, 46, 47]. This model suggests that the protease is bound to its receptor, forms a complex with associated proteins and then becomes activated. Subsequently, this protease is active on a specific area of the cell surface for a limited time, and thereafter it is inhibited and/or internalized. After internalization of the complex into endosomes, the protease can be regenerated and transferred back to the surface, or it is degraded in the lysosomes. Repeated small bouts of localized proteolytic activity are causing local detachment and allow generation of new cell–matrix interactions. The continuous de- and re-attachment of cell–matrix interactions enables the cell "to walk" along matrix fibers and to degrade other obstructing matrix fibers simultaneously. In addition, the mechanism provides a continuous mutual interplay between the proteolytic activity on the cell surface and the continuously adapting filamentous actin cytoskeleton, facilitating coordinated cell movement.

It is likely that proteases also play a role in the formation of anastomoses between the newly formed vascular sprouts and the existing perfused vessels. Anastomosis requires basement membrane degradation of the existing vessels and local migration of cells to enable proper fusion of the vessels. MT1-MMP was shown to play a role in Rac1- and Cdc42-mediated cell signaling that modulate such processes [48, 49]. Furthermore, monocytes have been indicated as important cells that control anastomosis formation [50]. In addition to their local action, their recruitment involves migration and invasion with a similar pericellular proteolysis participation as depicted above for the sprouting endothelial cell. The same applies to the recruitment of pericytes, which are necessary to stabilize the newly formed vessels.

4.5 Proteolytic Modification of Growth Factors and Receptors

During and preceding cell invasion into the extracellular matrix network, proteases can also activate or degrade growth factors and liberate matrix-bound growth factors. Plasminogen activators play an important role in the activation of latent transforming growth factor-β. Hepatocyt growth factor (HGF) is activated by HGF activating factor, a protease that is related to plasmin. VEGF can be liberated in an active state from a complex with connective tissue growth factor by the proteolytic action of MT-MMPs [51]. Stromal derived factor-1 (SDF-1) is trimmed by the carboxypeptidase thrombin activatable fibrinolysis inhibitor (TAFI) or the dipeptidyl peptidase DPPIV/CD26, which alter the affinity of SDF-1 for heparan sulfates. SDF-1 and its receptor C-X-C chemokine receptor type 4 (CXCR4) are cleaved and inactivated by MT1-MMP and MMP-1, -2, and -13 [52]. Degradation of the SDF-1/CXCR4 interaction and conversion of membrane-bound mKitL into soluble KitL by MMP-9 contribute to the mobilization of circulating angiogenesis stimulating cells from the bone marrow niche.

Proteolytic modification of cellular receptors is a rapidly growing field with many implications. Shedding of receptors such as VEGFR2 and the insulin receptor can cause microvascular rarefaction and—similar to VEGFR inhibition—contribute to hypertension [53, 54]. Shedding of IL-6 receptor (IL-6R) in septic conditions generates soluble IL-6R, which together with its adaptor protein gp130 can cause signaling in endothelial cells that are normally devoid of IL-6R [55]. ADAM-10 can generate soluble Ephrin (Eph) receptors by cleavage and shedding of EphA2 and EphA3, and these cleavage products can inhibit tumor angiogenesis in mice [56]. Indeed, a range of proteolytically modified proteins, in particular matrix proteins, have been recognized that modulate angiogenesis.

4.6 Proteolytic Modification of Matrix Proteins: Matrikines and Fibrin Forms

Proteins in the extracellular matrix, such as collagens and laminin, are degraded via multistep proteolysis. The intermediate products have received considerable attention, not only as parameters of proteolytic action, but also because some of them have angiogenesis inhibiting properties. Among these angiogenesis-modulating split products, which are nowadays indicated as matrikines, are fragments of collagens, like endostatin, tumstatin and canstatin, and laminin [57]. Several other circulating proteins, in particular plasminogen and fibrinogen, can also be degraded into anti-angiogenic fragments such as angiostatin and alphastatin, respectively. These inhibitory molecules act via specific receptors or via interaction with integrins [58].

Fibrinogen is an interesting molecule as it represents a temporary repair matrix. In addition to its role in hemostasis, it plays a facilitating role in the invasion of cells and capillary sprouts during the healing process [57, 59]. Multiple variants of fibrin were shown to occur in the circulation of healthy individuals; some of these variants are the result of partial proteolytic degradation [60–62]. Fibrin is primarily synthesized in the liver as intact high molecular weight (HMW) fibrinogen, which consists of six polypeptide chains, $A\alpha_2$, $B\beta_2$, and γ_2, respectively (Fig. 4.2). Partial proteolytic degradation of one fibrinogen $A\alpha$ chain results in low molecular weight fibrinogen (LMW), and partial degradation of both $A\alpha$ chains results in low molecular weight′ (LMW′) fibrinogen. The protease responsible for this conversion has not yet been identified [61, 63]. Physiological conditions may alter the ratio of the high and low molecular weight fibrinogen variants. Higher levels of HMW fibrinogen were found in pre-eclampsia patients, together with enhanced inflammatory responses and hypercoagulability [64]. In the acute phase of, e.g., stroke, pulmonary embolism, angina pectoris, or intracerebral hemorrhage, the fibrinogen levels in the circulation increase [65]. Specifically, the percentage of intact fibrinogen increases after surgery and myocardial infarction as shown by Holm and Godal [66]. The initial (3–4 days) increase in fibrinogen levels consists

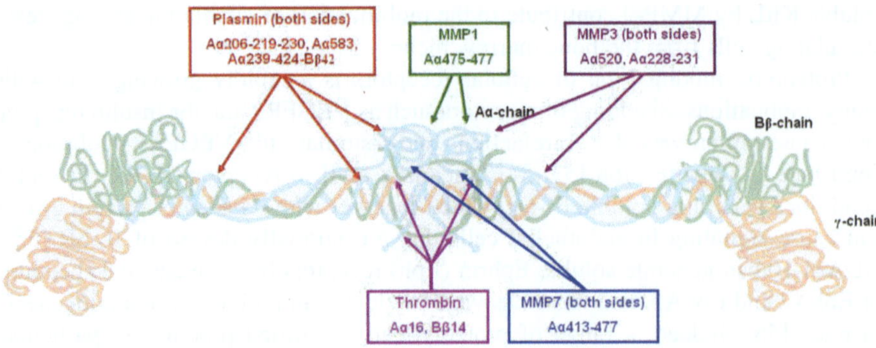

Fig. 4.2 Proteolytic degradation of native fibrinogen. Fibrinogen is composed of two similar molecule halves, each consisting of an Aα-, Bβ-, and γ-chain. Various proteases can split the fibrinogen molecule at different sites on both sides of the molecule. The conversion of HMW toward LMW and LMW′ occurs at positions Aα269, 297, and 309, but the protease(s) involved are not known yet

mostly of HMW fibrinogen, whereas only small and delayed increases in LMW fibrinogen levels were observed (8–11 days). Together with the yet unknown protease responsible for the HMW to LMW conversion, this implies that the formation of LMW fibrinogen is a slow and multistep process.

On the functional level the occurrence of fibrinogen molecular weight variants has substantial effects. When HMW and LMW fibrinogen are compared, LMW fibrinogen shows impaired clotability, thinner and denser fibrin fibers, and possibly slower fibrinolysis. These properties together with altered cell–matrix interactions have substantial effects on angiogenesis in HMW and LMW fibrin matrices. HMW fibrin matrices enhance the in vitro sprouting and in vivo angiogenesis when compared to the natural mixture of fibrin variants [67]. In contrast to HMW fibrin, the sprouting and in vivo angiogenesis are impaired in LMW fibrin when compared to the natural fibrin mixture. Thus, higher HMW fibrinogen levels in the acute phase might result in more optimal formation of vascular sprouts. Moreover, these fibrin variants show differential regulation of multiple genes in endothelial cells, representative for the influence of the matrix on cellular characteristics [68]. Fibrin heterogeneity via proteolytic degradations seems to give a delicate balance to angiogenesis.

4.7 Perspective

The protease family is one of many members in the fascinating complexity of molecules that regulate neovascularization. In this review we elaborate on protease action that is not limited to matrix degradation. In fact, proteases affect a wide range of steps in the process of angiogenesis.

Modulation of proteolytic activity during cell invasion and matrix remodeling has been proposed as potential therapeutic approach for tumors and rheumatoid arthritis. However, even though inhibition of proteases provides high expectations in various pathologies, modulation of proteolytic activity in pathological conditions appeared to be a complex process with yet limited success. Many antitumor MMP inhibitors failed in the early phase clinical trials, due to extensive homology between MMPs catalytic domains [69, 70]. In addition to MMP inhibition, the serine-protease family of uPA/plasmin appears as a potential target for anticancer therapy, and a few uPA inhibitors entered clinical trials [71]. In order to develop successful anti-angiogenic treatments, increased knowledge on growth factor, cytokine, and matrix modification by proteases are a necessity, together with their effect on trafficking, shedding, and signaling of receptors. Increased knowledge on the diversity of protease regulation and action, together with the development of specific and locally applicable inhibitors, may contribute to improved anti-angiogenic strategies. Recent developments using a degradomic discovery approach might reveal new and unexplored sets of proteases and targets of their action. In this degradomics approach, oligonucleotide microarrays, activity-based probes, and fluorescence-based assay are used to profile and monitor protease levels and activity [72].

A challenging new field is the role of proteases and their receptors in the recruitment and homing of stem and progenitor cells that contribute to angiogenesis. Substantial evidence exists on the contribution of MMPs to stem or progenitor cell recruitment. Active MMP-9 from bone marrow stromal cells causes shedding of soluble Kit ligand (sKitL) that upon binding to its receptor on endothelial progenitor cells results in migration of progenitor cells toward areas of vascularization [4, 73]. MT1-MMP has been implicated in progenitor cell mobilization via granulocyte colony-stimulating factor and CD44 on progenitor cells [27]. MMP-2 contributes to the release of stem/progenitor cells from the osteogenic niche by degrading the interaction between SDF-1 and its receptor CXCR4 that binds these cells to the local matrix [74]. Activation of tissue kallikreins (serine proteases) are involved in endothelial progenitor cell recruitment in rheumatoid arthritis, possibly via bradykinin type 2 receptor and CXCR4 [75]. Degradation of the extracellular matrix can also release molecules that contribute to the homing and invasion of stem and progenitor cells [76]. MMP-9 is involved in the incorporation of bone-marrow derived endothelial progenitor cells into ischemic areas to stimulate vascularization [77, 78]. In stem/progenitor cell migration into atherosclerotic lesions, an important role for MMP-8 was revealed [79]. Research toward the roles of proteases in stem and progenitor cell recruitment reveals a new aspect in the diversity of protease action that is involved in angiogenesis.

Finally, recognition of the angiogenesis-modulating effects of subtle changes in matrix proteins, as here illustrated for fibrin, bear the perspective of making matrix structures with a high or low stimulatory effect on angiogenesis and for improving tissue grafts that upon implantation in patients depend on an adequate vascularization.

Acknowledgments This work was supported by grants of The Netherlands Institute for Regenerative Medicine (NIRM, workpackage hypoxia and angiogenesis).

References

1. Carmeliet P, Collen D (1998) Development and disease in proteinase-deficient mice: role of the plasminogen, matrix metalloproteinase and coagulation system. Thromb Res 91 (6):255–285
2. Egeblad M, Werb Z (2002) New functions for the matrix metalloproteinases in cancer progression. Nat Rev Cancer 2(3):161–174
3. van Hinsbergh VWM, Engelse MA, Quax PH (2006) Pericellular proteases in angiogenesis and vasculogenesis. Arterioscler Thromb Vasc Biol 26(4):716–728
4. Verloop RE, Koolwijk P, van Zonneveld AJ, van Hinsbergh VW (2009) Proteases and receptors in the recruitment of endothelial progenitor cells in neovascularization. Eur Cytokine Netw 20(4):207–219
5. van Hinsbergh VWM, Koolwijk P (2008) Endothelial sprouting and angiogenesis: matrix metalloproteinases in the lead. Cardiovasc Res 78(2):203–212
6. Davis GE, Stratman AN, Sacharidou A, Koh W (2011) Molecular basis for endothelial lumen formation and tubulogenesis during vasculogenesis and angiogenic sprouting. Int Rev Cell Mol Biol 288:101–165
7. Lopez-Otin C, Bond JS (2008) Proteases: multifunctional enzymes in life and disease. J Biol Chem 283(45):30433–30437
8. Blasi F, Carmeliet P (2002) uPAR: a versatile signalling orchestrator. Nat Rev Mol Cell Biol 3 (12):932–943
9. Andreasen PA, Egelund R, Petersen HH (2000) The plasminogen activation system in tumor growth, invasion, and metastasis. Cell Mol Life Sci 57(1):25–40
10. Roy R, Zhang B, Moses MA (2006) Making the cut: protease-mediated regulation of angiogenesis. Exp Cell Res 312(5):608–622
11. Lijnen HR (2001) Plasmin and matrix metalloproteinases in vascular remodeling. Thromb Haemost 86(1):324–333
12. Van De Craen B, Scroyen I, Vranckx C, Compernolle G, Lijnen HR, Declerck PJ et al (2012) Maximal PAI-1 inhibition in vivo requires neutralizing antibodies that recognize and inhibit glycosylated PAI-1. Thromb Res 129(4):e126–e133
13. Syrovets T, Simmet T (2004) Novel aspects and new roles for the serine protease plasmin. Cell Mol Life Sci 61(7–8):873–885
14. Matsuno H (2006) Alpha2-antiplasmin on cardiovascular diseases. Curr Pharm Des 12 (7):841–847
15. Kessenbrock K, Plaks V, Werb Z (2010) Matrix metalloproteinases: regulators of the tumor microenvironment. Cell 141(1):52–67
16. Newby AC (2005) Dual role of matrix metalloproteinases (matrixins) in intimal thickening and atherosclerotic plaque rupture. Physiol Rev 85(1):1–31
17. Spinale FG, Janicki JS, Zile MR (2013) Membrane-associated matrix proteolysis and heart failure. Circ Res 112(1):195–208
18. Visse R, Nagase H (2003) Matrix metalloproteinases and tissue inhibitors of metalloproteinases: structure, function, and biochemistry. Circ Res 92(8):827–839
19. Nagase H, Woessner JFJ (1999) Matrix metalloproteinases. J Biol Chem 274 (31):21491–21494
20. Butler GS, Overall CM (2009) Updated biological roles for matrix metalloproteinases and new "intracellular" substrates revealed by degradomics. Biochemistry 48(46):10830–10845
21. Rodriguez D, Morrison CJ, Overall CM (2010) Matrix metalloproteinases: what do they not do? New substrates and biological roles identified by murine models and proteomics. Biochim Biophys Acta 1803(1):39–54

22. Handsley MM, Edwards DR (2005) Metalloproteinases and their inhibitors in tumor angiogenesis. Int J Cancer 115(6):849–860
23. Bauvois B (2012) New facets of matrix metalloproteinases MMP-2 and MMP-9 as cell surface transducers: outside-in signaling and relationship to tumor progression. Biochim Biophys Acta 1825(1):29–36
24. Williams KC, Coppolino MG (2011) Phosphorylation of membrane type 1-matrix metalloproteinase (MT1-MMP) and its vesicle-associated membrane protein 7 (VAMP7)-dependent trafficking facilitate cell invasion and migration. J Biol Chem 286 (50):43405–43416
25. Galvez BG, Matias-Roman S, Yanez-Mo M, Vicente-Manzanares M, Sanchez-Madrid F, Arroyo AG (2004) Caveolae are a novel pathway for membrane-type 1 matrix metalloproteinase traffic in human endothelial cells. Mol Biol Cell 15(2):678–687
26. Brew K, Nagase H (2010) The tissue inhibitors of metalloproteinases (TIMPs): an ancient family with structural and functional diversity. Biochim Biophys Acta 1803(1):55–71
27. Vagima Y, Avigdor A, Goichberg P, Shivtiel S, Tesio M, Kalinkovich A et al (2009) MT1-MMP and RECK are involved in human CD34+ progenitor cell retention, egress, and mobilization. J Clin Invest 119(3):492–503
28. Mascall KS, Small GR, Gibson G, Nixon GF (2012) Sphingosine-1-phosphate-induced release of TIMP-2 from vascular smooth muscle cells inhibits angiogenesis. J Cell Sci 125(Pt 9):2267–2275
29. Seo DW, Li H, Guedez L, Wingfield PT, Diaz T, Salloum R et al (2003) TIMP-2 mediated inhibition of angiogenesis: an MMP-independent mechanism. Cell 114(2):171–180
30. Kumar S, Rao N, Ruowen G (2012) Emerging roles of ADAMTSs in angiogenesis and cancer. Cancers 4(4):1252–1299
31. Kostoulas G, Lang A, Nagase H, Baici A (1999) Stimulation of angiogenesis through cathepsin B inactivation of the tissue inhibitors of matrix metalloproteinases. FEBS Lett 455(3):286–290
32. Urbich C, Heeschen C, Aicher A, Sasaki K, Bruhl T, Farhadi MR et al (2005) Cathepsin L is required for endothelial progenitor cell-induced neovascularization. Nat Med 11(2):206–213
33. Mohamed MM, Sloane BF (2006) Cysteine cathepsins: multifunctional enzymes in cancer. Nat Rev Cancer 6(10):764–775
34. Kobayashi H, Moniwa N, Sugimura M, Shinohara H, Ohi H, Terao T (1993) Effects of membrane-associated cathepsin B on the activation of receptor-bound prourokinase and subsequent invasion of reconstituted basement membranes. Biochim Biophys Acta 1178 (1):55–62
35. Sakamoto T, Seiki M (2010) A membrane protease regulates energy production in macrophages by activating hypoxia-inducible factor-1 via a non-proteolytic mechanism. J Biol Chem 285(39):29951–29964
36. Nishida C, Kusubata K, Tashiro Y, Gritli I, Sato A, Ohki-Koizumi M et al (2012) MT1-MMP plays a critical role in hematopoiesis by regulating HIF-mediated chemokine/cytokine gene transcription within niche cells. Blood 119(23):5405–5416
37. Chen TT, Luque A, Lee S, Anderson SM, Segura T, Iruela-Arispe ML (2010) Anchorage of VEGF to the extracellular matrix conveys differential signaling responses to endothelial cells. J Cell Biol 188(4):595–609
38. Lee S, Jilani SM, Nikolova GV, Carpizo D, Iruela-Arispe ML (2005) Processing of VEGF-A by matrix metalloproteinases regulates bioavailability and vascular patterning in tumors. J Cell Biol 169(4):681–691
39. Nakayama M, Nakayama A, van Lessen M, Yamamoto H, Hoffmann S, Drexler HCA et al (2013) Spatial regulation of VEGF receptor endocytosis in angiogenesis. Nat Cell Biol 15 (3):249–260
40. Jakobsson L, Franco CA, Bentley K, Collins RT, Ponsioen B, Aspalter IM et al (2010) Endothelial cells dynamically compete for the tip cell position during angiogenic sprouting. Nat Cell Biol 12(10):943–953

41. Potente M, Gerhardt H, Carmeliet P (2011) Basic and therapeutic aspects of angiogenesis. Cell 146(6):873–887
42. Funahashi Y, Shawber CJ, Sharma A, Kanamaru E, Choi YK, Kitajewski J (2011) Notch modulates VEGF action in endothelial cells by inducing matrix metalloprotease activity. Vasc Cell 3(1):2
43. Kamei M, Saunders WB, Bayless KJ, Dye L, Davis GE, Weinstein BM (2006) Endothelial tubes assemble from intracellular vacuoles in vivo. Nature 442(7101):453–456
44. Strilic B, Kucera T, Eglinger J, Hughes MR, McNagny KM, Tsukita S et al (2009) The molecular basis of vascular lumen formation in the developing mouse aorta. Dev Cell 17 (4):505–515
45. Mori H, Tomari T, Koshikawa N, Kajita M, Itoh Y, Sato H et al (2002) CD44 directs membrane-type 1 matrix metalloproteinase to lamellipodia by associating with its hemopexin-like domain. EMBO J 21(15):3949–3959
46. Tjwa M, Sidenius N, Moura R, Jansen S, Theunissen K, Andolfo A et al (2009) Membrane-anchored uPAR regulates the proliferation, marrow pool size, engraftment, and mobilization of mouse hematopoietic stem/progenitor cells. J Clin Invest 119(4):1008–1018
47. Uekita T, Itoh Y, Yana I, Ohno H, Seiki M (2001) Cytoplasmic tail-dependent internalization of membrane-type 1 matrix metalloproteinase is important for its invasion-promoting activity. J Cell Biol 155(7):1345–1356
48. Gonzalo P, Guadamillas MC, Hernandez-Riquer MV, Pollan A, Grande-Garcia A, Bartolome RA et al (2010) MT1-MMP is required for myeloid cell fusion via regulation of Rac1 signaling. Dev Cell 18(1):77–89
49. Sacharidou A, Koh W, Stratman AN, Mayo AM, Fisher KE, Davis GE (2010) Endothelial lumen signaling complexes control 3D matrix-specific tubulogenesis through interdependent Cdc42- and MT1-MMP-mediated events. Blood 115(25):5259–5269
50. Fantin A, Vieira JM, Gestri G, Denti L, Schwarz Q, Prykhozhij S et al (2010) Tissue macrophages act as cellular chaperones for vascular anastomosis downstream of VEGF-mediated endothelial tip cell induction. Blood 116(5):829–840
51. Hashimoto G, Inoki I, Fujii Y, Aoki T, Ikeda E, Okada Y (2002) Matrix metalloproteinases cleave connective tissue growth factor and reactivate angiogenic activity of vascular endothelial growth factor 165. J Biol Chem 277(39):36288–36295
52. McQuibban GA, Butler GS, Gong JH, Bendall L, Power C, Clark-Lewis I et al (2001) Matrix metalloproteinase activity inactivates the CXC chemokine stromal cell-derived factor-1. J Biol Chem 276(47):43503–43508
53. Mourad JJ, Des Guetz G, Debbabi H, Levy BI (2008) Blood pressure rise following angiogenesis inhibition by bevacizumab. A crucial role for microcirculation. Ann Oncol 19(5):927–934
54. van der Veldt AAM, de Boer MP, Boven E, Eringa EC, van den Eertwegh AJM, van Hinsbergh VW et al (2010) Reduction in skin microvascular density and changes in vessel morphology in patients treated with sunitinib. Anticancer Drugs 21(4):439–446
55. Romano M, Sironi M, Toniatti C, Polentarutti N, Fruscella P, Ghezzi P et al (1997) Role of IL-6 and its soluble receptor in induction of chemokines and leukocyte recruitment. Immunity 6 (3):315–325
56. Brantley DM, Cheng N, Thompson EJ, Lin Q, Brekken RA, Thorpe PE et al (2002) Soluble Eph A receptors inhibit tumor angiogenesis and progression in vivo. Oncogene 21 (46):7011–7026
57. Arroyo AG, Iruela-Arispe ML (2010) Extracellular matrix, inflammation, and the angiogenic response. Cardiovasc Res 86(2):226–235
58. Barczyk M, Carracedo S, Gullberg D (2010) Integrins. Cell Tissue Res 339(1):269–280
59. Laurens N, Koolwijk P, de Maat MP (2006) Fibrin structure and wound healing. J Thromb Haemost 4(5):932–939
60. Holm B, Nilsen DW, Kierulf P, Godal HC (1985) Purification and characterization of 3 fibrinogens with different molecular weights obtained from normal human plasma. Thromb Res 37(1):165–176

61. Henschen-Edman AH (2001) Fibrinogen non-inherited heterogeneity and its relationship to function in health and disease. Ann N Y Acad Sci 936:580–593
62. de Maat MP, Verschuur M (2005) Fibrinogen heterogeneity: inherited and noninherited. Curr Opin Hematol 12(5):377–383
63. Nakashima A, Sasaki S, Miyazaki K, Miyata T, Iwanaga S (1992) Human fibrinogen heterogeneity: the COOH-terminal residues of defective A alpha chains of fibrinogen II. Blood Coagul Fibrinolysis 3(4):361–370
64. Manten GTR, Sikkema JM, Franx A, Hameeteman TM, Visser GHA, de Groot PG et al (2003) Increased high molecular weight fibrinogen in pre-eclampsia. Thromb Res 111(3):143–147
65. Danesh J, Lewington S, Thompson SG, Lowe GDO, Collins R, Kostis JB et al (2005) Plasma fibrinogen level and the risk of major cardiovascular diseases and nonvascular mortality: an individual participant meta-analysis. JAMA 294(14):1799–1809
66. Holm B, Godal HC (1984) Quantitation of the three normally-occurring plasma fibrinogens in health and during so-called "acute phase" by SDS electrophoresis of fibrin obtained from EDTA-plasma. Thromb Res 35(3):279–290
67. Kaijzel EL, Koolwijk P, van Erck MG, van Hinsbergh VWM, de Maat MP (2006) Molecular weight fibrinogen variants determine angiogenesis rate in a fibrin matrix in vitro and in vivo. J Thromb Haemost 4(9):1975–1981
68. Weijers EM, Van Wijhe MH, Joosten L, Horrevoets AJ, de Maat MP, van Hinsbergh VWM et al (2010) Molecular weight fibrinogen variants alter gene expression and functional characteristics of human endothelial cells. J Thromb Haemost 8(12):2800–2809
69. Zucker S, Cao J (2009) Selective matrix metalloproteinase (MMP) inhibitors in cancer therapy: ready for prime time? Cancer Biol Ther 8(24):2371–2373
70. Coussens LM, Fingleton B, Matrisian LM (2002) Matrix metalloproteinase inhibitors and cancer: trials and tribulations. Science 295(5564):2387–2392
71. Ulisse S, Baldini E, Sorrenti S, D'Armiento M (2009) The urokinase plasminogen activator system: a target for anti-cancer therapy. Curr Cancer Drug Targets 9(1):32–71
72. Overall CM, Blobel CP (2007) In search of partners: linking extracellular proteases to substrates. Nat Rev Mol Cell Biol 8(3):245–257
73. Heissig B, Hattori K, Dias S, Friedrich M, Ferris B, Hackett NR et al (2002) Recruitment of stem and progenitor cells from the bone marrow niche requires MMP-9 mediated release of kit-ligand. Cell 109(5):625–637
74. Segers VFM, Tokunou T, Higgins LJ, MacGillivray C, Gannon J, Lee RT (2007) Local delivery of protease-resistant stromal cell derived factor-1 for stem cell recruitment after myocardial infarction. Circulation 116(15):1683–1692
75. Dai J, Agelan A, Yang A, Zuluaga V, Sexton D, Colman RW et al (2012) Role of plasma kallikrein-kinin system activation in synovial recruitment of endothelial progenitor cells in experimental arthritis. Arthritis Rheum 64(11):3574–3582
76. Reing JE, Zhang L, Myers-Irvin J, Cordero KE, Freytes DO, Heber-Katz E et al (2009) Degradation products of extracellular matrix affect cell migration and proliferation. Tissue Eng Part A 15(3):605–614
77. Iwakura A, Shastry S, Luedemann C, Hamada H, Kawamoto A, Kishore R et al (2006) Estradiol enhances recovery after myocardial infarction by augmenting incorporation of bone marrow-derived endothelial progenitor cells into sites of ischemia-induced neovascularization via endothelial nitric oxide synthase-mediated activation of matrix metalloproteinase-9. Circulation 113(12):1605–1614
78. Deryugina EI, Quigley JP (2010) Pleiotropic roles of matrix metalloproteinases in tumor angiogenesis: contrasting, overlapping and compensatory functions. Biochim Biophys Acta 1803(1):103–120
79. Xiao Q, Zhang F, Lin L, Fang C, Wen G, Tsai TN et al (2013) Functional role of matrix metalloproteinase-8 in stem/progenitor cell migration and their recruitment into atherosclerotic lesions. Circ Res 112(1):35–47

Serpins in Angiogenesis

5

Czeslaw S. Cierniewski and Joanna Boncela

Contents

5.1 Introduction 102
5.2 Structure of Serpins 103
5.3 Mechanism of Inhibition 103
5.4 PAI-1: An Exception to the Rule 104
5.5 Serpins and Angiogenesis 106
5.5.1 Pro- and Antiangiogenic Properties of PAI-1 106
5.5.2 Other Serpins and Angiogenesis 108
References 113

Abstract

Serpins (*ser*ine *p*roteinase *in*hibitors) are the largest superfamily of protease inhibitors. The serpins are structurally similar but functionally diverse proteins that fold into a conserved structure and employ a unique suicide substrate-like inhibitory mechanism. Most of them act as classical protease inhibitors, but there are also serpins that inhibit other types of proteinases, e.g., caspases, or have different, noninhibitory functions, e.g., hormone transport. Serpins are involved in regulation of numerous biological pathways that initiate inflammation, coagulation, fibrinolysis, complement activation responses, apoptosis, extracellular matrix composition, and angiogenesis. The following serpins have been identified as potential regulators of angiogenesis: plasminogen activator inhibitor type 1 (PAI-1), kallistatin, protein C inhibitor, angiotensinogen, maspin,

Professor Czeslaw Cierniewski has died on 24th October, 2013

C.S. Cierniewski
Institute of Medical Biology, Polish Academy of Sciences, Łódź, Poland

Department of Molecular and Medical Biophysics, Medical University of Łódź, Łódź, Poland

J. Boncela (✉)
Institute of Medical Biology, Polish Academy of Sciences, Łódź, Poland
e-mail: jboncela@cbm.pan.pl

J. Dulak et al. (eds.), *Angiogenesis and Vascularisation*,
DOI 10.1007/978-3-7091-1428-5_5, © Springer-Verlag Wien 2013

antithrombin, nexin-1, and pigment epithelial-derived factor. They exert mainly antiangiogenic activity, by inhibition of proteolytic processes in which serine proteases and matrix metalloproteinases (MMPs) are key players. Among them, PAI-1 appears to be the most controversial serpin in angiogenesis; it may act both as a pro- and antiangiogenic factor, depending upon the type of cells and existing conditions. Taken together, serpins remarkably contribute to vessels formation process; they are able to affect more than one of the angiogenic steps and their activity extend beyond the inhibition of target proteinases.

Keywords

Serine protease inhibitors • Plasminogen activator inhibitor type 1 • Extracellular matrix • Endothelial cells migration • Plasmin-dependent proteolysis

5.1 Introduction

Serpins constitute a family of *ser*ine *p*rotease *in*hibitor*s* that are identified by a conserved tertiary structure and a unique suicide substrate-like inhibitory mechanism. They are key regulators of numerous biological pathways that initiate inflammation, coagulation, fibrinolysis, angiogenesis, apoptosis, extracellular matrix composition, and complement activation responses. Serpins are found in viruses and various organisms, including vertebrates. Over 1,500 different serpins have been identified until now, with 36 serpins in humans [1, 2]. Based on the phylogenetic relationship of eukaryotic serpins, a consistent expandable nomenclature was established that includes 16 clades (termed A-I) and a group of highly diverged unclassified "orphans" [3]. However, many serpins are recognized, more frequently, by the names from before this classification was established, e.g., SerpinA1, the first serpin in the classification, is better known as antitrypsin. Human serpins exist extracellularly, and only members of clade B are present intracellularly, in different cellular compartments. Most of them act as classical protease inhibitors maintaining the homeostatic regulation of proteolysis. A decisive factor for their inhibitory activity is the surface-exposed reactive center loop (RCL) that provides a scissile peptide bond for the target protease. Once a protease cleaves a serpin, it can become trapped in an irreversible complex and be removed from the system [1, 2]. There are also serpins that inhibit other types of proteinases, i.e., caspases (proteinase inhibitor-9/SerpinB9) [4] and the pepsin-like cysteine proteases (α1 antitrypsin/SerpinA1) [5]. A few human serpins have different, noninhibitory functions, such as (a) hormone transport—cortisol-binding globulin (CBG/SerpinA6), thyroxine-binding globulin (TBG/SerpinA7) [6], (b) molecular chaperones—HSP47 (SerpinH1) [7], (c) tumor suppressors—maspin (SerpinB5) [8], (d) anti-angiogenic factor–pigment epithelium-derived factor (PEDF/SerpinF1) [9], and (e) blood pressure regulators—angiotensinogen (SerpinA8) [10]. The role of most human serpins is not yet fully understood and needs to be elucidated. They demonstrate a diverse set of biologic functions that extend beyond the ability for these molecules to inhibit target proteinases; thus a key challenge is to identify their partners and roles of noninhibitory activity.

5.2 Structure of Serpins

There is a high rate of conservation in structure among the members of the serpin family. The protein's average size is 350–400 amino acid residues, and the molecular weight is 40–50 kDa. The serpin molecule is comprised of three β-sheets (A, B, C), seven to nine α-helices, and the protease recognition site [11]. In the active conformation, RCL constitutes the 20–25 amino acid residue long flexible loop, which is positioned at the top of the molecule's body. The RCL binds to the active site of a target protease. The composition and length of the reactive center loop have been shown to be essential for the inhibition of proteases [1, 12]. Because the serpin is cleaved during inhibition, inhibitory serpins are called "suicide" or "single use" inhibitors [13].

5.3 Mechanism of Inhibition

The serpin fold represents a unique metastable conformation and a reaction with protease allows it to adopt the most stabilized conformation when RCL inserts itself, as the fourth strand, into the A β-sheet. Currently, a serpin can adopt at least four more stable conformations, termed *cleaved*, *delta*, *latent*, and *polymeric* states [14–16]. The mechanism of serpin inhibition has been demonstrated to be a suicide substrate-like one. In this inhibitory mechanism the proteases recognize RCL of serpin as "bait" and cleave between two residues known as P1 and P1′. This region's sequence is the primary determinant of inhibitory specificity. The initial recognition and reaction of serpin by protease are similar to that of serine protease acting on a substrate. Following cleavage, a covalent complex is formed with the target proteinase and the cut end of RCL begins to insert itself into β-sheet A and to transport the covalently bound protease with it. Upon complete loop insertion and a massive conformational change of serpin, the protease is translocated from the one pole of the serpin molecule to the other, its active site is distorted by being crushed against the serpin core, and an irreversible covalent serpin–enzyme complex arises, which is then a target for proteolytic degradation (the inhibitory pathway) [2, 11, 13, 17]. Such a "suicide cleavage" mechanism gives a rapid turnover of both inhibitor and proteinase. During the translocation event there is a branch point where the protease still has a chance to finish the proteolytic reaction and to be released from the serpin (the noninhibitory pathway or the substrate pathway) [11] (Fig. 5.1). The cleavage mechanism places particular constraints on the serpin fold, which is common for all serpin inhibitors. In particular, the suicide inhibitory mechanism requires a run of small side-chain amino acid residues in the proximal "hinge" region, located seven residues N-terminal to the scissile bond, to allow RCL insertion into β-sheet A. This flexible hinge region is absent in noninhibitory serpin-fold proteins [18]. One of the important features of serpins is their ability to bind various proteases and the nonprotease ligand. Binding the ligand to these

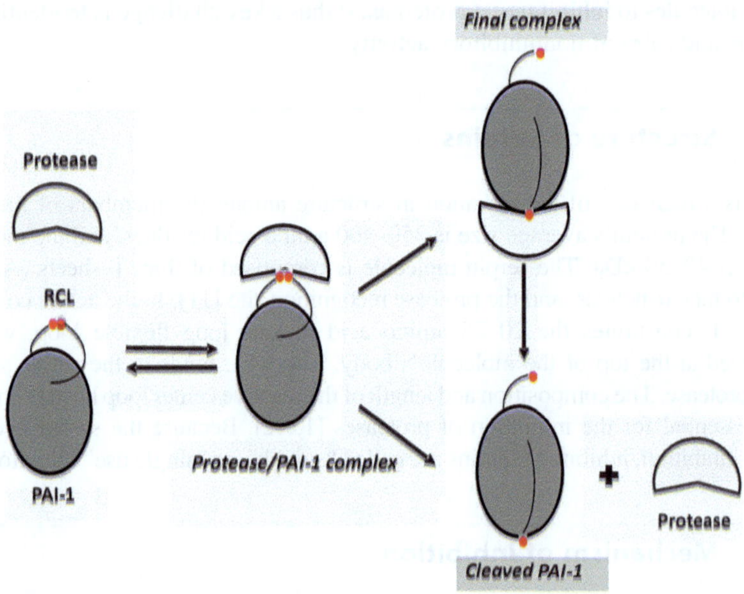

Fig. 5.1 Branched pathway mechanism of PAI-1's action. The serpin inhibition of protease proceeds through an initial noncovalent Michaelis complex and subsequently stable covalent complex of inhibitor and protease is formed (inhibitory pathway) or cleaved PAI-1 and active protease are released (noninhibitory pathway or substrate pathway). P1 and P1′ residues of RCL shown in *red*

serpins regulates their activity, which is similar to how PAI-1 is modified by interaction with vitronectin [19]. Several human serpins, including antithrombin, heparin cofactor II, protease nexin, PAI-1, and protein C inhibitor, have a binding site for sulfated polysaccharides [20]. Upon binding of heparin or other negatively charged polyanions, such as heparan sulfate and dermatan sulfate, the inhibitory activity of serpins is greatly enhanced; for example, antithrombin inhibitory efficiency increases 1,000-fold [21, 22]. This increased activity results from a substantial conformational change of the serpin molecule whereby the RCL can exist in an exposed enzyme-accessible conformation [1].

5.4 PAI-1: An Exception to the Rule

PAI-1, like most members of the serpin family, has a highly conserved structure consisting of three β-sheets (A–C), nine α-helices (A–I), and a 26-residue-long surface-exposed RCL. The residues of RCL designated as P16–P10′ provide a bait peptide bond, P1–P1′, that mimics the normal substrates of PAI-1, i.e., urokinase-type plasminogen inhibitor (u-PA) or tissue-type plasminogen activator (t-PA) [23]. In contrast to other serpins, the PAI-1 molecule demonstrates functional flexibility, and its inhibitory activity is regulated by a self-inactivation mechanism. Under normal physiological conditions, newly synthesized PAI-1 molecules are inherently unstable with an apparent half-life of approximately 2 h in 37 °C and readily

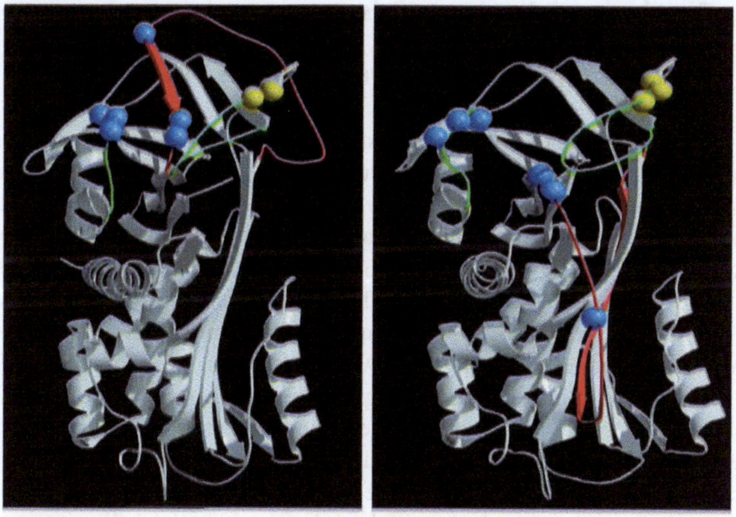

Fig. 5.2 Active and latent PAI-1. In both structures, the proteinase recognition site (RSL) shown in *red* represents a flexible stretch of 17 residues exposed in the active form which is spontaneously inserted into the middle of β-sheet A to give a fully antiparallel β-sheet. Image courtesy of Prof. Paul Declerck, University of Leuven, Belgium

convert from an active to a stable, inactive latent form. The latency transition of PAI-1 involves insertion of its RCL as a strand into β-sheet A and results in stabilization of the molecule structure (Fig. 5.2). The nonreactive latent form does not react with target proteases. The latent form can be partially reactivated in vitro by a denaturing agent [24] and also in vivo by mechanisms that still remain unidentified [25]. PAI-1 mainly occurs as the active form bound to the vitronectin that is present in the plasma or in ECM [26–29]. Vitronectin binds to PAI-1 with high affinity [30] and maintains the active conformation of the inhibitor, alone or with various metal ions [31]. The vitronectin-binding site on PAI-1 has been located on the central β-sheet, i.e., the one that incorporates a surface-exposed RCL and within the adjacent secondary structure [29, 32, 33]. Therefore, association of vitronectin with the region in the vicinity of the reactive center interferes with movements of the β-strands and thus inhibits a conversion of PAI-1 to the latent form. Recently, in addition to vitronectin, α1-acid glycoprotein (AGP) was identified as being able to bind to PAI-1. AGP is one of the major acute-phase proteins. Although AGP was bound to PAI-1 with a lower affinity than vitronectin, this reaction was strong enough to stabilize the active conformation of PAI-1 and to prolong its inhibitory activity towards u-PA and t-PA [34]. The AGP binding also stabilizes the active form of PAI-1 by restricting the movement of β-sheet A and thereby preventing insertion of the reactive center loop [35]. Therefore, the complex of PAI-1 with the α1-acid glycoprotein can play a role as an alternative reservoir of the inhibitor's physiologically active form, particularly during inflammation or another acute-phase reaction.

5.5 Serpins and Angiogenesis

5.5.1 Pro- and Antiangiogenic Properties of PAI-1

PAI-1 together with other components of the plasminogen activating system participates in carrying out the VEGF-induced "angiogenic switch," which involves pericellular proteolysis that increases vascular permeability and supports both endothelial cell proliferation and migration [36–40]. VEGF, a major positive regulator of angiogenesis [41], and bFGF synergically induce the expression of u-PA, t-PA, uPAR, and PAI-1, whereas TGF-β downregulates u-PA and enhances PAI-1 production [42]. Also, hypoxia, an essential stimulus for angiogenesis, has been reported to increase uPAR and PAI-1 expression in endothelial cells [43]. PAI-1 acts as an inhibitor of normal angiogenesis [44]. Paradoxically, PAI-1 has been shown to enhance tumor angiogenesis and progression [44]. In response to various cytokines in the tumor microenvironment, PAI-1 produced by adipocytes, fibroblasts, or macrophages can promote angiogenesis, which is crucial to tumor viability and dissemination. Interestingly, when endothelial cells are cocultured with fibroblasts, PAI-1 mRNA and promoter activity are induced only in a single row of fibroblasts adjacent to sprouting but not to quiescent endothelium [45]. This suggests that there is a paracrine induction which may be crucial during sprouting, i.e., during the stage in which the endothelial cells establish direct contact with the fibroblasts. Then the local increase in PAI-1 expression may provide a mechanism to counterbalance excessive pericellular proteolysis. However, it is not always clear if these functions depend on the antiproteolytic activity of PAI-1, on its binding to vitronectin, or on its interference with cellular migration or matrix binding.

PAI-1 is present in resting mature vessels, in which it averts both plasmin-mediated proteolysis and uPAR-dependent cell–ECM adhesions [46]. Quiescent endothelial cells express concomitant t-PA and PAI-1; thus net proteolysis is prevented [42, 47]. When the endothelial cells migrate, they upregulate u-PA, uPAR, and PAI-1, but the balance is shifted towards increased proteolysis [48]. Angiogenic endothelial cells require u-PA and plasmin to degrade ECM components and to migrate. Besides the endothelial cells, PAI-1 is also expressed by stromal and epithelial cells, where it is thought to play a role in the preservation of matrix integrity [42]. Thus, plasmin proteolysis is controlled by a physiological inhibitor such as PAI-1 to allow stabilization of the surrounding ECM and the assembly of endothelial cells into tubes.

The key role played by PAI-1 in angiogenesis and tumor invasion was demonstrated by studies using PAI-1 knockout mice which showed that tumor angiogenesis of transfected malignant cells is severely reduced [49]. A PAI-1 deficiency in mice led to abolished angiogenesis in two models of tumor transplantation [49, 50], in the mouse aortic ring assay [51, 52], and in the laser-induced choroidal neoangiogenesis assay [53]. Both tumoral and choroidal vascularization impaired in PAI-1 knockout mice could be restored in vivo by intravenous injection of recombinant adenovirus-expressing PAI-1 [49, 53]. Consistently, a microvessel outgrowth from the aortic rings was impaired when PAI-1$^{-/-}$ mice were tested using this quantitative in vitro model of angiogenesis. In this model, aortic explants derived from wild-type and gene-deficient mice (PAI-1−/−) were embedded in a three-dimensional matrix of collagen. Microvessel sprouting was restored by addition of exogenous PAI-1 as a purified active rPAI-1 or by adenovirus-mediated PAI-1 gene transfer. When other components of the plasminogen (Plg) activator system were tested in the same model using aortic explants of gene-deficient mice (Plg−/−, t-PA, u-PA, double t-PA−/−/u-PA−/−, uPAR) embedded in the collagen matrix, the importance of controlled plasmin-mediated proteolysis in this process was further confirmed. Angiogenesis was not remarkably affected in t-PA, u-PA, or double t-PA/u-PA-deficient mice. This may indicate that the deficiency of one plasminogen activator could be compensated for by upregulation of the other activator [50]. A delay of microvessel outgrowth was observed with aortic rings isolated from Plg−/− and a small acceleration in vessel outgrowth in the case of mice with uPAR deficiency confirmed that plasmin-mediated proteolysis in fact is a key event in angiogenesis. Further studies showed that tumor growth is inhibited in PAI-1-deficient mice and stimulated in mice that overexpress PAI-1 [54]. However, in a transgenic mouse model, metastasizing breast cancer primary tumor growth and vascular density were not different in PAI-1-deficient and wild-type mice [55]. Recent studies have highlighted the importance of host-derived PAI-1 as compared to tumor-derived PAI-1 for tumor growth and tumor angiogenesis [56, 57]. It was demonstrated that PAI-1 promotes tumor development at physiological concentrations in mice, but it inhibits invasion and vascularization at pharmacological concentrations [51]. Interestingly, high levels of PAI-1 in the primary tumor tissue of patients with various types of solid cancer correlate with disease recurrence and reduced survival [58]. PAI-1 was found to accelerate lung metastasis formation of human fibrosarcoma cells [59].

The precise mechanism by which PAI-1 promotes angiogenesis has remained largely indefinable. Originally it was suggested that the ability of PAI-1 to encourage angiogenesis is dependent on its protease inhibitor activity. This would imply that PAI-1 does not advance angiogenesis by promoting de-adhesion of endothelial cells from matrix proteins [50]. Based on other observations, PAI-1 has been proposed to be a molecular switch that governs cell detachment by dissociating vitronectin from uPAR, thereby promoting a migratory phenotype [60]. Some other mechanisms of the proangiogenic activity of PAI-1 were also proposed; e.g., PAI-1 has been found to induce the internalization of u-PA/uPAR/PAI-1 complexes, after which uPAR is recycled back to the plasma membrane, not randomly distributed but focused on new adhesion areas at the leading edge of the cell. Such uPAR relocalization was suggested as being crucial for endothelial as well as cancer cell motility [60, 61]. PAI-1 was also described as inhibiting pro-apoptotic cell signaling by blocking plasmin-mediated cleavage of the Fas ligand on the surface of endothelial cells, thus protecting them from apoptosis mediated by Fas [62]. Consistently, in PAI-1-deficient mice there was increased plasmin generation that stimulated the release of a soluble FasL fragment from the cell surface of the endothelial cells. This soluble FasL fragment activates Fas expressed on the endothelial cells and acts as a potent inducer of endothelial cell apoptosis. Therefore, by preventing apoptosis of endothelial cells, PAI-1 would promote angiogenesis. There are numerous observations showing that PAI-1 may also be implicated in controlling the generation of proteolytic fragments of ECM components or other molecules that display angioregulatory activities, which are either positive or negative. Particularly, angiostatin and endostatin, which are antiangiogenic protein fragments derived from Plg and collagen XVIII, respectively, upon limited proteolysis of Plg by plasmin, u-PA, and t-PA, were described to inhibit tumor angiogenesis [63, 64]. The number of protein fragments showing negative angiogenic regulator activity is increasing. Among them are fragments of collagen type XV (restin) [65] or different chains of collagen type IV, such as arresten, canstatin, and tumstatin derived from the α1 [66], α2 [67], and α3 chain [68], respectively. Fragments of non-ECM molecules such as MMP-2 (PEX), antithrombin [69], calreticulin (vasostatin) [70], and domain 5 of high-molecular-weight kininogen (kininostatin) [71] were also identified to possess angioinhibitory activity. Little is known, however, about the proteolytic pathway leading to their generation and about their biological significance in vivo.

Paradoxically, PAI-1 has been shown to be proangiogenic and antiangiogenic [50, 60, 72–74]. When utilizing PAI-1−/− mice it was shown that PAI-1 promotes angiogenesis in the tumor microenvironment at low concentrations (in the nanomolar range) but inhibits angiogenesis at extremely high concentrations (micromolar concentrations) [51]. At low concentrations, PAI-1 was described to act as a protease inhibitor, whereas at high concentrations it was postulated that it swayed the balance between adhesion/detachment and migration of normal and tumor cells [60, 75]. By blocking interaction between vitronectin, uPAR, and integrins, PAI-1 may induce cell detachment from the extracellular matrix and thus promote cellular migration and perhaps tumor invasion [60]. The relevance of the latter mechanism to overall tumor growth and angiogenesis has not been conclusively confirmed in vivo.

Conversely, there are data showing that PAI-1 can exhibit antiangiogenic activity. Administration of PAI-1 inhibits FGF-2-induced angiogenesis, which was tested using chick chorioallantoic membranes [74]. Tumor cells transduced with PAI-1 cDNA showed a reduction in primary tumor growth, tumor-associated angiogenesis, and metastasis [75].

5.5.2 Other Serpins and Angiogenesis

Several serpins (kallistatin, protein C inhibitor, angiotensinogen, maspin, antithrombin, nexin-1, and pigment epithelial-derived factor) have been recently shown to exert an antiangiogenic activity, suggesting a common mechanism of their effect on an endothelial cell proliferation and migration (Table 5.1).

Kallistatin (SerpinA4) is a specific inhibitor of tissue kallikrein [76, 77] but it has also multiple biologic functions independent of the tissue kallikrein–kinin system, such as blood pressure regulation [78], protection against inflammation [79], and vasculature relaxation [80]. It has been located in vascular smooth muscle cells and in endothelial cells of human blood vessels [81]. In contrast to kallikrein, kallistatin

Table 5.1 Serpins involved in angiogenesis regulation

Name	Gene symbol	Target proteinases	Number of aa	Chromosomal location	SWISS-PROT
Proteinase inhibitor 4, kallistatin	*SerpinA4*	Kallikrein	427	14q32.1	P29622
Protein C inhibitor, PAI-3	*SerpinA5*	Active protein C, thrombin, factors Xa and Xia, kalikrein, u-PA, t-PA	406	14q32.1	P05154
Angiotensinogen	*SerpinA8*	–	485	1q42–q43	P01019
Maspin	*SerpinB5*	–	375	18q21.3	P36952
Antithrombin	*SerpinC1*	Thrombin	464	1q23–q25	P01008
Plasminogen activator inhibitor type 1	*SerpinE1*	Thrombin, u-PA, t-PA, plasmin	402	7q21.3–q22	P05121
Nexin-1	*SerpinE2*	Thrombin, u-PA, t-PA, plasmin, trypsin	398	2q33–q35	P07093
Pigment epithelium-derived factor	*SerpinF1*	–	418	17p13.3	P36955

inhibits angiogenesis and the tumor growth. Native kallistatin inhibited VEGF- and bFGF-induced proliferation, migration, and adhesion of cultured endothelial cells as well as the microvessel formation in the Matrigel implants in mice [82]. The kallistatin gene delivery suppresses spontaneous angiogenesis in the ischemic hind limb and the tumor growth in nude mice [82]. It appears that the heparin-binding domain, but not the reactive site loop of kallistatin, is essential for an inhibiting VEGF-induced angiogenesis [83].

Protein C inhibitor (PCI), another member of serpins (SerpinA5), initially was found to be an inhibitor of the activated PCI and later shown to be a potent inhibitor of blood coagulation and fibrinolysis. Hence, it became known as the plasminogen activator inhibitor-3 (PAI-3). However, studies using transgenic mice expressing the human gene have suggested that PCI is also involved in regulation of lung remodeling, tissue regeneration, vascular permeability, proteolysis in the kidney, and tumor cell invasion [84]. A protease inhibitor-independent activity of PCI, the prevention of anti-angiogenesis and metastasis of tumor cells, has also been observed. In vivo angiogenesis in rat cornea and in vitro tube formation were inhibited by recombinant intact PCI, by reactive site-modified PCI (R354APCI), and by the N-terminal fragment of protease-cleaved PCI (NTPCI). Interestingly, the anti-angiogenic activity of PCI was comparable to that of the cleaved antithrombin and even stronger than that of PAI-1 or PEDF. It appears that, in addition to a reactive site-dependent mechanism, protein C inhibitor may also regulate tumor growth and metastasis independently of its protease inhibitory activity by inhibiting angiogenesis [85] (Fig. 5.3).

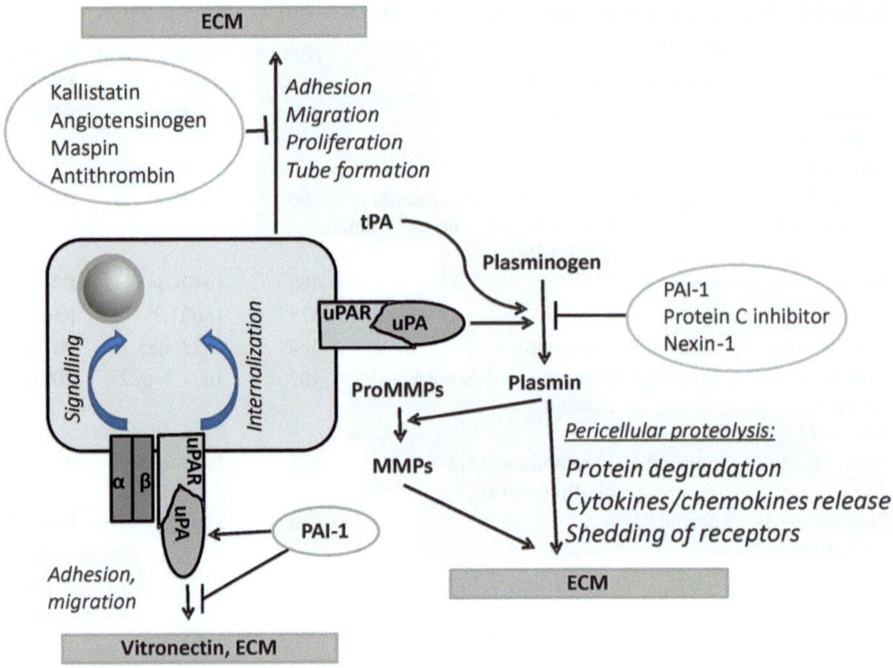

Fig. 5.3 Serpins in angiogenesis. Plasmin plays a central role by activating matrix metalloproteinases (MMPs) or by degrading extracellular matrix components. uPAR participates both in u-PA activation, cell adhesion, and migration and in signal transduction. The latter requires association of the uPAR/u-PA with integrins. The inhibitor PAI-1 and other serpins (protein C inhibitor, nexin-1) control the activity of plasminogen activators (t-PA and u-PA). In addition, PAI-1 exerts different effects by regulating cell adhesion and migration through its interaction with vitronectin and by internalization of the uPAR/u-PA complex. Serpins such as kallistatin, angiotensinogen, maspin, and antithrombin inhibit angiogenesis by undescribed mechanisms yet. However, all of them block adhesion, migration proliferation, and tube formation by endothelial cells

Angiotensinogen (AGT, SerpinA8) consists of a combination of an angiotensin I function, located at the N-terminal end, and the presence of a serpin structure at the opposite end. Angiotensin I, an inactive decapeptide that is converted by renin into Ang II, is the main effector of the renin–angiotensin system. This leaves a much larger fragment intact called des(angiotensin I)angiotensinogen [des(Ang I)AGT], which accounts for more than 97 % of the molecule and apparently has no function. Angiotensinogen and its renin-cleaved product, des(Ang I)AGT, are also angiogenesis inhibitors, both in vitro and in vivo at concentrations within the range of those observed in the plasma [86]. Angiotensinogen and des(Ang I)AGT exhibited similar anti-angiogenic effects when tested in different models of angiogenesis, that is, in the CAM angiogenic assay, endothelial cell proliferation and migration, and the formation of capillary-like structures on Matrigel.

Maspin is an epithelial specific member of serpins (SerpinB5). It is located in both cytoplasm and the nucleus. After secretion, it can also be found to be associated on cell membranes and in the extracellular matrix [87]. Its molecular structure shows a homology to PAI-1, PAI-2, and ovalbumin [88]. As a tumor suppressor, maspin functions to inhibit tumor cell migration and invasion, induces tumor cell apoptosis, and is a potent angiogenesis inhibitor [89]. Extracellularly, maspin seems to play a critical role in blocking the initiation step of the plasminogen activation [89–91], while intracellularly appears to be connected with a proapoptotic effect in tumor cells [92]. Recent studies showed that maspin inhibits angiogenesis through controlling endothelial cell adhesion, migration, and an adhesion-mediated cell signaling pathway. It activates β1 integrins, integrin-linked kinase (ILK), and the FAK signal transduction pathway. Subsequent changes in focal adhesion and cytoskeleton reorganization result in the attachment and spreading of endothelial cell on the matrix [93]. Nuclear maspin gives better differentiated epithelial phenotypes, a decreased tumor angiogenesis, the increased tumor sensitivity to drug-induced apoptosis, and a more favorable pro gnosis. The nuclear maspin correlated with better survival of lung adenocarcinoma [92], with significantly the lower proliferation index Ki67 in lung adenocarcinoma and inversely correlated with the VEGF-A level. It inhibited the oxidative stress-induced VEGF-A production in tumor cells [94] and blocked tumor angiogenesis in the xenograft tumor models [88, 89].

Antithrombin (SerpinC1) is an inhibitor of thrombin and other enzymes involved in the clotting cascade. Although its intact native molecule does not show antiangiogenic activity, after conversion into the cleaved, the stable locked, or the latent form, it becomes a potent antiangiogenic and antitumor inhibitor. All three forms are substantially similar to each other and differ significantly from the native molecule [95]. The cleaved antithrombin and the latent antithrombin inhibited a capillary endothelial cell proliferation and the growth of human SK-NAS neuroblastoma and Lewis lung carcinoma tumors in vivo. They acted selectively upon endothelial cells and the tumor vasculature, but the precise mechanism of the antiangiogenic activity of antithrombin has not been defined [96].

Nexin-1 (PN-1; Serpin E2) is a serine protease inhibitor produced by the most cell types [97]. In vascular system, it is expressed by fibroblasts [98], smooth muscle cells [99], and endothelial cells [100] in a tightly regulated manner. Normally, it is not present in blood plasma, but it may be rapidly released from the activated platelets which can be considered as its reservoir [101]. PN-1 is the most efficient tissue inhibitor of thrombin, plasminogen activators and plasmin, proteases largely involved in tissue remodeling. Recent study showed that PN-1 is antiangiogenic in vitro and ex vivo [102]. It inhibited the VEGF-induced endothelial cell responses such as proliferation, migration, and capillary tube formation and decreased cell spreading on the vitronectin. These effects did not require its antiprotease activity but involved PN-1 binding to glycosaminoglycans. Consistently, the aortic rings obtained from PN-1-deficient mice showed a significantly enhanced sprouting. There was also the increased neovessel formation in the Matrigel plug assay using PN-1-deficient mice as compared to the wild-type mice. These effects were reversed by the addition of recombinant PN-1, indicating that this serpin has direct antiangiogenic properties [102].

Pigment epithelial-derived factor (PEDF; SerpinF1) was discovered as the most potent inhibitor of angiogenesis. It was found to be twice as potent as angiostatin and seven times as potent as endostatin [103]. Widely expressed throughout the body, PEDF is a 50-kDa secreted glycoprotein that belongs to the noninhibitory serpin. The PEDF molecule contains an α/β core serine-protease inhibitor domain, three major β-sheets, and ten α-helices. Although PEDF does not inhibit either serine or cysteine proteinases, it exerts diverse physiological activities including anti-angiogenesis, anti-vasopermeability, antitumor, and neurotrophic activities [104]. Interestingly, a variety of peptides derived from PEDF possess activities similar to those of the parent molecule through interactions with the extracellular matrix, binding to PEDF receptors, nuclear localization, and phosphorylation. One of the advantages of using PEDF or its peptide fragments as a potential therapeutic agent against tumor angiogenesis and against the tumor cell is its ability to target new vessel growth and not disrupt the preexisting vasculature [105]. Mechanisms by which PEDF functions are very complex and involve an inducing apoptosis in endothelial cells, inhibiting their migration and the tube formation even in the presence of VEGF as well as the decrease in expression of the VEGF [106]. PEDF initiates cell apoptosis via the activation of caspase-3 and DNA fragmentation and induces expression of the peroxisome proliferator-activated receptor γ and p53 [107]. It also blocked the VEGF-induced endothelial cell migration and the tube formation by inhibiting the phosphorylation of P13K/Akt at Ser 473 [106]. Apart from apoptosis, PEDF inactivates the VEGF-dependent pathway via downregulating expression of the HIF-1α in tumor cells [108] and enhancing the γ secretase-dependent cleavage of VEGFR1 [109]. Furthermore, PEDF has been shown to be involved in differentiating cells into the mature phenotype as shown in the case of osteoblasts [110], leading to slowing of the tumor growth and spread. PEDF also increased adhesion of bone tumor cells to collagen type-1, the major protein in a mature bone. Thus, PEDF, as an antiangiogenic factor, targets multiple pathways playing a significant role in malignant tumor cells and in endothelial cells leading to the tumor cell proliferation and the initiation of apoptosis [111].

Concluding Remarks

Angiogenesis is an important process during normal physiology and pathologic conditions such as chronic inflammatory diseases and tumor growth. The expansion of normal vasculature depends on the homeostatic balance between pro- and antiangiogenic mediators. In tumor angiogenesis, a quiescent tumor may suddenly secrete and/or suppress angiogenic factors that will instantly produce blood vessel growth and invasion of the surrounding tissues due to inducing the imbalance between the levels of pro- and antiangiogenic signals [112]. Serpins are among many potent pro- and antiangiogenic molecules that are upregulated under these conditions. Interestingly, serpins are able to affect more than one of the angiogenic steps and the precise mechanism of their antiangiogenic activity in vivo has not been defined. It is known that some of them in addition to blocking serine proteases can express other protease-independent activities.

Acknowledgment This work was supported by Projects DEC-2011/01/B/NZ3/00194 (J.B.) and DEC-2011/02/A/NZ3/00068 (C.S.C.).

References

1. Irving JA, Pike RN, Lusk AM, Whisstock JC (2000) Phylogeny of the serpent super family: implications of patterns of amino acid conservation for structure and function. Genome Res 10:1845–1864
2. Law RH, Zhang Q, McGowan S, Buckle AM, Silverman GA, Wong W, Rosado CJ, Langendorf CG, Pike RN, Bird PI, Whisstock JC (2006) An overview of the serpin superfamily. Genome Biol 7:216–226
3. Silverman GA, Bird PI, Carrell RW, Church FC, Coughlin PB, Gettins PG, Irving JA, Lomas DA, Luke CJ, Moyer RW, Pemberton PA, Remold-O'Donnell E, Salvesen GS, Travis J, Whisstock JC (2001) The serpins are an expanding superfamily of structurally similar but functionally diverse proteins. Evolution, mechanism of inhibition, novel functions, and a revised nomenclature. J Biol Chem 276:33293–33296
4. Potempa J, Korzus E, Travis J (1994) The serpin superfamily of proteinase inhibitors: structure, function, and regulation. J Biol Chem 269:15957–15960
5. Irving JA, Pike RN, Dai W, Bromme D, Worrall DM, Silverman GA, Coetzer TH, Dennison C, Bottomley SP, Whisstock JC (2002) Evidence that serpin architecture intrinsically supports papain-like cysteine protease inhibition: engineering alpha(1)-antitrypsin to inhibit cathepsin proteases. Biochemistry 41:4998–5004
6. Pemberton PA, Stein PE, Pepys MB, Potter JM, Carrell RW (1988) Hormone binding globulins undergo serpin conformational change in inflammation. Nature 336:257–258
7. Nagata K (1996) Hsp47: a collagen-specific molecular chaperone. Trends Biochem Sci 21:22–26
8. Zou Z, Anisowicz A, Hendrix MJ, Thor A, Neveu M, Sheng S, Rafidi K, Seftor E, Sager R (1994) Maspin, a serpin with tumor-suppressing activity in human mammary epithelial cells. Science 263:526–529
9. Dawson DW, Volpert OV, Gillis P, Crawford SE, Xu HJ, Benedict W, Bouck NP (1999) Pigment epithelium-derived factor: a potent inhibitor of angiogenesis. Science 285:245–248
10. Corvol P, Lamande N, Cruz A, Celerier J, Gasc JM (2003) Inhibition of angiogenesis: a new function for angiotensinogen and des(angiotensin I)angiotensinogen. Curr Hypertens Rep 5:149–154
11. Gettins PGW (2002) Serpin structure, mechanism and function. Chem Rev 102:4751–4803
12. Zhou A, Carrell RW, Huntington JA (2001) The serpin inhibitory mechanism is critically dependent on the length of the reactive center loop. J Biol Chem 276:27541–27547
13. Huntington JA, Read RJ, Carrell RW (2000) Structure of a serpin-protease complex shows inhibition by deformation. Nature 407:923–926
14. Lomas DA, Mahadeva R (2002) Alpha1-antitrypsin polymerization and the serpinopathies: pathobiology and prospects for therapy. J Clin Invest 110:1585–1590
15. Chow MK, Lomas DA, Bottomley SP (2004) Promiscuous beta-strand interactions and the conformational diseases. Curr Med Chem 11:491–499
16. Mast AE, Enghild JJ, Salvesen G (1991) Conformation of the reactive site loop of alpha 1-proteinase inhibitor probed by limited proteolysis. Biochemistry 31:2720–2728
17. Whisstock J, Bottomley S (2006) Molecular gymnastics: serpin structure, folding and misfolding. Curr Opin Struct Biol 16:761–768
18. Hopkins PC, Carrell RW, Stone SR (1993) Effects of mutations in the hinge region of serpins. Biochemistry 32:7650–7657
19. Naski MC, Lawrence DA, Mosher DF, Podor TJ, Ginsburg D (1993) Kinetics of inactivation of α-thrombin by plasminogen activator inhibitor-1. Comparison of the effects of native and urea-treated forms of vitronectin. J Biol Chem 268:12367–12372

20. Patston PA, Church FC, Olson ST (2004) Serpin-ligand interactions. Methods 32:93–109
21. Jin L, Abrahams JP, Skinner R, Petitou M, Pike RN, Carrell RW (1997) The anticoagulant activation of antithrombin by heparin. Proc Natl Acad Sci U S A 94:14693–14698
22. Olson ST, Bjork I (1991) Predominant contribution of surface approximation to the mechanism of heparin acceleration of the antithrombin-thrombin reaction. Elucidation from salt concentration effects. J Biol Chem 266:6353–6354
23. Nar H, Bauer M, Stassen JM, Lang D, Gils A, Declerck PJ (2000) Plasminogen activator inhibitor 1. Structure of the native serpin, comparison to its other conformers and implications for serpin inactivation. J Mol Biol 297:683–695
24. Hekman CM, Loskutoff DJ (1985) Endothelial cells produce a latent inhibitor of plasminogen activators that can be activated by denaturants. J Biol Chem 260:11581–11587
25. Vaughan DE, Declerck PJ, Van Houtte E, De Mol M, Collen D (1990) Studies of recombinant plasminogen activator inhibitor- 1 in rabbits. Pharmacokinetics and evidence for reactivation of latent plasminogen activator inhibitor-1 in vivo. Circ Res 67:1281–1286
26. Declerck PJ, De Mol M, Alessi MC, Baudner S, Paques E-P, Preissner KT, Muller-Berghaus G, Collen D (1988) Purification and characterization of a plasminogen activator inhibitor 1 binding protein from human plasma. J Biol Chem 263:15454–15461
27. Preissner KT, Holzhtiter S, Justus C, Muller-Berghaus G (1989) Identification and partial characterization of platelet vitronectin: evidence for complex formation with platelet-derived plasminogen activator inhibitor-1. Blood 74:1989–1996
28. Seiffert D, Loskutoff DJ (1999) Kinetic analysis of the interaction between type 1 plasminogen activator inhibitor and vitronectin and evidence that the bovine inhibitor binds to a thrombin-derived amino-terminal fragment of bovine vitronectin. Biochim Biophys Acta 1078:23–30
29. Lawrence DA, Berkenpas MB, Palaniappan S, Ginsburg D (1994) Localization of vitronectin binding domain in plasminogen activator inhibitor-I. J Biol Chem 269:15223–15228
30. Lawrence DA, Palaniappani S, Stefansson S, Olson ST, Francis-Chmura AM, Shore JD, Ginsburg D (1997) Characterization of the binding of different conformational forms of plasminogen activator inhibitor-1 to vitronectin. J Biol Chem 272:7676–7680
31. Thompson LC, Goswami S, Peterson CB (2011) Metals affect the structure and activity of human plasminogen activator inhibitor-1. II. Modulation of stability and protease inhibition. Protein Sci 20:366–378
32. Mottonen J, Strand A, Symersky J, Sweet RM, Danley DE, Geogheggan KF, Gerard RD, Goldsmith EJ (1992) Structural basis of latency in plasminogen activator inhibitor-1. Nature 355:270–273
33. van Meijer M, Gebbink RK, Preissner KT, Pannekoek H (1994) Determination of the vitronectin binding site on plasminogen activator inhibitor 1 (PAI-1). FEBS Lett 352:342–346
34. Boncela J, Papiewska I, Fijalkowska I, Walkowiak B, Cierniewski CS (2001) Acute phase protein α1-acid glycoprotein interacts with plasminogen activator inhibitor type 1 and stabilizes its inhibitory activity. J Biol Chem 276:35305–35311
35. Smolarczyk K, Gils A, Boncela J, Declerck PJ, Cierniewski CS (2005) Function-stabilizing mechanism of plasminogen activator inhibitor type 1 induced upon binding to alpha1-acid glycoprotein. Biochemistry 44:12384–12390
36. Chen SC, Henry DO, Reczek PR, Wong MK (2008) Plasminogen activator inhibitor-1 inhibits prostate tumor growth through endothelial apoptosis. Mol Cancer Ther 7:1227–1236
37. Isogai C, Laug WE, Shimada H, Declerck PJ, Stins MF, Durden DL, Erdreich-Epstein A, DeClerck YA (2001) Plasminogen activator inhibitor-1 promotes angiogenesis by stimulating endothelial cell migration toward fibronectin. Cancer Res 61:5587–5594
38. Ploplis VA, Balsara R, Sandoval-Cooper MJ, Yin ZJ, Batten J, Modi N, Gadoua D, Donahue D, Martin JA, Castellino FJ (2004) Enhanced in vitro proliferation of aortic endothelial cells from plasminogen activator inhibitor-1-deficient mice. J Biol Chem 279:6143–6151

39. Soeda S, Oda M, Ochiai T, Shimeno H (2001) Deficient release of plasminogen activator inhibitor-1 from astrocytes triggers apoptosis in neuronal cells. Mol Brain Res 91:96–103
40. Degryse B, Neels JG, Czekay RP, Aertgeerts K, Kamikubo Y, Loskutoff DJ (2004) The low density lipoprotein receptor-related protein is a motogenic receptor for plasminogen activator inhibitor-1. J Biol Chem 279:22595–22604
41. Hanahan D, Folkman J (1996) Patterns and emerging mechanisms of the angiogenic switch during tumorigenesis. Cell 86:353–364
42. Carmeliet P, Collen D (2000) Transgenic mouse models in angiogenesis and cardiovascular disease. J Pathol 190:387–405
43. Uchiyama T, Kurabayashi M, Ohyama Y, Utsugi T, Akuzawa N, Sato M et al (2000) Hypoxia induces transcription of the plasminogen activator inhibitor-1 gene through genistein-sensitive tyrosine kinase pathways in vascular endothelial cells. Arterioscler Thromb Vasc Biol 20:1155–1161
44. Ulisse S, Baldini E, Sorrenti S, D'Armiento M (2009) The urokinase plasminogen activator system: a target for anticancer therapy. Curr Cancer Drug Targets 9:32–71
45. Bacharach E, Itin A, Keshet E (1998) Apposition-dependent induction of plasminogen activator inhibitor type 1 expression: a mechanism for balancing pericellular proteolysis during angiogenesis. Blood 92:939–945
46. Pepper MS (2001) Role of matrix metalloproteinases and plasminogen activator-plasmin system in angiogenesis. Arterioscler Thromb Vasc Biol 21:1104–1107
47. Simpson AJ, Booth NA, Moore NR, Bennett B (1991) Distribution of plasminogen activator inhibitor (PAI-1) in tissues. J Clin Pathol 44:139–143
48. Pepper MS, Sappino AP, Stocklin R, Montesano R, Orci L, Vassalli JD (1993) Upregulation of urokinase receptor expression on migrating endothelial cells. J Cell Biol 122:673–684
49. Bajou K, Noel A, Gerard RD, Masson V, Brunner N, Holst-Hansen C, Skobe M, Fusenig NE, Carmeliet P, Collen D, Foidart JM (1998) Absence of host plasminogen activator inhibitor 1 prevents cancer invasion and vascularization. Nat Med 4:923–928
50. Bajou K, Masson V, Gerard RD, Schmitt PM, Albert V, Praus M, Lund LR, Frandsen TL, Brunner N, Dano K, Fusenig NE, Weidle U, Carmeliet G, Loskutoff D, Collen D, Carmeliet P, Foidart JM, Noel A (2001) The plasminogen activator inhibitor PAI-1 controls in vivo tumor vascularization by interaction with proteases, not vitronectin. Implications for antiangiogenic strategies. J Cell Biol 152:777–784
51. Devy L, Blacher S, Grignet-Debrus C, Bajou K, Masson V, Gerard RD, Gils A, Carmeliet G, Carmeliet P, Declerck PJ, Noel A, Foidart JM (2002) The pro- or antiangiogenic effect of plasminogen activator inhibitor 1 is dose dependent. FASEB J 16:147–154
52. Masson V, Devy L, Grignet-Debrus C, Bernt S, Bajou K, Blacher S, Roland G, Chang Y, Fong T, Carmeliet P, Foidart JM, Noel A (2002) Mouse aortic ring assay: a new approach of the molecular genetics of angiogenesis. Biol Proced Online 4:24–31
53. Lambert V, Munaut C, Noel A, Frankenne F, Bajou K, Gerard R, Carmeliet P, Defresne MP, Foidart JM, Rakic JM (2001) Influence of plasminogen activator inhibitor type 1 on choroidal neovascularization. FASEB J 15:1021–1027
54. Deng G, Curriden SA, Wang S, Rosenberg S, Loskutoff DJ (1996) Is plasminogen activator inhibitor-1 the molecular switch that governs urokinase receptor-mediated cell adhesion and release? J Cell Biol 134:1563–1571
55. Andreasen PA, Kjoller L, Christensen L, Duffy MJ (1997) The urokinase-type plasminogen activator system in cancer metastasis: a review. Int J Cancer 72:1–22
56. Bajou K, Maillard C, Jost M, Lijnen HR, Gils A, Declerck P, Carmeliet P, Foidart JM, Noel A (2004) Host-derived plasminogen activator inhibitor-1 (PAI-1) concentration is critical for in vivo tumoral angiogenesis and growth. Oncogene 23:6986–6990
57. Gutierrez LS, Schulman A, Brito-Robinson T, Noria F, Ploplis VA, Castellino FJ (2000) Tumor development is retarded in mice lacking the gene for urokinase-type plasminogen activator or its inhibitor, plasmingen activator inhibitor-1. Cancer Res 60:5839–5847

58. Harbeck N, Alt U, Berger U, Kates R, Kruger A, Thomssen C, Janicke F, Graeff H, Schmitt M (2000) Long term follow-up confirms prognostic impact of PAI-1 and cathepsin D and L in primary breast cancer. Int J Biol Markers 15:79–83
59. Tsuchiya H, Sunayama C, Okada G, Matsuda E, Tomita K, Binder BR (1997) Plasminogen activator inhibitor-1 accelerates lung metastasis formation of human fibrosarcoma cells. Anticancer Res 17:313–316
60. Stefansson S, Petitclerc E, Wong MK, McMahon GA, Brooks PC, Lawrence DA (2001) Inhibition of angiogenesis in vivo by plasminogen activator inhibitor-1. J Biol Chem 276:8135–8141
61. Bacharach E, Itin A, Keshet E (1992) In vivo patterns of expression of urokinase and its inhibitor PAI-1 suggest a concerted role in regulating physiological angiogenesis. Proc Natl Acad Sci U S A 89:10686–10690
62. Bajou K, Peng H, Laug WE, Maillard C, Noel A, Foidart JM, Martial JA, DeClerck YA (2008) Plasminogen activator inhibitor-1 protects endothelial cells from FasL-mediated apoptosis. Cancer Cell 14:324–334
63. Stack MS, Gately S, Bafetti LM, Enghild J, Soff GA (1999) Angiostatin inhibitits endothelial and melanoma cellular invasion by blocking matrix-enhanced plasminogen activation. Biochem J 340:77–84
64. Gately S, Twardowski P, Stack MS, Cundiff DL, Grella D, Castellino FJ, Enghild J, Kwaan HC, Lee F, Kramer RA, Volpert O, Bouck N, Soff GA (1997) The mechanism of cancer mediated conversion of plasminogen to the angiogenesis inhibitor angiostatin. Proc Natl Acad Sci U S A 94:10868–10872
65. Ramchandran R, Dhanabal M, Volk R, Waterman MJ, Segal M, Lu H, Knelbelmann B, Sukhatme VP (1999) Antiangiogenic activity of restin, NC10 domain of human collagen XV: comparison to endostatin. Biochem Biophys Res Commun 255:735–739
66. Colorado PC, Torre A, Kamphaus G, Maeshima Y, Hopfer H, Takahashi K, Volk R, Zamborsky ED, Herman S, Sarkar PK, Ericksen MB, Dhanabal M, Simons M, Post M, Kufe DW, Weichselbaum RR, Sukhatme VP, Kalluri R (2000) Anti-angiogenic cues from vascular basement membrane collagen. Cancer Res 60:2520–2526
67. Kamphaus GD, Colorado PC, Panka DJ, Hopfer H, Ramchandran R, Torre A, Maeshima Y, Mier JW, Sukhatme VP, Kalluri R (2000) Canstatin, a novel matrix derived inhibitor of angiogenesis and tumor growth. J Biol Chem 275:1209–1215
68. Maeshima Y, Manfredi M, Reimer C, Holthaus KA, Hopfer H, Chandamuri BR, Kharbanda S, Kalluri R (2001) Identification of the antiangiogenic site within vascular basement membrane-derived tumstatin. J Biol Chem 276:15240–15248
69. O'Reilly MS, Pirie-Shepherd S, Lane WS, Folkman J (1999) Antiangiogenic activity of the cleaved conformation of the serpin antithrombin. Science 285:1926–1928
70. Pike SE, Yao L, Jones KD, Cherney B, Appella E, Sakaguchi K, Nakhasi H, Teruya-Feldstein J, Wirth P, Gupta G, Tosato G (1998) Vasostatin, a calreticulin fragment, inhibits angiogenesis and suppresses tumor growth. J Exp Med 188:2349–2356
71. Colman RW, Jameson BA, Lin Y, Johnson D, Mousa SA (2000) Domain 5 of high molecular weight kininogen (kininostatin) down-regulates endothelial cell proliferation and migration and inhibits angiogenesis. Blood 95:543–550
72. Degryse B, Sier CF, Resnati M, Conese M, Blasi F (2001) PAI-1 inhibits urokinase-induced chemotaxis by internalizing the urokinase receptor. FEBS Lett 505:249–254
73. McMahon GA, Petitclerc E, Stefansson S, Smith E, Wong MK, Westrick RJ, Ginsburg D, Brooks PC, Lawrence DA (2001) Plasminogen activator inhibitor-1 regulates tumor growth and angiogenesis. J Biol Chem 276:33964–33968
74. Loskutoff DJ, Curriden SA, Hu G, Deng G (1999) Regulation of cell adhesion by PAI-1. APMIS 107:54–61
75. Soff GA, Sanderowitz J, Gately S, Verrusio E, Weiss I, Brem S, Kwaan HC (1995) Expression of plasminogen activator inhibitor type 1 by human prostate carcinoma cells

inhibits primary tumor growth, tumor-associated angiogenesis, and metastasis to lung and liver in an athymic mouse model. J Clin Invest 96:2593–2600
76. Chao J, Tillman DM, Wang MY, Margolius HS, Chao L (1986) Identification of a new tissue-kallikrein binding protein. Biochem J 239:325–331
77. Chao J, Chai KX, Chen LM, Xiong W, Chao S, Woodley-Miller C, Chao L (1990) Tissue kallikrein-binding protein is a serpin, I: purification, characterization, and distribution in normotensive and spontaneously hypertensive rats. J Biol Chem 265:16394–16401
78. Chen LM, Ma JX, Liang YM, Chao L, Chao J (1996) Tissue kallikrein-binding protein reduces blood pressure in transgenic mice. J Biol Chem 271:27590–27594
79. Chen LM, Chao L, Chao J (1997) Beneficial effects of kallikrein-binding protein in transgenic mice during endotoxic shock. Life Sci 60:1431–1436
80. Chao J, Stallone JN, Liang YM, Chen LM, Chao L (1997) Kallistatin is a potent new vasodilator. J Clin Invest 100:11–17
81. Wolf WC, Harley RA, Sluce D, Chao L, Chao J (1999) Localization and expression of tissue kallikrein and kallistatin in human blood vessels. J Histochem Cytochem 47:221–228
82. Miao RQ, Agata J, Chao L, Chao J (2002) Kallistatin is new inhibitor of angiogenesis and tumor growth. Blood 100:3245–3252
83. Miao RQ, Chen V, Chao L, Chao J (2003) Structural elements of kallistatin required for inhibition of angiogenesis. Am J Physiol Cell Physiol 284:C1604–C1613
84. Suzuki K (2008) The multi-functional serpin, protein C inhibitor: beyond thrombosis and hemostasis. J Thromb Haemost 6:2017–2026
85. Asanuma K, Yoshikawa T, Hayashi T, Akita N, Nakagawa N, Hamada Y, Nishioka J, Kamada H, Gabazza EC, Ido M, Uchida A, Suzuki K (2007) Protein C inhibitor inhibits breast cancer cell growth, metastasis and angiogenesis independently of its protease inhibitory activity. Int J Cancer 121:955–965
86. Celerier J, Cruz A, Lamande N, Gasc JM, Corvol P (2002) Angiotensinogen and its cleaved derivatives inhibit angiogenesis. Hypertension 39:224–228
87. Sheng S (2006) A role of novel serpin maspin in tumor progression: the divergence revealed through efforts to converge. J Cell Physiol 209:631–635
88. Cher ML, Biliran HR Jr, Bhagat S, Meng Y, Che M, Lockett J, Abrams J, Fridman R, Zachareas M, Sheng S (2003) Maspin expression inhibits osteolysis, tumor growth, and angiogenesis in a model of prostate cancer bone metastasis. Proc Natl Acad Sci U S A 100:7847–7852
89. Biliran H Jr, Sheng S (2001) Pleiotrophic inhibition of pericellular urokinase-type plasminogen activator system by endogenous tumor suppressive maspin. Cancer Res 61:8676–8682
90. McGowen R, Biliran H Jr, Sager R, Sheng S (2000) The surface of prostate carcinoma DU145 cells mediates the inhibition of urokinase-type plasminogen activator by maspin. Cancer Res 60:4771–4778
91. Frey A, Soubani AO, Adam AK, Sheng S, Pass HI, Lonardo F (2009) Nuclear, compared with combined nuclear and cytoplasmic expression of maspin, is linked in lung adenocarcinoma to reduced VEGF-A levels and in Stage I, improved survival. Histopathology 54:590–597
92. Cella N, Contreras A, Latha K, Rosen JM, Zhang M (2006) Maspin is associated with beta1 integrin regulating cell adhesion in mammary epithelial cells. FASEB J 20:1510–1512
93. Yin S, Lockett J, Meng Y, Biliran H Jr, Blouse GE, Li X et al (2006) Maspin retards cell detachment *via* a novel interaction with the urokinase-type plasminogen activator/urokinase-type plasminogen activator receptor system. Cancer Res 66:4173–4181
94. Zhang M, Volpert O, Shi YH, Bouck N (2000) Maspin is an angiogenesis inhibitor. Nat Med 6:196–199
95. Azhar A, Singh P, Rashid Q, Naseem A, Khan MS, Jairajpuri MA (2013) Antiangiogenic function of antithrombin is dependent on its conformational variation: implication for other serpins. Protein Pept Lett 20(4):403–411
96. O'Reilly MS (2007) Antiangiogenic antithrombin. Semin Thromb Hemost 33:660–666

97. Eaton DL, Baker JB (1993) Evidence that a variety of cultured cells secrete protease nexin and produce a distinct cytoplasmic serine protease binding factor. J Cell Physiol 117:175–182
98. Baker JB, Low DA, Simmer RL, Cunningham DD (1980) Protease-nexin: a cellular component that links thrombin and plasminogen activator and mediates their binding to cells. Cell 21:37–45
99. Richard B et al (2004) Protease nexin-1: a cellular serpin down-regulated by thrombin in rat aortic smooth muscle cells. J Cell Physiol 201:138–145
100. Bouton MC et al (2007) Protease nexin-1 interacts with thrombomodulin and modulates its anticoagulant effect. Circ Res 100:1174–1181
101. Boulaftali Y et al (2010) Anticoagulant and antithrombotic properties of platelet protease nexin-1. Blood 115:97–106
102. Selbonne S, Azibani F, Iatmanen S, Boulaftali Y, Richard B, Jandrot-Perrus M, Bouton MC, Arocas V (2012) In vitro and in vivo antiangiogenic properties of the serpin protease nexin-1. Mol Cell 32:496–505
103. Phung M, Dass CR (2006) In-vitro and in-vivo assays for angiogenesis-modulating drug discovery and development. J Pharm Pharmacol 58:153–160
104. Kawaguchi T, Yamagishi SI, Sata M (2010) Structure-function relationships of PEDF. Curr Mol Med 10:302–311
105. Bouck N (2002) PEDF: anti-angiogenic guardian of ocular function. Trends Mol Med 8:330–334
106. Elayappan B, Ravinarayannan H, Pasha SP, Lee KJ, Gurunathan S (2009) PEDF inhibits VEGF- and EPO-induced angiogenesis in retinal endothelial cells through interruption of P13K/Akt phosphorylation. Angiogenesis 12:313–324
107. Ho TC, Chen SL, Yang YC, Liao CL, Cheng HC, Tsao YP (2007) PEDF induces p53-mediated apoptosis through PPAR gamma signaling in human umbilican vein endothelial cells. Cardiovasc Res 76:213–223
108. Yang H, Cheng R, Liu G et al (2009) PEDF inhibits growth of retinoblastoma by anti-angiogenic activity. Cancer Sci 100:2419–2425
109. Cai J, Jiang WG, Grant MB, Boulton M (2006) Pigment epithelium-derived factor inhibits angiogenesis via regulated intracellular proteolysis of vascular endothelial growth factor receptor-1. J Biol Chem 281:3604–3613
110. Ek ET, Dass C, Contreras KG, Choong PF (2007) Inhibition of orthotopic osteosarcoma growth and metastasis by multitargeted antitumor activities of pigment epithelium-derived factor. Clin Exp Metastasis 24:93–110
111. Manalo KB, Choong PFM, Dass CR (2011) Pigment epithelium-derived factor as an impending therapeutic agent against vascular epithelial growth factor-driven tumor-angiogenesis. Mol Carcinog 50:67–72
112. North S, Moenner M, Bikfalvi A (2005) Recent developments in the regulation of the angiogenic switch by cellular stress factors in tumors. Cancer Lett 218:1–14

Part II

Molecular Mechanisms

Endothelial Transcriptional Networks in the Control of Angiogenesis: the ETS Factor

6

Anna M. Randi

Contents

6.1 Introduction 122
6.2 The ETS Transcription Factor Family 122
6.2.1 ETS Factors Functional Domains 123
6.2.2 Control of ETS Protein Function by Alternative Splicing 125
6.2.3 ETS Factors as Targets of Signal Transduction Pathways 125
6.2.4 Transcriptional Networks Involving ETS Factors 126
6.3 The Role of ETS Transcription Factors in Endothelial Cells 127
6.3.1 ETS Target Genes in the Endothelium 127
6.3.2 Vascular Development 129
6.3.3 Angiogenesis 131
6.4 Two ETS Factors Regulating Angiogenesis: Ets-1 and Erg 131
6.4.1 Ets-1 131
6.4.2 Erg 133
References 135

Abstract

Endothelial gene expression is controlled by a few families of transcription factors (TF) which form regulatory networks. Amongst the most important are the ETS factors, a large family of which at least 19 members are expressed in human endothelial cells at some point throughout development. Through interaction with other TF families and co-regulators, ETS factors control most aspects of endothelial biology, from early differentiation in the embryo to homeostasis, angiogenesis and inflammation in the adult. Several ETS family members have been shown to be essential for vascular development and angiogenesis and to regulate processes from cell migration to survival and growth. Amongst the list of ETS targets genes involved in angiogenic pathways are

A.M. Randi (✉)
Vascular Sciences Unit, National Heart and Lung Institute (NHLI), Imperial College London, Hammersmith Hospital, Du Cane Road, London W12 0NN, UK
e-mail: a.randi@imperial.ac.uk

J. Dulak et al. (eds.), *Angiogenesis and Vascularisation*,
DOI 10.1007/978-3-7091-1428-5_6,

growth factor receptors for vascular endothelial growth factor and angiopoietins. ETS factors also control cell adhesion, matrix remodelling and cytoskeletal dynamics, processes essential for tissue homeostasis, which therefore influence the viability and stability of new vascular networks. Most ETS proteins have been shown to activate transcription; however, a few can also act as transcriptional repressors. Several ETS factors can control the same target genes, raising the question of possible redundancy in the system. Data from a variety of *in vivo* models are shedding new light on the different roles of ETS factors in vascular development, adult angiogenesis and endothelial homeostasis. In this review some general properties of ETS factors in the endothelium are discussed and then two of the best characterised ETS proteins in endothelial cells, Ets-1 and Erg, which exemplify the complex picture of distinct and overlapping targets and functions, are presented.

Keywords
ETS transcription factors • Angiogenesis • Endothelial homeostasis • Ets-1 • Erg

6.1 Introduction

By controlling gene expression, transcription factors (TF) are crucial in determining lineage identity and cellular phenotypes; therefore these proteins play essential roles in normal development and disease. TF act in complex, dynamic networks with other TF and transcriptional cofactors, integrating multiple signals. Based on motif analysis of the regulatory regions of endothelial-specific genes and on functional studies, several major families of TF have been identified as regulators of endothelial gene expression: the ETS, Sox, Forkhead, GATA and Kruppel-like families. This review will focus on the endothelial members of the ETS family and their role in angiogenesis.

6.2 The ETS Transcription Factor Family

ETS transcription factors are essential for basic cellular functions such as proliferation and survival [1]. Some members of the ETS family are ubiquitously expressed and are part of the basic cellular machinery of response to extracellular stimuli, such as the MAPK signalling pathway. Their dysfunction or dysregulated expression is associated with several human diseases, predominantly cancers [2, 3], but also chronic inflammatory diseases such as rheumatoid arthritis and systemic lupus erythematosus [4, 5]. Over the last 15 years, reports by many groups on the phenotypes of ETS-deficient mice have revealed that ETS factors are particularly important for lineage differentiation of the haematopoietic and vascular lineage (reviewed in [6]) and for neuronal development [7].

The founding member of this family, v-ets, was originally identified as a viral oncogene, part of a fusion protein expressed by a transforming retrovirus in the E26 avian leukaemia (hence ETS, for E-26 transformation specific [8]). Since then, 28 ETS genes have been identified in the human genome [9] and 26 in the mouse [10]; nearly two-thirds are ubiquitously expressed in adult tissue [11]. Whilst some appear to have fairly restricted expression profiles, no ETS factor is uniquely expressed in the endothelium.

6.2.1 ETS Factors Functional Domains

ETS factors are characterised by the presence of the ETS domain, a highly conserved 85 amino acids winged helix–turn–helix DNA-binding domain, consisting of three α-helices and four β-strands [12, 13]. On the basis of phylogenetic analysis of the highly conserved DNA-binding domains, the ETS family has been subdivided into subfamilies (reviewed in [14]) (Fig. 6.1).

The ETS domain mediates binding to a core DNA sequence 5′-GGA(A/T)-3′. NMR and X-ray crystallography studies of ETS domains have shown that helix-3 binds in the major groove of the basic DNA sequence [18, 19]. ETS binding sites have been identified in the promoter/regulatory regions of over 200 genes and ETS proteins have been estimated to bind between 5 and 15 % of gene promoters, including many housekeeping genes [9]. ETS proteins control expression of key regulators of cell proliferation, differentiation and survival (reviewed in [3, 6, 20]). The conservation of DNA-binding properties among ETS proteins raises the question of how individual family members' binding to the same sequence can elicit specific gene expression programmes. Flanking sequences to the core motif have been shown to influence binding affinities [14]. However, a genome-wide analysis of DNA-binding profiles for all human and mouse ETS factors has found that the DNA-binding specificities of ETS proteins fall into four distinct classes only [21], suggesting that DNA-binding specificity differences alone cannot explain the diversity of the ETS family.

The ETS domain also mediates protein binding: various ETS factors including Ets-1, Ets-2, Fli-1 and Erg (ETS-related gene) have been shown to interact directly with the Jun/Fos heterodimer via the ETS domain [17, 22] and evidence of transcriptional cooperation or repression between these two transcription families is available (see below). Partnerships with non-ETS proteins may also stabilise ternary complexes and help provide additional DNA-binding specificity and stability.

Almost one-third of ETS proteins contain a region of 65–100 residues named "Pointed" or PNT domain (Fig. 6.1), which has been implicated in signal transduction and protein–protein interactions (reviewed in [16]). The PNT Domain may have different functions in different ETS factors [23, 24]. For example, in Ets-1 and Ets-2, the monomeric PNT domain supports docking of the ERK2 kinase [25], which phosphorylates Ets-2, thus increasing its binding to the co-activator CBP [26]. In other ETS proteins, polymerisation through the PNT domain may favour

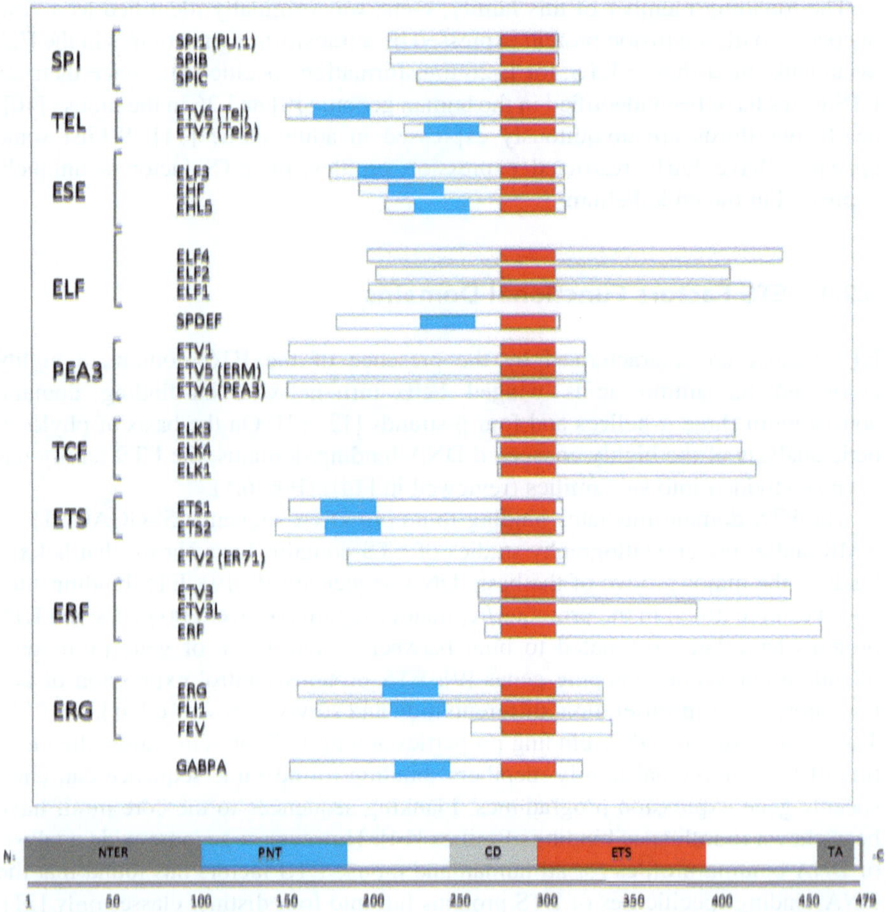

Fig. 6.1 Structural and functional domains of the ETS family of transcription factors and of the ETS factor Erg p55. (*Top*) domain organisation of the human ETS proteins, grouped according to Hollenhorst et al. [14]. (*Bottom*) domain structure of the Erg factor (isoform p55) and attributed functions. From the N-Terminus, the following domains have been described and functionally characterised: N-Terminal (NTER) (amino acids [aa] 1–114), interaction with the methyltrasferase ESET [15]; Pointed (PNT) domain (aa 108–201), protein interactions [16]; Central domain (CD) domain (aa 249–308), inhibition of cooperative Erg/AP-1 transactivation and protein–protein interaction [17]; ETS domain (aa 307–392): DNA binding, Erg interaction with AP-1 [17]; transactivation domain (TA) (aa 433–479), involved in cooperative Erg/AP-1 transactivation

association between ETS factors, as in the case of Etv-6/Tel [27]. Thus, through its diverse functions, the PNT domain may contribute to regulating specificity and activity of the ETS transcription factors.

The vast majority of ETS factors act as transactivators of gene transcription. Some ETS factors can also act as transcriptional repressors, depending on the target, on post-translational modifications or specific structural properties and possibly on the recruitment of different co-factors (reviewed in [28]). Only one

ETS factor, Etv-6/Tel, has been found to act exclusively as transcriptional repressor; Etv-6/Tel contains both an ETS and a PNT domain, and its ability to repress gene expression depends on its ability to self-associate [27]. However, several other ETS proteins which can act as transactivators also contain both domains [6], suggesting a further level of complexity in the regulation of ETS factors' function.

6.2.2 Control of ETS Protein Function by Alternative Splicing

Many ETS transcription factors have been shown to be present in multiple isoforms, mostly arising from alternative splicing, which is emerging as an important mechanism through which the function of these proteins is regulated. Alternatively spliced members of the ETS family include Ets-1 [29], Fli-1 [30], Etv-1/ER81 [31], Erg [32, 33], Etv-2/Tel [34] and Etv-2/ER71 [35]. In Ets-1, two inhibitory domains surrounding and interacting with the ETS domain cause autoinhibition of DNA binding [36]. Interestingly, this autoinhibition can be partly overcome by Ets-1 interaction with partners, thus enabling their cooperative binding to adjacent DNA elements [37]. An isoform of Ets-1, encoded by an alternatively spliced mRNA, lacks the intra-molecular inhibition of DNA binding [38].

Multiple isoforms have been described for the transcription factor Erg, produced through differential mRNA splicing and alternative use of translational start codons [32]. Although very limited data have been reported so far showing differences in the transcriptional activity of Erg isoforms [39], a tissue-specific role for distinct Erg splice variants was recently demonstrated by Vijayaraj et al. [40], suggesting that this may be an important mechanism for the regulation of Erg's function. This is supported by evidence from prostate cancer tissues, where Erg is frequently overexpressed due to its fusion to the androgen-responsive TMPRSS2 gene, and a multitude of splice variants have been described [39, 41–44]. The identification of specific *Erg* isoforms is currently being investigated as promising novel biomarker for prostate cancer [45].

6.2.3 ETS Factors as Targets of Signal Transduction Pathways

Proteins of the ETS family have been identified as key nuclear mediators of MAPK pathway activation [46]. For example, ERK-1/2 has been shown to phosphorylate Ets-2 in a number of cells lines, including NIH3T3 and RAW264 [47] and macrophages stimulated with colony-stimulating factor-1 [48]. Other enzymes such as Akt and JNK can also activate Ets-2 through phosphorylation [49].

Several kinases have been shown to phosphorylate Ets-1 and modulate its activity. Activation of the GTPase Ras leads to Ets-1 phosphorylation on Threonine 38 through the MAPK (ERK 1/2) pathway, resulting in increased Ets-1 binding to DNA [47].

Phosphorylation of ETS factors can influence their transactivation activity as well as their interaction with other transcription factors (reviewed in [50]).

For example, the EGF receptor can stimulate the transcriptional activity of Ets-1 and Ets-2, through phosphorylation of specific threonine residues within the PNT domains [47, 51].

6.2.4 Transcriptional Networks Involving ETS Factors

Cooperation between TF from different families is a well-established mechanism which has been described for several ETS proteins (reviewed in [24]). Ets-1 and HIF-2α, a crucial mediator of hypoxic signalling in the endothelium, have been shown to cooperate in the regulation of genes as diverse as vascular endothelial growth factor (VEGF)-receptor (VEGFR)-2 [52] and VE-cadherin [132]. A complex signalling network triggered by growth factors and involving these two transcription factors has been described in the regulatory axis EGF/basic FGF > ETS-1/HIF-2α > VEGF-R1 transcription [133].

Several studies dating back to the mid-1990s identified a synergy between GATA and ETS factors, analysing motifs in the regulatory regions of many endothelial genes including PECAM and VWF. Recently, chromatin immunoprecipitation (ChIP) sequencing and expression profiling in HUVEC revealed that 35 % of the GATA-2-occupied sites contain canonical motifs for ETS factors [134]. Transcription factors interacting with conserved ETS and composite GATA/Tal1 motifs were shown to confer enhancer activity to the human endothelial Protein C receptor gene [135]. Gata-2 and Ets-1 have also been shown to synergise in the activation of Angiopoietin-2 [53].

The AP-1 transcription complex consists of c-Jun/Fos family proteins and plays an important role in the signalling response to a vast array of stimuli [136]. Adjacent ETS and AP-1 binding sites occur in a large number of promoter/enhancer elements [54, 137]. Ets-1 binds to AP-1 proteins via the DNA-binding domain ([138], and see above); this results in transcriptional cooperation. Functional cooperation between Ets-1 and AP-1 is critical for the expression of many genes, including cytokines [139] and matrix metalloproteinases (MMP) [140]. Interestingly, the interaction between Ets-1 and AP-1 can also enhance transcription of tissue inhibitors of MMP (TIMP)-1 [55], suggesting that this transcription partnership controls the balance of tissue degradation. Other ETS factors that have been shown to functionally interact with AP-1 are Ets-2 and Erg [17, 22].

Finally, several ETS factors have been shown to interact with each other. A detailed study on Erg using a pull-down approach with GST-fusion recombinant proteins showed that the main endothelial isoform, Erg p55 (also called Erg-3), could form homodimers, as well as dimers, with other Erg isoforms and with other members of the ETS family [17]. There is also evidence of functional interaction between ETS factors: for example, Ets-2 function is influenced by Erg: in the case of the stromelysin promoter, Ets-2 was shown to activate the stromelysin-1 gene in a cell-based assay; however, Erg co-expression inhibited stromelysin-1 activation by Ets-2 [56]. These studies exemplify the complex relationship between ETS factors and the multiplicity of options they exploit in the control gene expression.

Defining the interaction between ETS factors may also provide some insight into the apparent redundancy of many ETS target genes (see Table 6.1).

6.3 The Role of ETS Transcription Factors in Endothelial Cells

6.3.1 ETS Target Genes in the Endothelium

ETS factors are crucial regulators of endothelial gene expression. Nearly all endothelial enhancers and promoters characterised so far contain multiple essential ETS binding sites. ETS target genes include several growth factor regulators such as VEGF and angiopoietin receptors, cell adhesion molecules such as intercellular adhesion molecule (ICAM)-2 and VE-cadherin, extracellular molecules such as von Willebrand factor (VWF), matrix remodelling enzymes such as several MMP and TIMPs and small GTPase and cytoskeletal regulators such as histone deacetylase (HDAC)-6 and Rho J [57, 58]. Table 6.1 reports a list of some of the genes regulated by ETS factors in the endothelium. From this partial summary, it is easy to see how ETS transcription factors control the expression of genes that define the endothelial lineage. Indeed, several ETS factors are involved in endothelial specification; a recent report by Rafii's group showed that the combination and co-ordinated expression of three ETS factors, namely Etv-2, Erg-1 and Fli1, in a complex cell culture environment, was able to reprogram amniotic cells into vascular endothelial cells ([94] and discussed below).

Despite the number of reports on specific ETS factors regulating individual genes, a GEO dataset search for genome-wide profiling of ETS targets in endothelial cells reveals a surprisingly small number of studies. Two transcriptome profiling studies carried out in human umbilical vein endothelial cells (HUVEC) to identify targets of Erg have revealed that a large number of pathways are under the control of this protein [57, 82]. Hopefully more of these experiments will be carried out on other ETS factors. These, in combination with genomic mapping of ETS factors' DNA-binding sites in endothelial cells by ChIP sequencing, will ultimately provide a comprehensive map of the gene targets specific for the different family member.

Table 6.1 also highlights the apparent redundancy in this transcription family: multiple ETS factors have been shown to transactivate, at least *in vitro*, the same target genes. This is the case for VE-cadherin, which has been described as a target for both Erg and Ets-1 [69, 70]. Interestingly, VE-cadherin expression also seems to be controlled by an ETS factor which functions as repressor, Etv-6/Tel: loss of Etv-6/Tel led to a significant increase in the expression of VE-cadherin *in vitro* [68]. Because ETS factors can form heterodimers with other ETS family members, it is possible to speculate that the regulation of some shared target genes may be under the control of ETS networks, whose composition may change depending on the differentiation and/or activation state of the cells. Clearly such models will have to be tested experimentally.

Table 6.1 ETS target genes

Target gene and category	Ets factor	References
Growth control		
Tie-1	Ets-1, Ets-2, NERF-2	[59]
Tie-2	Elf-1, NERF-2	[60, 61]
VEGF-receptor 1 (flt-1)	Elf-1 Ets-1, Ets-2, Erg	[62]
VEGF-receptor 2 (flk-1)	Erg Ets-1 Ets-1	[63] [64] [52]
Neuropilin-1	Ets-1	[65]
VEGF	Net	[66]
Angiopoietin-2	Ets-1	[53, 67]
Delta-like 4	Tel	[68]
Cell adhesion molecules and cytoskeleton		
VE-cadherin	Erg Ets-1	[69, 70]
ICAM-2	Erg	[71]
Robo-4	GABP	[72]
HDAC-6	Erg	[57]
Rho J	Erg	[58]
β3 integrin	Ets-1	[73]
Extracellular matrix and related proteases		
SPARC	Erg	[74]
Thrombospondin	Erg	[74]
MMP-1	Erg	[56]
MMP-2	Ets-1	[54]
MMP-9	Ets-1	[73]
MMP-10	Ets-1	[75]
TIMP	Ets-1	[55]
CD13/aminopeptidase N	Ets-2	[76]
Endothelial homeostasis and apoptosis		
Endoglin	Fli-1, Erg, Elf-1	[77]
Haem oxygenase-1 (HO-1)	Ets-1, Fli-1, Erg	[78]
Bcl2, Bcl-X_L, cIAP2	Ets-1 + Ets-2	[79]
Inflammation		
eNOS	Ets-1, Erg Elf (R)	[80]
MCP-1	Ets-1	[81]
IL-8	Erg (R)	[82, 83]
ICAM-1	Ets-1 Ets-2, ERM Erg (R)	[84] [85] [83, 86]
Transcription		
Egr-1	Ets-1, Net (R)	[82, 87]
Fli-1	Ets-1	[88]

(continued)

Table 6.1 (continued)

Target gene and category	Ets factor	References
Ets-1	Ets-1	[89]
JunB	Ets-1, Ets-2	[90]
LMO2	Fli1, Elf1, Ets1	[91]
Haemostasis		
Von Willebrand factor	Erg	[92]
Thrombomodulin	Ets-1	[93]

The table reports selected ETS target genes in endothelial cells, grouped in functional categories. (R) indicates repression of expression

Nineteen ETS factors have been found expressed at some stage during endothelial differentiation (reviewed in [6, 9]). Some ETS factors, such as Etv-2, are transiently expressed in the developing vasculature in the embryo but downregulated in the adult [95]. Other ETS factors, such as Erg and Fli-1, are expressed during embryonic development and in quiescent adult ECs [9, 96]. In adult endothelium, a small number of ETS factors are constitutively expressed, such as Erg, whilst others such as Ets-1 are expressed at very low levels in quiescent, resting cells and their levels can be upregulated by extracellular stimuli (reviewed in [6]). These patterns of expression reflect the different and multiple roles of ETS factors in endothelial cells, from driving differentiation and angiogenesis to controlling the response to inflammatory agents. It is also becoming evident that some ETS factors which are expressed throughout the life of the endothelium are essential for homeostasis and survival, as in the case of Erg (discussed below).

6.3.2 Vascular Development

In contrast with other tissues, no single transcription factor has been identified which dictates endothelial cell lineage. Instead, a number of TF that are not cell-type specific have been shown to act in a coordinate network to determine the endothelial lineage. The picture emerging from *in vivo* studies on ETS-deficient or mutant animals suggests that not all ETS factors are equally important for vascular development, adult angiogenesis and endothelial homeostasis. Gene knockdown experiments in *Xenopus* and Zebrafish have been carried out on most endothelial ETS factors, with variable consequences on angiogenesis (reviewed in [96–98]). In some cases, however, the phenotype of zebrafish knockdown was not as dramatic as that subsequently found in knockout (KO) mice. For example, morpholino inhibition of *Erg* resulted in a mild defect in intersomitic vessel (ISV) patterning, and a double knockdown with either *Fli-1* or *etsrp* (the fish homologue of *Etv-2*) was required to significantly disrupt angiogenesis [99, 100]. However, in the mouse, Erg global deficiency results in embryonic lethality with multiple defects, including vascular abnormalities [40]. This is confirmed by endothelial-specific deletion of

Erg (Randi et al., manuscript in preparation). These discrepancies indicate a level of redundancy in the Zebrafish model which is not present in mouse.

A recent review summarises the published phenotypes of ETS KO and transgenic mice, as well as Zebrafish and *Xenopus* models [101]. To note, none of the mouse KO reported in this list is endothelial specific. Because of the role of ETS factors in many other processes, particularly haematopoiesis, which shares early progenitors with endothelial differentiation, the phenotypes of global KO mice cannot be accurately attributed to vascular defects. Despite these limitations, some of the phenotypes seen in the KO mice clearly reflect defective vascular development and angiogenesis. Three of the most significant are discussed below.

Etv-2 (Etrsp/Er71): The most dramatic phenotype in vascular development caused by deletion or inactivation of an ETS protein is that of the Etv-2 KO. In mouse embryo, Etv-2 expression is detected from embryonic postcoital day 7.5 (E7.5) in the yolk sac blood islands, which contain both blood and endothelial precursors, and in the major blood vessels from E8.25 to E9.5 [102]. Etv-2 expression then decreases to disappear in adult endothelial tissue [95]. Homozygote deletion of Etv-2 in mice results in embryonic lethality at E9–E9.5, with yolk sac abnormality and dramatic loss of vascular structures [103]; both vascular and haematopoietic lineages are absent in these mutants. In line with these studies, morpholino-mediated knockdown of Etv-2 function in *Xenopus* and zebrafish embryos results in similar phenotypes, with near complete absence of endothelial markers [99]. Interestingly, Etv-2 was shown to cooperate with FoxO transcription factors in determining endothelial differentiation: co-expression of FoxC2 and Etv-2 resulted in ectopic vessel formation and strong induction of endothelial genes in *Xenopus* embryos [104]. FOX:ETS motifs were identified within many endothelial enhancers and promoters, including Tie-2, VEGF-R2 and VE-cadherin [104]. A recent study has provided more evidence for the crucial role of Etv-2 and shed new light on the relationship between different ETS factors in determining endothelial identity. Ginsberg et al. showed that lineage-committed amniotic cells can be reprogrammed into vascular endothelial cells by over-expression of three ETS factors, namely Etv-2, Erg-1 and Fli-1 [94]. Expression of these ETS factors was sufficient to switch on a panel of endothelial genes and, importantly, suppress expression of nonvascular genes. The study indicated that Etv-2 is transiently required to unlock the endothelial identity, whilst Erg and Fli-1 contribute to the maintenance of the mature EC phenotype. This is in line with the role of Erg in maintaining vascular stability in adult endothelium ([69] and see below). Together, these factors can trigger reprogramming towards the endothelial lineage. Interestingly, this phenotype required concomitant TGFβ inhibition [94].

Ets-1: At the other end of the spectrum, global deletion of Ets-1 has no effect on vascular development. Despite its expression in the vascular endothelium of the developing embryo, targeted deletion of the Ets-1 gene did not cause vascular defects or embryonic lethality [70, 105, 106]; however, perinatal mortality was about 50 %, depending on the genetic background. Crossing Ets-1 KO mice with mice either deficient or defective in Ets-2 resulted in embryonic lethality at day E14.5, with large, dilated blood vessels, oedema and haemorrhage suggesting compensatory mechanisms between Ets-1 and Ets-2 [79].

Erg: Between these two extremes are the phenotypes reported for several other ETS factors which have been implicated in vascular development. *Erg* expression is strongly linked with the development of the vascular tree. In the mouse embryo, Erg expression appears as early as E8.0, in both intra and extra-embryonic endothelial tissue such as the endothelium of dorsal aorta, the placental sinusoids and the yolk sack blood islands. By day E10.5, endothelial expression is found in all vascular structures [107, 108]. Global deletion of Erg results in embryonic lethality due to multiple causes, with some defects in vascular remodelling [40]. Unpublished data from our laboratory support these findings and shows that endothelial-specific deletion of Erg causes embryonic lethality with severe vascular abnormalities (Randi et al., manuscript in preparation).

6.3.3 Angiogenesis

Several studies have unequivocally demonstrated the importance of both constitutive and inducible ETS factors in angiogenesis. *In vitro* models of tube formation, cell migration and sprouting in co-culture have shown that most ETS factors play a role in these processes (reviewed in [6]). ETS factors have been associated with pathological angiogenesis, as in the case of tumour vascularisation [3] or in response to ischaemia [109]. Table 6.2 shows a partial list of ETS factor and their role in *in vivo* models of angiogenesis. However, in most of these models, ETS factors are likely to control angiogenesis through multiple mechanisms and cell types, not uniquely endothelial. As well as the lack of endothelial-specific models, *in vivo* angiogenesis studies for ETS transcription factors are complicated by embryonic lethality of several global ETS KO mice. Several approaches have been taken to circumvent this issue, from *in vivo* downregulation of the ETS protein through siRNA [69] to generation of transgenic animals with non-lethal mutations affecting the function of the selected ETS factor [117].

6.4 Two ETS Factors Regulating Angiogenesis: Ets-1 and Erg

6.4.1 Ets-1

The role of Ets-1, the founding member of the family, in endothelial cells has been extensively investigated. As mentioned above, Ets-1 is important for adult angiogenesis but is not essential for vascular development. Several Ets-1 target genes play a role in angiogenesis, and many are summarised in Table 6.1. In adult endothelial cells, Ets-1 expression is low and can be upregulated by pro-angiogenic stimuli, from growth factors such as basic fibroblast growth factor (FGF) and VEGF [118] to hypoxia [109] and pro-inflammatory cytokines [71]. Whilst for many ETS factors the data are mostly limited to a panel of target genes, in the case of Ets-1 complex pathways have been dissected, placing Ets-1 as the effector which modifies the expression of specific targets following extracellular signals. Ets-1

Table 6.2 ETS factors in *in vivo* models of angiogenesis

Ets factor	Ets factor manipulation	Tools	Angiogenesis model	Effect on angiogenesis	Reference
Ets-1	Over-expression	Liposome transfection of Ets-1 gene	Rat hindlimb ischaemia	Increased	[110]
Ets-1	Inhibition	Adenovirus Ets-1 DNA BD mutant	bFGF sponge (ear)	Decreased	[111]
Ets-1	Inhibition	Adenovirus Ets-1 TA mutant	Matrigel plug	Decreased	[112]
Ets-1	Inhibition	Adenovirus Ets-1 TA mutant; Intra-ocular injection	Proliferative retinopathy	Decreased	[113]
Ets-1	Inhibition	Adenovirus Ets-1 DNA BD	Pancreatic adenocarcinomas xenographs	Decreased	[114]
Ets-1	Inhibition	Antisense	Chick chorioallantois membrane	Decreased	[115]
GABPα/β	Over-expression	Liposome transfection of GAPB gene; Intra-ocular injection	Corneal neovascularisation	Decreased	[116]
Elf-1	Inhibition	Peptide	B16 melanoma tumour	Increased	[61]
Erg	Inhibition	siRNA	Matrigel plug	Decreased	[69]
Net (Elk-3/Sap-2/ERP)	Inhibition	Antisense	KSHV-GPCR tumours	Decreased	[66]

Summary of *in vivo* experiments on the role of ETS transcription factors in angiogenesis. All models have been carried out in adult mice, unless otherwise specified. For the phenotype of ETS-deficient mice which results in embryonic lethality, these have been recently reviewed [101]. *DNA BD* DNA binding domain, *TA* transactivation domain

appears to mediate pro-angiogenic signals stimulated by growth factors such as VEGF and acidic FGF [119], but also from less canonical pathways. For example, the coagulation protein tissue factor has been shown to promote angiogenesis by increasing Ets-1 DNA binding to its targets through the ERK1/2 pathway, which is known to phosphorylate Ets-1 and modify its transcriptional activity [120].

The role of Ets-1 in angiogenesis *in vitro* has been demonstrated using a variety of approaches, from over-expression via viral vectors, to ablation using antisense oligonucleotides or dominant negative mutants ([121] and reviewed in [122]). *In vivo*, Ets-1 has been shown to be a potent stimulator of angiogenesis: over-expression of Ets-1 in a rat hindlimb ischaemia model induced neovascularisation through the induction of angiogenic growth factors hepatocyte growth factor (HGF) and VEGF [110]. Ets-1 expression has been associated with angiogenesis in chronic inflammatory disease and tumour angiogenesis [123, 124]. Adenoviral-mediated expression of Ets-1 dominant negative significantly reduced tumour size and microvessel density in a xenograft model of human pancreatic adenocarcinomas [114] and suppressed retinal angiogenesis in mice [112, 113]. However, because of the extent of conservation of the DNA-binding domain in the ETS family, often used as dominant negative, concerns about the specificity of this approach ought to be considered. Some of the *in vivo* studies on Ets-1 and other ETS factors are summarised in Table 6.2.

6.4.2 Erg

Erg has become the focus of much attention in recent years because of its role in a number of human tumours, from prostate cancer to haematological malignancies. Erg over-expression in tumours has been shown to correlate with poor prognosis, and as a consequence strategies aimed at targeting Erg in these patients are being developed [125]. These approaches however often overlook the essential role of Erg in the endothelium, both during development and in adult life. Erg is involved in endothelial differentiation and lineage specification [74, 94, 126]. Over-expression of Erg in zebrafish promotes the proliferation of mesodermal angioblasts [100]. As discussed above, global as well as endothelial-specific deletion of Erg in the mouse result in vascular defects during development in the mouse and embryonic lethality. Vascular defects were also visible in Erg transgenic mice containing a single point mutation (Mld2) in the DNA-binding domain of *Erg* that significantly inhibits Erg transactivation [127]. This mutation is associated with multiple defects in definitive haematopoiesis and a failure to sustain self-renewal of haematopoietic stem cells (HSCs), ultimately leading to exhaustion of HSCs and death at E11.5.

At variance with other transcription factors required for vascular development, Erg expression is maintained throughout adulthood. Erg is the most abundantly expressed ETS factor in resting adult EC [82, 107, 128, 129] and its expression, although not uniquely endothelial, appears to be restricted to the haematopoietic/endothelial lineage and to chondrocytes [107]. In endothelial cells, Erg drives the expression of genes that define the endothelial lineage, such as endoglin, von Willebrand factor, ICAM-2 and VE-cadherin (reviewed in [6]). As well as being crucial during development, Erg plays a central role in adult endothelium, both in the maintenance of homeostasis and during angiogenesis. Figure 6.2 summarises the known functions of Erg in the endothelium. Several *in vitro* studies have shown that Erg is essential for endothelial homeostasis and angiogenesis [57, 69, 126], acting on a diverse array of endothelial processes and target genes. Many Erg target genes have been implicated in angiogenesis (reviewed in [122]). Erg is essential for endothelial junctional stability and endothelial survival [69], in established as well as new vessels. Erg controls cell–cell junction assembly, stability and function through expression of junctional adhesion molecules VE-cadherin, claudin-5 and ICAM-2 [69, 71, 130]. Another critical function of Erg in the endothelium is to control cell motility, migration and cytoskeletal dynamics. Microarray and network analysis identified multiple pathways as potential mediators of these effects, two of

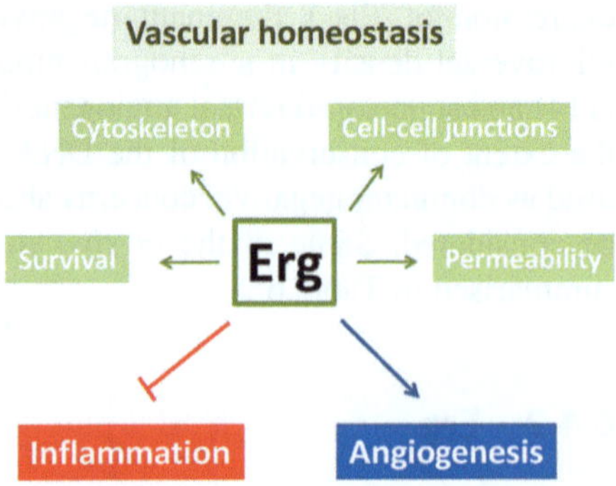

Fig. 6.2 The ETS-related gene Erg controls vascular homeostasis. In endothelial cells, Erg controls permeability and cell–cell junctions, the actin and tubulin cytoskeleton and survival, is required for angiogenesis and represses inflammation. See text for relevant references

which involve the histone deacetylase HDAC6 and the small GTPAse RhoJ [57, 58]. These targets and functions point to the essential role for this ETS factor in the maintenance of endothelial homeostasis. However, Erg has also been implicated in the regulation of growth factor receptors including VEGF-R1 [63] and VEGF-R2 [62], suggesting that this ETS factor may also play an important role in directly regulating the response to angiogenic stimuli.

In endothelial cells, Erg acts both as a transactivator and a repressor of gene expression. Erg binds to the promoters of pro-inflammatory molecules ICAM-1 and IL-8 and represses their basal expression in resting EC, thus preventing leukocyte adhesion to non-activated endothelium, an essential homeostatic property of healthy blood vessels [82, 83, 86]. The molecular pathways through which Erg mediates this repression involve inhibition of NF-κB p65 binding to the promoters of ICAM-1 and IL-8 and other pro-inflammatory genes [83]. These data suggest that Erg acts as a gatekeeper to inhibit the activity of low, constitutive levels of nuclear NF-κB p65, to protect the endothelium from the activation of NF-κB target genes and maintain homeostasis. The anti-inflammatory effects of Erg in endothelial cells appear to be exerted through multiple mechanisms. Over-expression of Erg in HUVEC via adenovirus resulted in inhibition of p65 phosphorylation by tumour necrosis factor (TNF), and blockade of TNF-induced leukocyte adhesion *in vitro* and acute inflammation *in vivo* [86]. Thus, the exciting potential for this ETS factor as a target to prevent vascular inflammation and maintain endothelial homeostasis is being explored. Structurally, Erg contains a PNT domain and several other functional domains (Fig. 6.1); the key to its multiple roles in EC may reside in the ability of these domains to mediate multiple interactions. For example, the N-terminal domain of Erg has been shown to interact with the histone methyltransferase ESET (for Erg-associated protein with a Set domain) [15]. Whether this interaction is crucial in mediating the repression of pro-inflammatory endothelial genes remains to be determined.

The ability of Erg to repress inflammation is in contrast to that of other ETS factors, including Ets-1, which has been shown to synergise with NF-κB in promoting inflammation [131]. Interestingly, expression of pro-inflammatory ETS factors, such as Ets-1, Ets-2 and ESE-1, is upregulated by pro-inflammatory agents, whilst expression of Erg is downregulated by agents such as TNF and LPS [71, 82]. The different expression levels and activities highlight the important role of ETS factors in regulating inflammation.

Conclusions

Over the last two decades, major advances in the understanding of the role of different ETS factors in the context of endothelial biology have been achieved. The use of model organisms coupled with ETS factor transgenic and KO mouse lines have highlighted the crucial role of ETS factors in vascular development. The study of the role of endothelial ETS factors in angiogenesis *in vivo* has been limited by the fact that, in many instances, deletion of an ETS factor in mice causes embryonic lethality. One notable exception is Ets-1, and in the case of this ETS family member, the evidence for its role in adult angiogenesis is indisputable. With the generation of lineage-specific, conditional KO mice, further data will become available on the contribution that various endothelial ETS factors make to physiological and pathological angiogenesis in the adult, a crucial step in the development of future therapeutic approaches. Despite the interest in ETS factors in angiogenesis, there is still much to be done to define the target genes and the transcriptional networks that control the function of these transcription factors. A novel chapter in the understanding of the role of ETS factors in endothelial biology and angiogenesis is likely to focus on microRNAs, which are rapidly being recognised as both targets and regulators of ETS factors' expression and function. Finally, the essential role of ETS proteins in the maintenance of vascular homeostasis will need to be considered when developing anti-tumour strategies aimed at targeting those ETS factors which are overexpressed in several types of cancer.

Acknowledgements The unpublished work mentioned in this chapter has been supported by grants from the British Heart Foundation. My thanks to Dr. Graeme Birdsey (Imperial College London, UK) for his invaluable contribution to my laboratory over many years, to Prof. Dorian Haskard (Imperial College London, UK) for his support and encouragement and to Prof. Justin Mason (Imperial College London, UK) for his continuous support and for critical reading of the manuscript.

References

1. Bartel FO, Higuchi T, Spyropoulos DD (2000) Mouse models in the study of the Ets family of transcription factors. Oncogene 19:6443–6454
2. Seth A, Watson DK (2005) ETS transcription factors and their emerging roles in human cancer. Eur J Cancer 41:2462–2478
3. Oettgen P (2010) The role of ets factors in tumor angiogenesis. J Oncol 2010:767384
4. Dittmer J, Nordheim A (1998) Ets transcription factors and human disease. Biochim Biophys Acta 1377:F1–F11
5. Oettgen P (2006) Regulation of vascular inflammation and remodeling by ETS factors. Circ Res 99:1159–1166
6. Randi AM, Sperone A, Dryden NH, Birdsey GM (2009) Regulation of angiogenesis by ETS transcription factors. Biochem Soc Trans 37:1248–1253

7. Vrieseling E, Arber S (2006) Target-induced transcriptional control of dendritic patterning and connectivity in motor neurons by the ETS gene Pea3. Cell 127:1439–1452
8. Leprince D, Gegonne A, Coll J, de Taisne C, Schneeberger A, Lagrou C, Stehelin D (1983) A putative second cell-derived oncogene of the avian leukaemia retrovirus E26. Nature 306:395–397
9. Hollenhorst PC, Shah AA, Hopkins C, Graves BJ (2007) Genome-wide analyses reveal properties of redundant and specific promoter occupancy within the ETS gene family. Genes Dev 21:1882–1894
10. Bult CJ, Eppig JT, Kadin JA, Richardson JE, Blake JA (2008) The mouse genome database (MGD): mouse biology and model systems. Nucleic Acids Res 36:D724–D728
11. Sharrocks AD (2001) The ETS-domain transcription factor family. Nat Rev Mol Cell Biol 2:827–837
12. Batchelor AH, Piper DE, de la Brousse FC, McKnight SL, Wolberger C (1998) The structure of GABPalpha/beta: an ETS domain- ankyrin repeat heterodimer bound to DNA. Science 279:1037–1041
13. Lamber EP, Vanhille L, Textor LC, Kachalova GS, Sieweke MH, Wilmanns M (2008) Regulation of the transcription factor Ets-1 by DNA-mediated homo-dimerization. EMBO J 27:2006–2017
14. Hollenhorst PC, McIntosh LP, Graves BJ (2011) Genomic and biochemical insights into the specificity of ETS transcription factors. Annu Rev Biochem 80:437–471
15. Yang L, Xia L, Wu DY, Wang H, Chansky HA, Schubach WH, Hickstein DD, Zhang Y (2002) Molecular cloning of ESET, a novel histone H3-specific methyltransferase that interacts with ERG transcription factor. Oncogene 21:148–152
16. Mackereth CD, Scharpf M, Gentile LN, MacIntosh SE, Slupsky CM, McIntosh LP (2004) Diversity in structure and function of the Ets family PNT domains. J Mol Biol 342:1249–1264
17. Carrere S, Verger A, Flourens A, Stehelin D, Duterque-Coquillaud M (1998) Erg proteins, transcription factors of the Ets family, form homo, heterodimers and ternary complexes via two distinct domains. Oncogene 16:3261–3268
18. Reddy SY, Obika S, Bruice TC (2003) Conformations and dynamics of Ets-1 ETS domain-DNA complexes. Proc Natl Acad Sci USA 100:15475–15480
19. Obika S, Reddy SY, Bruice TC (2003) Sequence specific DNA binding of Ets-1 transcription factor: molecular dynamics study on the Ets domain–DNA complexes. J Mol Biol 331:345–359
20. Hsu T, Trojanowska M, Watson DK (2004) Ets proteins in biological control and cancer. J Cell Biochem 91:896–903
21. Wei GH, Badis G, Berger MF, Kivioja T, Palin K, Enge M, Bonke M, Jolma A, Varjosalo M, Gehrke AR, Yan J, Talukder S, Turunen M, Taipale M, Stunnenberg HG, Ukkonen E, Hughes TR, Bulyk ML, Taipale J (2010) Genome-wide analysis of ETS-family DNA-binding in vitro and in vivo. EMBO J 29:2147–2160
22. Basuyaux JP, Ferreira E, Stehelin D, Buttice G (1997) The Ets transcription factors interact with each other and with the c-Fos/c-Jun complex via distinct protein domains in a DNA-dependent and -independent manner. J Biol Chem 272:26188–26195
23. Green SM, Coyne HJ III, McIntosh LP, Graves BJ (2010) DNA binding by the ETS protein TEL (ETV6) is regulated by autoinhibition and self-association. J Biol Chem 285:18496–18504
24. Li R, Pei H, Watson DK (2000) Regulation of Ets function by protein–protein interactions. Oncogene 19:6514–6523
25. Seidel JJ, Graves BJ (2002) An ERK2 docking site in the Pointed domain distinguishes a subset of ETS transcription factors. Genes Dev 16:127–137
26. Foulds CE, Nelson ML, Blaszczak AG, Graves BJ (2004) Ras/mitogen-activated protein kinase signaling activates Ets-1 and Ets-2 by CBP/p300 recruitment. Mol Cell Biol 24:10954–10964

27. Lopez RG, Carron C, Oury C, Gardellin P, Bernard O, Ghysdael J (1999) TEL is a sequence-specific transcriptional repressor. J Biol Chem 274:30132–30138
28. Lelievre E, Lionneton F, Soncin F, Vandenbunder B (2001) The Ets family contains transcriptional activators and repressors involved in angiogenesis. Int J Biochem Cell Biol 33:391–407
29. Leprivier G, Baillat D, Begue A, Hartmann B, Aumercier M (2009) Ets-1 p51 and p42 isoforms differentially modulate Stromelysin-1 promoter according to induced DNA bend orientation. Nucleic Acids Res 37:4341–4352
30. Sarrazin S, Starck J, Gonnet C, Doubeikovski A, Melet F, Morle F (2000) Negative and translation termination-dependent positive control of FLI-1 protein synthesis by conserved overlapping 5' upstream open reading frames in Fli-1 mRNA. Mol Cell Biol 20:2959–2969
31. Coutte L, Monte D, Imai K, Pouilly L, Dewitte F, Vidaud M, Adamski J, Baert JL, de Launoit Y (1999) Characterization of the human and mouse ETV1/ER81 transcription factor genes: role of the two alternatively spliced isoforms in the human. Oncogene 18:6278–6286
32. DuterqueCoquillaud M, Niel C, Plaza S, Stehelin D (1993) New human erg isoforms generated by alternative splicing are transcriptional activators. Oncogene 8:1865–1873
33. Owczarek CM, Portbury KJ, Hardy MP, O'Leary DA, Kudoh J, Shibuya K, Shimizu N, Kola I, Hertzog PJ (2004) Detailed mapping of the ERG-ETS2 interval of human chromosome 21 and comparison with the region of conserved synteny on mouse chromosome 16. Gene 324:65–77
34. Sasaki K, Nakamura Y, Maki K, Waga K, Nakamura F, Arai H, Imai Y, Hirai H, Mitani K (2004) Functional analysis of a dominant-negative DeltaETS TEL/ETV6 isoform. Biochem Biophys Res Commun 317:1128–1137
35. De Haro L, Janknecht R (2005) Cloning of the murine ER71 gene (Etsrp71) and initial characterization of its promoter. Genomics 85:493–502
36. Lee GM, Donaldson LW, Pufall MA, Kang HS, Pot I, Graves BJ, McIntosh LP (2005) The structural and dynamic basis of Ets-1 DNA binding autoinhibition. J Biol Chem 280:7088–7099
37. Pufall MA, Graves BJ (2002) Ets-1 flips for new partner Pax-5. Structure 10:11–14
38. Laitem C, Leprivier G, Choul-Li S, Begue A, Monte D, Larsimont D, Dumont P, Duterque-Coquillaud M, Aumercier M (2009) Ets-1 p27: a novel Ets-1 isoform with dominant-negative effects on the transcriptional properties and the subcellular localization of Ets-1 p51. Oncogene 28:2087–2099
39. Zammarchi F, Boutsalis G, Cartegni L (2013) 5' UTR control of native ERG and of Tmprss2: ERG variants activity in prostate cancer. PLoS One 8:e49721
40. Vijayaraj P, Le BA, Mitchell N, Kondo M, Juliao S, Wasserman M, Beeler D, Spokes K, Aird WC, Baldwin HS, Oettgen P (2012) Erg is a crucial regulator of endocardial-mesenchymal transformation during cardiac valve morphogenesis. Development 139:3973–3985
41. Clark J, Merson S, Jhavar S, Flohr P, Edwards S, Foster CS, Eeles R, Martin FL, Phillips DH, Crundwell M, Christmas T, Thompson A, Fisher C, Kovacs G, Cooper CS (2007) Diversity of TMPRSS2-ERG fusion transcripts in the human prostate. Oncogene 26:2667–2673
42. Sun C, Dobi A, Mohamed A, Li H, Thangapazham RL, Furusato B, Shaheduzzaman S, Tan SH, Vaidyanathan G, Whitman E, Hawksworth DJ, Chen Y, Nau M, Patel V, Vahey M, Gutkind JS, Sreenath T, Petrovics G, Sesterhenn IA, Mcleod DG, Srivastava S (2008) TMPRSS2-ERG fusion, a common genomic alteration in prostate cancer activates C-MYC and abrogates prostate epithelial differentiation. Oncogene 27:5348–5353
43. Tomlins SA, Rhodes DR, Perner S, Dhanasekaran SM, Mehra R, Sun XW, Varambally S, Cao XH, Tchinda J, Kuefer R, Lee C, Montie JE, Shah RB, Pienta KJ, Rubin MA, Chinnaiyan AM (2005) Recurrent fusion of TMPRSS2 and ETS transcription factor genes in prostate cancer. Science 310:644–648
44. Tomlins SA, Mehra R, Rhodes DR, Smith LR, Roulston D, Helgesson BE, Cao XH, Wei JT, Rubin MA, Shah RB, Chinnaiyan AM (2006) TMPRSS2: ETV4 gene fusions define a third molecular subtype of prostate cancer. Cancer Res 66:3396–3400

45. Hu Y, Dobi A, Sreenath T, Cook C, Tadase AY, Ravindranath L, Cullen J, Furusato B, Chen Y, Thangapazham RL, Mohamed A, Sun C, Sesterhenn IA, Mcleod DG, Petrovics G, Srivastava S (2008) Delineation of TMPRSS2-ERG splice variants in prostate cancer. Clin Cancer Res 14:4719–4725
46. Santen RJ, Song RX, McPherson R, Kumar R, Adam L, Jeng MH, Yue W (2002) The role of mitogen-activated protein (MAP) kinase in breast cancer. J Steroid Biochem Mol Biol 80:239–256
47. Yang BS, Hauser CA, Henkel G, Colman MS, Van BC, Stacey KJ, Hume DA, Maki RA, Ostrowski MC (1996) Ras-mediated phosphorylation of a conserved threonine residue enhances the transactivation activities of c-Ets1 and c-Ets2. Mol Cell Biol 16:538–547
48. Fowles LF, Martin ML, Nelsen L, Stacey KJ, Redd D, Clark YM, Nagamine Y, McMahon M, Hume DA, Ostrowski MC (1998) Persistent activation of mitogen-activated protein kinases p42 and p44 and ets-2 phosphorylation in response to colony-stimulating factor 1/c-fms signaling. Mol Cell Biol 18:5148–5156
49. Smith JL, Schaffner AE, Hofmeister JK, Hartman M, Wei G, Forsthoefel D, Hume DA, Ostrowski MC (2000) ets-2 is a target for an akt (Protein kinase B)/jun N-terminal kinase signaling pathway in macrophages of motheaten-viable mutant mice. Mol Cell Biol 20:8026–8034
50. Yordy JS, Muise-Helmericks RC (2000) Signal transduction and the Ets family of transcription factors. Oncogene 19:6503–6513
51. Rabault B, Roussel MF, Quang CT, Ghysdael J (1996) Phosphorylation of Ets1 regulates the complementation of a CSF-1 receptor impaired in mitogenesis. Oncogene 13:877–881
52. Elvert G, Kappel A, Heidenreich R, Englmeier U, Lanz S, Acker T, Rauter M, Plate K, Sieweke M, Breier G, Flamme I (2003) Cooperative interaction of hypoxia-inducible factor-2alpha (HIF-2alpha) and Ets-1 in the transcriptional activation of vascular endothelial growth factor receptor-2 (Flk-1). J Biol Chem 278:7520–7530
53. Simon MP, Tournaire R, Pouyssegur J (2008) The angiopoietin-2 gene of endothelial cells is up-regulated in hypoxia by a HIF binding site located in its first intron and by the central factors GATA-2 and Ets-1. J Cell Physiol 217:809–818
54. Wasylyk C, Gutman A, Nicholson R, Wasylyk B (1991) The c-Ets oncoprotein activates the stromelysin promoter through the same elements as several non-nuclear oncoproteins. EMBO J 10:1127–1134
55. Logan SK, Garabedian MJ, Campbell CE, Werb Z (1996) Synergistic transcriptional activation of the tissue inhibitor of metalloproteinases-1 promoter via functional interaction of AP-1 and Ets-1 transcription factors. J Biol Chem 271:774–782
56. Buttice G, DuterqueCoquillaud M, Basuyaux JP, Carrere S, Kurkinen M, Stehelin D (1996) Erg, an Ets-family member, differentially regulates human collagenase1 (MMP1) and stromelysin1 (MMP3) gene expression by physically interacting with the Fos/Jun complex. Oncogene 13:2297–2306
57. Birdsey GM, Dryden NH, Shah AV, Hanna RL, Hall MD, Haskard DO, Parsons M, Mason JC, Zvelebil M, Gottgens B, Ridley AJ, Randi AM (2012) The transcription factor Erg regulates expression of histone deacetylase 6 and multiple pathways involved in endothelial cell migration and angiogenesis. Blood 119:894–903, Ref Type: Generic
58. Yuan L, Sacharidou A, Stratman AN, Le Bras A, Zwiers PJ, Spokes K, Bhasin M, Shih SC, Nagy JA, Molema G, Aird WC, Davis GE, Oettgen P (2011) RhoJ is an endothelial cell-restricted Rho GTPase that mediates vascular morphogenesis and is regulated by the transcription factor ERG. Blood 118:1145–1153
59. Iljin K, Dube A, Kontusaari S, Korhonen J, Lahtinen I, Oettgen P, Alitalo K (1999) Role of ets factors in the activity and endothelial cell specificity of the mouse Tie gene promoter. FASEB J 13:377–386
60. Dube A, Akbarali Y, Sato TN, Libermann TA, Oettgen P (1999) Role of the Ets transcription factors in the regulation of the vascular-specific Tie2 gene. Circ Res 84:1177–1185

61. Huang X, Brown C, Ni W, Maynard E, Rigby AC, Oettgen P (2006) Critical role for the Ets transcription factor ELF-1 in the development of tumor angiogenesis. Blood 107:3153–3160
62. Wakiya K, Begue A, Stehelin D, Shibuya M (1996) A cAMP response element and an Ets motif are involved in the transcriptional regulation of flt-1 tyrosine kinase (vascular endothelial growth factor receptor 1) gene. J Biol Chem 271:30823–30828
63. Meadows SM, Salanga MC, Krieg PA (2009) Kruppel-like factor 2 cooperates with the ETS family protein ERG to activate Flk1 expression during vascular development. Development 136:1115–1125
64. Kappel A, Schlaeger TM, Flamme I, Orkin SH, Risau W, Breier G (2000) Role of SCL/Tal-1, GATA, and ets transcription factor binding sites for the regulation of flk-1 expression during murine vascular development. Blood 96:3078–3085
65. Teruyama K, Abe M, Nakano T, Takahashi S, Yamada S, Sato Y (2001) Neurophilin-1 is a downstream target of transcription factor Ets-1 in human umbilical vein endothelial cells. FEBS Lett 504:1–4
66. Zheng H, Wasylyk C, Ayadi A, Abecassis J, Schalken JA, Rogatsch H, Wernert N, Maira SM, Multon MC, Wasylyk B (2003) The transcription factor Net regulates the angiogenic switch. Genes Dev 17:2283–2297
67. Hasegawa Y, Abe M, Yamazaki T, Niizeki O, Shiiba K, Sasaki I, Sato Y (2004) Transcriptional regulation of human angiopoietin-2 by transcription factor Ets-1. Biochem Biophys Res Commun 316:52–58
68. Roukens MG, Alloul-Ramdhani M, Baan B, Kobayashi K, Peterson-Maduro J, van Dam H, Schulte-Merker S, Baker DA (2010) Control of endothelial sprouting by a Tel-CtBP complex. Nat Cell Biol 12:933–942
69. Birdsey GM, Dryden NH, Amsellem V, Gebhardt F, Sahnan K, Haskard DO, Dejana E, Mason JC, Randi AM (2008) Transcription factor Erg regulates angiogenesis and endothelial apoptosis through VE-cadherin. Blood 111:3498–3506
70. Lelievre E, Mattot V, Huber P, Vandenbunder B, Soncin F (2000) ETS1 lowers capillary endothelial cell density at confluence and induces the expression of VE-cadherin. Oncogene 19:2438–2446
71. McLaughlin F, Ludbrook VJ, Kola I, Campbell CJ, Randi AM (1999) Characterisation of the tumour necrosis factor (TNF)-(alpha) response elements in the human ICAM-2 promoter. J Cell Sci 112(Pt 24):4695–4703
72. Okada Y, Jin E, Nikolova-Krstevski V, Yano K, Liu J, Beeler D, Spokes K, Kitayama M, Funahashi N, Doi T, Janes L, Minami T, Oettgen P, Aird WC (2008) A GABP-binding element in the Robo4 promoter is necessary for endothelial expression in vivo. Blood 112:2336–2339
73. Oda N, Abe M, Sato Y (1999) ETS-1 converts endothelial cells to the angiogenic phenotype by inducing the expression of matrix metalloproteinases and integrin beta3. J Cell Physiol 178:121–132
74. McLaughlin F, Ludbrook VJ, Cox J, von Carlowitz I, Brown S, Randi AM (2001) Combined genomic and antisense analysis reveals that the transcription factor Erg is implicated in endothelial cell differentiation. Blood 98:3332–3339
75. Heo SH, Choi YJ, Ryoo HM, Cho JY (2010) Expression profiling of ETS and MMP factors in VEGF-activated endothelial cells: role of MMP-10 in VEGF-induced angiogenesis. J Cell Physiol 224:734–742
76. Petrovic N, Bhagwat SV, Ratzan WJ, Ostrowski MC, Shapiro LH (2003) CD13/APN transcription is induced by RAS/MAPK-mediated phosphorylation of Ets-2 in activated endothelial cells. J Biol Chem 278:49358–49368
77. Pimanda JE, Chan WY, Donaldson IJ, Bowen M, Green AR, Gottgens B (2006) Endoglin expression in the endothelium is regulated by Fli-1, Erg, and Elf-1 acting on the promoter and a −8-kb enhancer. Blood 107:4737–4745
78. Deramaudt BM, Remy P, Abraham NG (1999) Upregulation of human heme oxygenase gene expression by Ets-family proteins. J Cell Biochem 72:311–321

79. Wei G, Srinivasan R, Cantemir-Stone CZ, Sharma SM, Santhanam R, Weinstein M, Muthusamy N, Man AK, Oshima RG, Leone G, Ostrowski MC (2009) Ets1 and Ets2 are required for endothelial cell survival during embryonic angiogenesis. Blood 114:1123–1130
80. Karantzoulis-Fegaras F, Antoniou H, Lai SLM, Kulkarni G, D'Abreo C, Wong GKT, Miller TL, Chan Y, Atkins J, Wang Y, Marsden PA (1999) Characterization of the human endothelial nitric-oxide synthase promoter. J Biol Chem 274:3076–3093
81. Zhan Y, Brown C, Maynard E, Anshelevich A, Ni W, Ho IC, Oettgen P (2005) Ets-1 is a critical regulator of Ang II-mediated vascular inflammation and remodeling. J Clin Invest 115:2508–2516
82. Yuan L, Nikolova-Krstevski V, Zhan Y, Kondo M, Bhasin M, Varghese L, Yano K, Carman CV, Aird WC, Oettgen P (2009) Antiinflammatory effects of the ETS factor ERG in endothelial cells are mediated through transcriptional repression of the interleukin-8 gene. Circ Res 104:1049–1057
83. Dryden NH, Sperone A, Martin-Almedina S, Hannah RL, Birdsey GM, Khan ST, Layhadi JA, Mason JC, Haskard DO, Gottgens B, Randi AM (2012) The transcription factor Erg controls endothelial cell quiescence by repressing activity of nuclear factor (NF)-kappaB p65. J Biol Chem 287:12331–12342
84. Yockell-Lelievre J, Spriet C, Cantin P, Malenfant P, Heliot L, de Launoit Y, Audette M (2009) Functional cooperation between Stat-1 and ets-1 to optimize icam-1 gene transcription. Biochem Cell Biol 87:905–918
85. de Launoit Y, Audette M, Pelczar H, Plaza S, Baert JL (1998) The transcription of the intercellular adhesion molecule-1 is regulated by Ets transcription factors. Oncogene 16:2065–2073
86. Sperone A, Dryden NH, Birdsey GM, Madden L, Johns M, Evans PC, Mason JC, Haskard DO, Boyle JJ, Paleolog EM, Randi AM (2011) The transcription factor Erg inhibits vascular inflammation by repressing NF-kappaB activation and proinflammatory gene expression in endothelial cells. Arterioscler Thromb Vasc Biol 31:142–150
87. Ayadi A, Zheng H, Sobieszczuk P, Buchwalter G, Moerman P, Alitalo K, Wasylyk B (2001) Net-targeted mutant mice develop a vascular phenotype and up-regulate egr-1. EMBO J 20:5139–5152
88. Lelievre E, Lionneton F, Mattot V, Spruyt N, Soncin F (2002) Ets-1 regulates fli-1 expression in endothelial cells. Identification of ETS binding sites in the fli-1 gene promoter. J Biol Chem 277:25143–25151
89. Majerus MA, Bibollet-Ruche F, Telliez JB, Wasylyk B, Bailleul B (1992) Serum, AP-1 and Ets-1 stimulate the human ets-1 promoter. Nucleic Acids Res 20:2699–2703
90. Coffer P, de Jonge M, Mettouchi A, Binetruy B, Ghysdael J, Kruijer W (1994) junB promoter regulation: Ras mediated transactivation by c-Ets-1 and c-Ets-2. Oncogene 9:911–921
91. Landry JR, Kinston S, Knezevic K, Donaldson IJ, Green AR, Gottgens B (2005) Fli1, Elf1, and Ets1 regulate the proximal promoter of the LMO2 gene in endothelial cells. Blood 106:2680–2687
92. Schwachtgen JL, Janel N, Barek L, Duterque-Coquillaud M, Ghysdael J, Meyer D, Kerbiriou-Nabias D (1997) Ets transcription factors bind and transactivate the core promoter of the von Willebrand factor gene. Oncogene 15:3091–3102
93. von der Ahe D, Nischan C, Kunz C, Otte J, Knies U, Oderwald H, Wasylyk B (1993) Ets transcription factor binding site is required for positive and TNF alpha-induced negative promoter regulation. Nucleic Acids Res 21:5636–5643
94. Ginsberg M, James D, Ding BS, Nolan D, Geng F, Butler JM, Schachterle W, Pulijaal VR, Mathew S, Chasen ST, Xiang J, Rosenwaks Z, Shido K, Elemento O, Rabbany SY, Rafii S (2012) Efficient direct reprogramming of mature amniotic cells into endothelial cells by ETS factors and TGFbeta suppression. Cell 151:559–575
95. Rasmussen TL, Shi X, Wallis A, Kweon J, Zirbes KM, Koyano-Nakagawa N, Garry DJ (2012) VEGF/Flk1 signaling cascade transactivates Etv2 gene expression. PLoS One 7: e50103

96. De Val S (2011) Key transcriptional regulators of early vascular development. Arterioscler Thromb Vasc Biol 31:1469–1475
97. Patterson LJ, Patient R (2006) The "Ets" factor: vessel formation in zebrafish – the missing link? PLoS Biol 4:21–24
98. De Val S, Black BL (2009) Transcriptional control of endothelial cell development. Dev Cell 16:180–195
99. Liu F, Patient R (2008) Genome-wide analysis of the zebrafish ETS family identifies three genes required for hemangioblast differentiation or angiogenesis. Circ Res 103:1147–1154
100. Ellett F, Kile BT, Lieschke GJ (2009) The role of the ETS factor erg in zebrafish vasculogenesis. Mech Dev 126:220–229
101. Meadows SM, Myers CT, Krieg PA (2011) Regulation of endothelial cell development by ETS transcription factors. Semin Cell Dev Biol 22:976–984
102. Ristevski S, Tam PP, Hertzog PJ, Kola I (2002) Ets2 is expressed during morphogenesis of the somite and limb in the mouse embryo. Mech Dev 116:165–168
103. Yamamoto H, Flannery ML, Kupriyanov S, Pearce J, McKercher SR, Henkel GW, Maki RA, Werb Z, Oshima RG (1998) Defective trophoblast function in mice with a targeted mutation of Ets2. Genes Dev 12:1315–1326
104. De Val S, Chi NC, Meadows SM, Minovitsky S, Anderson JP, Harris IS, Ehlers ML, Agarwal P, Visel A, Xu SM, Pennacchio LA, Dubchak I, Krieg PA, Stainier DY, Black BL (2008) Combinatorial regulation of endothelial gene expression by ets and forkhead transcription factors. Cell 135:1053–1064
105. Bories JC, Willerford DM, Grevin D, Davidson L, Camus A, Martin P, Stehelin D, Alt FW (1995) Increased T-cell apoptosis and terminal B-cell differentiation induced by inactivation of the Ets-1 proto-oncogene. Nature 377:635–638
106. Muthusamy N, Barton K, Leiden JM (1995) Defective activation and survival of T cells lacking the Ets-1 transcription factor. Nature 377:639–642
107. Vlaeminck-Guillem V, Carrere S, Dewitte F, Stehelin D, Desbiens X, Duterque-Coquillaud M (2000) The Ets family member Erg gene is expressed in mesodermal tissues and neural crests at fundamental steps during mouse embryogenesis. Mech Dev 91:331–335
108. Mohamed AA, Tan SH, Mikhalkevich N, Ponniah S, Vasioukhin V, Bieberich CJ, Sesterhenn IA, Dobi A, Srivastava S, Sreenath TL (2010) Ets family protein, erg expression in developing and adult mouse tissues by a highly specific monoclonal antibody. J Cancer 1:197–208
109. Oikawa M, Abe M, Kurosawa H, Hida W, Shirato K, Sato Y (2001) Hypoxia induces transcription factor ETS-1 via the activity of hypoxia-inducible factor-1. Biochem Biophys Res Commun 289:39–43
110. Hashiya N, Jo N, Aoki M, Matsumoto K, Nakamura T, Sato Y, Ogata N, Ogihara T, Kaneda Y, Morishita R (2004) In vivo evidence of angiogenesis induced by transcription factor Ets-1: Ets-1 is located upstream of angiogenesis cascade. Circulation 109:3035–3041
111. Pourtier-Manzanedo A, Vercamer C, Van BE, Mattot V, Mouquet F, Vandenbunder B (2003) Expression of an Ets-1 dominant-negative mutant perturbs normal and tumor angiogenesis in a mouse ear model. Oncogene 22:1795–1806
112. Nakano T, Abe M, Tanaka K, Shineha R, Satomi S, Sato Y (2000) Angiogenesis inhibition by transdominant mutant Ets-1. J Cell Physiol 184:255–262
113. Watanabe D, Takagi H, Suzuma K, Suzuma I, Oh H, Ohashi H, Kemmochi S, Uemura A, Ojima T, Suganami E, Miyamoto N, Sato Y, Honda Y (2004) Transcription factor Ets-1 mediates ischemia- and vascular endothelial growth factor-dependent retinal neovascularization. Am J Pathol 164:1827–1835
114. Lefter LP, Dima S, Sunamura M, Furukawa T, Sato Y, Abe M, Chivu M, Popescu I, Horii A (2009) Transcriptional silencing of ETS-1 efficiently suppresses angiogenesis of pancreatic cancer. Cancer Gene Ther 16:137–148
115. Wernert N, Stanjek A, Kiriakidis S, Hugel A, Jha HC, Mazitschek R, Giannis A (1999) Inhibition of angiogenesis in vivo by ets-1 antisense oligonucleotides-inhibition of Ets-1

transcription factor expression by the antibiotic fumagillin. Angew Chem Int Ed Engl 38:3228–3231
116. Yoon KC, Bae JA, Park HJ, Im SK, Oh HJ, Lin XH, Kim MY, Lee JH, Lee SE, Ahn KY, Kim KK (2009) Subconjunctival gene delivery of the transcription factor GA-binding protein delays corneal neovascularization in a mouse model. Gene Ther 16:973–981
117. Man AK, Young LJ, Tynan JA, Lesperance J, Egeblad M, Werb Z, Hauser CA, Muller WJ, Cardiff RD, Oshima RG (2003) Ets2-dependent stromal regulation of mouse mammary tumors. Mol Cell Biol 23:8614–8625
118. Iwasaka C, Tanaka K, Abe M, Sato Y (1996) Ets-1 regulates angiogenesis by inducing the expression of urokinase-type plasminogen activator and matrix metalloproteinase-1 and the migration of vascular endothelial cells. J Cell Physiol 169:522–531
119. Stamatovic SM, Keep RF, Mostarica-Stojkovic M, Andjelkovic AV (2006) CCL2 regulates angiogenesis via activation of Ets-1 transcription factor. J Immunol 177:2651–2661
120. Arderiu G, Pena E, Aledo R, Espinosa S, Badimon L (2012) Ets-1 transcription is required in tissue factor driven microvessel formation and stabilization. Angiogenesis 15:657–669
121. Mattot V, Vercamer C, Soncin F, Calmels T, Huguet C, Fafeur V, Vandenbunder B (2000) Constitutive expression of the DNA-binding domain of Ets1 increases endothelial cell adhesion and stimulates their organization into capillary-like structures. Oncogene 19:762–772
122. Dejana E, Taddei A, Randi AM (2007) Foxs and Ets in the transcriptional regulation of endothelial cell differentiation and angiogenesis. Biochim Biophys Acta 1775:298–312
123. Wernert N, Raes MB, Lassalle P, Dehouck MP, Gosselin B, Vandenbunder B, Stehelin D (1992) c-Ets1 proto-oncogene is a transcription factor expressed in endothelial cells during tumor vascularization and other forms of angiogenesis in humans. Am J Pathol 140:119–127
124. Tokuhara K, Ogata Y, Nakagawa M, Shirouzu K (2003) Ets-1 expression in vascular endothelial cells as an angiogenic and prognostic factor in colorectal carcinoma. Int Surg 88:25–33
125. Tomlins SA, Bjartell A, Chinnaiyan AM, Jenster G, Nam RK, Rubin MA, Schalken JA (2009) ETS gene fusions in prostate cancer: from discovery to daily clinical practice. Eur Urol 56:275–286
126. Nikolova-Krstevski V, Bhasin M, Otu HH, Libermann T, Oettgen P (2008) Gene expression analysis of embryonic stem cells expressing VE-cadherin (CD144) during endothelial differentiation. BMC Genomics 9:240
127. Loughran SJ, Kruse EA, Hacking DF, de Graaf CA, Hyland CD, Willson TA, Henley KJ, Ellis S, Voss AK, Metcalf D, Hilton DJ, Alexander WS, Kile BT (2008) The transcription factor Erg is essential for definitive hematopoiesis and the function of adult hematopoietic stem cells. Nat Immunol 9:810–819
128. Hollenhorst PC, Jones DA, Graves BJ (2004) Expression profiles frame the promoter specificity dilemma of the ETS family of transcription factors. Nucleic Acids Res 32:5693–5702
129. Hewett PW, Nishi K, Daft EL, Murray JC (2001) Selective expression of erg isoforms In human endothelial cells. Int J Biochem Cell Biol 33:347–355
130. Yuan L, Le BA, Sacharidou A, Itagaki K, Zhan Y, Kondo M, Carman CV, Davis GE, Aird WC, Oettgen P (2012) ETS-related gene (ERG) controls endothelial cell permeability via transcriptional regulation of the claudin 5 (CLDN5) gene. J Biol Chem 287:6582–6591
131. Song KS, Yoon JH, Kim KS, Ahn DW (2012) c-Ets1 inhibits the interaction of NF-kappaB and CREB, and downregulates IL-1beta-induced MUC5AC overproduction during airway inflammation. Mucosal Immunol 5:207–215
132. Le Bras A, Lionneton F, Mattot V, Lelièvre E, Caetano B, Spruyt N, Soncin F (2007) HIF-2 specifically activates the VE-cadherin promoter independently of hypoxia and in synergy with Ets-1 through two essential ETS-binding sites. Oncogene 26:7480–7489
133. Dutta D, Ray S, Vivian JL, Paul S (2008) Activation of the VEGFR1 chromatin domain: an angiogenic signal-ETS1/HIF-2alpha regulatory axis. J Biol Chem 283(37):25404–25413

134. Linnemann AK, O'Geen H, Keles S, Farnham PJ, Bresnick EH (2011) Genetic framework for GATA factor function in vascular biology. Proc Natl Acad Sci U S A 108(33):13641–13646
135. Mollica LR, Crawley JT, Liu K, Rance JB, Cockerill PN, Follows GA, Landry JR, Wells DJ, Lane DA (2006) Role of a 5′-enhancer in the transcriptional regulation of the human endothelial cell protein C receptor gene. Blood 108(4):1251–1259
136. Wisdom R (1999) AP-1: one switch for many signals. Exp Cell Res 253(1):180–185
137. Westermarck J, Kähäri VM (1999) Regulation of matrix metalloproteinase expression in tumor invasion. FASEB J 13(8):781–792
138. Bassuk AG, Leiden JM (2007) A direct physical association between ETS and AP-1 transcription factors in normal human T cells. Immunity 27(4):561–571
139. Gottschalk LR, Giannola DM, Emerson SG (1995) Molecular regulation of the human IL-3 gene: inducible T cell-restricted expression requires intact AP-1 and Elf-1 nuclear protein binding sites. Immunity 3(2):223–237
140. Crawford HC, Matrisian LM (1996) Mechanisms controlling the transcription of matrix metalloproteinase genes in normal and neoplastic cells. Enzyme Protein 49(1–3):20–37

Insights into Roles of Immediate-Early Genes in Angiogenesis

7

Leonel Prado-Lourenço, Ahmad M.N. Alhendi, and Levon M. Khachigian

Contents

7.1 Immediate-Early Genes 146
7.2 Angiogenesis 146
7.3 Immediate-Early Genes and Angiogenesis 146
7.3.1 Transcription Factors 148
7.3.2 Cytokines 152
7.3.3 Growth Factors 153
7.3.4 Enzymes 153
7.3.5 Secreted Factors/Proteins 154
7.3.6 Transporters 154
7.3.7 Cytoskeletal Proteins 155
7.3.8 Anti-apoptotic Proteins 155
References 156

Abstract

Immediate-early genes are those that are rapidly induced in response to a cellular stimulus in the absence of protein synthesis and can influence the biology and pathobiology of the cell. These span transcription factors, cytokines, growth factors, enzymes, secreted factors, cytoskeletal proteins, transporters and anti-apoptotic proteins that are attractive targets for the control of pathologic angiogenesis. This chapter focuses on immediate-early genes and specifically their regulation of processes underpinning angiogenesis.

Keywords

Angiogenesis • Immediate-early gene • Vascular • Endothelial cell

L. Prado-Lourenço • A.M.N. Alhendi • L.M. Khachigian (✉)
Centre for Vascular Research, University of New South Wales, Sydney, NSW 2052, Australia
e-mail: l.khachigian@unsw.edu.au

J. Dulak et al. (eds.), *Angiogenesis and Vascularisation*,
DOI 10.1007/978-3-7091-1428-5_7, © Springer-Verlag Wien 2013

7.1 Immediate-Early Genes

Immediate-early genes (IEGs) are genes whose interphasic expression is transient and rapid in response to many cellular stimuli (e.g. growth factors, mitogens, stress) before de novo protein synthesis [1–3]. A large number of IEGs have been identified representing master regulators that are involved in a multitude of cellular functions including proliferation, differentiation, metabolism and immunity (for review see [4–6]). Even though a majority of them are transcription factors, these are also represented by cytokines, growth factors, enzymes, secreted factors, cytoskeletal proteins, transporters and anti-apoptotic proteins [5]. They are a diverse group of mediators and exquisitely sensitive to changes in the local cellular environment [5]. Their expression is activated by a range of signalling pathways including mitogen-activated protein kinase (MAPK) and p38-MAPK [7–10]. At the genetic level, chromatin structure may govern their rapid activation. Furthermore, mRNA transcript length is generally shorter and more labile compared to other genes, and, at the protein level, IEG products are generally unstable and rapidly degraded via the proteasome, which may account for poor basal expression.

7.2 Angiogenesis

Angiogenesis is the formation of new blood vessels from the pre-existing vasculature and is required in normal growth, development and other physiological situations. This process is also a feature of pathological situations such as in ocular disease, inflammatory disease, cardiovascular disease, tumour growth and metastasis [11, 12]. Blood vessel formation is normally controlled by a delicate balance of growth and inhibitory factors that can be disrupted and the consequence is either an excess or lack of angiogenesis [13]. As a result, angiogenic processes and the genes controlling these have become a target of choice for combating neovascular diseases. Such therapeutic interventions have also allowed dissection of mechanistic roles for these genes [14]. Greater understanding of the molecular and cellular pathways involved in angiogenesis has facilitated the development of genomedicines. Among these approaches are gene-silencing technologies, such as ribozymes, antisense oligonucleotides, small interference RNA (siRNA) and DNAzymes. Many targets have been evaluated in different disease models *in vitro* and *in vivo*. Below we discuss the diverse range of IEGs that have effects on angiogenesis.

7.3 Immediate-Early Genes and Angiogenesis

IEGs are known to be widely induced during stress responses, exposure of cells to mitogens and hormones. We review here (also see Fig. 7.1) several of these IEGs which have been shown to be major mediators of angiogenesis and they represent a class of proteins that are able to coordinate the plasticity of the cell and an integrated rapid response.

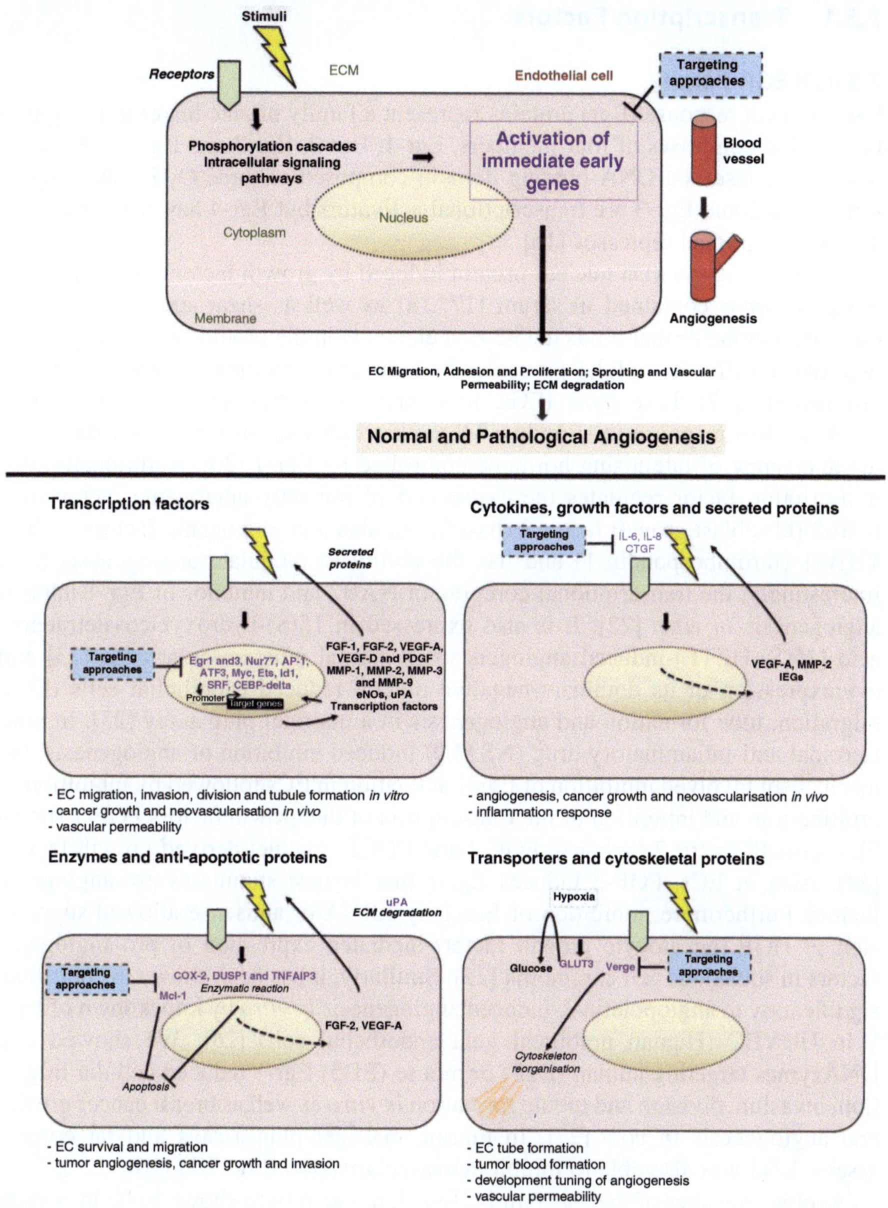

Fig. 7.1 Mechanisms of action of immediate early genes in angiogenesis. Figure shows a general concept of the regulation of angiogenesis following activation of IEGs (*top panel*) as well as specific mechanistic examples of these proteins involved in blood vessel formation (*bottom panel*)

7.3.1 Transcription Factors

7.3.1.1 EGR Family

Early growth response (Egr) proteins represent a family of zinc finger transcription factors that comprises of four members, Egr-1, Egr-2, Egr-3 and Egr-4 [15]. They possess a conserved DNA-binding domain composed of three C_2H_2 zinc fingers. Egr-1, Egr-2 and Egr-3 are transcriptional activators but Egr-4 has been described as a transcriptional repressor [16].

Egr-1 is a 60–80 kDa nuclear protein induced by growth factors, hormones and other mitogens contained in serum [17, 18] as well as shear stress [19]. It is a transcription factor that binds to GC-rich elements in the promoters of many genes required in different cellular processes including proliferation, differentiation and mitogenesis [17]. Like most IEGs, its expression is transiently induced within 15–30 min in response to stimulus. Moreover, deficient mice are infertile due to the deficiency of luteinising hormone controlled by Egr-1 [20]. Additionally, this transcription factor regulates the expression of not only angiogenic factors like FGF-2 [fibroblast growth factor 2 (basic)] but also anti-angiogenic factors such as THBS1 (thrombospondin-1) and has the ability to modulate angiogenesis [21]. Interestingly, the transcriptional corepressor NAB2 and inhibitor of Egr-1 inhibits angiogenesis *in vitro* [22]. It is also expressed in 15(S)-hydroxyeicosatetraenoic acid [15(S)-HETE]-induced angiogenesis in dermal microvascular cells [23] and overexpression of its dominant-negative mutant reduced endothelial cells (ECs) migration, tube formation and angiogenesis in a matrigel plug assay [23]. In non-steroidal anti-inflammatory drug (NSAID)-induced inhibition of angiogenesis, the mechanism involved inhibition of Egr-1 activation in ECs followed by inhibition of proliferation and inhibition of the transcription of the growth factors FGF-1 [fibroblast growth factor 2 (acidic)], FGF-2 and PDGF (platelet-derived growth factor) [24]. Also in ECs, FGF-2 induces Egr-1 that in turn stimulates pro-angiogenic factors. Furthermore, inhibition of Egr-1 by siRNA or antisense allowed suppression of HGF (hepatocyte growth factor)-mediated expression of pro-angiogenic factors in squamous cell carcinoma [25]. Similarly, it has been shown to contribute significantly to angiopoietin-1-induced angiogenesis *in vitro* by knockdown of Egr-1 in HUVEC (Human umbilical vein endothelial cell) [26]. We showed that DNAzymes targeting human (DzF) or mouse (ED5) Egr-1 reduced cellular migration, invasion, division and tubule formation *in vitro* as well as breast cancer growth and angiogenesis *in vivo* [27]. In murine matrigel plug assays and rat corneal assays, ED5 was also able to inhibit neovascularisation (Fig. 7.2) [26].

Another member of the Egr family, Egr-3, has also been shown to be important in angiogenesis [28]. Like Egr-1, it is up-regulated by various stimuli, including serum, growth factors such as VEGF (vascular endothelial growth factor) [29] and hormones [30]. Knockdown of Egr-3 by siRNA reduced VEGF-inducible cell growth, migration and tubulogenesis of ECs [31].

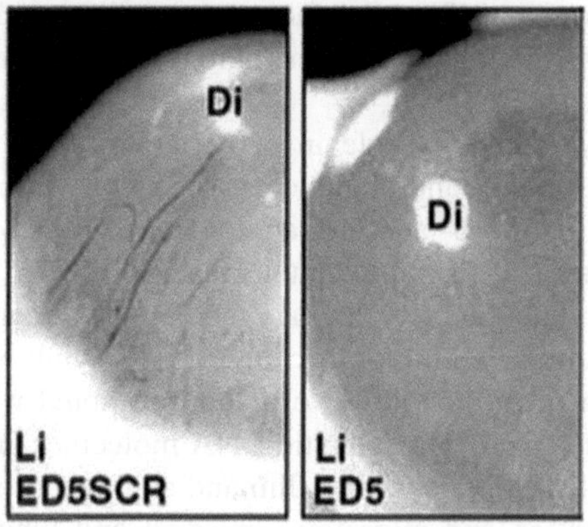

Fig. 7.2 ED5 blocks neovascularisation in rat corneas implanted with $VEGF_{165}$-containing filter discs. Figure shows representative corneas 5 days after treatment with an Egr-1 targeting DNAzyme (ED5) or the same DNAzyme with its hybridising arms scrambled (ED5SCR). *Di* disk, *Li* limbus. Compared with ED5SCR, ED5 inhibits the angiogenic response from the limbus toward the disc [reproduced from Fahmy et al., *Nature Medicine* **9**, 1026–1032 (2003) with permission]

7.3.1.2 NR4A Family of Nuclear Receptors

The NR4A (nuclear receptor subfamily 4 group A) subfamily includes Nur77 (NR4A1), Nurr1 (NR4A2) and Nor1 (NR4A3). These are transcription factors with roles in differentiation, cell growth and cell death in different cell types [32]. Nur77 also known as nerve growth factor IB (NGFIB) has been studied in angiogenesis and is induced by VEGF [33]. This induction by VEGF can occur in a PKD (protein kinase D)-HDAC5 (histone deacetylase 5) pathway-dependent manner [34]. Consequently, inhibition by antisense or siRNA of its expression suppresses VEGF-inducible angiogenesis, whereas adenoviral overexpression in ECs leads to cell cycle arrest [35, 36]. Nur77−/− mice have no apparent vascular defects, but in adult B16F1 melanoma growth and angiogenesis are greatly reduced [35]. Additionally, histamine and serotonin, which increase vascular permeability, stimulate angiogenesis through Nur77 [37]. Finally, it is also capable of regulating microvessel permeability by targeting eNOS (endothelial NO synthase) [38].

7.3.1.3 AP-1 Family

Activator protein-1 (AP-1) is a transcription factor composed of certain proteins including JunB, c-Jun, JunD, c-Fos, FosB, Fra-1 and Fra-2 or ATFs or CREB proteins [39, 40]. To form active complexes, Jun proteins (c-Jun, JunB, JunD) can both homo- or heterodimerise with Fos proteins (c-Fos, FosB, Fra-1, Fra-2), whereas Fos proteins can only heterodimerise with Jun proteins. Jun homodimers are less stable than a Jun–Fos heterodimers. Binding site(s) for these complexes are present in the promoter or enhancer region of specific target genes. Deficiency in c-Jun, JunB, Fra-1 and Fra-2 results in embryonic or early postnatal lethality [41]. AP-1 controls the expression of multiple genes and is linked with cell proliferation, migration and cell survival [42]. AP-1 factors generally regulate the expression of metalloproteinases (MMPs) implicated in angiogenesis and tumoural angiogenesis [41].

c-Jun plays a key regulatory role in neovascularisation and its activation is associated with proliferation and angiogenesis in invasive breast cancer [43] and known to deregulate both pro- and anti-angiogenic factors [44]. Moreover, c-Jun can induce angiogenesis in rabbit skeletal muscle [45] and it has been recently suggested that c-src induction by c-Jun may mediate angiogenic cell invasion [46]. c-Jun also induces cellular migration and invasion through induction of stem cell factor [47]. Finally a peptide inhibitor of c-Jun promoted wound healing *in vivo* [48]. Almost a decade ago, using catalytic DNA molecules targeting c-Jun (Dz13) we demonstrated a clear link between c-Jun and angiogenesis [49]. *In vitro*, Dz13 inhibits ECs proliferation and migration, invasion and tubule formation. It also suppresses the growth of solid tumours and angiogenesis, in part through its inhibition of MMP-2 and MMP-9 [50]. Recent studies by our group have determined that Dz13 inhibits expression of FGF-2 and VEGF-A expression and increases immune and inflammatory cell infiltration [51].

c-Fos can stimulate VEGF-D expression and induce angiogenesis *in vivo* and *in vitro* [52]. It can also regulate the expression of stromelysin (MMP-3), collagenase-1 (MMP-1) and MMP-9 suggesting a role in cellular invasion that is indispensable to angiogenesis [53]. Additionally, endothelial morphogenesis in collagen gels is inhibited by EGCG (epigallocatechin gallate) through down-regulation of c-Jun, Ets-1 and c-Fos [54].

Similarly, Fra-1 induces the expression of extracellular matrix proteinases and of the urokinase plasminogen activator system [urokinase-type plasminogen activator (uPA) and uPA receptor (uPAR)] and mediates MMP-2 expression in angiogenesis [55, 56]. Unexpectedly, studies using mice deficient in JunD demonstrated its anti-angiogenic properties on cardiac capillary [57], whereas tumour angiogenesis is impaired in JunB-deficient mice [58].

Investigators have attempted to inhibit AP-1 using a range of other approaches but some of these studies can be considered incomplete because the composition of AP-1 is not taken into account. For example TAM67, a dominant-negative mutant of c-Jun unable to activate transcription [59], was used widely to demonstrate by overexpression *in vitro* and *in vivo* inhibition of AP-1 [59] and angiogenesis blockade in experimental tumours but its clinical use has had limitations [60]. More recent strategies include siRNA, decoys and peptides. c-Fos and c-jun siRNA inhibit cell motility in epidermoid carcinoma cells [61]. Fra-1 siRNA was able to inhibit invasion, angiogenesis and cell proliferation in breast cancer cells [55]. Decoys for AP-1 were also developed and found to reduce its activity in cancer [62].

Chemotherapeutic and chemopreventive agents have also been used to inhibit AP-1 activity and angiogenesis such as quercetin, taxol and resveratrol. Natural inhibitors have been discovered, one of which is the compound curcumin, one of the principle components of the spice turmeric [63]. Curcumin also has anti-proliferative properties, inhibiting tubule formation by rat lymphatic ECs [64].

Hence there is evidence that inhibitors of this transcription factor could possess active anti-angiogenic properties. AP-1 has also been the subject of previous drug design attempts [65–67]. These studies identified several candidate compounds, but there are no reports of *in vivo* studies. Recently the feasibility of inhibiting selectively c-Fos/AP-1 *in vivo* with a small molecule inhibitor has been tested in a preclinical model of arthritis with positive outcomes [68].

7.3.1.4 ATF3

Cyclic AMP (Adenosine monophosphate)-dependent transcription factor ATF-3 is a member of the mammalian cAMP responsive element-binding (CREB) protein family. ATF3 isoforms can heterodimerise with each other and with other transcription factors such as c-Jun, ATF2 or Smad3. The protein functions either as a transcriptional activator or repressor depending on the target gene and the situation [69] and has a direct role in angiogenesis through the transcriptional repression of Id1 that changes the expression levels of THBSP1 and VEGF [70]. Interestingly, its expression was also found in atherosclerotic lesions, where ECs are under programmed cell death [71]. Moreover, *in vitro* oxidative stress-induced angiogenesis in matrigel can be inhibited by ATF3 siRNA [72].

7.3.1.5 MYC

Myc (c-Myc) is a multifunctional nuclear transcription factor and mutations in this gene have been associated with cancers [73] and tumour angiogenesis [74]. It is induced by EPO (erythropoietin) in ECs [75] and studies in deficient mice have revealed that it is essential for vasculogenesis, as well as tumoural and developmental angiogenesis [76]. Furthermore, it has an essential role in cardiac angiogenesis leading to cardiac remodelling [77]. Conditioned media from c-Myc-overexpressing cells promote ECs migration *in vitro* and neovascularisation of the cornea *in vivo* [78]. Additionally, the mRNA of THBSP-1, an important regulator in angiogenesis, was decreased after overexpression of c-Myc [79] and an increase in tumour angiogenesis was associated with a Myc-activated miRNA cluster [80].

7.3.1.6 Ets Family

The ETS (E-26) family is one of the largest families of transcription factors and is defined by an ETS DNA-binding domain. This family has been involved in activation or repression of angiogenesis [81, 82] and c-ets-1 (Ets-1) has been widely implicated in angiogenesis [83]. Indeed, Ets-1 controls angiogenesis by activating uPA and MMP-1 [84] and target genes also include MMP-9 implicated in ECs migration [85]. Both Ets-1 and Ets-2 are essential in embryonic development, with lethality due to vascular defects [86].

7.3.1.7 Id1

The DNA-binding protein inhibitor Id1 is a helix–loop–helix (HLH) protein that can form heterodimers with members of the basic HLH family of transcription factors [87]. The encoded protein has no DNA-binding activity and therefore can inhibit the capacity of basic HLH proteins to transactivate gene expression. Id1- and Id3-deficient mice exhibit vascular malformations in the forebrain, lack of vascular branching and sprouting in the neuroectoderm. Impaired angiogenesis in transplanted and spontaneous tumours was also observed [88] and overexpression of Id1 in prostate cancer cells promotes angiogenesis through activation of VEGF [89]. It has been proposed as a potential anti-angiogenic target for endothelial progenitor cells [90].

7.3.1.8 SRF

Serum response factor (SRF) is a member of the MADS (MCM1, Agamous, Deficiens, and SRF) box superfamily of transcription factors [91]. It binds to the serum response element (SRE) in the promoter regions of target genes [92]. Such elements are often present in IEGs promoters such as c-Fos and Egr-1 [93]. SRF is essential in VEGF-induced angiogenesis [94]. A knockdown of SRF proteins in ECs with antisense oligonucleotides impairs cell migration and proliferation [94]. Moreover, *in vivo* injection of an SRF antisense expression plasmid into gastric ulcers in rats inhibits angiogenesis [94] and SRF ablation in ECs using a Tie1-Cre transgenic line leads to embryonic lethality due to decreased vascular density and oedema [95]. Deficient mice exhibit impaired expression of IEGs [96]. In a recent study, mir483-5p, which targets SRF, is able to regulate angiogenesis [97].

7.3.1.9 CEBP-delta

CCAAT/enhancer-binding protein delta (CEBP-delta) is an intronless bZIP transcription factor binding to certain promoters as a homodimer or as heterodimers with CEBP-alpha [98]. It plays an important role in inflammation and has implications in rheumatoid arthritis with reduced collagen-induced arthritis in deficient mice for CEBP-delta compared to WT mice [98]. In addition a decreased pannus proliferation and angiogenesis was also observed in deficient mice [99]. In lymphangiogenesis and metastasis this transcription factor regulates VEGF-C autocrine signalling through HIF-1alpha [99].

7.3.2 Cytokines

Interleukin-6 (IL-6) is both a pro-inflammatory and anti-inflammatory secreted cytokine produced at sites of acute and chronic inflammation [100]. IL-6-deficient mice have impaired liver regeneration and abnormalities in IEGs expression [101]. This cytokine also activates VEGF expression and angiogenesis [102, 103]. Another member of this family, IL-8 is a chemokine produced and secreted by macrophages and other cell types including ECs that store it in Weibel–Palade bodies [104]. It is a member of the CXC chemokine family and is a major mediator of the inflammatory response functioning as a chemoattractant and angiogenic factor. Additionally, it has been implicated in cancer and tumoral angiogenesis [105]. IL-8 induces migration and proliferation of HUVEC and can induce angiogenesis when injected in rat cornea [105]. *In vitro*, treatment of melanoma cells by humanised ABX-IL8 antibodies reduced MMP-2 expression and metastasis [106]. Interestingly, a cross talk between IL-8 and c-myc (see above Transcriptions factors) involving the hypoxia-inducible factors (HIFs) HIF-1 and HIF-2 has been recently suggested [107]. However this transcriptional response activated by hypoxia differs depending on the isoform: HIF-1 diminished the expression of IL-8, whereas HIF-2 augmented its expression [107].

7.3.3 Growth Factors

Connective tissue growth factor (CTGF), also known as CCN2, is a member of the CCN family of extracellular matrix-associated heparin-binding proteins that are found at the cell surface or extracellular matrix [108]. Deletion of this gene is lethal with defects in angiogenesis, with impaired interactions between ECs, pericytes and collagen IV deficiency in the endothelial basement membrane [109]. This protein has important roles in many biological processes, including cell adhesion, migration, proliferation and angiogenesis [110, 111]. It is a mitogen secreted by vascular ECs and it can be up-regulated after VEGF stimulation [112]. However, a controversial study showed that CTGF could be anti-angiogenic by binding to VEGF [113] and this binding terminates when CTGF is digested by various MMPs [114, 115]. Nevertheless, a specific CTGF antibody was able to reduce tumour growth and angiogenesis in an orthoptic mouse model of pancreatic cancer [116].

7.3.4 Enzymes

7.3.4.1 COX-2

Prostaglandin-endoperoxide synthase 2 or cyclooxygenase-2 (COX-2) is an enzyme rapidly induced by various cytokines that catalyses prostaglandin production from arachidonic acid [117]. This enzyme is pro-angiogenic, increasing VEGF and FGF-2 production and inhibiting ECs apoptosis [118]. COX-2 was also found to play a role in tumour angiogenesis [119] and it has been proposed as a therapeutic target for cancer, chronic inflammatory disease and retinopathies [120]. Indeed, COX-2 inhibitors can directly block angiogenesis [120].

7.3.4.2 DUSP1

Dual specificity protein phosphatase 1 (DUSP1) is a nuclear phosphatase member of a large family and contains a highly conserved C-terminal catalytic domain and an N-terminal Cdc25-like (CH2) domain. It is transcriptionally induced by growth factors, oxidative and heat stress and specifically inactivates the MAPK pathway *in vitro* by dephosphorylation of phosphothreonine and phosphotyrosine residues [121]. Moreover, it plays a role in the regulation of stress responses and suppression of apoptosis [122]. In ECs, VEGF can induce DUSP1 expression and play a role in migration [123]. This enzyme is known to be a thrombin-responsive gene in ECs [123] and to promote angiogenesis in lung cancer [124].

7.3.4.3 TNFAIP3

Tumour necrosis factor alpha-induced protein 3 (TNFAIP3) is a zinc finger protein and an ubiquitin-editing enzyme that can inhibit NF-kappa B activation and TNF-mediated apoptosis [125]. TNFAIP3 was shown to be rapidly induced after TNF exposure in ECs [126] and it was known that TNF play an important role in the development and maintenance of inflammation. It has been associated with rheumatoid arthritis [127] and angiogenesis [128, 129].

7.3.4.4 Urokinase-type Plasminogen Activator

Urokinase-type plasminogen activator (uPA) is a serine protease involved in the degradation of the extracellular matrix (ECM) necessary for the migration and proliferation of ECs [130]. It was confirmed with siRNA approaches that down-regulation of uPA reduces tumour angiogenesis and invasiveness in breast cancer cells involving MMP-9 [131]. RNA interference using a bi-cistronic construct targeting both uPAR and MMP-9 results in decreased angiogenesis in glioma [131].

7.3.5 Secreted Factors/Proteins

7.3.5.1 EGFL7/ZNeu1

EGF-like domain-containing protein 7 (EGFL7) is a secreted protein in ECs that contains two epidermal growth factor-like domains. EGFL7/ZNeu1 regulates both vasculogenesis and angiogenesis [132–134]. Loss of function in zebrafish shows that it plays a role in tubulogenesis [135]. When EGFL7 was specifically overexpressed in ECs in mice, it induced embryonic lethality with head haemorrhage, cardiac and heartbeat defects, and head and yolk sac vasculature defects [133].

7.3.5.2 Cyr61

Cysteine-rich angiogenic inducer 61 (CYR61) or CCN family member 1 (CCN1) is a secreted matricellular protein from the same CCN family as CTGF/CCN2 (see above). This ECM-associated signalling protein regulates many cellular activities including cell adhesion, migration and angiogenesis [111, 136, 137]. Cyr61-deficient mice have been generated but are unable to survive more than 24 h after birth due to severe vascular defects [138, 139].

7.3.6 Transporters

7.3.6.1 SLC2A3/GLUT3

Glucose transporter 3 (GLUT3) or solute carrier family 2, facilitated glucose transporter member 3 (SLC2A3) is a protein facilitating the transport of glucose across the plasma membranes that can be induced by hypoxia, a major angiogenesis activator [140]. It also has been associated with tumour blood formation in epithelial ovarian carcinoma [141]. Supporting its implication in angiogenesis, HUVEC treated with a SLC2A3 siRNA showed a 27 % reduction in tube formation compared with control siRNA [142].

7.3.7 Cytoskeletal Proteins

Verge (vascular early response gene) or APOLD1(apolipoprotein L domain containing 1) is a protein playing a role in cytoskeleton reorganisation and vascular permeability [143]. It is one of the most recent IEGs that has been implicated in vascular function and is induced in cultured ECs by TNFα, FGF-2 and hypoxia and developmentally regulated in the rat heart [143]. In addition to regulate angiogenesis, a role in barrier function was demonstrated with an adaptation to hypertonicity [144]. Verge is rapidly induced after focal stroke and studies in deficient mice revealed that this protein affects brain vasculature and blood–brain permeability [145]. Moreover, after brain ischemia there is decreased oedema formation but no change in stroke outcomes in the deficient mice [145].

7.3.8 Anti-apoptotic Proteins

Induced myeloid leukaemia cell differentiation protein Mcl-1 is a protein belonging to the Bcl-2 (B-cell lymphoma 2) family and different isoforms have apparently different activities. Indeed, isoform 1 inhibits apoptosis, whereas, isoforms 2/3 induce apoptosis. Morever, 3,6-di(2,3-epoxypropoxy)xanthone (EPOX) is a drug that inhibits angiogenesis by ERK-dependent degradation of Mcl-1 suggesting that novel Mcl-1 targeting drugs may be used to inhibit angiogenesis [146].

Concluding Remarks

Altogether, a range of strategies has been used with success to dissect the roles of IEGs in angiogenesis even though some of them are transcription factors and are usually considered difficult to target [147, 148]. These targeting approaches include dominant-negative mutants, small molecule inhibitors, decoys, siRNA, DNAzymes, peptides/antibodies and natural products. The mechanisms involved in angiogenesis including degradation of the extracellular matrix, ECs migration and proliferation, tube formation and vessel maturation are complex and depend on tissue type. By impacting directly and rapidly on endothelial growth and angiogenesis, IEGs represent an important class of molecular targets. Nevertheless, it remains to be seen whether IEGs targeting approaches will serve as useful tools in pathologic angiogenesis and related human diseases.

References

1. Sweeney C et al (2001) Growth factor-specific signaling pathway stimulation and gene expression mediated by ErbB receptors. J Biol Chem 276(25):22685–22698
2. Khachigian LM, Collins T (1997) Inducible expression of Egr-1-dependent genes. A paradigm of transcriptional activation in vascular endothelium. Circ Res 81(4):457–461
3. Lau LF, Nathans D (1987) Expression of a set of growth-related immediate early genes in BALB/c 3T3 cells: coordinate regulation with c-fos or c-myc. Proc Natl Acad Sci USA 84 (5):1182–1186
4. Healy S, Khan P, Davie JR (2013) Immediate early response genes and cell transformation. Pharmacol Ther 137(1):64–77
5. Tullai JW et al (2007) Immediate-early and delayed primary response genes are distinct in function and genomic architecture. J Biol Chem 282(33):23981–23995
6. Herschman HR (1991) Primary response genes induced by growth factors and tumor promoters. Annu Rev Biochem 60:281–319
7. O'Donnell A, Odrowaz Z, Sharrocks AD (2012) Immediate-early gene activation by the MAPK pathways: what do and don't we know? Biochem Soc Trans 40(1):58–66
8. McKay MM, Morrison DK (2007) Integrating signals from RTKs to ERK/MAPK. Oncogene 26(22):3113–3121
9. Dunn KL et al (2005) The Ras-MAPK signal transduction pathway, cancer and chromatin remodeling. Biochem Cell Biol 83(1):1–14
10. Yang SH, Sharrocks AD, Whitmarsh AJ (2003) Transcriptional regulation by the MAP kinase signaling cascades. Gene 320:3–21
11. Carmeliet P (2005) Angiogenesis in life, disease and medicine. Nature 438(7070):932–936
12. Carmeliet P, Jain RK (2000) Angiogenesis in cancer and other diseases. Nature 407 (6801):249–257
13. Hanahan D, Folkman J (1996) Patterns and emerging mechanisms of the angiogenic switch during tumorigenesis. Cell 86(3):353–364
14. Ferrara N, Kerbel RS (2005) Angiogenesis as a therapeutic target. Nature 438(7070):967–974
15. O'Donovan KJ et al (1999) The EGR family of transcription-regulatory factors: progress at the interface of molecular and systems neuroscience. Trends Neurosci 22(4):167–173
16. Madden SL, Rauscher FJ 3rd (1993) Positive and negative regulation of transcription and cell growth mediated by the EGR family of zinc-finger gene products. Ann N Y Acad Sci 684:75–84
17. Gashler A, Sukhatme VP (1995) Early growth response protein 1 (Egr-1): prototype of a zinc-finger family of transcription factors. Prog Nucleic Acid Res Mol Biol 50:191–224
18. Cao XM et al (1990) Identification and characterization of the Egr-1 gene product, a DNA-binding zinc finger protein induced by differentiation and growth signals. Mol Cell Biol 10 (5):1931–1939
19. Khachigian LM et al (1997) Egr-1 is activated in endothelial cells exposed to fluid shear stress and interacts with a novel shear-stress-response element in the PDGF A-chain promoter. Arterioscler Thromb Vasc Biol 17(10):2280–2286
20. Lee SL et al (1996) Luteinizing hormone deficiency and female infertility in mice lacking the transcription factor NGFI-A (Egr-1). Science 273(5279):1219–1221
21. Silverman ES, Collins T (1999) Pathways of Egr-1-mediated gene transcription in vascular biology. Am J Pathol 154(3):665–670
22. Lucerna M et al (2003) NAB2, a corepressor of EGR-1, inhibits vascular endothelial growth factor-mediated gene induction and angiogenic responses of endothelial cells. J Biol Chem 278(13):11433–11440
23. Kundumani-Sridharan V et al (2010) 15(S)-hydroxyeicosatetraenoic acid-induced angiogenesis requires Src-mediated Egr-1-dependent rapid induction of FGF-2 expression. Blood 115 (10):2105–2116

24. Szabo IL et al (2001) NSAIDs inhibit the activation of egr-1 gene in microvascular endothelial cells. A key to inhibition of angiogenesis? J Physiol Paris 95(1–6):379–383
25. Worden B et al (2005) Hepatocyte growth factor/scatter factor differentially regulates expression of proangiogenic factors through Egr-1 in head and neck squamous cell carcinoma. Cancer Res 65(16):7071–7080
26. Abdel-Malak NA et al (2009) Early growth response-1 regulates angiopoietin-1-induced endothelial cell proliferation, migration, and differentiation. Arterioscler Thromb Vasc Biol 29(2):209–216
27. Fahmy RG et al (2003) Transcription factor Egr-1 supports FGF-dependent angiogenesis during neovascularization and tumor growth. Nat Med 9(8):1026–1032
28. Abe M, Sato Y (2001) cDNA microarray analysis of the gene expression profile of VEGF-activated human umbilical vein endothelial cells. Angiogenesis 4(4):289–298
29. Suehiro J et al (2010) Vascular endothelial growth factor activation of endothelial cells is mediated by early growth response-3. Blood 115(12):2520–2532
30. Patwardhan S et al (1991) EGR3, a novel member of the Egr family of genes encoding immediate-early transcription factors. Oncogene 6(6):917–928
31. Liu D et al (2008) The zinc-finger transcription factor, early growth response 3, mediates VEGF-induced angiogenesis. Oncogene 27(21):2989–2998
32. Moll UM, Marchenko N, Zhang XK (2006) p53 and Nur77/TR3 – transcription factors that directly target mitochondria for cell death induction. Oncogene 25(34):4725–4743
33. Liu D et al (2003) Vascular endothelial growth factor-regulated gene expression in endothelial cells: KDR-mediated induction of Egr3 and the related nuclear receptors Nur77, Nurr1, and Nor1. Arterioscler Thromb Vasc Biol 23(11):2002–2007
34. Ha CH et al (2008) Protein kinase D-dependent phosphorylation and nuclear export of histone deacetylase 5 mediates vascular endothelial growth factor-induced gene expression and angiogenesis. J Biol Chem 283(21):14590–14599
35. Zeng H et al (2006) Orphan nuclear receptor TR3/Nur77 regulates VEGF-A-induced angiogenesis through its transcriptional activity. J Exp Med 203(3):719–729
36. Arkenbout EK et al (2003) TR3 orphan receptor is expressed in vascular endothelial cells and mediates cell cycle arrest. Arterioscler Thromb Vasc Biol 23(9):1535–1540
37. Qin L et al (2013) The vascular permeabilizing factors histamine and serotonin induce angiogenesis through TR3/Nur77 and subsequently truncate it through thrombospondin-1. Blood 121(11):2154–2164
38. Zhao D et al (2011) Orphan nuclear transcription factor TR3/Nur77 regulates microvessel permeability by targeting endothelial nitric oxide synthase and destabilizing endothelial junctions. Proc Natl Acad Sci USA 108(29):12066–12071
39. Eferl R, Wagner EF (2003) AP-1: a double-edged sword in tumorigenesis. Nat Rev Cancer 3 (11):859–868
40. Shaulian E, Karin M (2001) AP-1 in cell proliferation and survival. Oncogene 20 (19):2390–2400
41. Shaulian E, Karin M (2002) AP-1 as a regulator of cell life and death. Nat Cell Biol 4(5): E131–E136
42. Angel P, Karin M (1991) The role of Jun, Fos and the AP-1 complex in cell-proliferation and transformation. Biochim Biophys Acta 1072(2–3):129–157
43. Vleugel MM et al (2006) c-Jun activation is associated with proliferation and angiogenesis in invasive breast cancer. Hum Pathol 37(6):668–674
44. Kraemer M et al (1999) Rat embryo fibroblasts transformed by c-Jun display highly metastatic and angiogenic activities *in vivo* and deregulate gene expression of both angiogenic and antiangiogenic factors. Cell Growth Differ 10(3):193–200
45. Michel JB et al (1994) Biphasic induction of immediate early gene expression accompanies activity-dependent angiogenesis and myofiber remodeling of rabbit skeletal muscle. J Clin Invest 94(1):277–285

46. Jiao X et al (2008) Disruption of c-Jun reduces cellular migration and invasion through inhibition of c-Src and hyperactivation of ROCK II kinase. Mol Biol Cell 19(4):1378–1390
47. Katiyar S et al (2007) Somatic excision demonstrates that c-Jun induces cellular migration and invasion through induction of stem cell factor. Mol Cell Biol 27(4):1356–1369
48. Giles N et al (2008) A peptide inhibitor of c-Jun promotes wound healing in a mouse full-thickness burn model. Wound Repair Regen 16(1):58–64
49. Zhang G et al (2004) Effect of deoxyribozymes targeting c-Jun on solid tumor growth and angiogenesis in rodents. J Natl Cancer Inst 96(9):683–696
50. Fahmy RG et al (2006) Suppression of vascular permeability and inflammation by targeting of the transcription factor c-Jun. Nat Biotechnol 24(7):856–863
51. Cai H et al (2012) DNAzyme targeting c-jun suppresses skin cancer growth. Sci Transl Med 4 (139):139ra82
52. Marconcini L et al (1999) c-Fos-induced growth factor/vascular endothelial growth factor D induces angiogenesis *in vivo* and *in vitro*. Proc Natl Acad Sci USA 96(17):9671–9676
53. Saez E et al (1995) c-Fos is required for malignant progression of skin tumors. Cell 82 (5):721–732
54. Lai HC et al (2004) Effect of EGCG, a major component of green tea, on the expression of Ets-1, c-Fos, and c-Jun during angiogenesis *in vitro*. Cancer Lett 213(2):181–188
55. Belguise K et al (2005) FRA-1 expression level regulates proliferation and invasiveness of breast cancer cells. Oncogene 24(8):1434–1444
56. Kustikova O et al (1998) Fra-1 induces morphological transformation and increases *in vitro* invasiveness and motility of epithelioid adenocarcinoma cells. Mol Cell Biol 18 (12):7095–7105
57. Hilfiker-Kleiner D et al (2005) Lack of JunD promotes pressure overload-induced apoptosis, hypertrophic growth, and angiogenesis in the heart. Circulation 112(10):1470–1477
58. Schmidt D et al (2007) Critical role for NF-kappaB-induced JunB in VEGF regulation and tumor angiogenesis. EMBO J 26(3):710–719
59. Brown PH, Chen TK, Birrer MJ (1994) Mechanism of action of a dominant-negative mutant of c-Jun. Oncogene 9(3):791–799
60. Kang MI et al (2012) Targeting of noncanonical Wnt5a signaling by AP-1 blocker dominant-negative Jun when it inhibits skin carcinogenesis. Genes Cancer 3(1):37–50
61. Andersen H et al (2005) Immediate and delayed effects of E-cadherin inhibition on gene regulation and cell motility in human epidermoid carcinoma cells. Mol Cell Biol 25 (20):9138–9150
62. Shiratsuchi T, Ishibashi H, Shirasuna K (2002) Inhibition of epidermal growth factor-induced invasion by dexamethasone and AP-1 decoy in human squamous cell carcinoma cell lines. J Cell Physiol 193(3):340–348
63. Weber WM et al (2006) TPA-induced up-regulation of activator protein-1 can be inhibited or enhanced by analogs of the natural product curcumin. Biochem Pharmacol 72(8):928–940
64. Matsuo M et al (2007) Curcumin inhibits the formation of capillary-like tubes by rat lymphatic endothelial cells. Cancer Lett 251(2):288–295
65. Tsuchida K et al (2004) Design, synthesis, and biological evaluation of new cyclic disulfide decapeptides that inhibit the binding of AP-1 to DNA. J Med Chem 47(17):4239–4246
66. Tsuchida K et al (2006) Discovery of nonpeptidic small-molecule AP-1 inhibitors: lead hopping based on a three-dimensional pharmacophore model. J Med Chem 49(1):80–91
67. Ruocco KM et al (2007) A high-throughput cell-based assay to identify specific inhibitors of transcription factor AP-1. J Biomol Screen 12(1):133–139
68. Aikawa Y et al (2008) Treatment of arthritis with a selective inhibitor of c-Fos/activator protein-1. Nat Biotechnol 26(7):817–823
69. Hai T et al (1999) ATF3 and stress responses. Gene Expr 7(4–6):321–335
70. Volpert OV et al (2002) Id1 regulates angiogenesis through transcriptional repression of thrombospondin-1. Cancer Cell 2(6):473–483

71. Nawa T et al (2002) Expression of transcriptional repressor ATF3/LRF1 in human atherosclerosis: colocalization and possible involvement in cell death of vascular endothelial cells. Atherosclerosis 161(2):281–291
72. Okamoto A, Iwamoto Y, Maru Y (2006) Oxidative stress-responsive transcription factor ATF3 potentially mediates diabetic angiopathy. Mol Cell Biol 26(3):1087–1097
73. Nesbit CE, Tersak JM, Prochownik EV (1999) MYC oncogenes and human neoplastic disease. Oncogene 18(19):3004–3016
74. Shanmugham R et al (2004) Tumour angiogenesis and C-myc expression in breast carcinomas. Indian J Pathol Microbiol 47(3):340–342
75. Fodinger M et al (2000) Erythropoietin-inducible immediate-early genes in human vascular endothelial cells. J Investig Med 48(2):137–149
76. Baudino TA et al (2002) c-Myc is essential for vasculogenesis and angiogenesis during development and tumor progression. Genes Dev 16(19):2530–2543
77. Souders CA et al (2012) c-Myc is required for proper coronary vascular formation via cell- and gene-specific signaling. Arterioscler Thromb Vasc Biol 32(5):1308–1319
78. Ngo CV et al (2000) An *in vivo* function for the transforming Myc protein: elicitation of the angiogenic phenotype. Cell Growth Differ 11(4):201–210
79. Janz A et al (2000) Activation of the myc oncoprotein leads to increased turnover of thrombospondin-1 mRNA. Nucleic Acids Res 28(11):2268–2275
80. Dews M et al (2006) Augmentation of tumor angiogenesis by a Myc-activated microRNA cluster. Nat Genet 38(9):1060–1065
81. Lelievre E et al (2001) The Ets family contains transcriptional activators and repressors involved in angiogenesis. Int J Biochem Cell Biol 33(4):391–407
82. Hashiya N et al (2004) *In vivo* evidence of angiogenesis induced by transcription factor Ets-1: Ets-1 is located upstream of angiogenesis cascade. Circulation 109(24):3035–3041
83. Sato Y (1998) Transcription factor ETS-1 as a molecular target for angiogenesis inhibition. Hum Cell 11(4):207–214
84. Iwasaka C et al (1996) Ets-1 regulates angiogenesis by inducing the expression of urokinase-type plasminogen activator and matrix metalloproteinase-1 and the migration of vascular endothelial cells. J Cell Physiol 169(3):522–531
85. Oettgen P (2010) The role of ets factors in tumor angiogenesis. J Oncol 2010:767384
86. Wei G et al (2009) Ets1 and Ets2 are required for endothelial cell survival during embryonic angiogenesis. Blood 114(5):1123–1130
87. Pesce S, Benezra R (1993) The loop region of the helix-loop-helix protein Id1 is critical for its dominant-negative activity. Mol Cell Biol 13(12):7874–7880
88. Lyden D et al (1999) Id1 and Id3 are required for neurogenesis, angiogenesis and vascularization of tumour xenografts. Nature 401(6754):670–677
89. Ling MT et al (2005) Overexpression of Id-1 in prostate cancer cells promotes angiogenesis through the activation of vascular endothelial growth factor (VEGF). Carcinogenesis 26 (10):1668–1676
90. Mellick AS et al (2010) Using the transcription factor inhibitor of DNA binding 1 to selectively target endothelial progenitor cells offers novel strategies to inhibit tumor angiogenesis and growth. Cancer Res 70(18):7273–7282
91. Nurrish SJ, Treisman R (1995) DNA-binding specificity determinants in mads-box transcription factors. Mol Cell Biol 15(8):4076–4085
92. Vialou V et al (2012) Serum response factor and camp response element binding protein are both required for cocaine induction of delta FosB. J Neurosci 32(22):7577–7584
93. Lee SM, Vasishtha M, Prywes R (2010) Activation and repression of cellular immediate early genes by serum response factor cofactors. J Biol Chem 285(29):22036–22049
94. Chai J, Jones MK, Tarnawski AS (2004) Serum response factor is a critical requirement for VEGF signaling in endothelial cells and VEGF-induced angiogenesis. FASEB J 18 (11):1264–1266

95. Franco CA et al (2008) Serum response factor is required for sprouting angiogenesis and vascular integrity. Dev Cell 15(3):448–461
96. Schratt G et al (2001) Serum response factor is required for immediate-early gene activation yet is dispensable for proliferation of embryonic stem cells. Mol Cell Biol 21(8):2933–2943
97. Qiao Y et al (2011) MiR-483-5p controls angiogenesis *in vitro* and targets serum response factor. FEBS Lett 585(19):3095–3100
98. Chang LH et al (2012) Role of macrophage CCAAT/enhancer binding protein delta in the pathogenesis of rheumatoid arthritis in collagen-induced arthritic mice. PLoS One 7(9): e45378
99. Min Y et al (2011) C/EBP-delta regulates VEGF-C autocrine signaling in lymphangiogenesis and metastasis of lung cancer through HIF-1alpha. Oncogene 30(49):4901–4909
100. Gabay C (2006) Interleukin-6 and chronic inflammation. Arthritis Res Ther 8:S3
101. Li W et al (2001) Global changes in interleukin-6-dependent gene expression patterns in mouse livers after partial hepatectomy. Hepatology 33(6):1377–1386
102. Cohen T et al (1996) Interleukin 6 induces the expression of vascular endothelial growth factor. J Biol Chem 271(2):736–741
103. Huang SP et al (2004) Interleukin-6 increases vascular endothelial growth factor and angiogenesis in gastric carcinoma. J Biomed Sci 11(4):517–527
104. Koch AE et al (1992) Interleukin-8 as a macrophage-derived mediator of angiogenesis. Science 258(5089):1798–1801
105. Bancroft CC et al (2001) Coexpression of proangiogenic factors IL-8 and VEGF by human head and neck squamous cell carcinoma involves coactivation by MEK-MAPK and IKK-NF-kappaB signal pathways. Clin Cancer Res 7(2):435–442
106. Huang S et al (2002) Fully humanized neutralizing antibodies to interleukin-8 (ABX-IL8) inhibit angiogenesis, tumor growth, and metastasis of human melanoma. Am J Pathol 161 (1):125–134
107. Florczyk U et al (2011) Opposite effects of HIF-1alpha and HIF-2alpha on the regulation of IL-8 expression in endothelial cells. Free Radic Biol Med 51(10):1882–1892
108. Nishida T et al (2007) CCN2 (connective tissue growth factor) is essential for extracellular matrix production and integrin signaling in chondrocytes. J Cell Commun Signal 1(1):45–58
109. Hall-Glenn F et al (2012) CCN2/connective tissue growth factor is essential for pericyte adhesion and endothelial basement membrane formation during angiogenesis. PLoS One 7 (2):e30562
110. Markiewicz M et al (2011) Connective tissue growth factor (CTGF/CCN2) mediates angiogenic effect of S1P in human dermal microvascular endothelial cells. Microcirculation 18 (1):1–11
111. Brigstock DR (2002) Regulation of angiogenesis and endothelial cell function by connective tissue growth factor (CTGF) and cysteine-rich 61 (CYR61). Angiogenesis 5(3):153–165
112. Suzuma K et al (2000) Vascular endothelial growth factor induces expression of connective tissue growth factor via KDR, Flt1, and phosphatidylinositol 3-kinase-akt-dependent pathways in retinal vascular cells. J Biol Chem 275(52):40725–40731
113. Inoki I et al (2002) Connective tissue growth factor binds vascular endothelial growth factor (VEGF) and inhibits VEGF-induced angiogenesis. FASEB J 16(2):219–221
114. Hashimoto G et al (2002) Matrix metalloproteinases cleave connective tissue growth factor and reactivate angiogenic activity of vascular endothelial growth factor 165. J Biol Chem 277 (39):36288–36295
115. Kondo S et al (2002) Connective tissue growth factor increased by hypoxia may initiate angiogenesis in collaboration with matrix metalloproteinases. Carcinogenesis 23(5):769–776
116. Aikawa T et al (2006) Connective tissue growth factor-specific antibody attenuates tumor growth, metastasis, and angiogenesis in an orthotopic mouse model of pancreatic cancer. Mol Cancer Ther 5(5):1108–1116
117. Nantel F et al (1999) Distribution and regulation of cyclooxygenase-2 in carrageenan-induced inflammation. Br J Pharmacol 128(4):853–859

118. Tsujii M et al (1998) Cyclooxygenase regulates angiogenesis induced by colon cancer cells. Cell 93(5):705–716
119. Gately S (2000) The contributions of cyclooxygenase-2 to tumor angiogenesis. Cancer Metastasis Rev 19(1–2):19–27
120. Iniguez MA et al (2003) Cyclooxygenase-2: a therapeutic target in angiogenesis. Trends Mol Med 9(2):73–78
121. Taylor DM et al (2013) MAP kinase phosphatase 1 (MKP-1/DUSP1) is neuroprotective in Huntington's disease via additive effects of JNK and p38 inhibition. J Neurosci 33 (6):2313–2325
122. Denkert C et al (2002) Expression of mitogen-activated protein kinase phosphatase-1 (MKP-1) in primary human ovarian carcinoma. Int J Cancer 102(5):507–513
123. Kinney CM et al (2008) VEGF and thrombin induce MKP-1 through distinct signaling pathways: role for MKP-1 in endothelial cell migration. Am J Physiol Cell Physiol 294(1): C241–C250
124. Moncho-Amor V et al (2011) DUSP1/MKP1 promotes angiogenesis, invasion and metastasis in non-small-cell lung cancer. Oncogene 30(6):668–678
125. Song HY, Rothe M, Goeddel DV (1996) The tumor necrosis factor-inducible zinc finger protein A20 interacts with TRAF1/TRAF2 and inhibits NF-kappaB activation. Proc Natl Acad Sci USA 93(13):6721–6725
126. Opipari AW, Boguski MS, Dixit VM (1990) The A20 cDNA induced by tumor necrosis factor-alpha encodes a novel type of zinc finger protein. J Biol Chem 265(25):14705–14708
127. Elsby LM et al (2010) Functional evaluation of TNFAIP3 (A20) in rheumatoid arthritis. Clin Exp Rheumatol 28(5):708–714
128. Chng HW et al (2006) A new role for the anti-apoptotic gene A20 in angiogenesis. Exp Cell Res 312(15):2897–2907
129. Daniel S et al (2004) A20 protects endothelial cells from TNF-, Fas-, and NK-mediated cell death by inhibiting caspase 8 activation. Blood 104(8):2376–2384
130. Balsara RD, Castellino FJ, Ploplis VA (2006) A novel function of plasminogen activator inhibitor-1 in modulation of the AKT pathway in wild-type and plasminogen activator inhibitor-1-deficient endothelial cells. J Biol Chem 281(32):22527–22536
131. Lakka SS et al (2005) Specific interference of urokinase-type plasminogen activator receptor and matrix metalloproteinase-9 gene expression induced by double-stranded RNA results in decreased invasion, tumor growth, and angiogenesis in gliomas. J Biol Chem 280 (23):21882–21892
132. Soncin F et al (2003) VE-statin, an endothelial repressor of smooth muscle cell migration. EMBO J 22(21):5700–5711
133. Nichol D et al (2010) Impaired angiogenesis and altered Notch signaling in mice overexpressing endothelial Egfl7. Blood 116(26):6133–6143
134. Campagnolo L et al (2005) EGFL7 is a chemoattractant for endothelial cells and is up-regulated in angiogenesis and arterial injury. Am J Pathol 167(1):275–284
135. Parker LH et al (2004) The endothelial-cell-derived secreted factor Egfl7 regulates vascular tube formation. Nature 428(6984):754–758
136. Babic AM et al (1998) CYR61, a product of a growth factor-inducible immediate early gene, promotes angiogenesis and tumor growth. Proc Natl Acad Sci USA 95(11):6355–6360
137. Fataccioli V et al (2002) Stimulation of angiogenesis by Cyr61 gene: a new therapeutic candidate. Hum Gene Ther 13(12):1461–1470
138. Mo FE et al (2002) CYR61 (CCN1) is essential for placental development and vascular integrity. Mol Cell Biol 22(24):8709–8720
139. Mo FE, Lau LF (2006) The matricellular protein CCN1 is essential for cardiac development. Circ Res 99(9):961–969
140. Airley RE, Mobasheri A (2007) Hypoxic regulation of glucose transport, anaerobic metabolism and angiogenesis in cancer: novel pathways and targets for anticancer therapeutics. Chemotherapy 53(4):233–256

141. Tsukioka M et al (2007) Expression of glucose transporters in epithelial ovarian carcinoma: correlation with clinical characteristics and tumor angiogenesis. Oncol Rep 18(2):361–367
142. Mimura I et al (2012) Dynamic change of chromatin conformation in response to hypoxia enhances the expression of GLUT3 (SLC2A3) by cooperative interaction of hypoxia-inducible factor 1 and KDM3A. Mol Cell Biol 32(15):3018–3032
143. Regard JB et al (2004) Verge: a novel vascular early response gene. J Neurosci 24 (16):4092–4103
144. Maallem S et al (2008) Gene expression profiling in brain following acute systemic hypertonicity: novel genes possibly involved in osmoadaptation. J Neurochem 105(4):1198–1211
145. Liu F et al (2012) Loss of vascular early response gene reduces edema formation after experimental stroke. Exp Transl Stroke Med 4(1):12
146. Sun HL et al (2009) EPOX inhibits angiogenesis by degradation of Mcl-1 through ERK inactivation. Clin Cancer Res 15(15):4904–4914
147. Abbott A (2002) On the offensive. Nature 416(6880):470–474
148. Gibbs JB (2000) Mechanism-based target identification and drug discovery in cancer research. Science 287(5460):1969–1973

Molecular Mechanisms of Hypoxia-Regulated Angiogenesis

8

Agnieszka Łoboda, Alicja Józkowicz, and Józef Dulak

Contents

8.1 Introduction 164
8.2 Stabilization of the Hypoxia Inducible Factor 165
8.2.1 The Hydroxylation of HIF-α as a Major Mechanism for Its Destabilization ... 165
8.2.2 Not Only Hydroxylation: ROS-Mediated HIFs Stabilization 168
8.2.3 Not Only Hydroxylation: The Role of (De)acetylation in HIFs (De)stabilization 169
8.3 The Cross Talk Between HIFs and NFκB: The Implications for Inflammatory Angiogenesis 170
8.4 Nitric Oxide: Important Regulator of Hypoxia-Induced Angiogenesis 171
8.5 microRNA and Hypoxia-Induced Angiogenesis 173
8.6 Hypoxia and Angiogenic Regulators 173
8.6.1 Regulation of VEGF by HIFs: A Classic Example of Pro-angiogenic Gene Regulation in Hypoxic Conditions 173
8.6.2 Complex Effect of Hypoxia on Angiogenesis Stimulators: Not Only Induction Is Observed 175
8.6.3 Hypoxia-Regulated Angiogenic Inhibitors 177
References 178

Abstract

Hypoxia is an important regulator of angiogenesis and it stimulates neovascularization from existing blood vessels. At the molecular level, it occurs mostly through transcriptional regulation of genes which contain a core consensus sequence called hypoxia response element (HRE) via hypoxia inducible factors (HIFs) action. The discovery of HIFs hydroxylases as a family of dioxygenases that regulate HIFs dependently on the oxygen availability have significantly

A. Łoboda (✉) • A. Józkowicz • J. Dulak
Department of Medical Biotechnology, Faculty of Biochemistry, Biophysics and Biotechnology, Jagiellonian University, Gronostajowa 7, 30-387 Kraków, Poland
e-mail: agnieszka.loboda@uj.edu.pl

J. Dulak et al. (eds.), *Angiogenesis and Vascularisation*,
DOI 10.1007/978-3-7091-1428-5_8, © Springer-Verlag Wien 2013

improved our understanding of the mechanisms of hypoxia signaling. Moreover, a broad number of factors have been shown to influence HIF stability and their effects could be mediated via several possible mechanisms including nitrosylation, deacetylation, or oxidation. The induction of HIF leads to the complex regulation of the pro- and antiangiogenic factors and extensive research will be essential for thorough understanding of the role of hypoxia in disease development and will help to identify new therapeutic targets for treatment of such hypoxia-dependent disorders.

Keywords
Hypoxia inducible factor • HIF-1 • HIF-2 • HRE • MicroRNA • Oxidative stress • Vascular endothelial growth factor • IL-8

8.1 Introduction

Oxygen is vital for living cells and it plays a fundamental role in their metabolism. An inadequate oxygen supply to tissues and cells causes hypoxia, a state which can restrict their functions. The ability to maintain oxygen homeostasis is essential to the survival of all vertebrate species. On the other hand, several pathological conditions like stroke (cerebral ischemia) and heart infarction (myocardial ischemia) may result from the detrimental effect of hypoxia. Moreover, hypoxia has been recognized as one of the fundamentally important features of tumor growth and metastasis.

The hypoxic response is primarily mediated by the family of hypoxia inducible transcription factors (HIFs). Their activation leads to transcription of HIFs target genes—nowadays more than 150 HIFs-dependent genes controlling different processes including energy metabolism, cell growth and apoptosis, or vasomotor control was identified (Fig. 8.1). Moreover, to counteract the undesirable effects of low oxygen concentration, the new blood vessels providing oxygen to the ischemic tissues are formed. The expression of pro-angiogenic factors like vascular endothelial growth factor (VEGF) or interleukin-8 (IL-8) was thought to be strongly increased in hypoxic conditions; however, recent studies underline the complexity of angiogenic response after HIFs induction. The complication of cellular response to low oxygen concentration may be also related to context specific effects of individual HIF isoforms. Latest research underlined that HIF-1 and HIF-2 possess both overlapping and unique target genes and may also trigger specific roles: HIF-1 predominantly drives the initial response to hypoxia (<24 h), whereas HIF-2 is responsible for chronic response (>24 h).

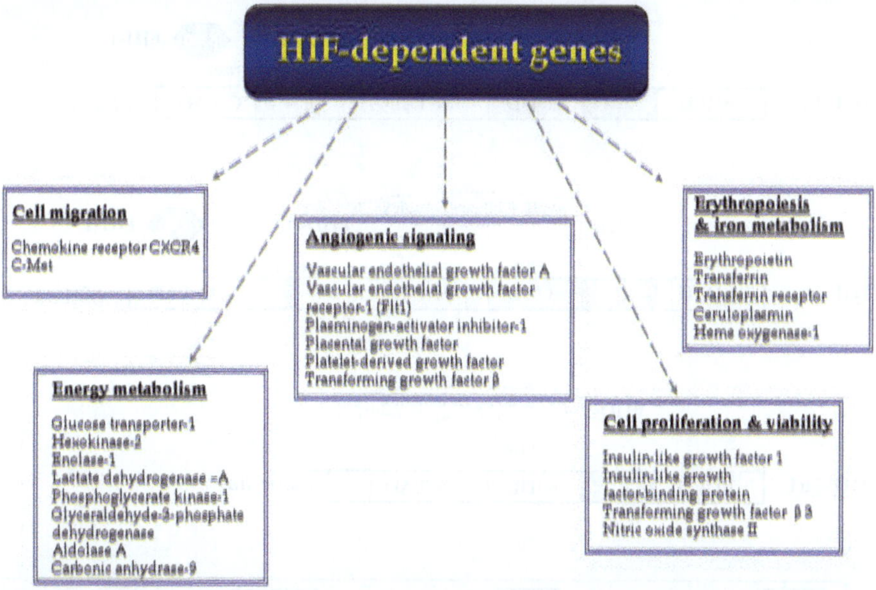

Fig. 8.1 Selected HIF-dependent genes. HIFs transactivate genes involved in a number of adaptive processes in response to hypoxia including angiogenesis, energy metabolism, cell migration, proliferation, etc.

8.2 Stabilization of the Hypoxia Inducible Factor

8.2.1 The Hydroxylation of HIF-α as a Major Mechanism for Its Destabilization

The discovery of the hypoxia response element (HRE), the oxygen-regulated sequence, in the erythropoietin gene and the identification of the hypoxia inducible factor-1 (HIF-1) as a master transcription factor binding this sequence [1, 2] had started the intensive research in hypoxia field. Further studies performed in other than erythropoietin-producing cells have revealed that the same regulatory element is involved in the hypoxic regulation of gene expression indicating the presence of the general oxygen-sensing system in mammalian cells.

It is now well established that the regulatory sequence, called HRE, comprises of consensus-binding site (HBS, HIF-1-binding site) with a core motif (RCGTG) [3] and HIF-1 ancillary sequence (HAS), located eight to nine nt down- or upstream of HBS, which facilitates HIF-1-mediated transcription activation [4]. The existence of HRE was confirmed in hundreds of genes regulated in hypoxic conditions.

HRE is recognized by HIFs transcription factors, a heterodimers consisting of oxygen-labile α (HIF-α) and oxygen-independent β (HIF-β, also called aryl hydrocarbon receptor nuclear translocator, ARNT) subunits. Both α and β subunits are

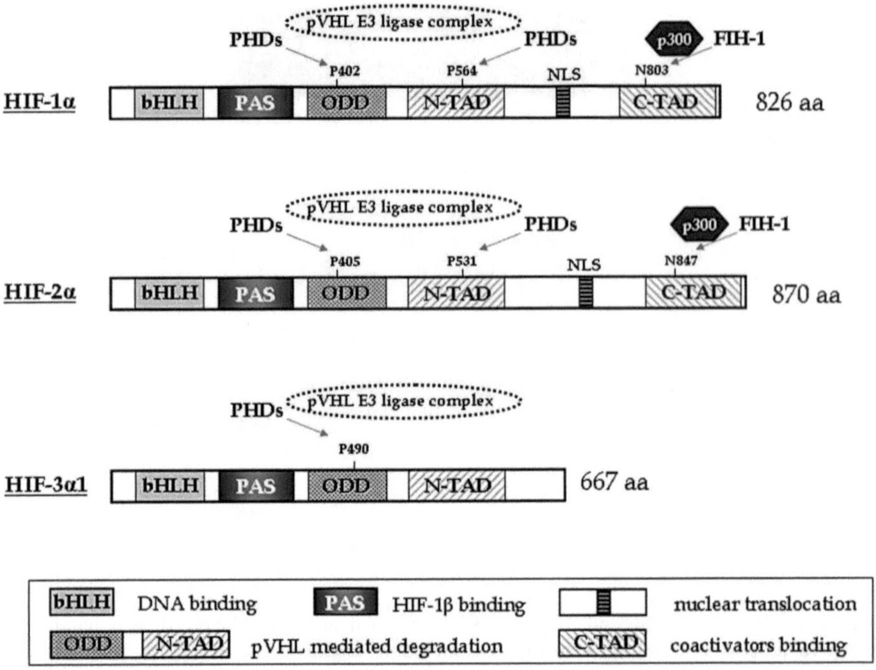

Fig. 8.2 The schematic structure of HIF-α isoforms. HIF-1α is closely related to HIF-2α, whereas the variant of HIF-3α, HIF-3α1 lacks C-terminal transactivation domain (C-TAD), which is required for transcriptional activation. However all α subunits are prone to oxygen-dependent degradation by the von Hippel–Lindau protein (pVHL) E3 ligase complex. The modifications of specific residues are highlighted above each protein, and the proteins that perform those modifications are shown (PHDs, FIH-1)

members of the bHLH/PAS (Basic Helix–Loop–Helix/PER-ARNT-SIM) domain family [5]. Three genes encoding distinct HIF-α isoforms exist in humans: *HIF1A*, encoding HIF-1α; *EPAS1*, encoding HIF-2α; and *HIF3A*, which is expressed as multiple HIF-3α splice variants [5–7]. Mainly HIF-1α and HIF-2α mediate the hypoxia-dependent signaling, whereas HIF-3α was shown to inhibit transcriptional activity of HIF-1α [8]. Such difference in isoform-specific effects may be the result of their structure: HIF-3α has high similarity to HIF-1α and HIF-2α in the bHLH and PAS domains but it lacks the C-terminal transactivation domain (C-TAD), which is required for transcriptional activation (Fig. 8.2).

Only HIF-β subunit is constitutively expressed and not regulated by oxygen concentration. On the other hand, the expression of HIF-α subunits is strictly controlled at the protein level. This occurs, predominantly, through the hydroxylation of specific proline residues (Pro402 and Pro564 in HIF-1, Pro405 and Pro531 in HIF-2, Pro490 in HIF-3) within oxygen-dependent degradation domain (ODDD) present in α subunits [7, 9, 10]. The hydroxylation is performed by a class of

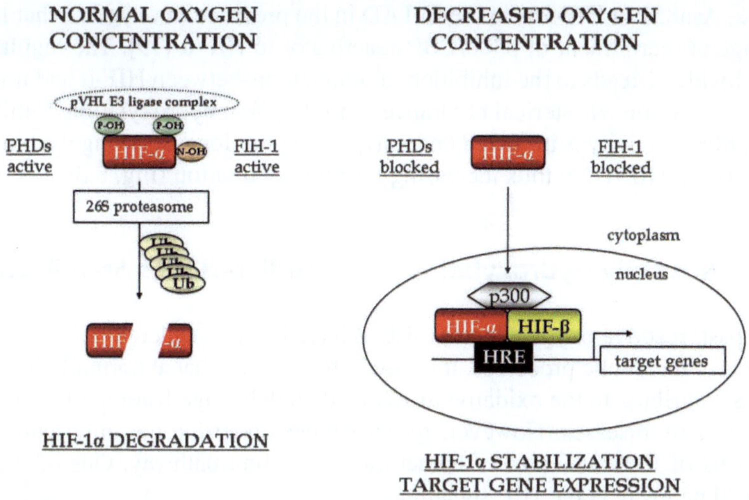

Fig. 8.3 HIFs stabilization during hypoxia. Under normoxia, HIF-α is subjected to oxygen-dependent prolyl hydroxylation by PHDs leading to pVHL E3 ligase complex binding and ubiquitination and degradation by the 26S proteasome. Moreover, asparagine hydroxylation by FIH-1 inhibits binding of the HIFα coactivator p300/CBP. In hypoxic conditions, both PHDs and FIH-1 are not active and non-hydroxylated HIF-1α associates with HIFβ in nucleus. HIFα/HIFβ heterodimer binds to HRE sequence at the promoters of HIF-responsive genes and upon binding to the coactivators p300/CBP initiates their transcription

enzymes called prolyl hydroxylases (PHDs). They require molecular O_2, 2-oxoglutarate, iron ions (Fe^{2+}), and ascorbic acid to be fully active, and they are inhibited in hypoxic conditions [11]. From three PHDs identified so far and shown to hydroxylate HIF-1α in vitro (PHD1, 2, and 3) [12], in vivo PHD2 isoform plays a major role in normoxic HIF-1α regulation [13, 14]. Hydroxylated HIF-α is recognized by the β-domain of von Hippel–Lindau tumor suppressor protein (pVHL) and is targeted for degradation by the 26S proteasome [15–17]. This process is preceded by the ubiquitination of the pVHLα domain by the elongin-C/elongin-B/cullin-2 E3-ubiquitin-ligase complex. The requirement of oxygen for PHDs functioning suggests that under hypoxic condition their activity is abrogated. In fact, when the oxygen level drops down, HIF-α subunit escapes recognition by the pVHL ubiquitin–ligase complex and proteasomal degradation and could be transported to the nucleus. There, after dimerization with HIF-1β [18], the active complex binds to HRE to induce transcription of numerous hypoxia-responsive genes (Fig. 8.3).

Importantly, there are also other mechanisms responsible for the controlling of the HIF-mediated transcription. To form an active HIF-α/β complex able to transactivate numerous genes, the binding of specific coactivators, like p300/CREB-binding protein (CBP) is necessary. Other than PHDs, oxygen-regulated hydroxylase-domain protein termed factor-inhibiting HIF-1 (FIH-1) regulates the binding of p300/CBP. FIH-1 modifies asparagine residue (Asn803 in human

HIF-1α, Asn851 in HIF-2α) in the C-TAD in the presence of oxygen, what leads to blockage of recruitment of p300/CBP coactivator to HIF-α [19]. The regulation of HIF-α by FIH-1 leads to the inhibition of interactions between HIF-α and transcriptional coactivators via sterical hindrance caused by Asn hydroxylation. Similarly to PHDs, FIH-1 activity is inhibited under hypoxic conditions, allowing the binding of p300/CBP to HIF-1/2α, thus increasing HIF transactivation (Fig. 8.3).

8.2.2 Not Only Hydroxylation: ROS-Mediated HIFs Stabilization

In the past, reactive oxygen species (ROS) have been considered only as toxic by-products of metabolic processes. It is also believed now that abnormally high levels of ROS contribute to the oxidative stress-mediated damage leading to the development of many diseases. However, recent studies underline the importance of the low levels of ROS as a part of homeostatic signaling pathway. One of the ROS-regulated pathway is the HIF signaling.

A numerous studies have presented so-called ROS hypothesis and underlined the significance of ROS for HIF-1*α* stabilization [20–22]. It was proposed that hypoxia-mediated production of superoxide at complex III of the mitochondrial electron transport chain and its conversion to H_2O_2 by the superoxide dismutase (SOD) directly inhibits PHDs enzymes by oxidizing the essential nonheme-bound iron. However, recently Masson and colleagues have shown that PHDs have low sensitivity to inhibition by H_2O_2. In contrast, the other HIF-hydroxylated enzyme, FIH-1, was much more susceptible to H_2O_2-dependent inactivation [23]. These results suggest that molecular mechanisms responsible for the interaction between ROS and HIFs still have to be investigated.

On the other hand, there are other observations linking ROS and PHDs/HIF. Sirtuins are NAD-dependent deacetylases involved in metabolism, stress response, and longevity [24]. It was recently reported that the member of sirtuins family, SIRT3, destabilizes HIF-1α by inhibiting ROS production and keeping PHDs in the active state, thereby leading to HIF-1α degradation. In SIRT3 null cells, ROS level is increased and it contributed to increased HIF-1α stabilization/activity and finally to increased glycolysis and cellular growth [25]. Additionally, Bell et al. [26] proved that SIRT3 modulates the progression of tumors in ROS- and HIF-dependent way. The absence of SIRT3 led to increased ROS level and to activation of pro-tumorigenic HIF-1α in normoxia and its hyperactivation in hypoxia. After injection of HCT116 colorectal carcinoma cells with stable knockdown of SIRT3 to immunodeficient Nu/Nu mice, the growing tumors were bigger than tumors derived from control HCT116 cells. Additionally, when injected mice were subjected to potent antioxidant, N-acetylcysteine (NAC), the rate of tumor growth and its size was comparable to control conditions [26]. Moreover, there are other studies suggesting the effect of SIRT3 on ROS production. SIRT3 can directly target isocitrate dehydrogenase 2 (IDH2), a major source of NADPH in the mitochondria [27] and SOD2, activating mitochondrial ROS scavenging [28] and consequently influence cellular redox status and affect HIF signaling.

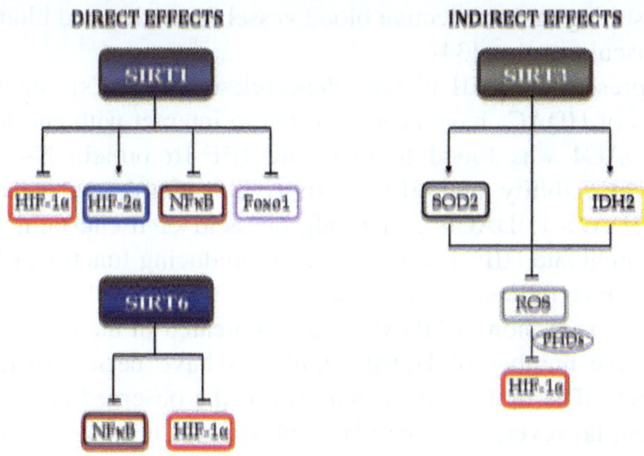

Fig. 8.4 The regulation of angiogenic factors by sirtuins. SIRT1 and SIRT6 have direct, complex effect on several factors involved in the angiogenic response, as they inhibit pro-angiogenic HIF-1α, NFκB, as well as anti-angiogenic Foxo1 with concomitant upregulation of HIF-2α. SIRT3 indirectly, via inhibition of ROS, through several mechanisms including deacetylation of SOD2 and IDH2 proteins, activates PHDs leading to inhibition of HIF-1α

8.2.3 Not Only Hydroxylation: The Role of (De)acetylation in HIFs (De)stabilization

As mentioned above, SIRT3, a member of deacetylases family may regulate HIF stabilization through ROS- and PHD-dependent signaling. However, the action of SIRT3 and other members of sirtuins family may also rely on the direct deacetylation of HIFs protein (Fig. 8.4).

Interestingly, it was shown that specific members of sirtuins family may exert different effects on HIF transcription factors. Deacetylation of HIF-1α at Lys674 by SIRT1 led to blocking p300 recruitment and consequently this repressed HIF-1 target genes [29]. On the other hand, the activity of HIF-2α increases after deacetylation by the same nuclear SIRT1 [30]. Moreover, the expression of HIF-2α-dependent genes, namely, VEGF, SOD2, and erythropoietin was also induced in response to SIRT1 action [30]. SIRT1 was also able to downregulate PHD2 protein level through deacetylase activity [31]. Importantly, SIRT1 is regulated by hypoxia in HIF-1α- and HIF-2α-dependent manner what suggests both positive and negative feedback loops between SIRT1 and HIFs [32]. Moreover, SIRT6 was also shown to be a negative regulator of HIF-1α stability and protein as SIRT6-deficient cells exhibit increased HIF-α activity [33].

Additionally, the direct involvement of sirtuins in the controlling angiogenic response was underlined. SIRT1 was found to be highly expressed in the vasculature during blood vessel growth, where via deacetylation of the forkhead transcription factor Foxo1, a crucial negative regulator of blood vessel development inhibits its anti-angiogenic activity [34]. Moreover, the knockout of SIRT1 led to dysregulation of number of genes essential for vascular growth, maturation, and remodeling resulting in the defective blood vessel formation and blunted ischemia-induced neovascularization [34].

Sirtuins represent class III histone deacetylases (HDACs) enzymes and also other members of HDACs have been reported to interact with and influence HIF-1 activity. HDAC4 was found to modulate HIF-1α protein N-terminal lysine acetylation level, stability, and HIF-1 activity [35]. On the other hand, inhibitors of class I/II HDACs (HDACIs), like valproic acid or trichostatin A, have been shown to downregulate HIF signaling by either reducing functional HIF-1α levels or repressing HIF-α transactivation activity (reviewed in [36]).

The aberrant expression of HDACs was implicated in the cancer development and the extensive number of HDAC inhibitors have been shown to exert the antitumor effects. The molecular mechanism of the observed action may involve HIFs regulation, however, as deacetylases play a role in numerous processes like transcriptional regulation, epigenetic programming, chromosomal remodeling and they regulate plethora of factors, HIFs modification represents only one possibility.

8.3 The Cross Talk Between HIFs and NFκB: The Implications for Inflammatory Angiogenesis

Inflammatory reaction is not only characterized by the excessive expression of inflammatory cytokines like TNF-α or IL-1β but is also associated with reduced oxygen tension [37]. It was shown that various inflammatory stimuli like lipopolysaccharide (LPS) [38] or pro-inflammatory cytokines [39–41] are able to induce HIF system under normoxic conditions. Such situation may be especially important during tumor progression as inflammation can facilitate the development of cancer [42, 43]. The cross talk between HIFs and inflammatory machinery may lead to the upregulation of the expression of pro-angiogenic factors, resulting in the increased vascular leakage, tumor vascularization, and finally metastasis. Moreover, a growing body of evidence indicates the presence and activation of HIF signaling in various inflammatory diseases, like rheumatoid arthritis [44], asthma [45], or atherosclerosis [46].

The regulation of HIF-1α by inflammatory cytokines was shown to occur both at the transcriptional and translational level. Especially important seems to be the interaction of HIFs and nuclear factor κB (NFκB) as both transcription factors are induced by hypoxia and inflammation and influence each other. Importantly, not only hypoxia may regulate HIF-1α mRNA level through NFκB [38, 47] but also in normoxic conditions NFκB is a direct modulator of HIF-1α expression [48]. In fact, the functional NFκB response element is present in the HIF-1α promoter at −197/−188 base pairs upstream of the transcriptional start site [48]. First indication that an active NFκB-binding site is present in HIF-1α promoter was shown in 2006, when Frede and colleagues studied the effect of LPS on HIF-1α mRNA in human monocytes and macrophages [38]. Additionally, during hypoxic induction of HIF-1α in pulmonary artery smooth muscle cells, the binding of the NFκB subunits p50 and p65 to the HIF-1α promoter have been shown [47]. Finally, direct interaction (although with different affinity) of all NFκB family members—p65, RelB, c-Rel, p52, and p50 with HIF-1α gene/promoter at an NFκB consensus site has been reported, whereas a truncated version of the HIF-1α promoter construct which lacks the NFκB site was not activated by the NFκB subunits [48].

Of note, the regulation of HIF by NFκB is evolutionary conserved—NFκB-binding element was found to be conserved across different species [47]. Moreover, recent work by van Uden et al. showed that even in *Drosophila*, NFκB regulates HIF-1β (tango) and HIF-α (sima) levels and activity both in normoxia and hypoxia [49].

As discussed earlier, sirtuins have been shown to associate, deacetylate, and regulate the activity of HIF-1α. Similarly, NFκB acetylation status and transcriptional activity could be modified by sirtuins. Sirtuins, although with different mechanisms of action, act as a negative regulators of NFκB activity (reviewed in [50]) (Fig. 8.4).

8.4 Nitric Oxide: Important Regulator of Hypoxia-Induced Angiogenesis

As discussed above, the activation of HIF transcription factors has been increasingly implicated in inflammatory diseases. A common hallmark of inflammatory disorders is the increased synthesis of nitric oxide (NO) by induction and activation of inducible nitric oxide synthase (iNOS). The implication of NO in the regulation of angiogenic mediators as well as HIF transcription factor has been suggested in a great number of papers.

In the past, several studies have shown the discrepant data concerning the effect of NO on HIF-1α accumulation—both stabilization [51, 52] and destabilization [53, 54] have been reported. Such confusing results have been nicely explained by Mateo and colleagues who have shown that NO may exert biphasic effect on HIF-α stabilization, dependently on the concentration used. At low concentration, up to 400 nM diminishment in HIF-1α stability was observed, whereas higher NO concentrations (above 1 μM) increased HIF-1α stability [55].

Moreover, the discrepant data about NO-mediated effects could be the result of various NO donors used in the experiments. Different factors which release NO are used, mostly SNAP (S-nitroso-*N*-acetylpenicillamine), SIN-1 (S-morpholinosydnonimine), GSNO (S-nitrosoglutathione), and DETA (diethylenetriamine NONOate). Unfortunately, sodium nitroprusside (SNP), a complex compound which releases NO together with iron ions and cyanides, was also used in many studies. SNP was shown to inhibit HIF-1 accumulation in human glioblastoma A-172 cells, whereas in the same experiment SNAP or GSNO increased HIF-1α accumulation [56]. Similarly, in the human bladder cancer and in the human prostate cancer cell lines SNP decreased hypoxia-induced HIF-1α protein level [57]. As mentioned earlier, SNP, after releasing NO, is converted to ferrocyanide and ferricyanide. Cyanides, inhibitors of cytochrome c oxidase, are cytotoxic molecules as well as iron which could be released from SNP is a possible toxic factor generating highly reactive radicals, such as hydroxyl radicals via the Fenton reaction.

Importantly, NO-dependent accumulation of HIF-α might be a possible mechanism responsible for increase in HIF-dependent gene expression in tumors. Both increased NOS expression and NO level led to stabilization of HIF-1α protein in human oral squamous cell carcinoma [58]. Similarly, in prostate cancer, Nanni and colleagues have identified the nuclear co-localization between endothelial NOS, estrogen receptor β, HIF-1α, and HIF-2α and they found that these proteins cooperate to activate transcription of protumorigenic genes [59].

The induction of HIF by NO may lead to the increased transcription of angiogenic gene expression, and nowadays it is well proven that NO is an important mediator of angiogenesis. We [60, 61] and others [62] have shown that NO is significant regulator of VEGF signaling both in vitro and in vivo. In rat vascular smooth muscle cells, IL-1β-induced NO production, NO donors, or plasmid delivery of NOS led to the upregulation of VEGF synthesis, which was abolished when inhibitor of NOS, like L-NAME was added [60, 61]. In rat ischemic hindlimb injected with eNOS cDNA increase in peripheral blood flow in ischemic tissue, better vascularization (increased number of CD31-positive cells in muscles), and increased VEGF level have been detected [62].

Similarly to discrepant data about HIF-1 accumulation, also there are results showing inhibition of VEGF expression after SNP treatment [56, 63] again indicating the importance of the appropriate NO donor usage. In our hands, direct comparison of SNAP, SIN-1, DETA, GSNO, and SNP on VEGF production in rat or human vascular smooth muscle cells showed that all donors, but not SNP, potently upregulate VEGF release [61, 64]. These results strongly indicate that NO increases VEGF level, but SNP cannot be used as a NO donor.

In summary, it seems that different effects of NO on HIF-1 or VEGF level described in the literature may be the result of various methodological approaches including different NO donors and various concentrations used.

The mechanism responsible for NO-mediated HIF modification relies mostly on S-nitrosylation of critical cysteine residues of several key proteins involved in HIF regulation, including HIF-1α itself but not on cGMP involvement. For example, NO was shown to have inhibitory effect on PHDs 1–3 in vitro. In human embryonic kidney cells stimulated with exogenous NO donor (GSNO) dose dependently inhibition of PHD activity and reduction of HIF-1α hydroxylation leading to its stabilization was observed [65]. Moreover, SNAP was shown to inhibit FIH-1 leading to enhanced HIF-1α C-TAD activity. Additionally, NO may S-nitrosylate the VHL blocking its ubiquitination activity [66].

Interesting findings linking NO and mitochondrial respiratory chain were published recently [67]. The authors based on the information that NO could be produced in the mitochondria in the reaction catalyzed by cytochrome c oxidase (Cco/NO) [68]. To find if NO produced in mitochondria could be responsible for HIF-α stabilization, HEK 293.7 cells without NOS activity have been used. Interestingly, the cells used Cco/NO activity to generate NO and further for the stabilization of HIF-1α [67].

8.5 microRNA and Hypoxia-Induced Angiogenesis

Identified in 1993 as small, noncoding RNA molecules which bind to the 3′ UTRs of target mRNAs to negatively regulate gene expression [69], microRNAs have been shown to be implicated in many different processes including angiogenesis. To date, more than 1,500 human miRNAs have been reported (http://www.mirbase.org) and it is suggested that they could regulate more than one-third of the mRNAs produced.

In 2010, Chan and Loscalzo have introduced the term "hypoxamirs" to stress the existence of a number of hypoxia-regulated miRNAs [70]. Several HIF-related miRNAs have been identified till now, with the most prominent miR-210. Moreover, miR-20a, miR-20b, miR-199a, miR-424, miR-130a, miR-130b, and miR-155 have been shown to affect HIF expression, although it seems that only miR-20a (encoded by miR-17-92 cluster), miR-20b, and miR-199a directly target the 3′UTR of HIF-1α [31, 71, 72].

From well-described hypoxamirs, miR-210 was shown to be regulated by both HIF-1 and HIF-2 [73] and is able to target several angiogenic factors. Overexpression of miR-210 in human umbilical vein endothelial cells increases the expression of VEGF and its receptor, VEGF-R2 leading to increased renal angiogenesis [74]. The receptor tyrosine kinase ligand Ephrin-A3, an important regulator of endothelial cells survival, migration, and differentiation is another example of angiogenic gene regulated by miR-210 [75]. Moreover, Notch signaling pathway was suggested to be activated by miR-210 in a model of cerebral ischemia [76].

Importantly, the overexpression of miR-210 has been detected in a variety of cardiovascular diseases and solid tumors. Particularly, its high level characterizes the renal clear cell carcinomas (RCCs), in which also HIF is overexpressed as a consequence of pVHL inactivation. Its important role in tumor development may be related to the fact that in response to hypoxia this miRNA regulates a wide spectrum of genes involved not only in angiogenesis but also in the mitochondrial metabolism, DNA repair, or cell survival.

8.6 Hypoxia and Angiogenic Regulators

8.6.1 Regulation of VEGF by HIFs: A Classic Example of Pro-angiogenic Gene Regulation in Hypoxic Conditions

Angiogenesis is controlled through the equilibrium of pro- and anti-angiogenic factors. However, under special conditions, e.g., during tumor development, the level of pro-angiogenic factors increases and this facilitates the blood vessel formation.

HIFs regulate the expression of more than 150 genes, including several, involved in angiogenesis (Fig. 8.1), like VEGF, the major pro-angiogenic factor which is known to enhance proliferation, survival, and tube formation by endothelial cells.

The mechanism of VEGF regulation in hypoxic conditions is well known. It was shown already in 1995 that the promoter of the VEGF gene contains HRE and binding of HIF to this region increases the transcription of VEGF [77]. Moreover, the stability of VEGF mRNA, which is very low under normoxia is dramatically increased in hypoxia [78]. It is interesting, that whereas the average half-life of eukaryotic mRNAs is 10–12 h, the half-life of VEGF mRNA is less than 1 h, e.g., in PC12 rat pheochromocytoma cells it is about 40 min [79].

Post-transcriptional VEGF regulation may be controlled at least in part by RNA-binding proteins (RBPs) and miRNAs. The *VEGF* 3′UTR contains several important cis-acting elements including CA-rich element (CARE) or AU-rich element (ARE) and is regulated by its RBPs such as AUF1, tristetraprolin (TTP), heterogeneous nuclear ribonucleoprotein (hnRNP L), and HuR [80–83]. One of the recently discovered factor with strong binding preference for the VEGF mRNA 5′UTR is the DEAD-box RNA helicase DDX6 [84]. Some proteins like TTP and AUF destabilize VEGF mRNA in macrophages and tumor cells, respectively [81, 82]. In contrast, the hypoxia-induced inhibition of DDX6 positively affects VEGF expression [84].

The stabilization of labile VEGF mRNA relies largely on the activity of specific RNA-binding protein HuR, a member of the Elav family of proteins found in *Drosophila*. HuR binds to ARE in the VEGF 3′UTR, forming an RNA–protein complex in a hypoxia-inducible fashion [83]. Till now, the molecular details of how HuR regulates VEGF expression are not well understood. It was suggested that the mechanism of HuR action may be related to its protective effect against endonucleases as a 40 bp region adjacent to the HuR-binding site in the VEGF stability region has been identified to be susceptible to ribonucleases in the absence of HuR [85].

Data accumulated over the last years have highlighted the significant roles for miRNAs, small noncoding RNAs, for the regulation of VEGF signaling. Not only VEGF but also its receptors and components of the intracellular signaling pathway are controlled by different miRNAs, including miR-15, -16, -23, -27, -93, -200b, -221, or miR-424 (reviewed in [86]). Recently, numerous studies indicate also the cross talk between RBPs and miRNAs in the post-transcriptional regulation of gene expression, like the interplay between HuR and miRNAs which associate with the HuR-regulated mRNAs. Chang et al. indicated that HuR suppressed miR-200b expression and antagonizes its anti-angiogenic effects, leading to increase in VEGF level [87].

The transcriptional regulation of VEGF by HIFs can be modulated by other transcription factors. Interestingly, the complex regulation is achieved by the members of E2F transcription factors family. This family consists of eight members which may have the opposite effect on VEGF transcription. E2F7 and E2F8 were shown to stimulate angiogenesis via transcriptional induction of VEGF [88]. The key event in this regulation is the formation of an E2F7/8–HIF-1α transcriptional complex that directly binds and stimulates the activity of VEGF promoter. The positive regulation of VEGF transcription by E2F7/8 is quite surprising as these factors are classified as repressors, and activation of VEGF transcription is unique mechanism. However, the significance of these factors for angiogenic process is also provided by in vivo studies using E2F7$^{-/-}$E2F8$^{-/-}$ mice. Deletion of the E2F7 and E2F8 genes causes serious vascular defects and lethality around embryonic day 10.5 [89]. In contrast, another member of E2F family, E2F1 negatively regulates hypoxia-induced VEGF expression [90]. In hypoxic conditions, E2F1 associates with p53 and downregulates VEGF expression exclusively, without affecting other hypoxia-inducible genes. Importantly, E2F1$^{-/-}$ mice display enhanced angiogenesis, endothelial cell proliferation, and reperfusion in a hindlimb ischemia model, resulting from enhanced VEGF expression [90].

Additional transcription factors play important role in the hypoxic regulation of VEGF-driven angiogenesis. Related transcription enhancer factor-1 (RTEF-1) and early growth response 1 (EGR-1) can both target VEGF to enhance angiogenesis [91, 92]. RTEF-1 was shown to regulate HIF-1α both in normoxic and hypoxic conditions and to increase capillary density and blood recovery after the hindlimb ischemia [93]. Not only VEGF (VEGF-A) is a target gene of RTEF-1. Several studies have shown that other genes involved in the regulation of vascular growth, like fibroblast growth factor receptor-1, FGFR-1 [94], or VEGF-B [95] are regulated by RTEF-1. Similarly, EGR-1 acts not only on VEGF. This transcription factor was suggested to be a master regulator of angiogenic, inflammatory, procoagulant, and permeability-related genes [92].

Finally, translation of VEGF mRNA is increased under low oxygen tension. It was shown that even in such unfavorable stress conditions like hypoxia, when cap-dependent initiation of translation is compromised, the synthesis of VEGF protein is maintained at high level. This is because VEGF mRNA contains the internal ribosomal entry site (IRES) in the 5′UTR sequence and when cap-dependent translation is suppressed under hypoxia this IRES-dependent translation predominates [96].

8.6.2 Complex Effect of Hypoxia on Angiogenesis Stimulators: Not Only Induction Is Observed

The hypoxia-induced upregulation of VEGF expression at mRNA and protein level was shown in the multiple cell lines, including cancer cells. However, although very important, VEGF is not the sole-acting angiogenesis factor. Among others, there are platelet-derived growth factor (PDGF), placental growth factor (PlGF), transforming growth factor-β (TGFβ), fibroblast growth factor (FGF), angiopoietins (Ang), matrix metalloproteinases (MMP), or interleukin-8 (IL-8) (Fig. 8.5).

A lot of studies have found the correlation between HIF expression and intratumoral microvascular density (reviewed in [97]). Since hypoxia is one of the major inducers of tumor vasculature, HIFs and VEGF were suggested to be the attractive targets for anticancer therapy. However, clinical data suggest that the benefit of anti-angiogenic therapies with anti-VEGF antibodies does not last long, as many patients encounter progression of cancers. This can be explained by the recurrence of tumor angiogenesis through the compensatory production of

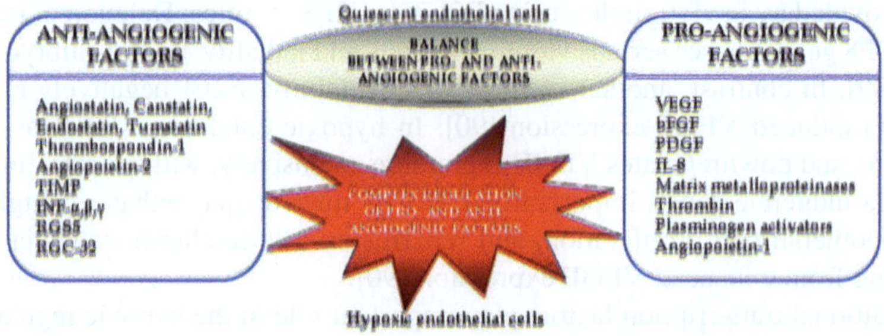

Fig. 8.5 Angiogenesis regulation—the balance between pro-and anti-angiogenic factors. The balance between pro- and antiangiogenic factors is important for the quiescent endothelial cells. Generally, the expression of proangiogenic factors increases whereas inhibitors of angiogenesis decreases in hypoxic conditions; however, recent reports suggest the complex regulation of factors regulating angiogenesis

mediators other than VEGF, indicating for the complex regulation of angiogenic factors in response to hypoxia.

In relation to the above observations we demonstrated that 24 h incubation of human microvascular endothelial cell line (HMEC-1) in the atmosphere of 1 % O_2 led to decrease of PlGF expression, the stimulator of blood vessel growth [98]. Moreover, we have also shown that the expression of another pro-angiogenic agent, IL-8 was downregulated by hypoxia/HIF-1 in endothelial cells [98, 99]. Importantly, in such conditions VEGF was induced, showing opposite effect of hypoxia conditions on the expression of different pro-angiogenic factors. Next we reported that in contrast to HIF-1, overexpression of HIF-2 isoform resulted in increased expression of IL-8 [100]. The mechanism of the opposite effect of HIF-1 and HIF-2 on IL-8 expression involves the recruitment of different transcription factors—IL-8 inhibition by HIF-1 is mediated via the down-regulation of the NF-E2-related factor 2 (Nrf2) transcription factor [99], whereas stimulatory effect of HIF-2 on IL-8 relies on the induction of Sp-1 [100]. Moreover, c-Myc transcription factor plays a crucial role in the regulation of HIF-1/HIF-2 mediated IL-8 expression. We have shown that HIF-1α not only downregulates the level of c-Myc but concomitantly it also increases the production of Mxi-1, c-Myc antagonist. On the other hand, HIF-2α increases c-Myc activity [100].

There are a growing number of studies underlying the complexity of the regulation of hypoxia-dependent angiogenesis regulators. Interestingly, in HIF-1-deficient colon cancer cells, the hypoxic induction of VEGF was only partially blocked, whereas NFκB-dependent increase in IL-8 expression was observed. What is more, HIF-1 inhibition did not influence vascularization of tumors—the microvessel density was identical in tumors that had wild-type or knock-down HIF-1α [101]. The unexpected and surprising results have been recently presented by Yu and Hales [102]. In lung cancer tumor, long-term exposure to hypoxia repressed the growth of tumor and decreased microvessel density in the tumor tissues. Importantly, the same conditions led to acceleration of tumor progression and microvasculature in the colon cancer. The mechanism responsible for the opposite effect of hypoxia on angiogenesis and tumorigenesis in different tumor types observed in this study was not investigated in details, however, the differences in the expression of Na^{+}–K^{+} ATPase, a versatile signal transducer, between these two tumor types was underlined. The inhibition of the sodium pump expression was observed in lung cancer tumors but not in colon cancer tumors. It is possible that the basal and hypoxia-induced expression of pro-angiogenic mediators was different in these two models, but it was not checked in this study. In another work [103], in the transgenic polyomavirus middle T breast cancer mouse model, the exposure of mice to acute cyclic hypoxia did not influence the primary tumor growth as well as lung metastasis. Interestingly, the number of CD31-positive cells in hypoxic tumors was lower than in control tumors.

Importantly, several studies tried to correlate HIF-1 expression with the survival rate of cancer patients. Although the general idea indicates that HIF-1 promotes angiogenic-dependent tumor progression, there are also studies presenting opposite results.

Volm and Koomagi have analyzed 96 paraffin-embedded sections obtained from patients with non-small cell lung carcinomas (NSCLCs) [104] and reported no relationship between HIF-1α or HIF-1β and proliferation. On the other hand, significant correlation between HIF-1 expression and apoptotic markers was detected. Surprisingly, patients with HIF-positive carcinomas had significantly longer median survival times than patients with HIF-negative carcinomas. These interesting results prompted the authors to check the expression of oncogene and tumor suppressor products and proliferative, apoptotic, and angiogenic factors in 216 patients with long-term surviving NSCLCs patients. Again, from several factors differentially expressed between control and studied patients, high level of HIF-1β was observed in carcinomas of long-term survivors [105]. Finally, recent data suggest also that activation of HIFs due to the inhibition of PHD2, can paradoxically improve the effectiveness of antitumor therapy due to tumor vessel normalization, increasing the delivery of chemotherapeutics [106].

8.6.3 Hypoxia-Regulated Angiogenic Inhibitors

As mentioned before, the regulation of pro-angiogenic factors by hypoxia/HIFs is complicated and still not fully discovered process. Moreover, not only activators but also inhibitors of angiogenesis (Fig. 8.5) could be regulated in HIF-dependent manner. Generally, the results of different studies indicate the inhibitory effect of HIFs activation on the regulation of angiogenic inhibitors. Thrombospondin-1 (TSP-1), a glycoprotein with major roles in cellular adhesion and vascular smooth muscle proliferation and migration, was shown to be downregulated in hypoxic conditions in different in vitro and in vivo models [98, 107, 108]. In the human microvascular endothelial cells and pericytes, the expression of other potent anti-angiogenic factor, endostatin was shown to be inhibited by hypoxia [109]. Additionally, the expression of angiostatin was decreased in a swine model of neonatal hypoxia [110]. In contrast to the above studies, several work underline the importance of hypoxia-induced expression of anti-angiogenic and anti-apoptotic factors.

One of the newly recognized hypoxia-overexpressed angiogenic inhibitor is regulator of G protein signaling 5 (RGS5) [111]. The acceleration of RGS5 by hypoxia is HIF-1 dependent and occurs also after treatment with cobalt chloride ($CoCl_2$), hypoxia-mimicking compound, stabilizing HIF as well as with prolyl hydroxylase inhibitors, dimethyloxalylglycine (DMOG), or ethyl-3,4-dihydroxy-benzoate (DHB). Importantly, RGS5 attenuates growth and induces apoptosis of endothelial cells in vitro and hinders angiogenesis in vivo [111].

An and colleagues [112] reported hypoxic induction of response gene to complement-32 (RGC-32), important modulator of cell proliferation. Increase in RGC-32 expression is mediated by HIF-1, both at the transcriptional and posttranscriptional levels, and moreover, also VEGF stimulates RGC-32 expression. Notably, the overexpression of RGC-32 in endothelial cells inhibits cell proliferation and migration via downregulation of another major angiogenic protein, FGF-2. In vivo, anti-angiogenic effect of RGC-32 was tested in two different models: this factor suppressed SW480 melanoma tumor growth as well as attenuated the blood flow recovery after hindlimb ischemia. Additionally, the reduction of CD-31-positive endothelial cells was observed in both experimental models.

Conclusions

Although extensive studies on the identifying molecular mechanisms of hypoxia-regulated angiogenesis are performed in many laboratories worldwide, still much work remains to be done to discover fully the mechanisms of hypoxia/HIF-regulated expression of angiogenic factors and to transfer this knowledge to translational medicine. Based on this information, optimal HIF-acting therapeutic agents could be developed what will result in an improved clinical outcome.

Acknowledgments The authors apologize for the inability to cite all of the important studies that have been performed in the field due to space limitations.

AL is supported by the Foundation for Polish Science—PARENT-BRIDGE Programme co-financed by the European Union within European Regional Development Fund (POMOST/2010-2/8) and she is the recipient of L'Oreal Poland for Women in Science Scholarship. The Faculty of Biochemistry, Biophysics and Biotechnology of the Jagiellonian University is a beneficiary of the structural funds from the European Union and the Polish Ministry of Science and Higher Education (grants No: POIG.02.01.00-12 064/08, POIG 01.01.02-00-109/09, POIG.02.02.00-014/08 and 01.01.02-00-069/09).

References

1. Semenza GL, Nejfelt MK, Chi SM, Antonarakis SE (1991) Hypoxia-inducible nuclear factors bind to an enhancer element located 3′ to the human erythropoietin gene. Proc Natl Acad Sci USA 88(13):5680–5684
2. Pugh CW, Tan CC, Jones RW, Ratcliffe PJ (1991) Functional analysis of an oxygen-regulated transcriptional enhancer lying 3′ to the mouse erythropoietin gene. Proc Natl Acad Sci USA 88(23):10553–10557

3. Semenza GL, Jiang BH, Leung SW, Passantino R, Concordet JP, Maire P, Giallongo A (1996) Hypoxia response elements in the aldolase A, enolase 1, and lactate dehydrogenase A gene promoters contain essential binding sites for hypoxia-inducible factor 1. J Biol Chem 271(51):32529–32537
4. Kimura H, Weisz A, Ogura T, Hitomi Y, Kurashima Y, Hashimoto K, D'Acquisto F, Makuuchi M, Esumi H (2001) Identification of hypoxia-inducible factor 1 ancillary sequence and its function in vascular endothelial growth factor gene induction by hypoxia and nitric oxide. J Biol Chem 276(3):2292–2298
5. Wang GL, Jiang BH, Rue EA, Semenza GL (1995) Hypoxia-inducible factor 1 is a basic-helix-loop-helix-PAS heterodimer regulated by cellular O_2 tension. Proc Natl Acad Sci USA 92(12):5510–5514
6. Makino Y, Kanopka A, Wilson WJ, Tanaka H, Poellinger L (2002) Inhibitory PAS domain protein (IPAS) is a hypoxia-inducible splicing variant of the hypoxia-inducible factor-3alpha locus. J Biol Chem 277(36):32405–32408
7. Maynard MA, Qi H, Chung J, Lee EH, Kondo Y, Hara S, Conaway RC, Conaway JW, Ohh M (2003) Multiple splice variants of the human HIF-3 alpha locus are targets of the von Hippel-Lindau E3 ubiquitin ligase complex. J Biol Chem 278(13):11032–11040
8. Makino Y, Cao R, Svensson K, Bertilsson G, Asman M, Tanaka H, Cao Y, Berkenstam A, Poellinger L (2001) Inhibitory PAS domain protein is a negative regulator of hypoxia-inducible gene expression. Nature 414(6863):550–554
9. Jaakkola P, Mole DR, Tian YM, Wilson MI, Gielbert J, Gaskell SJ, von Kriegsheim A, Hebestreit HF, Mukherji M, Schofield CJ, Maxwell PH, Pugh CW, Ratcliffe PJ (2001) Targeting of HIF-alpha to the von Hippel-Lindau ubiquitylation complex by O_2-regulated prolyl hydroxylation. Science 292(5516):468–472
10. Ivan M, Kondo K, Yang H, Kim W, Valiando J, Ohh M, Salic A, Asara JM, Lane WS, Kaelin WG Jr (2001) HIFalpha targeted for VHL-mediated destruction by proline hydroxylation: implications for O2 sensing. Science 292(5516):464–468
11. Epstein AC, Gleadle JM, McNeill LA, Hewitson KS, O'Rourke J, Mole DR, Mukherji M, Metzen E, Wilson MI, Dhanda A, Tian YM, Masson N, Hamilton DL, Jaakkola P, Barstead R, Hodgkin J, Maxwell PH, Pugh CW, Schofield CJ, Ratcliffe PJ (2001) *C. elegans* EGL-9 and mammalian homologs define a family of dioxygenases that regulate HIF by prolyl hydroxylation. Cell 107(1):43–54
12. Bruick RK, McKnight SL (2001) A conserved family of prolyl-4-hydroxylases that modify HIF. Science 294(5545):1337–1340
13. Berra E, Benizri E, Ginouves A, Volmat V, Roux D, Pouyssegur J (2003) HIF prolyl-hydroxylase 2 is the key oxygen sensor setting low steady-state levels of HIF-1alpha in normoxia. EMBO J 22(16):4082–4090
14. Appelhoff RJ, Tian YM, Raval RR, Turley H, Harris AL, Pugh CW, Ratcliffe PJ, Gleadle JM (2004) Differential function of the prolyl hydroxylases PHD1, PHD2, and PHD3 in the regulation of hypoxia-inducible factor. J Biol Chem 279(37):38458–38465
15. Maxwell PH, Wiesener MS, Chang GW, Clifford SC, Vaux EC, Cockman ME, Wykoff CC, Pugh CW, Maher ER, Ratcliffe PJ (1999) The tumour suppressor protein VHL targets hypoxia-inducible factors for oxygen-dependent proteolysis. Nature 399(6733):271–275
16. Ohh M, Park CW, Ivan M, Hoffman MA, Kim TY, Huang LE, Pavletich N, Chau V, Kaelin WG (2000) Ubiquitination of hypoxia-inducible factor requires direct binding to the beta-domain of the von Hippel-Lindau protein. Nat Cell Biol 2(7):423–427
17. Cockman ME, Masson N, Mole DR, Jaakkola P, Chang GW, Clifford SC, Maher ER, Pugh CW, Ratcliffe PJ, Maxwell PH (2000) Hypoxia inducible factor-alpha binding and ubiquitylation by the von Hippel-Lindau tumor suppressor protein. J Biol Chem 275(33):25733–25741
18. Chilov D, Camenisch G, Kvietikova I, Ziegler U, Gassmann M, Wenger RH (1999) Induction and nuclear translocation of hypoxia-inducible factor-1 (HIF-1): heterodimerization with ARNT is not necessary for nuclear accumulation of HIF-1alpha. J Cell Sci 112(Pt 8): 1203–1212

19. Lando D, Peet DJ, Gorman JJ, Whelan DA, Whitelaw ML, Bruick RK (2002) FIH-1 is an asparaginyl hydroxylase enzyme that regulates the transcriptional activity of hypoxia-inducible factor. Genes Dev 16(12):1466–1471
20. Brunelle JK, Bell EL, Quesada NM, Vercauteren K, Tiranti V, Zeviani M, Scarpulla RC, Chandel NS (2005) Oxygen sensing requires mitochondrial ROS but not oxidative phosphorylation. Cell Metab 1(6):409–414
21. Guzy RD, Hoyos B, Robin E, Chen H, Liu L, Mansfield KD, Simon MC, Hammerling U, Schumacker PT (2005) Mitochondrial complex III is required for hypoxia-induced ROS production and cellular oxygen sensing. Cell Metab 1(6):401–408
22. Mansfield KD, Guzy RD, Pan Y, Young RM, Cash TP, Schumacker PT, Simon MC (2005) Mitochondrial dysfunction resulting from loss of cytochrome c impairs cellular oxygen sensing and hypoxic HIF-alpha activation. Cell Metab 1(6):393–399
23. Masson N, Singleton RS, Sekirnik R, Trudgian DC, Ambrose LJ, Miranda MX, Tian YM, Kessler BM, Schofield CJ, Ratcliffe PJ (2012) The FIH hydroxylase is a cellular peroxide sensor that modulates HIF transcriptional activity. EMBO Rep 13(3):251–257
24. Finkel T, Deng CX, Mostoslavsky R (2009) Recent progress in the biology and physiology of sirtuins. Nature 460(7255):587–591
25. Finley LW, Carracedo A, Lee J, Souza A, Egia A, Zhang J, Teruya-Feldstein J, Moreira PI, Cardoso SM, Clish CB, Pandolfi PP, Haigis MC (2011) SIRT3 opposes reprogramming of cancer cell metabolism through HIF1alpha destabilization. Cancer Cell 19(3):416–428
26. Bell EL, Emerling BM, Ricoult SJ, Guarente L (2011) SirT3 suppresses hypoxia inducible factor 1alpha and tumor growth by inhibiting mitochondrial ROS production. Oncogene 30 (26):2986–2996
27. Yu W, Dittenhafer-Reed KE, Denu JM (2012) SIRT3 protein deacetylates isocitrate dehydrogenase 2 (IDH2) and regulates mitochondrial redox status. J Biol Chem 287(17): 14078–14086
28. Tao R, Coleman MC, Pennington JD, Ozden O, Park SH, Jiang H, Kim HS, Flynn CR, Hill S, Hayes McDonald W, Olivier AK, Spitz DR, Gius D (2010) Sirt3-mediated deacetylation of evolutionarily conserved lysine 122 regulates MnSOD activity in response to stress. Mol Cell 40(6):893–904
29. Lim JH, Lee YM, Chun YS, Chen J, Kim JE, Park JW (2010) Sirtuin 1 modulates cellular responses to hypoxia by deacetylating hypoxia-inducible factor 1alpha. Mol Cell 38(6): 864–878
30. Dioum EM, Chen R, Alexander MS, Zhang Q, Hogg RT, Gerard RD, Garcia JA (2009) Regulation of hypoxia-inducible factor 2alpha signaling by the stress-responsive deacetylase sirtuin 1. Science 324(5932):1289–1293
31. Rane S, He M, Sayed D, Vashistha H, Malhotra A, Sadoshima J, Vatner DE, Vatner SF, Abdellatif M (2009) Downregulation of miR-199a derepresses hypoxia-inducible factor-1alpha and Sirtuin 1 and recapitulates hypoxia preconditioning in cardiac myocytes. Circ Res 104(7):879–886
32. Chen R, Dioum EM, Hogg RT, Gerard RD, Garcia JA (2011) Hypoxia increases sirtuin 1 expression in a hypoxia-inducible factor-dependent manner. J Biol Chem 286(16): 13869–13878
33. Zhong L, D'Urso A, Toiber D, Sebastian C, Henry RE, Vadysirisack DD, Guimaraes A, Marinelli B, Wikstrom JD, Nir T, Clish CB, Vaitheesvaran B, Iliopoulos O, Kurland I, Dor Y, Weissleder R, Shirihai OS, Ellisen LW, Espinosa JM, Mostoslavsky R (2010) The histone deacetylase Sirt6 regulates glucose homeostasis via Hif1alpha. Cell 140(2):280–293
34. Potente M, Ghaeni L, Baldessari D, Mostoslavsky R, Rossig L, Dequiedt F, Haendeler J, Mione M, Dejana E, Alt FW, Zeiher AM, Dimmeler S (2007) SIRT1 controls endothelial angiogenic functions during vascular growth. Genes Dev 21(20):2644–2658
35. Geng H, Harvey CT, Pittsenbarger J, Liu Q, Beer TM, Xue C, Qian DZ (2011) HDAC4 protein regulates HIF1alpha protein lysine acetylation and cancer cell response to hypoxia. J Biol Chem 286(44):38095–38102

36. Chen S, Sang N (2011) Histone deacetylase inhibitors: the epigenetic therapeutics that repress hypoxia-inducible factors. J Biomed Biotechnol 2011:197946
37. Karhausen J, Haase VH, Colgan SP (2005) Inflammatory hypoxia: role of hypoxia-inducible factor. Cell Cycle 4(2):256–258
38. Frede S, Stockmann C, Freitag P, Fandrey J (2006) Bacterial lipopolysaccharide induces HIF-1 activation in human monocytes via p44/42 MAPK and NF-kappaB. Biochem J 396(3): 517–527
39. Jung YJ, Isaacs JS, Lee S, Trepel J, Neckers L (2003) IL-1beta-mediated up-regulation of HIF-1alpha via an NFkappaB/COX-2 pathway identifies HIF-1 as a critical link between inflammation and oncogenesis. FASEB J 17(14):2115–2117
40. Frede S, Freitag P, Otto T, Heilmaier C, Fandrey J (2005) The proinflammatory cytokine interleukin 1beta and hypoxia cooperatively induce the expression of adrenomedullin in ovarian carcinoma cells through hypoxia inducible factor 1 activation. Cancer Res 65(11): 4690–4697
41. Jung Y, Isaacs JS, Lee S, Trepel J, Liu ZG, Neckers L (2003) Hypoxia-inducible factor induction by tumour necrosis factor in normoxic cells requires receptor-interacting protein-dependent nuclear factor kappa B activation. Biochem J 370(Pt 3):1011–1017
42. Balkwill F, Mantovani A (2001) Inflammation and cancer: back to Virchow? Lancet 357(9255): 539–545
43. Hanahan D, Weinberg RA (2011) Hallmarks of cancer: the next generation. Cell 144(5): 646–674
44. Hollander AP, Corke KP, Freemont AJ, Lewis CE (2001) Expression of hypoxia-inducible factor 1alpha by macrophages in the rheumatoid synovium: implications for targeting of therapeutic genes to the inflamed joint. Arthritis Rheum 44(7):1540–1544
45. Huerta-Yepez S, Baay-Guzman GJ, Bebenek IG, Hernandez-Pando R, Vega MI, Chi L, Riedl M, Diaz-Sanchez D, Kleerup E, Tashkin DP, Gonzalez FJ, Bonavida B, Zeidler M, Hankinson O (2011) Hypoxia inducible factor promotes murine allergic airway inflammation and is increased in asthma and rhinitis. Allergy 66(7):909–918
46. Parathath S, Mick SL, Feig JE, Joaquin V, Grauer L, Habiel DM, Gassmann M, Gardner LB, Fisher EA (2011) Hypoxia is present in murine atherosclerotic plaques and has multiple adverse effects on macrophage lipid metabolism. Circ Res 109(10):1141–1152
47. Belaiba RS, Bonello S, Zahringer C, Schmidt S, Hess J, Kietzmann T, Gorlach A (2007) Hypoxia up-regulates hypoxia-inducible factor-1alpha transcription by involving phosphatidylinositol 3-kinase and nuclear factor kappaB in pulmonary artery smooth muscle cells. Mol Biol Cell 18(12):4691–4697
48. van Uden P, Kenneth NS, Rocha S (2008) Regulation of hypoxia-inducible factor-1alpha by NF-kappaB. Biochem J 412(3):477–484
49. van Uden P, Kenneth NS, Webster R, Muller HA, Mudie S, Rocha S (2011) Evolutionary conserved regulation of HIF-1beta by NF-kappaB. PLoS Genet 7(1):e1001285
50. Tafani M, Pucci B, Russo A, Schito L, Pellegrini L, Perrone GA, Villanova L, Salvatori L, Ravenna L, Petrangeli E, Russo MA (2013) Modulators of HIF1alpha and NFkB in cancer treatment: is it a rational approach for controlling malignant progression? Front Pharmacol 4:13
51. Sandau KB, Faus HG, Brune B (2000) Induction of hypoxia-inducible-factor 1 by nitric oxide is mediated via the PI 3K pathway. Biochem Biophys Res Commun 278(1):263–267
52. Kimura H, Weisz A, Kurashima Y, Hashimoto K, Ogura T, D'Acquisto F, Addeo R, Makuuchi M, Esumi H (2000) Hypoxia response element of the human vascular endothelial growth factor gene mediates transcriptional regulation by nitric oxide: control of hypoxia-inducible factor-1 activity by nitric oxide. Blood 95(1):189–197
53. Sogawa K, Numayama-Tsuruta K, Ema M, Abe M, Abe H, Fujii-Kuriyama Y (1998) Inhibition of hypoxia-inducible factor 1 activity by nitric oxide donors in hypoxia. Proc Natl Acad Sci USA 95(13):7368–7373

54. Yin JH, Yang DI, Ku G, Hsu CY (2000) iNOS expression inhibits hypoxia-inducible factor-1 activity. Biochem Biophys Res Commun 279(1):30–34
55. Mateo J, Garcia-Lecea M, Cadenas S, Hernandez C, Moncada S (2003) Regulation of hypoxia-inducible factor-1alpha by nitric oxide through mitochondria-dependent and -independent pathways. Biochem J 376(Pt 2):537–544
56. Kimura H, Ogura T, Kurashima Y, Weisz A, Esumi H (2002) Effects of nitric oxide donors on vascular endothelial growth factor gene induction. Biochem Biophys Res Commun 296(4): 976–982
57. Wang F, Zhang R, Xia T, Hsu E, Cai Y, Gu Z, Hankinson O (2007) Inhibitory effects of nitric oxide on invasion of human cancer cells. Cancer Lett 257(2):274–282
58. Quintero M, Brennan PA, Thomas GJ, Moncada S (2006) Nitric oxide is a factor in the stabilization of hypoxia-inducible factor-1alpha in cancer: role of free radical formation. Cancer Res 66(2):770–774
59. Nanni S, Benvenuti V, Grasselli A, Priolo C, Aiello A, Mattiussi S, Colussi C, Lirangi V, Illi B, D'Eletto M, Cianciulli AM, Gallucci M, De Carli P, Sentinelli S, Mottolese M, Carlini P, Strigari L, Finn S, Mueller E, Arcangeli G, Gaetano C, Capogrossi MC, Donnorso RP, Bacchetti S, Sacchi A, Pontecorvi A, Loda M, Farsetti A (2009) Endothelial NOS, estrogen receptor beta, and HIFs cooperate in the activation of a prognostic transcriptional pattern in aggressive human prostate cancer. J Clin Invest 119(5):1093–1108
60. Dulak J, Jozkowicz A, Dembinska-Kiec A, Guevara I, Zdzienicka A, Zmudzinska-Grochot D, Florek I, Wojtowicz A, Szuba A, Cooke JP (2000) Nitric oxide induces the synthesis of vascular endothelial growth factor by rat vascular smooth muscle cells. Arterioscler Thromb Vasc Biol 20(3):659–666
61. Jozkowicz A, Cooke JP, Guevara I, Huk I, Funovics P, Pachinger O, Weidinger F, Dulak J (2001) Genetic augmentation of nitric oxide synthase increases the vascular generation of VEGF. Cardiovasc Res 51(4):773–783
62. Namba T, Koike H, Murakami K, Aoki M, Makino H, Hashiya N, Ogihara T, Kaneda Y, Kohno M, Morishita R (2003) Angiogenesis induced by endothelial nitric oxide synthase gene through vascular endothelial growth factor expression in a rat hindlimb ischemia model. Circulation 108(18):2250–2257
63. Ghiso N, Rohan RM, Amano S, Garland R, Adamis AP (1999) Suppression of hypoxia-associated vascular endothelial growth factor gene expression by nitric oxide via cGMP. Invest Ophthalmol Vis Sci 40(6):1033–1039
64. Dulak J, Jozkowicz A (2003) Regulation of vascular endothelial growth factor synthesis by nitric oxide: facts and controversies. Antioxid Redox Signal 5(1):123–132
65. Metzen E, Zhou J, Jelkmann W, Fandrey J, Brune B (2003) Nitric oxide impairs normoxic degradation of HIF-1alpha by inhibition of prolyl hydroxylases. Mol Biol Cell 14(8): 3470–3481
66. Park YK, Ahn DR, Oh M, Lee T, Yang EG, Son M, Park H (2008) Nitric oxide donor, (+/−)-S-nitroso-N-acetylpenicillamine, stabilizes transactive hypoxia-inducible factor-1alpha by inhibiting von Hippel-Lindau recruitment and asparagine hydroxylation. Mol Pharmacol 74(1): 236–245
67. Ball KA, Nelson AW, Foster DG, Poyton RO (2012) Nitric oxide produced by cytochrome c oxidase helps stabilize HIF-1alpha in hypoxic mammalian cells. Biochem Biophys Res Commun 420(4):727–732
68. Castello PR, David PS, McClure T, Crook Z, Poyton RO (2006) Mitochondrial cytochrome oxidase produces nitric oxide under hypoxic conditions: implications for oxygen sensing and hypoxic signaling in eukaryotes. Cell Metab 3(4):277–287
69. Lee RC, Feinbaum RL, Ambros V (1993) The C. elegans heterochronic gene lin-4 encodes small RNAs with antisense complementarity to lin-14. Cell 75(5):843–854
70. Chan SY, Loscalzo J (2010) MicroRNA-210: a unique and pleiotropic hypoxamir. Cell Cycle 9(6):1072–1083

71. Lei Z, Li B, Yang Z, Fang H, Zhang GM, Feng ZH, Huang B (2009) Regulation of HIF-1alpha and VEGF by miR-20b tunes tumor cells to adapt to the alteration of oxygen concentration. PLoS One 4(10):e7629
72. Taguchi A, Yanagisawa K, Tanaka M, Cao K, Matsuyama Y, Goto H, Takahashi T (2008) Identification of hypoxia-inducible factor-1 alpha as a novel target for miR-17-92 microRNA cluster. Cancer Res 68(14):5540–5545
73. Zhang Z, Sun H, Dai H, Walsh RM, Imakura M, Schelter J, Burchard J, Dai X, Chang AN, Diaz RL, Marszalek JR, Bartz SR, Carleton M, Cleary MA, Linsley PS, Grandori C (2009) MicroRNA miR-210 modulates cellular response to hypoxia through the MYC antagonist MNT. Cell Cycle 8(17):2756–2768
74. Liu F, Lou YL, Wu J, Ruan QF, Xie A, Guo F, Cui SP, Deng ZF, Wang Y (2012) Upregulation of microRNA-210 regulates renal angiogenesis mediated by activation of VEGF signaling pathway under ischemia/perfusion injury in vivo and in vitro. Kidney Blood Press Res 35(3):182–191
75. Fasanaro P, D'Alessandra Y, Di Stefano V, Melchionna R, Romani S, Pompilio G, Capogrossi MC, Martelli F (2008) MicroRNA-210 modulates endothelial cell response to hypoxia and inhibits the receptor tyrosine kinase ligand Ephrin-A3. J Biol Chem 283(23): 15878–15883
76. Lou YL, Guo F, Liu F, Gao FL, Zhang PQ, Niu X, Guo SC, Yin JH, Wang Y, Deng ZF (2012) *m*iR-210 activates notch signaling pathway in angiogenesis induced by cerebral ischemia. Mol Cell Biochem 370(1–2):45–51
77. Levy AP, Levy NS, Wegner S, Goldberg MA (1995) Transcriptional regulation of the rat vascular endothelial growth factor gene by hypoxia. J Biol Chem 270(22):13333–13340
78. Shima DT, Deutsch U, D'Amore PA (1995) Hypoxic induction of vascular endothelial growth factor (VEGF) in human epithelial cells is mediated by increases in mRNA stability. FEBS Lett 370(3):203–208
79. Levy AP, Levy NS, Goldberg MA (1996) Post-transcriptional regulation of vascular endothelial growth factor by hypoxia. J Biol Chem 271(5):2746–2753
80. Ray PS, Jia J, Yao P, Majumder M, Hatzoglou M, Fox PL (2009) A stress-responsive RNA switch regulates VEGFA expression. Nature 457(7231):915–919
81. Essafi-Benkhadir K, Onesto C, Stebe E, Moroni C, Pages G (2007) Tristetraprolin inhibits Ras-dependent tumor vascularization by inducing vascular endothelial growth factor mRNA degradation. Mol Biol Cell 18(11):4648–4658
82. Fellows A, Griffin ME, Petrella BL, Zhong L, Parvin-Nejad FP, Fava R, Morganelli P, Robey RB, Nichols RC (2012) AUF1/hnRNP D represses expression of VEGF in macrophages. Mol Biol Cell 23(8):1414–1422
83. Levy NS, Chung S, Furneaux H, Levy AP (1998) Hypoxic stabilization of vascular endothelial growth factor mRNA by the RNA-binding protein HuR. J Biol Chem 273(11):6417–6423
84. de Vries S, Naarmann de Vries IS, Urlaub H, Lue H, Bernhagen J, Ostareck DH, Ostareck-Lederer A (2013) Identification of DEAD-box RNA helicase 6 (DDX6) as a cellular modulator of vascular endothelial growth factor expression under hypoxia. J Biol Chem 288(8):5815–5827
85. Goldberg-Cohen I, Furneauxb H, Levy AP (2002) A 40-bp RNA element that mediates stabilization of vascular endothelial growth factor mRNA by HuR. J Biol Chem 277(16): 13635–13640
86. Dang LT, Lawson ND, Fish JE (2013) MicroRNA control of vascular endothelial growth factor signaling output during vascular development. Arterioscler Thromb Vasc Biol 33(2): 193–200
87. Chang SH, Lu YC, Li X, Hsieh WY, Xiong Y, Ghosh M, Evans T, Elemento O, Hla T (2013) Antagonistic function of the RNA-binding protein HuR and miR-200b in post-transcriptional regulation of VEGF-A expression and angiogenesis. J Biol Chem 288:4908–4921
88. Weijts BG, Bakker WJ, Cornelissen PW, Liang KH, Schaftenaar FH, Westendorp B, de Wolf CA, Paciejewska M, Scheele CL, Kent L, Leone G, Schulte-Merker S, de Bruin A (2012) E2F7 and

E2F8 promote angiogenesis through transcriptional activation of VEGFA in cooperation with HIF1. EMBO J 31(19):3871–3884

89. Li J, Ran C, Li E, Gordon F, Comstock G, Siddiqui H, Cleghorn W, Chen HZ, Kornacker K, Liu CG, Pandit SK, Khanizadeh M, Weinstein M, Leone G, de Bruin A (2008) Synergistic function of E2F7 and E2F8 is essential for cell survival and embryonic development. Dev Cell 14(1):62–75
90. Qin G, Kishore R, Dolan CM, Silver M, Wecker A, Luedemann CN, Thorne T, Hanley A, Curry C, Heyd L, Dinesh D, Kearney M, Martelli F, Murayama T, Goukassian DA, Zhu Y, Losordo DW (2006) Cell cycle regulator E2F1 modulates angiogenesis via p53-dependent transcriptional control of VEGF. Proc Natl Acad Sci USA 103(29):11015–11020
91. Shie JL, Wu G, Wu J, Liu FF, Laham RJ, Oettgen P, Li J (2004) RTEF-1, a novel transcriptional stimulator of vascular endothelial growth factor in hypoxic endothelial cells. J Biol Chem 279(24):25010–25016
92. Yan SF, Fujita T, Lu J, Okada K, Shan Zou Y, Mackman N, Pinsky DJ, Stern DM (2000) Egr-1, a master switch coordinating upregulation of divergent gene families underlying ischemic stress. Nat Med 6(12):1355–1361
93. Jin Y, Wu J, Song X, Song Q, Cully BL, Messmer-Blust A, Xu M, Foo SY, Rosenzweig A, Li J (2011) RTEF-1, an upstream gene of hypoxia-inducible factor-1alpha, accelerates recovery from ischemia. J Biol Chem 286(25):22699–22705
94. Messmer-Blust AF, Zhang C, Shie JL, Song Q, He P, Lubenec I, Liu Y, Sellke F, Li J (2012) Related transcriptional enhancer factor 1 increases endothelial-dependent microvascular relaxation and proliferation. J Vasc Res 49(3):249–259
95. Xu M, Jin Y, Song Q, Wu J, Philbrick MJ, Cully BL, An X, Guo L, Gao F, Li J (2011) The endothelium-dependent effect of RTEF-1 in pressure overload cardiac hypertrophy: role of VEGF-B. Cardiovasc Res 90(2):325–334
96. Stein I, Itin A, Einat P, Skaliter R, Grossman Z, Keshet E (1998) Translation of vascular endothelial growth factor mRNA by internal ribosome entry: implications for translation under hypoxia. Mol Cell Biol 18(6):3112–3119
97. Brahimi-Horn MC, Pouyssegur J (2005) The hypoxia-inducible factor and tumor progression along the angiogenic pathway. Int Rev Cytol 242:157–213
98. Loboda A, Jazwa A, Jozkowicz A, Molema G, Dulak J (2006) Angiogenic transcriptome of human microvascular endothelial cells: Effect of hypoxia, modulation by atorvastatin. Vascul Pharmacol 44(4):206–214
99. Loboda A, Stachurska A, Florczyk U, Rudnicka D, Jazwa A, Wegrzyn J, Kozakowska M, Stalinska K, Poellinger L, Levonen AL, Yla-Herttuala S, Jozkowicz A, Dulak J (2009) HIF-1 induction attenuates Nrf2-dependent IL-8 expression in human endothelial cells. Antioxid Redox Signal 11(7):1501–1517
100. Florczyk U, Czauderna S, Stachurska A, Tertil M, Nowak W, Kozakowska M, Poellinger L, Jozkowicz A, Loboda A, Dulak J (2011) Opposite effects of HIF-1alpha and HIF-2alpha on the regulation of IL-8 expression in endothelial cells. Free Radic Biol Med 51(10):1882–1892
101. Mizukami Y, Jo WS, Duerr EM, Gala M, Li J, Zhang X, Zimmer MA, Iliopoulos O, Zukerberg LR, Kohgo Y, Lynch MP, Rueda BR, Chung DC (2005) Induction of interleukin-8 preserves the angiogenic response in HIF-1alpha-deficient colon cancer cells. Nat Med 11(9):992–997
102. Yu L, Hales CA (2011) Long-term exposure to hypoxia inhibits tumor progression of lung cancer in rats and mice. BMC Cancer 11:331
103. Kalliomaki TM, McCallum G, Lunt SJ, Wells PG, Hill RP (2008) Analysis of the effects of exposure to acute hypoxia on oxidative lesions and tumour progression in a transgenic mouse breast cancer model. BMC Cancer 8:151
104. Volm M, Koomagi R (2000) Hypoxia-inducible factor (HIF-1) and its relationship to apoptosis and proliferation in lung cancer. Anticancer Res 20(3A):1527–1533
105. Volm M, Koomagi R, Mattern J, Efferth T (2002) Expression profile of genes in non-small cell lung carcinomas from long-term surviving patients. Clin Cancer Res 8(6):1843–1848

106. Leite de Oliveira R, Deschoemaeker S, Henze AT, Debackere K, Finisguerra V, Takeda Y, Roncal C, Dettori D, Tack E, Jonsson Y, Veschini L, Peeters A, Anisimov A, Hofmann M, Alitalo K, Baes M, D'Hooge J, Carmeliet P, Mazzone M (2012) Gene-targeting of Phd2 improves tumor response to chemotherapy and prevents side-toxicity. Cancer Cell 22(2): 263–277
107. Fleitas T, Martinez-Sales V, Vila V, Reganon E, Mesado D, Martin M, Gomez-Codina J, Montalar J, Reynes G (2013) VEGF and TSP1 levels correlate with prognosis in advanced non-small cell lung cancer. Clin Transl Oncol 15(11):897–902
108. Tenan M, Fulci G, Albertoni M, Diserens AC, Hamou MF, El Atifi-Borel M, Feige JJ, Pepper MS, Van Meir EG (2000) Thrombospondin-1 is downregulated by anoxia and suppresses tumorigenicity of human glioblastoma cells. J Exp Med 191(10):1789–1798
109. Wu P, Yonekura H, Li H, Nozaki I, Tomono Y, Naito I, Ninomiya Y, Yamamoto H (2001) Hypoxia down-regulates endostatin production by human microvascular endothelial cells and pericytes. Biochem Biophys Res Commun 288(5):1149–1154
110. Emara M, Obaid L, Johnson S, Bigam DL, Cheung PY (2007) The effect of hypoxia on plasma angiostatin and related factors in newborn pigs. Proc West Pharmacol Soc 50:47–52
111. Jin Y, An X, Ye Z, Cully B, Wu J, Li J (2009) RGS5, a hypoxia-inducible apoptotic stimulator in endothelial cells. J Biol Chem 284(35):23436–23443
112. An X, Jin Y, Guo H, Foo SY, Cully BL, Wu J, Zeng H, Rosenzweig A, Li J (2009) Response gene to complement 32, a novel hypoxia-regulated angiogenic inhibitor. Circulation 120(7): 617–627

MicroRNA Regulation of Angiogenesis

9

Chandan K. Sen and Yuk Cheung Cyrus Chan

Contents

9.1	Introduction	188
	9.1.1 microRNAs and Angiogenesis	190
9.2	Role of microRNAs in Developmental Angiogenesis	190
	9.2.1 Global Loss of miR: Dicer Deficiency	190
	9.2.2 Specific miRs in Developmental Angiogenesis	191
9.3	microRNA-Dependent Tissue Regeneration Angiogenesis	193
	9.3.1 Developmental AngiomiRs	193
	9.3.2 Hypoxia-Responsive miRs (HypoxamiRs)	193
	9.3.3 miRs Regulating HIF Signaling	196
	9.3.4 Ischemia-Associated miRs	197
9.4	Tumor Angiogenesis	198
	9.4.1 miRs Regulating VEGF Signaling	198
	9.4.2 miRs Regulating Secretory Factors	201
	9.4.3 miRs as Secretory Factors	201
	9.4.4 miRs Regulating Metastatic Angiogenesis	202
References		205

Abstract

Angiogenesis is a key biological response regulating embryonic development, tissue repair, tumor growth, and metastasis. It requires complex regulation of gene expression in a temporal-spatial manner. microRNAs (miRs) are 18–24 nucleotide-containing endogenous RNAs that exert substantial gene regulatory effects via a posttranscriptional mechanism. Recent advance in basic research revealed that miRs exert potent angiogenic control via direct targeting certain critical secretory factors and transcription factors in a cell autonomous and

C.K. Sen (✉) • Y.C.C. Chan
Department of Surgery, Davis Heart and Lung Research Institute, The Ohio State University Medical Center, 473 West 12th Avenue, Columbus, OH 43210, USA
e-mail: chandan.sen@osumc.edu

J. Dulak et al. (eds.), *Angiogenesis and Vascularisation*,
DOI 10.1007/978-3-7091-1428-5_9, © Springer-Verlag Wien 2013

non-cell autonomous manner. This chapter comprehensively summarizes the current understanding in miR regulation of angiogenesis during embryonic stage, tissue injury, as well as tumorigenesis. Approaches to deliver anti-miR entity have been successfully developed in silencing miR level in vivo. miR-based therapy might provide specific and effective solution for managing ischemic diseases or controlling tumor growth and metastasis.

Keywords
MicroRNAs • AngiomiRs • HypoxamiRs • Tissue repair • miR-210 • miR-200b • miR-126

9.1 Introduction

The central dogma of molecular biology states that the flow of genetic information is from DNA to RNA and then to protein [1]. Transcription, followed by translation, has been thought as the rule of thumb for eliciting the biological actions in the cells. The remaining nonprotein coding regions, which constitute enormous proportion of the whole genome, were once disregarded as sequence having no discernible functions and no selective advantage to the organism [2, 3]. In 1993, however, the findings from Dr. Ambros and his colleagues had entirely shifted the paradigm and redefined the function of these so-called "junk DNA." They reported that endogenous nonprotein coding short RNAs silence gene expression and dictate the phenotype of the cells in *Caenorhabditis elegans* (*C. elegans*) [4]. Later findings from the same group and others indicated that similar mechanisms are present in plants and wide range of mammalian cells which are phylogenetically conserved [5, 6]. As the size of these RNAs is extraordinarily small, consisting of around 18–24 nucleotides, they were thus named "microRNAs" (miRs).

According to the miRBase (http://www.miRbas.org), more than 1,500 human miRs have been identified [7] and were predicted to control the expression of over 30 % of genes in the entire genome [8]. In general, miRs silence gene expression via complementarity binding with the 3′ untranslated region (UTR) of target messenger RNAs (mRNAs). Binding of miR to 3′ UTR suppresses protein translation via hindering translation initiation, elongation, and ribosomal protein assembly [9]. In some cases, the mRNA–miR duplex triggers deadenylation and decapping of the 5′ cap structure, leading to transcript instability [9]. The sequence responsible for 3′ UTR interaction comprising of seven to eight nucleotides is known as "seed sequence." Such short sequence requirement allows a single miR to target multiple mRNAs, thus exerting robust gene regulatory effects. Moreover, some of the targets are transcription factors or repressors which further amplify particular biological functions via transcriptional control of an array of genes.

The biogenesis of miRs is a highly orchestrated process, essentially requiring the coordination of ribonucleases, RNA-binding proteins, and the miR genes [10]. Transcription and appropriate truncation of the nucleic acid, which is known as

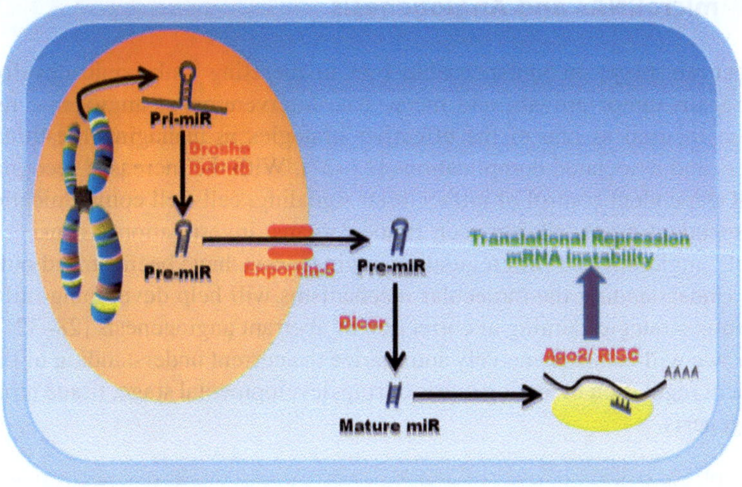

Fig. 9.1 An overview of miR biogenesis. miRs are transcribed as primary miRs (pri-miRs) which are subsequently cleaved by Drosha/DGCR8 microprocessor to form premature miRs (pre-miRs). The pre-miRs are transported from nucleus to cytoplasm via exportin-5. In the cytosol, pre-miRs are further processed by dicer to form 22 nucleotide mature miRs, which are then loaded into Argonaute-2 (Ago2)-associated RNA-inducing silencing complex (RISC) to elucidate gene silencing effects

miR maturation, are vital for the correct miR strand synthesis. These small RNAs are encoded as miR genes, locating in either intragenic or intergenic sequence in the genome. Most miRs are generated by polymerase II, but certain large miR transcripts can be transcribed by polymerase III [11]. Primary miR (pri-miR) firstly interacts with a complex known as "microprocessor" consisting of DiGeorge critical region 8 (DGCR8) and Drosha [12–14], resulting in cleavage to form a premature miRs (pre-miRs). These pre-miRs are transported out from nucleus via a double-stranded RNA-binding protein exportin-5 [15–17]. In the cytosol, pre-miRs are further processed by forming double-stranded RNAs containing approximately 22 nucleotides. In order to elucidate the gene silencing effects, the double-stranded RNA are unwound, correct mature miRs (known as guide strand) being selected and incorporated into RNA-Induced Silencing Complex (RISC) [18]. Once loaded, the guide strand is brought in close contact with the target mRNA, exerting gene silencing effect via complementarity binding. RISC effector protein Argonaute-2 (Ago2), via binding to cap region, conceals the interaction between eukaryotic initiation factor (EIF) 4E and thus suppresses initiation of translation [19, 20]. Ago2 possesses endonuclease activity which exposes the 3′ end of the mRNA, subsequently facilitating the degradation in an exonuclease-dependent mechanism [21]. An overview of the key processes involved in miR biogenesis is illustrated in Fig. 9.1.

9.1.1 microRNAs and Angiogenesis

Angiogenesis plays an indispensable role in directing embryonic development, tissue repair, tumor growth, and metastasis. Intervention of angiogenic response has been focused as one of the effective strategies in managing life-threatening disorders and associated complications [22–27]. With the increased recognition of the miR-dependent control of intracellular signaling, cell–cell communication, and miR-dependent cross talk between tissues/organs, investigations focusing on the study of "angiomiRs," miRs regulating angiogenesis, have been carried out extensively. Understanding the molecular mechanisms will help developing miR-based therapeutic strategies aiming at correction of aberrant angiogenesis [28–32]. In this chapter, we will comprehensively summarize the current understanding of the miR-dependent regulation of angiogenesis during developmental stage, tissue repair, and tumorigenesis.

9.2 Role of microRNAs in Developmental Angiogenesis

9.2.1 Global Loss of miR: Dicer Deficiency

The first line of evidence supporting the notion of miR regulation of developmental angiogenesis stemmed from the findings of Dr. Zhao's group in 2005 [33]. The striking phenotypic defect in angiogenesis of mouse embryo after homozygous deletion of dicer pinpointed the critical role of miRs in dictating angiogenesis during development [33]. Global depletion of miRs resulting from dicer knockout leads to severe impairment in blood vessel formation [34] and reduces expression of PECAM, an endothelial marker, in yolk sac [33]. Reminiscent of dicer-deficient mouse embryo, genetic ablation of dicer in zebrafish results in perturbation of blood circulation [35], suggesting that miR regulation of developmental angiogenesis is highly conserved among species. In fact, this is consistent with the finding from our group and others reporting that dicer knockdown in endothelial cells results in severe impairment in angiogenic response [36–40]. Loss of dicer stalls endothelial proliferation and migration [37–39]. Efforts in dissecting the underlying mechanisms revealed that endothelial cells with dicer deficiency fails to effectively produce reactive oxygen species (ROS), a critical intracellular signal promoting endothelial proliferation and tubulogenesis, in response to angiogenic stimuli [39]. These findings are further supported by abrupt phenotypic defect in endothelial-specific dicer knockout mice. Loss of dicer in endothelial cells, but not in other non-endothelial region, is sufficient enough to compromise blood flow reestablishment and endothelial proliferation in murine model of ischemic hindlimb [36]. Interestingly, the honeycomb-like primary vascular structure is intact in dicer-deficient mouse embryo [33], indicating that vasculogenesis and the initial steps of angiogenesis are not affected in the mutants. It should be noted that embryonic lethality of global dicer knockout animals [33, 35] is not likely due to miR deficiency in endothelial cells since endothelial-specific dicer-deficient mice exhibit normal development and are indistinguishable from their littermate during embryonic stage [36]. Loss of dicer in non-endothelial cells confers the embryonic lethality which is independent of compromised angiogenesis during development.

9.2.2 Specific miRs in Developmental Angiogenesis

Since the discovery of dicer-dependent regulation of angiogenesis, scientists have been trying to look for specific miRs regulating angiogenesis during development. One of the first miRs identified as a key player in developmental angiogenesis is miR-126 [41–43], a miR encoded from the intronic region of endothelial cell-specific peptide EGF-like domain 7 (Egfl7). Injection of morpholinos (MOs) targeting pri-miR-126 into fertilized eggs of zebrafish reduces the abundance of endothelial cells in the developing vasculature, narrowed lumen diameter in branchial arch vessels, and occasional cranial hemorrhages [42]. Similarly, genetic ablation of miR-126 in mouse embryo results in 40–50 % of embryonic lethality characterized by severe systemic edema, multifocal hemorrhages, ruptured blood vessels, abnormal thickening of endothelial sprouts, retarded growth of the cranial vessels, and impaired vascularization of retina [41, 43]. The pro-angiogenic effects of miR-126 depend on the positive regulation of vascular endothelial growth factor (VEGF) pathways. miR-126 targets sprouty-related, EVH1 domain-containing protein 1 (SPRED1) [41–43], vascular cell adhesion molecule-1 (VCAM-1) [42, 44], and phosphoinositide 3-kinase regulatory subunit 2 (PI3KR2) [42, 43]. Of noted, both SPRED1 and PI3KR2 exert negative effects on VEGF signaling via repressing the activation of extracellular-regulated kinase 1/2 (ERK1/2) and Akt, respectively. Thus, miR-126 induction in developing embryo silences SPRED1 and PI3KR2, enabling VEGF signaling and directing proper angiogenesis.

Apart from miR-126, numerous VEGF-regulating miRs have been characterized to play a key role in developmental angiogenesis. miR-10, for example, is highly expressed in endothelium during blood vessel development. Anti-miR-10 injection to zebrafish embryo significantly impairs the sprouting of intersegmental vessels (ISVs) and displayed defective formation of the dorsal longitudinal anastomotic vessels [45]. miR-10 directly targets Flt-1 [45], a decoy receptor for VEGF. In this regard, elevated miR-10 expression during development silences Flt-1, resulting in enhanced bioavailability of VEGF for activation of VEGFR2 (or KDR), the major pro-angiogenic VEGF receptor in endothelial cells [45]. miR-30 family, another miR cluster regulating VEGF signaling, has been shown to modulate the vasculature development during embryonic stage. Bridge et al. reported that the expression of miR-30b and -30c is downregulated from 18 to 30 h postfertilization of zebrafish embryo, which corresponds to the temporal window of angiogenic sprouting [46]. Injection of synthetic miR-30 to fertilized eggs results in hyperbranching of ISVs in a dose-dependent manner [46]. Experiments aiming at dissecting the underlying mechanisms revealed that delta-like 4 (Dll4), a negative regulator of VEGF signaling, serves as a direct miR-30 target. While overexpression of miR-30 mitigates Dll4 expression in zebrafish embryo and endothelial cells, co-injection of Dll4

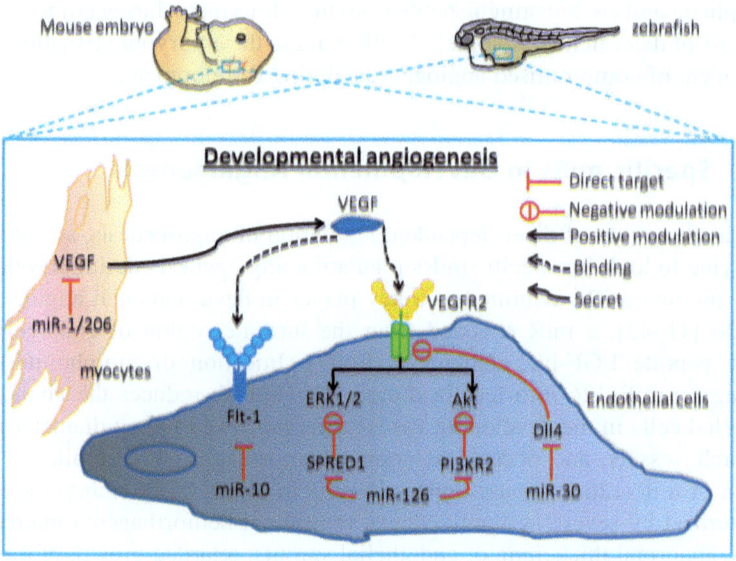

Fig. 9.2 Role of miRs in developmental angiogenesis. Findings from basic research using mouse embryo and zebrafish as models indicated that developmental angiogenesis is controlled by miR-126, which directly targets SPRED-1 and PI3KR2, supporting angiogenesis via activation of both ERK1/2 and Akt. miR-10 enables angiogenesis by targeting Flt-1 which enhances VEGF bioavailability. miR-30 positively regulates angiogenesis via targeting Dll4, a negative regulator of VEGFR2. Muscle-derived miR-1/206 modulates developmental angiogenesis by restricting VEGF bioavailability

target protector or KDR antisense partially attenuates the miR-30-induced excessive ISV branching, indicating that Dll4/VEGF signaling plays an obligatory role in mediating pro-angiogenic response of miR-30 [46]. The miR-dependent regulation of developmental angiogenesis does not only confine in endothelial cells but also occurs in non-endothelial compartment. An interesting observation in the findings from Drs. Giraldez and Mishima's group indicated that muscle-enriched miR-1/206 negatively regulates developmental angiogenesis via silencing VEGF [47]. Delivery of antisense MOs to zebrafish targeting miR-1/206 results in enlargement of ISVs and enrichment of endothelial cells. Transplantation of ISVs from a miR-1/206 MO-injected zebrafish to wild-type recipient, however, displays a normal vessel phenotype. These data pinpointed that non-vessel-derived miR-1/206 is a critical angiostatic signal. This miR cluster at physiological level prevents the overproduction of VEGF, thus directing proper vasculature growth during development in a non-cell-autonomous manner [47]. These findings not only revealed the complex cross talk between vasculature and non-vessel compartment but also underscored the indirect angiogenic control by non-endothelial miRs during developmental stage. Figure 9.2 summarizes the key miR-dependent pathways regulating developmental angiogenesis.

9.3 microRNA-Dependent Tissue Regeneration Angiogenesis

9.3.1 Developmental AngiomiRs

Reestablishment of blood flow by angiogenesis is critical in dictating adult tissue repair in response to injury. Sufficient blood supply not only delivers oxygen and nutrients essential for reparative procedure but also enables mobilization of bone marrow-derived stem cells for regeneration, a process known as stem cell homing. Tissue repair revascularization shares similarities with developmental (both embryonic and postnatal) angiogenesis in terms of cellular morphological changes, production of angiogenic factors, and mitogenic effects of vascular cells [48]. Notably, recent publications revealed that angiomiRs regulate angiogenesis in both developmental and adult stage. miR-126, for example, not only supports developmental angiogenesis (see above section) but also controls angiogenic response during ischemic injury. Treatment with anti-miR-126 reduces capillary density in the gastrocnemius muscle in ischemic hindlimb model [49]. Angiogenic early outgrowth cells (EOCs) isolated from chronic heart failure (CHF) patients reveal a loss of miR-126 [50]. While ex vivo treatment of anti-miR-126 blocks angiogenic behavior of EOCs impairing cardiac neovascularization, miR-126 delivery substantially promotes tubulogenesis and facilitates the angiogenesis in the murine model of myocardial infarction [50]. Transplantation of mesenchymal stem cells (MSCs) overexpressing miR-126 improves myocardial blood flow and microvessel density in ischemic heart [51, 52]. Similar to developmental angiogenesis, miR-126 induction results in activation of ERK1/2 and Akt pathways promoting cardioprotection [52]. High level of miR-126 in circulating endothelial progenitor cells (EPCs) correlates with better survival rate of CHF patients and is an independent predictor for cardiac death in human ischemic cardiomyopathy [53].

9.3.2 Hypoxia-Responsive miRs (HypoxamiRs)

Tissue injury is either triggered by (ischemia) or associated with (hemorrhagic or wounding) insufficient oxygen supply. As an immediate self-protective response, cells acutely exposed to hypoxia (a state where partial pressure of oxygen is lower than the physiological one) activate pathways essential for survival as well as signals for reestablishing oxygen supply. Indeed, numerous angiogenic factors, namely, VEGF, fibroblast growth factor (FGF), and hepatocyte growth factor (HGF), are sensitive to low oxygen environment. Certain hypoxia-sensitive miRs, known as hypoxamiRs, have been characterized to serve as key mediators in tissue repair angiogenesis [54]. Among all candidates, miR-210 is one of those which merit investigation owing to its robust induction under hypoxia in nearly all kind of cells [25]. Both mature and primary miR-210 are upregulated under the condition of low oxygen environment [55]. Given the fact that nearly 90 % of gene expression is regulated by hypoxia-inducing factor (HIF) when oxygen is depleted, it is not surprising that the transcription of miR-210 depends on HIF transactivation.

Indeed, miR-210 promoter region contains HIF binding sites [55, 56]. Constitutively active HIF-1 or HIF-2 induces miR-210 expression under normoxic conditions [56]. Two genes located within 33 kb downstream of the miR-210 gene, namely, RASSF7 and HRAS, are hypoxia inducible, suggesting that the local chromatin structure might facilitate the expression of genes within the loci via a de novo synthesis mechanism [57, 58]. Fasanaro et al. firstly reported that hypoxia-induced miR-210 enables angiogenesis via silencing receptor tyrosine kinase ligand Ephrin-A3 [58]. Knockdown of miR-210 or overexpression of Ephrin A-3 in endothelial cells blunts hypoxia-associated pro-angiogenic response [58]. Upregulation of miR-210 is associated with VEGF induction in renal ischemia/reperfusion (I/R) injury [59] and Notch activation during cerebral ischemia [60]. Treatment with minicircle DNA carrying miR-210 significantly enhances angiogenesis and improves cardiac functions during myocardiac infarction by targeting protein tyrosine phosphatase 1B (Ptp1b) [61], a phosphatase inactivating VEGFR2 signal. Similarly, intra-articular injection of double-stranded (ds) miR-210 augments angiogenesis via upregulation of VEGF and FGF2 in a rodent model of anterior cruciate ligament injury [62]. In this regard, hypoxia-induced miR-210 upregulation serves as an endogenous self-protective response compensating insufficient blood supply during ischemia. The pro-angiogenic effects of miR-210 were not only evidenced in endothelial cells but also validated in stem (progenitor) cell-mediated vasculogenesis. VEGF treatment induces differentiation of progenitor cells towards an endothelial lineage with the concomitant induction of miR-210 [63]. Induction of miR-210 in stem cells phenocopies the effect of VEGF treatment on tubulogenesis [63]. Interestingly, ex vivo miR-210 knockdown partially attenuates the angiogenic ability of VEGF-treated stem cells, indicating that VEGF-induced angiogenesis is partly due to miR-210 induction [63]. Apart from direct modulation of angiogenic effectors, recent investigations unveiled that miR-210 serves as a potent self-feed-forward signal further amplifying the hypoxic response. Knockdown of miR-210 in hypoxia attenuates the HIF activity [64]. miR-210 directly silences glycerol-3-phosphate dehydrogenase 1-like (GPD1L) [65], a positive regulator of prolyl hydroxylases (PHD2). In light of the fact that PHD2 promotes the degradation of HIF, hypoxia-associated induction of miR-210 downregulates GPD1L and PHD2, thus preventing HIF degradation. Stabilized HIF allows miR-210 promoter transactivation, supporting HIF signaling and therefore establishing self-feed-forward loop.

miR-23∼27∼24 cluster is another well-characterized hypoxamiRs regulating angiogenesis during tissue repair [66]. Kulshreshtha et al. reported that hypoxia upregulates the expression of miR-24 via HIF activation [56]. During cardiac ischemia, miR-24 is specifically upregulated in endothelial cells [67]. Overexpression of miR-24 induces endothelial cells apoptosis and impairs angiogenesis in vitro [67]. Treatment with anti-miR-24 improves neovascularization and cardiac function in a murine model of myocardial infarction [67]. The authors identified that GATA binding protein 2 (GATA2) and p21-activated kinase PAK4 serve as direct targets of miR-24 which accounts for the angiostatic effects [67]. On the contrary, miR-23 and -27 appear to exert pro-angiogenic effects during tissue injury. Laser-induced choroidal injury upregulates the expression of miR-23a, -23b, -27a, -27b, and -24 [66]. Anti-miR-23/27 treatment blocks the choroidal neovascularization (CNV) [66]. The pro-angiogenic effect of miR-23/27 is due to direct silencing of angiostatic protein Sprouty-2 and soluble repulsive factor semaphorin 6A (SEMA6A) [66]. This is in line with the findings from another group reporting that miR-27a/b promotes embryonic vessel formation and inhibits angiogenic response in human adult endothelial cells via silencing SEMA6A [68]. The presence of both pro-angiogenic and anti-angiogenic miRs in the same cluster implies a possible self-regulatory mechanism aiming at achieving optimal angiogenic signal and prevention of exaggerated angiogenic outcome.

On the other hand, certain miRs such as miR-16 family members (miR-503, -15a/b, and -16) are downmodulated upon hypoxia exposure or HIF stabilization [69, 70]. Forced overexpression of miR-503 in endothelial cells retards proliferation, compromises migratory capacity, and impairs tubulogenesis in vitro. miR-503 exerts anti-angiogenic effects via silencing multiple transcripts including cyclin E1 (CCNE1) [71], cell division cycle 25 homolog A (CDC25a) [71], VEGF [70], FGF2 [70, 72], and FGFR1 [72]. Elevation of miR-503 was evidenced in myocardial microvascular endothelial cells from diabetic rat [73] and diabetic ischemic limb muscles from both mouse and human [71]. Inhibition of miR-503 normalizes post-ischemic recovery and muscular neovascularization in diabetic mice, with the concomitant restoration of CDC25a and CCNE1 [71]. Similarly, the expression of miR-16 and -15a/b is upregulated in blood or EPCs from patients suffering from ischemic limb. Forced overexpression of miR-16 in endothelial cells stalls proliferation, impairs migration, and inhibits tubulogenesis [74]. miR-15a delivery mitigates endothelial cell migration and differentiation [75]. Mesenchymal cells with miR-16 overexpression attenuate the angiogenic commitment of endothelial cells in a coculture system [76]. The pro-angiogenic capacity is improved when the isolated EPCs are treated ex vivo with anti-miR15a/16. These findings are further supported by the observation from a recent study reporting that endothelium-selective miR-15a transgenic mice exhibit impaired angiogenesis and blood flow in ischemic hindlimb [75]. Both miR-15a and -16 directly silence the angiogenic factors and their receptors including VEGF [75], FGF2 [75], VEGFR2 [74, 77], and FGFR1 [74, 77] to elucidate the angiostatic effects.

miR-200b is another angiostatic miR which is downregulated upon hypoxic challenge and forced HIF stabilization [78]. Our group identified that HIF-associated transcription factor v-ets erythroblastosis virus E26 oncogene homolog 1 (Ets-1—*see Chap. 6 for review of Ets*) serves as a direct target of miR-200b [78]. Indeed, hypoxia-associated induction of Ets-1 and pro-angiogenic behavior is blunted upon delivery of exogenous miR-200b [78]. miR-200b not only targets Ets-1 but also silences proteins essential for VEGF signaling such as VEGF [79] and VEGFR2 [80, 81]. In human and murine diabetic retina, miR-200b is significantly downmodulated, which is associated with VEGF overproduction. Treatment with synthetic miR-200b blocks the overexpression of VEGF as well as attenuates aberrant murine choroidal angiogenesis. Our group reported that dermal microvascular endothelial cells exhibits downregulation of miR-200b acutely in response to injury. Forced overexpression of miR-200b in vivo not only blocks injury-associated blood flow and endothelial abundance but also stalls wound closure [81]. The downregulation of miR-200b is associated with an induction of endothelial VEGFR2 and GATA2. While miR-200b inhibits tubulogenesis in vitro [78–81], overexpression of GATA2/VEGFR2 attenuates the anti-angiogenic effects brought by miR-200b delivery. Moreover, VEGF-induced ERK1/2 activation, a well-known pro-angiogenic signal, is blunted in miR-200b overexpressing endothelial cells [80], suggesting that downregulation of miR-200b supports VEGF signaling during cutaneous wound injury. Interestingly, we further identified that miR-200b-VEGFR2/GATA2 signaling is disrupted in diabetic skin wounds [81]. The aberrant miR-200b expression is associated with exaggerated inflammatory response as neutralization of tumor necrosis factor (TNF) α is able to restore wound angiogenesis and rectifies the corresponding downstream pathways [81]. It should be noted that miR-200b is not only regulated by HIF but also modulates HIF activity. PHD2 is a predicted target of miR-200 family [82]. Overexpression of miR-200b silences the expression of PHD2 with the concomitance stabilization of HIF [82]. However, whether this feedback mechanism exists in endothelial cells remains elusive.

9.3.3 miRs Regulating HIF Signaling

Owing to their ability to interfere gene expression, miRs can directly act on HIF protein (or proteins regulating HIF stability), thus fine-tuning hypoxic signaling. One of the most well-characterized examples is the interaction of miR-17-92 cluster on HIF-1α transcript. miR-17-92 cluster directly binds to the 3′ UTR of HIF-1α and negatively regulates its expression [83]. Interestingly, miR-17-92 cluster is downregulated during hypoxic challenge via p53-dependent mechanism [84]. The miR-17-92 cluster serves as an endogenous signal blocking angiogenesis. Delivery of synthetic miR-17-92 cluster to endothelial cells blocks sprout formation in the spheroid assay [85]. In particular, the potent angiostatic effect is partially attributed by miR-17/20 as intravenous injection of antisense targeting these miRs (but not miR-18a, -19a, or -20a alone) promotes endothelial cell proliferation in mouse [85]. The authors further characterized that Jak1, a key tyrosine kinase involved in STAT3 pathway, serves as a direct miR-17 target [85]. miR-17 is highly specific for pro-angiogenic control under physiological condition but not tumor angiogenesis. Anti-miR-17 treatment neither increases perfusion of the implanted tumors nor affects tumor size [85], indicating a distinct discrepancy between miR regulation of physiological and pathological angiogenesis. Another miR-17-92 cluster member involved in angiostatic control is miR-92a. miR-92a not only controls the sprouting of endothelial cells in vitro but also implicates in tissue repair angiogenesis [86]. Delivery of anti-miR-92a improves blood flow reestablishment in ischemic hindlimb and promotes neovascularization during myocardial infarction [86]. miR-92a targets integrin αV (ITGA5) [86], which is a critical transmembrane protein for endothelial migration.

miR-199a is another hypoxamiR regulating tissue regeneration angiogenesis. This intronic miR not only directly targets HIF but also silences SIRT1, a deacetylase inhibiting PHD2 expression [87]. Interestingly, miR-199a is downregulated during cardiac ischemia–reperfusion injury and upon hypoxia exposure [87]. In this regard, ischemic or hypoxic challenge drives miR-199a downregulation, leading to derepression of HIF. On the other hand, loss of miR-199a results in elevation of SIRT1 expression, subsequently inhibiting PHD2 expression. These mechanisms act simultaneously enabling HIF stabilization. Interestingly, we identified that miR-199a expression is downregulated in dermal microvascular endothelial cells during cutaneous wounding [88]. Induction of miR-199a blocks tubulogenesis in vitro while inhibition of endogenous miR-199a enhances the aggressiveness of endothelial cells [88]. We further characterized that Ets-1-matrix metalloproteinase (MMP1) pathway is involved in miR-199a-dependent angiogenic control [88]. Not only miR-199a itself but also another member of the same miR cluster exerts angiostatic effect on endothelial cells. miR-214 lies in the same cluster with miR-199a and the expression is under the control of the same promoter activity [89]. It was reported that miR-214 is highly expressed in endothelial cells, vascular smooth muscle cells (VSMCs), and vascular fibroblasts (vFBs) [90, 91]. Induction of miR-214 inhibits murine retinal developmental angiogenesis and exerts potent angiostatic effect on embryonic endothelial cells (MEECs) and human adult microvascular endothelial cells (HMECs). The underlying mechanism involving anti-angiogenic effects of miR-214 depends on direct targeting Quaking (QKI) [90], a RNA-binding protein regulating VEGF secretion.

9.3.4 Ischemia-Associated miRs

Using a high-throughput microarray analysis, Grundmann et al. reported that miR-100 is significantly downregulated in adductor muscle and isolated endothelial cells during ischemia [92]. Overexpression of miR-100 blocks the angiogenic effects and proliferation of endothelial cells [92]. Experiments aiming at dissecting the underlying mechanism revealed that mammalian target of rapamycin (mTOR) serves a direct target of miR-100 [92]. While intravenous injection of anti-miR-100 improves blood flow and endothelial cell proliferation in ischemic hindlimb, co-treatment of rapamycin significantly reverses the protective effect of anti-miR-100 [92], indicating that miR-100 mediates angiostatic effect via rapamycin-dependent pathway. The miR-dependent angiogenic control is not only restricted to endothelial cells but also evidenced in bone marrow-derived stromal cells during ischemic injury. A recent report from Dr. George's research team indicated that miR-106b-25 cluster, a miR cluster that is elevated in murine ischemic hindlimb model and myocardial infarction, modulates bone marrow-derived stromal cell-dependent tissue repair angiogenesis [93]. Mice with genetic deletion of miR-106-25 cluster (miR-106-25 KO mice) reveal blunted blood flow reestablishment and neovascularization in ischemic hindlimb model [93]. The bone marrow-derived cells from miR-106-25 KO mice exhibit impaired tubulogenesis, with the concomitant

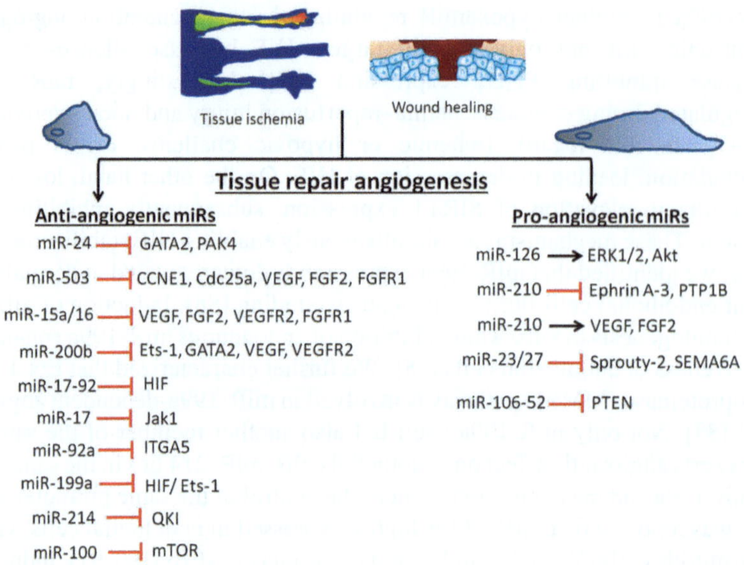

Fig. 9.3 An overview of angiomiRs in regulating tissue repair angiogenesis. miR-126 supports angiogenic response via activation of ERK1/2 and Akt in response to injury. miR-210 promotes angiogenesis via secretion of FGF2 and VEGF and repressing the expression of angiostatic proteins such as Ephrin-A3 and PTP1B. Pro-angiogenic miRs including miR-23/27 and -106/25 cluster enable angiogenic response via Sprouty-2 as well as SEMA6A and PTEN, respectively, during tissue injury. Angiostatic miRs, namely, miR-24, -503, -15a/16, -200b, -17-92 cluster, -17, -92a, -199a, -214, -100, block angiogenesis by directly targeting numerous pro-angiogenic transcripts including GATA2, PAK4, CCNE1, Cdc25a, VEGF, VEGFR2, FGF2, FGFR1, Ets-1, HIF, Jak1, ITGA5, QKI, and mTOR

reduction in secretion of pro-angiogenic factors [93]. Similar angiostatic effect of miR-106-25 deficiency was observed in residential endothelial cells, suggesting that this miR cluster serves as an angiogenic switch in both cell types by silencing similar downstream targets. The authors further demonstrated that miR-106-25 targets phosphatase and tensin homolog (PTEN) [93], a negative regulator of Akt pathway, thus supporting the survival and proliferation pathway during tissue repair angiogenesis. Figure 9.3 summarizes the key miR-dependent pathways regulating tissue repair angiogenesis.

9.4 Tumor Angiogenesis

9.4.1 miRs Regulating VEGF Signaling

Tumor angiogenesis is a key step for expansion of cancer cells as they require enormous nutrient supporting proliferation and development. It is also critical for malignant progression of a tumor. Of noted, metastasis accounts for more than 90 %

of mortality associated with cancer [94]. It is thus of paramount importance to understand the biological clues attributing to this pathological angiogenesis in order to develop intervention inhibiting tumor growth and blocking the initiation of metastasis. As mentioned in the early section, VEGF is the key soluble factor supporting angiogenesis. Interestingly, a number of miRs that bind directly to VEGF 3′ UTR have been implicated in tumor angiogenesis. miR-15/16, for example, is lost in relapsed/refractory multiple myeloma (MM) patients [95, 96]. Similarly, downregulation of these miRs was observed in sample from human nasopharyngeal carcinoma [69]. Stable overexpression of miR-15a/-16 in the human myeloma cell mitigates the VEGF expression and reduces the number of vessel in the tumor, with the concomitant reduction in tumor volume [95]. After delivery of miR-15a and -16, mice bearing myeloma tumor burden reveal significant improvement in survival rate [95]. Another study reported that lower level of miR-29b is associated with metastatic behavior with human luminal breast cancer [97]. The authors identified that miR-29 directly interacts with the 3′ UTR of VEGF to elucidate the angiogenic control of tumor cells [97]. The angiostatic effect of miR-29b is also contributed by direct silencing of other soluble factors such as platelet-derived growth factor (PDGF) A, PDGFB, and PDGFC [97]. On the other hand, downregulation of miR-195, a miR that directly interacts with VEGF 3′ UTR, is associated with poor prognosis of human hepatocarcinoma as well as higher microvessel density in the tumor [98]. Conditioned medium from miR-195 overexpressing hepatocarcinoma cells (HCCs) exhibits a blunted endothelial migration and capillary tube formation [98]. This is associated with a reduced VEGF secretion from the HCCs in vitro and impaired VEGF expression in xenografts from the miR-195 overexpressing HCCs in vivo [98]. miR-195 not only targets VEGF but also silences VAV2, a guanine nucleotide exchange factor (GEF) regulating Rac1 signaling, and CDC42 to negatively modulate the metastatic behavior of HCCs.

Apart from targeting VEGF 3′ UTR, a number of miRs regulate VEGF-directed tumor angiogenesis via silencing transcription factors or cell surface receptors essential for VEGF production. miR-107, a hypoxia-upregulated miR, directly silences HIF1β in colon cancer cells. While blocking miR-107 induces HIF1β expression in cervical cancer cells and colon adenocarcinoma, delivery of miR-107 significantly downregulates HIF1β under normoxic and hypoxic condition [99]. Destabilization of HIF1β is associated with loss of VEGF secretion, reduced tumor volume, and blunted vessel density in tumor xenograft [99]. Similarly, neuroblastoma cells reveal a loss of miR-145, a miR targeting HIF-2α, compared with healthy dorsal ganglia [100]. Synthetic miR-145 delivery to neuroblastoma cells not only blocks proliferation and migration but also inhibits angiogenic response of endothelial cells. The angiostatic control is accompanied with the blunted VEGF expression and reduced number of intratumoral vessels [100]. Another miR controlling HIF-dependent VEGF production during tumorigenesis is miR-155. Upregulation of miR-155 enables tumor angiogenesis by silencing Von Hippel–Lindau (VHL), a protein facilitating HIF degradation, and is associated with poor prognosis in triple negative breast cancer [101]. Forced overexpression of miR-155 induces loss of VHL and accumulation of both HIF-1α and HIF-2α, enhances VEGF production, increases tumor size, and facilitates endothelial cell accumulation in tumor xenograft [101]. miR-9, a miR that directly target E-Cadherin, is enriched in breast cancer cells [94]. Elevated miR-9 leads to downregulation of E-Cadherin, resulting in impaired β-catenin activity and subsequent compromised VEGF expression [94]. Kumar et al. reported that loss of miR-34a was evidenced in head and neck squamous cell carcinoma (HNSCCs) [102]. Downregulation of miR-34a leads to derepression of E2F3, a E2F family of transcription factors regulating cell cycle progression, resulting in elevated secretion of VEGF [102]. In glioma cells, loss of miR-125b [103] and -128 [104] was evidenced compared with normal brain tissues. Both miR-125b and -128 negatively regulate VEGF expression via silencing Myc-associated zinc finger protein (MAZ) [103] and p70S6K1 [104], respectively. Besides MAZ, miR-125b inhibits VEGF expression via direct silencing of HER3, a receptor tyrosine kinase of the EGFR family, in ovarian cancer [105]. miR-148a/152 modulates VEGF expression via direct silencing of insulin-like growth factor 1 receptor (IGF1R) in breast cancer cells [106].

Not only VEGF production but also the expression of VEGF receptor is under the control of miR during tumor angiogenesis. Würdinger et al. reported that coculture of endothelial cells with glioma cells promotes the accumulation of endothelial miR-296 [107]. Similarly, grade II and IV glioma endothelial cells are enriched with miR-296 compared with endothelial cells from normal brain [107]. In silico prediction and target validation experiments indicated that hepatocyte growth factor-regulated tyrosine kinase substrate (HGS) serves as a direct miR-296 target [107]. Given the fact that HGS potentiates the degradation of VEGFR2 [108, 109], miR-296 elevation in endothelial cells blocks HGS signaling, which in turn increases functional VEGFR2 for VEGF to elucidate the biological activity [107]. Indeed, intravenous injection of anti-miR-296 substantially blocks tumor vasculature in the xenograft [107], indicating that it might serve as a potential therapy to treat glioma in the patients.

Recently, it was reported that VEGF-mediated tumor angiogenesis is subjected to be modulated by multiple miR regulation via a posttranscriptional gene silencing mechanism. Chen et al. reported that hypoxia-sensitive miRs, namely let-7, miR-103, and -107, are enriched in endothelial cells upon exposure to low oxygen [110]. The authors characterized that Ago1, an Argonaute protein involved in mammalian transcriptional silencing mechanism, serves as direct target of let-7 and miR-103/-107 [110]. Using an immunoprecipitation method, the authors identified that hypoxia exposure abrogates the physical interaction between Ago1 and VEGF transcript [110]. In this regard, hypoxia-induced upregulation of let-7 and miR-103/-107 inhibits Ago1 expression, resulting in blunted interaction between Ago1 and VEGF transcript. A loss of Ago1-VEGF transcript association leads to relief of posttranscriptional silencing of VEGF under the low oxygen environment and subsequently results in elevation of VEGF secretion from tumor. Anti-miR treatment blocks while Ago1 siRNA delivery facilitates angiogenic response in vivo in a model of Matrigel-plug assay [110]. The elevation of let-7 and miR-103 is associated with the loss of Ago1 in human breast cancer sample [110].

9.4.2 miRs Regulating Secretory Factors

VEGF is not the sole soluble factor directing tumor angiogenesis. Hepatoma-derived growth factor (HDGF), also known as high mobility group protein 1-like 2 (HMG-1 L2), exerts pro-angiogenic effect on endothelial cells supporting tumor angiogenesis [111]. miR-214 directly targets HDGF 3′ UTR and is downregulated in human HCCs. Loss of miR-214 is associated with poor prognosis [112]. Conditioned medium from miR-214 overexpressing HCCs reveals blunted angiogenic response of endothelial cells [112]. Ectopic expression of miR-214 significantly suppresses orthotopic xenograft tumor growth and reduces microvessel density [112]. It should be noted that miR-214 not only silences HDGF but also directly targets endothelial nitric oxide synthase (eNOS) [112] and QKI [90] (see above), implying that the miR-214-dependent angiostatic action relies on both endothelial cells and cancer cells. Angiogenin is another soluble factor promoting tumor angiogenesis [113, 114]. miR-221 [113] and miR-409-3p [114] negatively modulate the expression of this factor in hepatocarcinoma and fibrosarcoma, respectively, thus serving as an effective suppressor of tumor angiogenesis. Thrombospondin 1 (TSP-1) is a glycoprotein that is secreted from cancer cells and negatively regulates angiogenic response of endothelial cells. Sundaram et al. reported that miR-194 targets TSP-1 in colon cancer cells, resulting in exaggerated tumor angiogenesis [115]. While delivery of miR-194 in colon cancer cells results in enhanced microvessel density and enlargement of vessel lumen in xenograft tumor, overexpression of TSP-1 without the 3′ UTR, but not the wild type, significantly abrogates the pro-angiogenic effects [115]. This observation indicated that miR-194 promotes tumor angiogenesis via repression of TSP-1 expression [115]. Besides miR-194, miR-17-92 is another miR cluster that modulates TSP-1 expression. Delivery of anti-miR17-92 to colon cancer cells results in reduction of TSP-1 expression [116]. The pro-angiogenic effect of miR-17-92 in tumor cells not only is attributed by TSP-1 downregulation but also acts through a transforming growth factor receptor (TGF) β-dependent mechanism. TGFβ receptor 2 (TGFBR2) serves as a direct target of miR-17-92 [117]. Loss of TGFBR2 leads to impaired expression of clusterin [117]. Indeed, overexpression of clusterin in colon cancer cells is sufficient to exert negative effect on tumor angiogenesis [117]. It should be noted that miR-17-92 cluster in cancer cells promotes tumor angiogenesis [116, 117] but on the other hand serves as a potent endogenous angiostatic signal in endothelial cells [85]. A number of investigations revealed that the action of miR in cancer cells under tumor microenvironment is completely different from that in endothelial cells during tissue repair (see below).

9.4.3 miRs as Secretory Factors

Recently, it was reported that miRs are not only retained in the intracellular compartment but also actively secreted out from the cells [25]. They are protected from RNase degradation by enclosing in microvesicles or bound by proteins [118, 119].

More importantly, the secreted miRs act on neighbor cells and elucidate potent biological effects including angiogenic response in a paracrine manner. Zhuang et al. reported the first time that tumor cell-secreted miR-9 stimulates endothelial cells promoting tumor angiogenesis. Coculture of endothelial cells with melanoma cells, but not normal skin melanocytes or fibroblasts, specifically induces miR-9 accumulation in endothelial cells [120]. The cultured medium contains microvesicles which are enriched in mature miR-9 but not the pri-miR-9 [120]. While conditioned medium induces endothelial cell migration and tubulogenesis, miR-9 knockdown in tumor cells or in endothelial cells significantly abrogates this effect, indicating that miR-9 secreted from tumor cells to endothelial cells contributes to the pro-angiogenic effects [120]. This is in line with the in vivo observation indicating miR-9 knockdown in murine lung carcinoma reduces microvessel density in xenograft tumor [120]. Efforts in understanding the underlying mechanisms uncovered that miR-9 directly targets suppressor of cytokine signaling 5 (SOCS5), a negative regulator of JAK-STAT pathway cells [120]. Pharmacological inhibition of JAK-STAT activation alone is sufficient to block endothelial cell migration and angiogenesis [120]. Another study reveals that metastatic cancer cell-derived miR-210 actively participates in tumor angiogenesis. Data from miR microarray analysis revealed that miR-210 is highly enriched in exosomes from metastatic breast cancer cell-conditioned medium compared with that from non-metastatic or normal mammary epithelial cells [121]. Treatment of endothelial cells with miR-210-containing exosomes induces the accumulation of mature miR-210 but not pri-miR-210 in endothelial cells, with the concomitant enhancement of endothelial cell migration and capillary formation in vitro [121]. The pro-angiogenic effect of miR-210 is attributed by downregulation of Ephrin-A3 [121], which is similar to the observation in tissue repair angiogenesis (see above section). Table 9.1 summarizes the current understanding of miR regulation of tumor angiogenesis.

9.4.4 miRs Regulating Metastatic Angiogenesis

Cancer cell-derived miRs modulate the recruitment of endothelial cells mediating tumor metastasis. Loss of miR-126 has been characterized in a number of human cancers [122–124]. Downregulation of miR-126 results in induction of key downstream mediators including VEGF [124], insulin-like growth factor-binding protein 2 (IGFBP2) [125], and proto-oncogene tyrosine-protein kinase MER (MERTK) [125]. IGFBP2 promotes the action of insulin growth factor (IGF)-1 on endothelial cells leading to migratory and chemotatic phenotype [125]. MERTK captures and lowers the bioavailability of growth arrest-specific 6 (Gas6), an anti-chemotatic soluble factor, thus promoting endothelial cell recruitment [125]. While overexpression of miR-126 attenuates metastatic behavior of cancer cells, co-injection of endothelial cells with breast cancer cells to null mice reverses the miR-126-associated metastatic defect [125], suggesting that miR-126-dependent metastatic control relies on endothelial recruitment. Similarly, upregulation of miR-

Table 9.1 Summarized table of miR regulating tumor angiogenesis

miRs	Target(s)	Cancer	References
miRs regulating VEGF signaling			
miR-15a/-16	VEGF	Multiple myeloma cells	[97, 98]
		Nasopharyngeal carcinoma cells	[71]
miR-29b	VEGF, PDGF	Breast cancer cells	[99]
miR-195	VEGF, VAV2, CDC42	Hepatocarcinoma cells	[100]
miR-107	HIF-1β	Cervical cancer cells Colon cancer cells	[101]
miR-145	HIF-2α	Neuroblastoma cells	[102]
miR-155	VHL	Breast cancer cells	[103]
miR-9	E-Cadherin	Breast cancer cells	[96]
miR-34a	E2F3	Head and neck squamous cell carcinoma cells	[104]
miR-128	p70S6K1	Glioma cells	[106]
miR-125b	MAZ	Glioma/endothelial cells	[105]
	HER3	Ovarian cancer	[107]
miR-148a/-152	IGF1R	Breast cancer	[108]
miR-296	HGS	Glioma/endothelial cells	[109]
let-7, miR-103/-107	Agol	Breast cancer cells	[112]
miRs regulating secretory factors			
miR-214	HDGF, eNOS	Hepatocarcinoma cells	[114]
miR-221	Angiogenin	Hepatocarcinoma cells	[115]
miR-409-3p		Fibrosarcoma cells	[116]
miR-194	TSP1	Colorectal cancer cells	[117]
miR-17-92	TSP-1	Colon cancer cells	[118]
	Clusterin via TGFBRII		[119]
miRs as secretory factors			
miR-9	SOCS5	Melanoma cells to endothelial cells	[122]
miR-210	Ephrin-A3	Breast cancer cell to endothelial cells	[123]

1908/199a-5p/199a-3p has been characterized in metastasized melanoma [126]. These miRs directly target the protein apolipoprotein E (Apo E) and its upstream protein DNAJA4 [126]. While genetic inactivation of Apo E promotes metastasis, treatment of recombinant Apo E rescues the miR-199a-5p-dependent metastasis of melanoma [126]. Intriguingly, the miR-dependent metastatic regulation is likely to be mediated by endothelial recruitment via Apo E-dependent activation of LRP8, a low-density lipoprotein receptor that suppresses endothelial migration [126]. In this regard, elevated miR-1908/199a-5p/199a-3p inhibits the expression of Apo E and DNAJA4 in cancer cells, leading to loss of activation of LRP8 receptor in endothelial cells and promotion of endothelial cell recruitment to the tumor cells. It is noteworthy to mention that both endogenous miR-126 and miR-199a-5p exert angiogenic control in endothelial cells [41–43, 88] (see above sections). However,

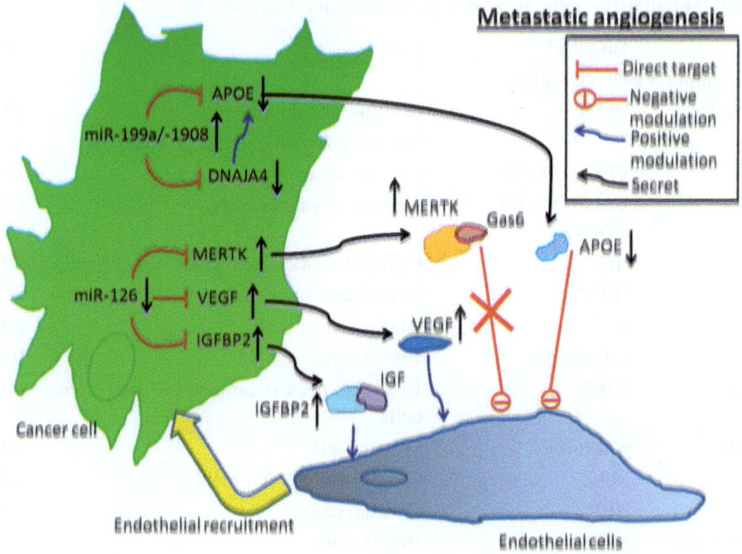

Fig. 9.4 Involvement of miR in regulation of tumor angiogenesis during metastatic transformation. Loss of miR-126 stimulates endothelial cells in the tumor microenvironment via derepression of VEGF. Reduction of miR-126 promotes the expression of IGFBP2 facilitating the IGF pathway supporting tumor angiogenesis. Blunted miR-126 expression in cancer cells enables the expression of MERTK, leading to the capture of anti-chemotatic factor Gas6 and subsequent switching of endothelial cells to migratory phenotype. Elevated miR-199a/-1908 targets both Apo E and DNAAJ4 resulting in reduction of Apo E. Loss of Apo E leads to impaired angiostatic signal resulting in endothelial cell recruitment towards cancer cells

the miR-dependent effects on endothelial cell recruitment to cancer cells are entirely distinct from that on developmental/tissue repair angiogenesis. These findings pinpointed the fact that miRs exert different biological actions in a cell context-dependent manner. It also uncovered the complexity of different regulatory mechanisms in physiological angiogenesis versus tumor angiogenic control. Figure 9.4 illustrates a schematic diagram summarizing the major miR-dependent regulation of metastatic angiogenesis.

Conclusions

In summary, miRs are the critical signaling molecules regulating angiogenesis during developmental stage, tissue repair, and tumorigenesis through targeting soluble factors, transcription factors, and surface receptors in cell autonomous and non-cell autonomous manner. In light of the fact that miRs serve as master angiogenic switches, miR-based therapy becomes an attractive strategy combating aberrant angiogenesis. Indeed, a number of investigations have provided compelling evidence showing that antisense miR or miR supplementation in vivo can effectively rectify the inappropriate angiogenesis during pathological conditions. Insufficient angiogenesis in ischemic diseases can be treated with angiomiR delivery aiming at facilitating perfusion. Excessive angiogenic response during tumorigenesis can be attenuated by treating modified stable form of antisense angiomiRs (known as antagomiRs or lock-nucleic acid antisense). With the advancement of technology improving stability and efficiency of synthetic miR/antisense miR, miR-based drugs targeting aberrant angiogenesis will be developed in the future. Indeed, the first miR-targeting drug has already entered Phase 2 clinical trials in multiple centers treating hepatitis C [127]. However, precaution needs to be taken concerning the specificity of the miR-based delivery. As mentioned in the above section, some miRs, for example, induce anti-angiogenic signals in cancer cells but exert potent pro-angiogenic effects on endothelial cells. The choice of target miR(s), dosage, and delivery approach are essential to determine the therapeutic value of the miR-based intervention. Further investigations are required to figure out an optimal dosage and appropriate route of administration of angiomiR-based drugs targeting pathogenic angiogenesis.

Acknowledgements This work was supported by National Institute of Health Grants GM 069589 and GM 077185 to C.K.S.

References

1. Crick F (1970) Central dogma of molecular biology. Nature 227:561–563
2. Ohno S (1972) So much "junk" DNA in our genome. Brookhaven Symp Biol 23:366–370
3. Orgel LE, Crick FH (1980) Selfish DNA: the ultimate parasite. Nature 284:604–607
4. Lee RC, Feinbaum RL, Ambros V (1993) The *C. elegans* heterochronic gene lin-4 encodes small RNAs with antisense complementarity to lin-14. Cell 75:843–854
5. Llave C, Kasschau KD, Rector MA, Carrington JC (2002) Endogenous and silencing-associated small RNAs in plants. Plant Cell 14:1605–1619
6. Lee RC, Ambros V (2001) An extensive class of small RNAs in *Caenorhabditis elegans*. Science 294:862–864
7. Schee K, Boye K, Abrahamsen TW, Fodstad O, Flatmark K (2012) Clinical relevance of microRNA miR-21, miR-31, miR-92a, miR-101, miR-106a and miR-145 in colorectal cancer. BMC Cancer 12:505
8. Rajewsky N (2006) microRNA target predictions in animals. Nat Genet 38(Suppl):S8–S13
9. Filipowicz W, Bhattacharyya SN, Sonenberg N (2008) Mechanisms of post-transcriptional regulation by microRNAs: are the answers in sight? Nat Rev Genet 9:102–114
10. Slezak-Prochazka I, Durmus S, Kroesen BJ, van den Berg A (2010) MicroRNAs, macrocontrol: regulation of miRNA processing. RNA 16:1087–1095
11. Lee Y, Kim M, Han J, Yeom KH, Lee S, Baek SH, Kim VN (2004) MicroRNA genes are transcribed by RNA polymerase II. EMBO J 23:4051–4060
12. Lee EJ, Baek M, Gusev Y, Brackett DJ, Nuovo GJ, Schmittgen TD (2008) Systematic evaluation of microRNA processing patterns in tissues, cell lines, and tumors. RNA 14:35–42
13. Han J, Lee Y, Yeom KH, Kim YK, Jin H, Kim VN (2004) The Drosha-DGCR8 complex in primary microRNA processing. Genes Dev 18:3016–3027
14. Landthaler M, Yalcin A, Tuschl T (2004) The human DiGeorge syndrome critical region gene 8 and its *D. melanogaster* homolog are required for miRNA biogenesis. Curr Biol 14:2162–2167

15. Bohnsack MT, Czaplinski K, Gorlich D (2004) Exportin 5 is a RanGTP-dependent dsRNA-binding protein that mediates nuclear export of pre-miRNAs. RNA 10:185–191
16. Lund E, Guttinger S, Calado A, Dahlberg JE, Kutay U (2004) Nuclear export of microRNA precursors. Science 303:95–98
17. Yi R, Qin Y, Macara IG, Cullen BR (2003) Exportin-5 mediates the nuclear export of pre-microRNAs and short hairpin RNAs. Genes Dev 17:3011–3016
18. Schwarz DS, Hutvagner G, Du T, Xu Z, Aronin N, Zamore PD (2003) Asymmetry in the assembly of the RNAi enzyme complex. Cell 115:199–208
19. Choe J, Cho H, Chi SG, Kim YK (2011) Ago2/miRISC-mediated inhibition of CBP80/20-dependent translation and thereby abrogation of nonsense-mediated mRNA decay require the cap-associating activity of Ago2. FEBS Lett 585:2682–2687
20. Kiriakidou M, Tan GS, Lamprinaki S, De Planell-Saguer M, Nelson PT, Mourelatos Z (2007) An mRNA m7G cap binding-like motif within human Ago2 represses translation. Cell 129:1141–1151
21. Meister G, Landthaler M, Patkaniowska A, Dorsett Y, Teng G, Tuschl T (2004) Human Argonaute2 mediates RNA cleavage targeted by miRNAs and siRNAs. Mol Cell 15:185–197
22. Emanueli C, Madeddu P (2006) Therapeutic angiogenesis: translating experimental concepts to medically relevant goals. Vascul Pharmacol 45:334–339
23. Zachary I, Morgan RD (2011) Therapeutic angiogenesis for cardiovascular disease: biological context, challenges, prospects. Heart 97:181–189
24. Duan L, Yang G, Zhang R, Feng L, Xu C (2012) Advancement in the research on vascular endothelial growth inhibitor (VEGI). Target Oncol 7:87–90
25. Chan YC, Banerjee J, Choi SY, Sen CK (2012) miR-210: the master hypoxamir. Microcirculation 19:215–223
26. Sen CK, Gordillo GM, Khanna S, Roy S (2009) Micromanaging vascular biology: tiny microRNAs play big band. J Vasc Res 46:527–540
27. Sen CK, Roy S (2012) OxymiRs in cutaneous development, wound repair and regeneration. Semin Cell Dev Biol 23:971–980
28. Caporali A, Emanueli C (2011) MicroRNA-503 and the extended microRNA-16 family in angiogenesis. Trends Cardiovasc Med 21:162–166
29. Staszel T, Zapala B, Polus A, Sadakierska-Chudy A, Kiec-Wilk B, Stepien E, Wybranska I, Chojnacka M, Dembinska-Kiec A (2011) Role of microRNAs in endothelial cell pathophysiology. Pol Arch Med Wewn 121:361–366
30. Caporali A, Emanueli C (2011) MicroRNA regulation in angiogenesis. Vascul Pharmacol 55:79–86
31. Yang W, Lee DY, Ben-David Y (2011) The roles of microRNAs in tumorigenesis and angiogenesis. Int J Physiol Pathophysiol Pharmacol 3:140–155
32. Hartmann D, Thum T (2011) MicroRNAs and vascular (dys)function. Vascul Pharmacol 55: 92–105
33. Yang WJ, Yang DD, Na S, Sandusky GE, Zhang Q, Zhao G (2005) Dicer is required for embryonic angiogenesis during mouse development. J Biol Chem 280:9330–9335
34. Otsuka M, Jing Q, Georgel P, New L, Chen J, Mols J, Kang YJ, Jiang Z, Du X, Cook R, Das SC, Pattnaik AK, Beutler B, Han J (2007) Hypersusceptibility to vesicular stomatitis virus infection in Dicer1-deficient mice is due to impaired miR24 and miR93 expression. Immunity 27:123–134
35. Giraldez AJ, Cinalli RM, Glasner ME, Enright AJ, Thomson JM, Baskerville S, Hammond SM, Bartel DP, Schier AF (2005) MicroRNAs regulate brain morphogenesis in zebrafish. Science 308:833–838
36. Suarez Y, Fernandez-Hernando C, Yu J, Gerber SA, Harrison KD, Pober JS, Iruela-Arispe ML, Merkenschlager M, Sessa WC (2008) Dicer-dependent endothelial microRNAs are necessary for postnatal angiogenesis. Proc Natl Acad Sci USA 105:14082–14087
37. Suarez Y, Fernandez-Hernando C, Pober JS, Sessa WC (2007) Dicer dependent microRNAs regulate gene expression and functions in human endothelial cells. Circ Res 100:1164–1173
38. Kuehbacher A, Urbich C, Zeiher AM, Dimmeler S (2007) Role of Dicer and Drosha for endothelial microRNA expression and angiogenesis. Circ Res 101:59–68

39. Shilo S, Roy S, Khanna S, Sen CK (2008) Evidence for the involvement of miRNA in redox regulated angiogenic response of human microvascular endothelial cells. Arterioscler Thromb Vasc Biol 28:471–477
40. Poliseno L, Tuccoli A, Mariani L, Evangelista M, Citti L, Woods K, Mercatanti A, Hammond S, Rainaldi G (2006) MicroRNAs modulate the angiogenic properties of HUVECs. Blood 108: 3068–3071
41. Wang S, Aurora AB, Johnson BA, Qi X, McAnally J, Hill JA, Richardson JA, Bassel-Duby R, Olson EN (2008) The endothelial-specific microRNA miR-126 governs vascular integrity and angiogenesis. Dev Cell 15:261–271
42. Fish JE, Santoro MM, Morton SU, Yu S, Yeh RF, Wythe JD, Ivey KN, Bruneau BG, Stainier DY, Srivastava D (2008) miR-126 regulates angiogenic signaling and vascular integrity. Dev Cell 15:272–284
43. Kuhnert F, Mancuso MR, Hampton J, Stankunas K, Asano T, Chen CZ, Kuo CJ (2008) Attribution of vascular phenotypes of the murine Egfl7 locus to the microRNA miR-126. Development 135:3989–3993
44. Harris TA, Yamakuchi M, Ferlito M, Mendell JT, Lowenstein CJ (2008) MicroRNA-126 regulates endothelial expression of vascular cell adhesion molecule 1. Proc Natl Acad Sci USA 105:1516–1521
45. Hassel D, Cheng P, White MP, Ivey KN, Kroll J, Augustin HG, Katus HA, Stainier DY, Srivastava D (2012) MicroRNA-10 regulates the angiogenic behavior of zebrafish and human endothelial cells by promoting vascular endothelial growth factor signaling. Circ Res 111: 1421–1433
46. Bridge G, Monteiro R, Henderson S, Emuss V, Lagos D, Georgopoulou D, Patient R, Boshoff C (2012) The microRNA-30 family targets DLL4 to modulate endothelial cell behavior during angiogenesis. Blood 120:5063–5072
47. Stahlhut C, Suarez Y, Lu J, Mishima Y, Giraldez AJ (2012) miR-1 and miR-206 regulate angiogenesis by modulating VegfA expression in zebrafish. Development 139:4356–4365
48. Risau W (1991) Embryonic angiogenesis factors. Pharmacol Ther 51:371–376
49. van Solingen C, Seghers L, Bijkerk R, Duijs JM, Roeten MK, van Oeveren-Rietdijk AM, Baelde HJ, Monge M, Vos JB, de Boer HC, Quax PH, Rabelink TJ, van Zonneveld AJ (2009) Antagomir-mediated silencing of endothelial cell specific microRNA-126 impairs ischemia-induced angiogenesis. J Cell Mol Med 13:1577–1585
50. Jakob P, Doerries C, Briand S, Mocharla P, Krankel N, Besler C, Mueller M, Manes C, Templin C, Baltes C, Rudin M, Adams H, Wolfrum M, Noll G, Ruschitzka F, Luscher TF, Landmesser U (2012) Loss of angiomiR-126 and 130a in angiogenic early outgrowth cells from patients with chronic heart failure: role for impaired in vivo neovascularization and cardiac repair capacity. Circulation 126:2962–2975
51. Huang F, Zhu X, Hu XQ, Fang ZF, Tang L, Lu XL, Zhou SH (2012) Mesenchymal stem cells modified with miR-126 release angiogenic factors and activate Notch ligand Delta-like-4, enhancing ischemic angiogenesis and cell survival. Int J Mol Med 31:484–492
52. Chen JJ, Zhou SH (2011) Mesenchymal stem cells overexpressing MiR-126 enhance ischemic angiogenesis via the AKT/ERK-related pathway. Cardiol J 18:675–681
53. Qiang L, Hong L, Ningfu W, Huaihong C, Jing W (2013) Expression of miR-126 and miR-508-5p in endothelial progenitor cells is associated with the prognosis of chronic heart failure patients. Int J Cardiol 168:2082–2088
54. Suarez Y, Sessa WC (2009) MicroRNAs as novel regulators of angiogenesis. Circ Res 104: 442–454
55. Huang X, Ding L, Bennewith KL, Tong RT, Welford SM, Ang KK, Story M, Le QT, Giaccia AJ (2009) Hypoxia-inducible mir-210 regulates normoxic gene expression involved in tumor initiation. Mol Cell 35:856–867
56. Kulshreshtha R, Ferracin M, Wojcik SE, Garzon R, Alder H, Agosto-Perez FJ, Davuluri R, Liu CG, Croce CM, Negrini M, Calin GA, Ivan M (2007) A microRNA signature of hypoxia. Mol Cell Biol 27:1859–1867

57. Huang X, Le QT, Giaccia AJ (2010) MiR-210–micromanager of the hypoxia pathway. Trends Mol Med 16:230–237
58. Fasanaro P, D'Alessandra Y, Di Stefano V, Melchionna R, Romani S, Pompilio G, Capogrossi MC, Martelli F (2008) MicroRNA-210 modulates endothelial cell response to hypoxia and inhibits the receptor tyrosine kinase ligand Ephrin-A3. J Biol Chem 283: 15878–15883
59. Liu F, Lou YL, Wu J, Ruan QF, Xie A, Guo F, Cui SP, Deng ZF, Wang Y (2012) Upregulation of microRNA-210 regulates renal angiogenesis mediated by activation of VEGF signaling pathway under ischemia/perfusion injury in vivo and in vitro. Kidney Blood Press Res 35:182–191
60. Lou YL, Guo F, Liu F, Gao FL, Zhang PQ, Niu X, Guo SC, Yin JH, Wang Y, Deng ZF (2012) miR-210 activates notch signaling pathway in angiogenesis induced by cerebral ischemia. Mol Cell Biochem 370:45–51
61. Hu S, Huang M, Li Z, Jia F, Ghosh Z, Lijkwan MA, Fasanaro P, Sun N, Wang X, Martelli F, Robbins RC, Wu JC (2010) MicroRNA-210 as a novel therapy for treatment of ischemic heart disease. Circulation 122:S124–S131
62. Shoji T, Nakasa T, Yamasaki K, Kodama A, Miyaki S, Niimoto T, Okuhara A, Kamei N, Adachi N, Ochi M (2012) The effect of intra-articular injection of microRNA-210 on ligament healing in a rat model. Am J Sports Med 40:2470–2478
63. Alaiti MA, Ishikawa M, Masuda H, Simon DI, Jain MK, Asahara T, Costa MA (2012) Up-regulation of miR-210 by vascular endothelial growth factor in ex vivo expanded CD34+ cells enhances cell-mediated angiogenesis. J Cell Mol Med 16:2413–2421
64. Puissegur MP, Mazure NM, Bertero T, Pradelli L, Grosso S, Robbe-Sermesant K, Maurin T, Lebrigand K, Cardinaud B, Hofman V, Fourre S, Magnone V, Ricci JE, Pouyssegur J, Gounon P, Hofman P, Barbry P, Mari B (2011) miR-210 is overexpressed in late stages of lung cancer and mediates mitochondrial alterations associated with modulation of HIF-1 activity. Cell Death Differ 18:465–478
65. Kelly TJ, Souza AL, Clish CB, Puigserver P (2011) A hypoxia-induced positive feedback loop promotes hypoxia-inducible factor 1alpha stability through miR-210 suppression of glycerol-3-phosphate dehydrogenase 1-like. Mol Cell Biol 31:2696–2706
66. Zhou Q, Gallagher R, Ufret-Vincenty R, Li X, Olson EN, Wang S (2011) Regulation of angiogenesis and choroidal neovascularization by members of microRNA-23 27 24 clusters. Proc Natl Acad Sci USA 108:8287–8292
67. Fiedler J, Jazbutyte V, Kirchmaier BC, Gupta SK, Lorenzen J, Hartmann D, Galuppo P, Kneitz S, Pena JT, Sohn-Lee C, Loyer X, Soutschek J, Brand T, Tuschl T, Heineke J, Martin U, Schulte-Merker S, Ertl G, Engelhardt S, Bauersachs J, Thum T (2011) MicroRNA-24 regulates vascularity after myocardial infarction. Circulation 124:720–730
68. Urbich C, Kaluza D, Fromel T, Knau A, Bennewitz K, Boon RA, Bonauer A, Doebele C, Boeckel JN, Hergenreider E, Zeiher AM, Kroll J, Fleming I, Dimmeler S (2012) MicroRNA-27a/b controls endothelial cell repulsion and angiogenesis by targeting semaphorin 6A. Blood 119:1607–1616
69. Hua Z, Lv Q, Ye W, Wong CK, Cai G, Gu D, Ji Y, Zhao C, Wang J, Yang BB, Zhang Y (2006) MiRNA-directed regulation of VEGF and other angiogenic factors under hypoxia. PLoS One 1:e116
70. Zhou B, Ma R, Si W, Li S, Xu Y, Tu X, Wang Q (2013) MicroRNA-503 targets FGF2 and VEGFA and inhibits tumor angiogenesis and growth. Cancer Lett 333:159–169
71. Caporali A, Meloni M, Vollenkle C, Bonci D, Sala-Newby GB, Addis R, Spinetti G, Losa S, Masson R, Baker AH, Agami R, le Sage C, Condorelli G, Madeddu P, Martelli F, Emanueli C (2011) Deregulation of microRNA-503 contributes to diabetes mellitus-induced impairment of endothelial function and reparative angiogenesis after limb ischemia. Circulation 123: 282–291
72. Kim J, Kang Y, Kojima Y, Lighthouse JK, Hu X, Aldred MA, McLean DL, Park H, Comhair SA, Greif DM, Erzurum SC, Chun HJ (2013) An endothelial apelin-FGF link

mediated by miR-424 and miR-503 is disrupted in pulmonary arterial hypertension. Nat Med 19:74–82
73. Wang XH, Qian RZ, Zhang W, Chen SF, Jin HM, Hu RM (2009) MicroRNA-320 expression in myocardial microvascular endothelial cells and its relationship with insulin-like growth factor-1 in type 2 diabetic rats. Clin Exp Pharmacol Physiol 36:181–188
74. Chamorro-Jorganes A, Araldi E, Penalva LO, Sandhu D, Fernandez-Hernando C, Suarez Y (2011) MicroRNA-16 and microRNA-424 regulate cell-autonomous angiogenic functions in endothelial cells via targeting vascular endothelial growth factor receptor-2 and fibroblast growth factor receptor-1. Arterioscler Thromb Vasc Biol 31:2595–2606
75. Yin KJ, Olsen K, Hamblin M, Zhang J, Schwendeman SP, Chen YE (2012) Vascular endothelial cell-specific microRNA-15a inhibits angiogenesis in hindlimb ischemia. J Biol Chem 287:27055–27064
76. Wang Y, Fan H, Zhao G, Liu D, Du L, Wang Z, Hu Y, Hou Y (2012) miR-16 inhibits the proliferation and angiogenesis-regulating potential of mesenchymal stem cells in severe pre-eclampsia. FEBS J 279:4510–4524
77. Musumeci M, Coppola V, Addario A, Patrizii M, Maugeri-Sacca M, Memeo L, Colarossi C, Francescangeli F, Biffoni M, Collura D, Giacobbe A, D'Urso L, Falchi M, Venneri MA, Muto G, De Maria R, Bonci D (2011) Control of tumor and microenvironment cross-talk by miR-15a and miR-16 in prostate cancer. Oncogene 30:4231–4242
78. Chan YC, Khanna S, Roy S, Sen CK (2011) miR-200b targets Ets-1 and is down-regulated by hypoxia to induce angiogenic response of endothelial cells. J Biol Chem 286:2047–2056
79. McArthur K, Feng B, Wu Y, Chen S, Chakrabarti S (2011) MicroRNA-200b regulates vascular endothelial growth factor-mediated alterations in diabetic retinopathy. Diabetes 60:1314–1323
80. Choi YC, Yoon S, Jeong Y, Yoon J, Baek K (2011) Regulation of vascular endothelial growth factor signaling by miR-200b. Mol Cells 32:77–82
81. Chan YC, Roy S, Khanna S, Sen CK (2012) Downregulation of endothelial microRNA-200b supports cutaneous wound angiogenesis by desilencing GATA binding protein 2 and vascular endothelial growth factor receptor 2. Arterioscler Thromb Vasc Biol 32:1372–1382
82. Lee ST, Chu K, Jung KH, Yoon HJ, Jeon D, Kang KM, Park KH, Bae EK, Kim M, Lee SK, Roh JK (2010) MicroRNAs induced during ischemic preconditioning. Stroke 41:1646–1651
83. Taguchi A, Yanagisawa K, Tanaka M, Cao K, Matsuyama Y, Goto H, Takahashi T (2008) Identification of hypoxia-inducible factor-1 alpha as a novel target for miR-17-92 microRNA cluster. Cancer Res 68:5540–5545
84. Yan HL, Xue G, Mei Q, Wang YZ, Ding FX, Liu MF, Lu MH, Tang Y, Yu HY, Sun SH (2009) Repression of the miR-17-92 cluster by p53 has an important function in hypoxia-induced apoptosis. EMBO J 28:2719–2732
85. Doebele C, Bonauer A, Fischer A, Scholz A, Reiss Y, Urbich C, Hofmann WK, Zeiher AM, Dimmeler S (2010) Members of the microRNA-17-92 cluster exhibit a cell-intrinsic antiangiogenic function in endothelial cells. Blood 115:4944–4950
86. Bonauer A, Carmona G, Iwasaki M, Mione M, Koyanagi M, Fischer A, Burchfield J, Fox H, Doebele C, Ohtani K, Chavakis E, Potente M, Tjwa M, Urbich C, Zeiher AM, Dimmeler S (2009) MicroRNA-92a controls angiogenesis and functional recovery of ischemic tissues in mice. Science 324:1710–1713
87. Rane S, He M, Sayed D, Vashistha H, Malhotra A, Sadoshima J, Vatner DE, Vatner SF, Abdellatif M (2009) Downregulation of miR-199a derepresses hypoxia-inducible factor-1alpha and Sirtuin 1 and recapitulates hypoxia preconditioning in cardiac myocytes. Circ Res 104:879–886
88. Chan YC, Roy S, Huang Y, Khanna S, Sen CK (2012) The MicroRNA miR-199a-5p down-regulation switches on wound angiogenesis by derepressing the v-ets erythroblastosis virus E26 oncogene homolog 1-matrix metalloproteinase-1 pathway. J Biol Chem 287:41032–41043
89. Lee YB, Bantounas I, Lee DY, Phylactou L, Caldwell MA, Uney JB (2009) Twist-1 regulates the miR-199a/214 cluster during development. Nucleic Acids Res 37:123–128

90. van Mil A, Grundmann S, Goumans MJ, Lei Z, Oerlemans MI, Jaksani S, Doevendans PA, Sluijter JP (2012) MicroRNA-214 inhibits angiogenesis by targeting Quaking and reducing angiogenic growth factor release. Cardiovasc Res 93:655–665
91. Smits M, Mir SE, Nilsson RJ, van der Stoop PM, Niers JM, Marquez VE, Cloos J, Breakefield XO, Krichevsky AM, Noske DP, Tannous BA, Würdinger T (2011) Down-regulation of miR-101 in endothelial cells promotes blood vessel formation through reduced repression of EZH2. PLoS One 6:e16282
92. Grundmann S, Hans FP, Kinniry S, Heinke J, Helbing T, Bluhm F, Sluijter JP, Hoefer I, Pasterkamp G, Bode C, Moser M (2011) MicroRNA-100 regulates neovascularization by suppression of mammalian target of rapamycin in endothelial and vascular smooth muscle cells. Circulation 123:999–1009
93. Semo J, Sharir R, Afek A, Avivi C, Barshack I, Maysel-Auslender S, Krelin Y, Kain D, Entin-Meer M, Keren G, George J (2013) The 106b ~ 25 microRNA cluster is essential for neovascularization after hindlimb ischaemia in mice. Eur Heart J [Epub ahead of print]
94. Ma L, Young J, Prabhala H, Pan E, Mestdagh P, Muth D, Teruya-Feldstein J, Reinhardt F, Onder TT, Valastyan S, Westermann F, Speleman F, Vandesompele J, Weinberg RA (2010) miR-9, a MYC/MYCN-activated microRNA, regulates E-cadherin and cancer metastasis. Nat Cell Biol 12:247–256
95. Sun CY, She XM, Qin Y, Chu ZB, Chen L, Ai LS, Zhang L, Hu Y (2013) miR-15a and miR-16 affect the angiogenesis of multiple myeloma by targeting VEGF. Carcinogenesis 34:426–435
96. Roccaro AM, Sacco A, Thompson B, Leleu X, Azab AK, Azab F, Runnels J, Jia X, Ngo HT, Melhem MR, Lin CP, Ribatti D, Rollins BJ, Witzig TE, Anderson KC, Ghobrial IM (2009) MicroRNAs 15a and 16 regulate tumor proliferation in multiple myeloma. Blood 113: 6669–6680
97. Chou J, Lin JH, Brenot A, Kim JW, Provot S, Werb Z (2013) GATA3 suppresses metastasis and modulates the tumour microenvironment by regulating microRNA-29b expression. Nat Cell Biol 15:201–213
98. Wang R, Zhao N, Li S, Fang JH, Chen MX, Yang J, Jia WH, Yuan Y, Zhuang SM (2013) MicroRNA-195 suppresses angiogenesis and metastasis of hepatocellular carcinoma by inhibiting the expression of VEGF, VAV2 and CDC42. Hepatology 58:642–653
99. Yamakuchi M, Lotterman CD, Bao C, Hruban RH, Karim B, Mendell JT, Huso D, Lowenstein CJ (2010) P53-induced microRNA-107 inhibits HIF-1 and tumor angiogenesis. Proc Natl Acad Sci USA 107:6334–6339
100. Zhang H, Pu J, Qi T, Qi M, Yang C, Li S, Huang K, Zheng L, Tong Q (2012) MicroRNA-145 inhibits the growth, invasion, metastasis and angiogenesis of neuroblastoma cells through targeting hypoxia-inducible factor 2 alpha. Oncogene [Epub ahead of print]
101. Kong W, He L, Richards EJ, Challa S, Xu CX, Permuth-Wey J, Lancaster JM, Coppola D, Sellers TA, Djeu JY, Cheng JQ (2013) Upregulation of miRNA-155 promotes tumour angiogenesis by targeting VHL and is associated with poor prognosis and triple-negative breast cancer. Oncogene [Epub ahead of print]
102. Kumar B, Yadav A, Lang J, Teknos TN, Kumar P (2012) Dysregulation of microRNA-34a expression in head and neck squamous cell carcinoma promotes tumor growth and tumor angiogenesis. PLoS One 7:e37601
103. Smits M, Würdinger T, van het Hof B, Drexhage JA, Geerts D, Wesseling P, Noske DP, Vandertop WP, de Vries HE, Reijerkerk A (2012) Myc-associated zinc finger protein (MAZ) is regulated by miR-125b and mediates VEGF-induced angiogenesis in glioblastoma. FASEB J 26:2639–2647
104. Shi ZM, Wang J, Yan Z, You YP, Li CY, Qian X, Yin Y, Zhao P, Wang YY, Wang XF, Li MN, Liu LZ, Liu N, Jiang BH (2012) MiR-128 inhibits tumor growth and angiogenesis by targeting p70S6K1. PLoS One 7:e32709
105. He J, Jing Y, Li W, Qian X, Xu Q, Li FS, Liu LZ, Jiang BH, Jiang Y (2013) Roles and mechanism of miR-199a and miR-125b in tumor angiogenesis. PLoS One 8:e56647

106. Xu Q, Jiang Y, Yin Y, Li Q, He J, Jing Y, Qi YT, Xu Q, Li W, Lu B, Peiper SS, Jiang BH, Liu LZ (2013) A regulatory circuit of miR-148a/152 and DNMT1 in modulating cell transformation and tumor angiogenesis through IGF-IR and IRS1. J Mol Cell Biol 5:3–13
107. Würdinger T, Tannous BA, Saydam O, Skog J, Grau S, Soutschek J, Weissleder R, Breakefield XO, Krichevsky AM (2008) miR-296 regulates growth factor receptor overexpression in angiogenic endothelial cells. Cancer Cell 14:382–393
108. Bache KG, Raiborg C, Mehlum A, Stenmark H (2003) STAM and Hrs are subunits of a multivalent ubiquitin-binding complex on early endosomes. J Biol Chem 278:12513–12521
109. Stern KA, Visser Smit GD, Place TL, Winistorfer S, Piper RC, Lill NL (2007) Epidermal growth factor receptor fate is controlled by Hrs tyrosine phosphorylation sites that regulate Hrs degradation. Mol Cell Biol 27:888–898
110. Chen Z, Lai TC, Jan YH, Lin FM, Wang WC, Xiao H, Wang YT, Sun W, Cui X, Li YS, Fang T, Zhao H, Padmanabhan C, Sun R, Wang DL, Jin H, Chau GY, Huang HD, Hsiao M, Shyy JY (2013) Hypoxia-responsive miRNAs target argonaute 1 to promote angiogenesis. J Clin Invest 123:1057–1067
111. Okuda Y, Nakamura H, Yoshida K, Enomoto H, Uyama H, Hirotani T, Funamoto M, Ito H, Everett AD, Hada T, Kawase I (2003) Hepatoma-derived growth factor induces tumorigenesis in vivo through both direct angiogenic activity and induction of vascular endothelial growth factor. Cancer Sci 94:1034–1041
112. Shih TC, Tien YJ, Wen CJ, Yeh TS, Yu MC, Huang CH, Lee YS, Yen TC, Hsieh SY (2012) MicroRNA-214 downregulation contributes to tumor angiogenesis by inducing secretion of the hepatoma-derived growth factor in human hepatoma. J Hepatol 57:584–591
113. Santhekadur PK, Das SK, Gredler R, Chen D, Srivastava J, Robertson C, Baldwin AS Jr, Fisher PB, Sarkar D (2012) Multifunction protein staphylococcal nuclease domain containing 1 (SND1) promotes tumor angiogenesis in human hepatocellular carcinoma through novel pathway that involves nuclear factor kappaB and miR-221. J Biol Chem 287:13952–13958
114. Weng C, Dong H, Chen G, Zhai Y, Bai R, Hu H, Lu L, Xu Z (2012) miR-409-3p inhibits HT1080 cell proliferation, vascularization and metastasis by targeting angiogenin. Cancer Lett 323:171–179
115. Sundaram P, Hultine S, Smith LM, Dews M, Fox JL, Biyashev D, Schelter JM, Huang Q, Cleary MA, Volpert OV, Thomas-Tikhonenko A (2011) p53-responsive miR-194 inhibits thrombospondin-1 and promotes angiogenesis in colon cancers. Cancer Res 71:7490–7501
116. Dews M, Homayouni A, Yu D, Murphy D, Sevignani C, Wentzel E, Furth EE, Lee WM, Enders GH, Mendell JT, Thomas-Tikhonenko A (2006) Augmentation of tumor angiogenesis by a Myc-activated microRNA cluster. Nat Genet 38:1060–1065
117. Dews M, Fox JL, Hultine S, Sundaram P, Wang W, Liu YY, Furth E, Enders GH, El-Deiry W, Schelter JM, Cleary MA, Thomas-Tikhonenko A (2010) The myc-miR-17 92 axis blunts TGF{beta} signaling and production of multiple TGF{beta}-dependent antiangiogenic factors. Cancer Res 70:8233–8246
118. Xu J, Zhao J, Evan G, Xiao C, Cheng Y, Xiao J (2012) Circulating microRNAs: novel biomarkers for cardiovascular diseases. J Mol Med 90:865–875
119. Dimmeler S, Zeiher AM (2010) Circulating microRNAs: novel biomarkers for cardiovascular diseases? Eur Heart J 31:2705–2707
120. Zhuang G, Wu X, Jiang Z, Kasman I, Yao J, Guan Y, Oeh J, Modrusan Z, Bais C, Sampath D, Ferrara N (2012) Tumour-secreted miR-9 promotes endothelial cell migration and angiogenesis by activating the JAK-STAT pathway. EMBO J 31:3513–3523
121. Kosaka N, Iguchi H, Hagiwara K, Yoshioka Y, Takeshita F, Ochiya T (2013) Neutral sphingomyelinase 2 (nSMase2)-dependent exosomal transfer of angiogenic microRNAs regulate cancer cell metastasis. J Biol Chem 288:10849–10859
122. Guo C, Sah JF, Beard L, Willson JK, Markowitz SD, Guda K (2008) The noncoding RNA, miR-126, suppresses the growth of neoplastic cells by targeting phosphatidylinositol 3-kinase signaling and is frequently lost in colon cancers. Genes Chromosomes Cancer 47:939–946
123. Feng R, Chen X, Yu Y, Su L, Yu B, Li J, Cai Q, Yan M, Liu B, Zhu Z (2010) miR-126 functions as a tumour suppressor in human gastric cancer. Cancer Lett 298:50–63

124. Zhu N, Zhang D, Xie H, Zhou Z, Chen H, Hu T, Bai Y, Shen Y, Yuan W, Jing Q, Qin Y (2011) Endothelial-specific intron-derived miR-126 is down-regulated in human breast cancer and targets both VEGFA and PIK3R2. Mol Cell Biochem 351:157–164
125. Png KJ, Halberg N, Yoshida M, Tavazoie SF (2012) A microRNA regulon that mediates endothelial recruitment and metastasis by cancer cells. Nature 481:190–194
126. Pencheva N, Tran H, Buss C, Huh D, Drobnjak M, Busam K, Tavazoie SF (2012) Convergent multi-miRNA targeting of ApoE drives LRP1/LRP8-dependent melanoma metastasis and angiogenesis. Cell 151:1068–1082
127. Janssen HL, Reesink HW, Lawitz EJ, Zeuzem S, Rodriguez-Torres M, Patel K, van der Meer AJ, Patick AK, Chen A, Zhou Y, Persson R, King BD, Kauppinen S, Levin AA, Hodges MR (2013) Treatment of HCV infection by targeting microRNA. N Engl J Med 368:1685–1694

Nrf2 Transcription Factor and Heme Oxygenase-1 as Modulators of Vascular Injury and Angiogenesis

10

Urszula Florczyk, Alicja Józkowicz, and Józef Dulak

Contents

10.1	Introduction	214
10.2	Nrf2/HO-1 as a Key Path Against Vascular Injury	215
10.3	HO-1 Functions in Blood Vessel Formation	216
10.3.1	An Overall Effect of HO-1 Knockout	216
10.3.2	HO-1 Is Involved in Angiogenic Signaling Cascades: Link to VEGF and SDF-1	217
10.3.3	HO-1 Function in Therapeutic Angiogenesis and Progenitor Cells	220
10.3.4	HO-1 Function in Tumor Angiogenesis and Growth	224
10.4	Novel Proangiogenic Function of Nrf2	225
10.4.1	The Effect of Nrf2 Knockout	225
10.4.2	Nrf2 Place in Angiogenic Signaling Pathways	226
10.4.3	Nrf2 in Regenerative Neovascularization of Damaged Tissue	229
10.4.4	Nrf2 in Tumor Angiogenesis	231
References		232

Abstract

The cytoprotective actions of Nrf2 transcription factor against vascular injuries associated with oxidative stress and tissue ischemia are widely reported. Amongst Nrf2 target genes, heme oxygenase-1 (HO-1) is responsible for at least a part of such protective effects by playing strong antioxidant and anti-inflammatory roles in the vascular system. Nonetheless, the area of Nrf2/HO-1 functioning could be extended to the control of vessel growth. Although the angiogenic involvement of HO-1 in physiological states and under conditions of tissue damage is well demonstrated, a direct role of Nrf2 in the process of blood vessel formation is just coming to light. Nrf2 has been linked to known angiogenic signaling pathways, comprising not only HO-1, but also vascular

U. Florczyk • A. Józkowicz • J. Dulak (✉)
Department of Medical Biotechnology, Faculty of Biochemistry, Biophysics and Biotechnology, Jagiellonian University, Gronostajowa 7, 30-387 Kraków, Poland
e-mail: jozef.dulak@uj.edu.pl

J. Dulak et al. (eds.), *Angiogenesis and Vascularisation*,
DOI 10.1007/978-3-7091-1428-5_10, © Springer-Verlag Wien 2013

endothelial growth factor and hypoxia-inducible factor-1, and suggested to act in normal vascular development as well as in the formation of blood vessels nourishing tumor. Strikingly, Nrf2 deficiency may promote oxidative stress-related inflammatory neovascularization accompanying damaged/ischemic tissue regeneration. Thus, further studies are definitely required for a thorough assessment of Nrf2 place in the processes of blood vessel formation.

Keywords

Heme oxygenase-1 • NF-E2-related factor 2 • Angiogenesis • Neovascularization • Ischemia • Proangiogenic cells • Endothelial progenitor cells

10.1 Introduction

NF-E2-related factor 2 (Nrf2) is considered as the major player in cellular protection against insults coming from heightened production of reactive oxygen species (ROS) or electrophilic metabolites of xenobiotics/carcinogens. Under such stresses Nrf2 is released from cytoplasmic repression of Keap1 protein, undergoes nuclear translocation and activates transcription of detoxifying and antioxidant genes such as NAD(P)H:quinone oxidoreductase (NQO1), peroxiredoxin-1, or glutathione S-transferases (GSTs) (reviewed in [1]). However, broader spectrum of Nrf2 functions is suggested since it also regulates genes of crucial importance in blood vessel formation, such as interleukin-8 (IL-8) [2] and cytoprotective enzyme, heme oxygenase-1 (HO-1) [3]. Proangiogenic effects of the latter are well defined and comprise control of normal vascularization (vascular cells proliferation, migration, and angiogenic capacity) as well as the involvement in tissue regeneration/revascularization and proangiogenic cells homing in response to ischemic/oxidative injuries. On the other hand, HO-1 contribution in tumor angiogenesis, a hallmark of cancer growth and progression [4], seems to be more complex and rather tissue-dependent. Nonetheless, for majority of the reported proangiogenic actions of HO-1, especially important could be reciprocal regulation of HO-1 and proangiogenic factors, such as vascular endothelial growth factor (VEGF) and stromal cell-derived factor-1 (SDF-1).

Regulation of HO-1 by Nrf2 provides grounds to assume that Nrf2 may not only be involved in cellular protection but in the mechanisms of neovascularization as well. Indeed, studies aiming at understanding of the proangiogenic function of Nrf2 are gathering pace. In addition to Nrf2 link to HO-1, its involvement in VEGF signaling and cross talk with hypoxia-inducible factor-1 (HIF-1), a major mediator of hypoxia-driven neovascularization, was suggested. Specifically, Nrf2 may contribute to the angiogenic capacity of endothelial cells as well as functioning and mobilization of proangiogenic cells. The role of Nrf2 in normal vascular development, in inflammation-related neovascularization, as well as in growing tumor is postulated.

10.2 Nrf2/HO-1 as a Key Path Against Vascular Injury

Being a bZip transcription factor Nrf2 is a member of the Cap 'n' Collar family of regulatory proteins and structurally consists of six functional Neh domains (Neh1–Neh6). Among them, the amino-terminal Neh2 domain is responsible for Keap1 binding leading to Nrf2 sequestration in the cytoplasm. The latter implies the repression of antioxidant response element (ARE)-containing cytoprotective and anti-inflammatory genes. Such situation occurs until electrophiles and oxidants will switch on Nrf2-dependent cellular defense mechanisms (reviewed in [1]).

It is commonly believed that oxidative stress occurs in the course or is at the root of cardiovascular diseases particularly by inducing endothelial cell dysfunction. Increased ROS production may result, e.g., from tissue ischemia/reperfusion as well as an occurrence of atherosclerosis risk factors, such as oxidized low-density lipoproteins (OxLDLs) or hyperglycemia, the latter condition attributed to diabetes, a metabolic disease with long-term vascular complications (angiopathies) [5–7]. In such conditions Nrf2 protection system is considered to play pivotal role in vascular cells. For example, phytochemical activation of Nrf2 protected human coronary artery endothelial cells against hydrogen peroxide-induced apoptosis [8]. Similarly we observed that in human microvascular endothelial cells (HMEC-1) as well as murine bone marrow-derived proangiogenic cells (PACs) H_2O_2-dependent increase in mortality is correlated with Nrf2 deficiency (Florczyk et al. in press). Moreover, studies of Kim et al. revealed that in response to laminar flow the activation of Nrf2 by ERK5 protein exerts atheroprotective effect in vascular endothelium [9]. Similarly Ashino et al. suggesting that Nrf2 system may be of importance in counteracting atherosclerosis showed that Nrf2 depletion enhances vascular smooth muscle cell (VSMC) migration in response to platelet-derived growth factor (PDGF) or wound scratch and in Nrf2-deficient mice stimulates neointimal hyperplasia in a wire vascular injury model [10]. On the other hand, a defect in Nrf2 signaling was shown to constitute a mechanism for cellular stress hypersensitivity in a genetic rat model of type 2 diabetes [11].

It is reasonable to assume that at least a part of the protective effects of Nrf2 could be related to HO-1 activation, the proangiogenic actions of which, but also strong cytoprotective, antioxidant, and anti-inflammatory functions in the vascular system, are well understood. Interestingly, HO-1-assigned protection may come from each of the enzyme activity products: biliverdin, ferrous ions, and carbon monoxide (CO), released as a result of heme degradation. Noteworthy, a set of HO-1 inducers, including ROS/H_2O_2, OxLDLs, or 15d-PGJ2 (15-deoxy-D12,14-prostaglandin J2, a natural ligand of peroxisome proliferator-activated receptor-gamma transcription factor), act through Nrf2 (reviewed in [12]). Indeed, a wide spectrum of direct effects of HO-1 against vascular injury, inflammation, and oxidative stress have been demonstrated. For example, HO-1 protected retinal endothelial cells against high glucose- and oxidative/nitrosative stress-induced toxicity [13]. The activation of HO-1 was also shown to inhibit atherosclerotic lesion formation in LDL receptor knockout mice [14]. Accordingly, in another model of atherosclerosis, namely, apolipoprotein E (apoE)-deficient mice, an absence of HO-1 (HO-1$^{-/-}$/apoE$^{-/-}$

mice) exacerbated atherosclerotic lesion formation and vascular remodeling [15]. Confirming HO-1 involvement in Nrf2-dependent protection Kim et al. reported, e.g., that sulfasalazine induces HO-1 via ROS-dependent Nrf2 signaling leading to suppression of neointimal growth [16]. On the other hand the induction of HO-1 partially reversed hampered in diabetes and oxidant damage-related vascular recruitment of endothelial progenitor cells in rats [17]. Moreover, pharmacologic induction of HO-1 was shown to play a protective role in diabetic retinopathy in rats [18]. Accordingly, the absence of HO-1 exacerbated myocardial ischemia/reperfusion injury in diabetic mice [19].

It is suggested that the protective effect of Nrf2/HO-1 against vascular injuries such as atherosclerosis is especially due to anti-inflammatory response and antioxidant actions in the vascular wall. Nonetheless, the area of Nrf2/HO-1 functioning could be extended to the control of vessel growth in physiological states and under conditions of tissue damage. Although the proangiogenic function of HO-1 is already almost a dogma, the direct role of Nrf2 in the processes of blood vessel formation is just coming to light.

10.3 HO-1 Functions in Blood Vessel Formation

10.3.1 An Overall Effect of HO-1 Knockout

The high significance of HO-1 has been underlined by studies on HO-1 knockout animals and was confirmed by single human cases of HO-1 deficiency. In first experiments HO-1-deficient mice were characterized mostly by splenomegaly, high CD4+:CD8+ T-cell ratios, increased lipid peroxidation, fibrosis and hepatic injury, late-onset weight loss, decreased mobility, and premature mortality [20]. Further studies revealed reduced stress defense in HO-1-deficient cells, including vascular endothelial and smooth muscle cells [15, 21, 22]. So far only two human cases of HO-1 deficiency have been recognized and observed to involve endothelial cells more severely, resulting in hemolysis and disseminated intravascular coagulation [23, 24]. Noteworthy, it was reported that HO-1 promoter contains a (GT)n microsatellite DNA and such polymorphism can significantly modulate a proangiogenic, cytoprotective, and anti-inflammatory function of HO-1 in human endothelium [25]. Moreover, apart from the difference in the occurrence of asplenia in the human HO-1-deficient patient and splenomegaly in mice counterpart one can find many similarities between both species including prenatal death (usually stillbirth/abortion in humans and 20 % birth rate in mice), developmental failure, iron deficiency anemia, and chronic systemic inflammatory disorders [26]. Thus, one can sum up that HO-1 deficiency may have broad spectrum of severe side effects, which could affect many organs (reviewed in [12]).

Since the first generation of HO-1-deficient mice in 1997 [20] a variety of studies, as described more closely below, particularly revealed the importance of HO-1 in vascular development under physiological and pathological conditions, underlying its influence on proangiogenic capacity of endothelial cells and bone marrow-derived progenitors.

10.3.2 HO-1 Is Involved in Angiogenic Signaling Cascades: Link to VEGF and SDF-1

10.3.2.1 Proangiogenic Function of HO-1

There is a broad spectrum of HO-1 functions in the vascular system comprising its protective, anti-apoptotic effects towards endothelial cells, attenuation of inflammatory response in the vessel wall, and regulation of vascular tone through vessel relaxation (reviewed in [12, 27]). However, it is particularly famous as one of the key players in the processes of the formation of blood vessels. As first it was shown in 1998 by Deramaudt et al. that gene transfer of human HO-1 into rabbit coronary endothelial cells potentially promotes angiogenesis [28]. Further in vitro studies revealed that HO-1 is significant also in prolactin-mediated cell proliferation and angiogenesis in human endothelial cells [29]. Recently our group showed also that both constitutive and hypoxia-induced HO-1 overexpression increases migratory properties of HMEC-1 [30]. Strikingly, from 2003, when Suzuki et al. observed facilitated angiogenesis after intra-arterial delivery of adenoviral vectors encoding HO-1 in a rat model of hind limb ischemia [31], the growing body of evidence confirmed stimulatory effects of HO-1 on neovascularization in many in vivo models, such as myocardial infarction, wound healing, cancer, or hind limb ischemia [32–36].

The importance of HO-1 and/or its activity products in the vasculogenesis and angiogenesis can be attributed to its role in regulation of proangiogenic mediators, such as VEGF, SDF-1, monocyte chemotactic protein-1 (MCP-1), or transforming growth factor-β (TGF-β). On the other hand, HO-1 may negatively affect the production of anti-angiogenic factors such as soluble VEGF receptor-1 (sVEGFR-1), soluble endoglin (sEng), or CXCL-10 chemokine [12, 37].

In addition to the involvement of HO-1 in the regulation of VEGF and SDF-1 expression in different cell types (see below), it seems that HO-1 functions as a downstream mediator in proliferative and pro-migratory response of endothelial cells and progenitors to VEGF and SDF-1 may be especially important for the overall proangiogenic role of HO-1 [38, 39].

10.3.2.2 Interactions with VEGF Signaling

The fact that VEGF is one of the major or even the most crucial mediator of blood vessel formation was strikingly proved in 1996 by two independent groups [40, 41]. Ferrara et al. reported that loss of even a single VEGF allele is lethal in the mouse embryo between days 11 and 12 [41]. They demonstrated developmental anomalies coming from impaired angiogenesis and blood island formation. Moreover, VEGF-null embryonic stem cells exhibited a dramatically reduced ability to form tumors in nude mice [41]. Similarly, Carmeliet et al. showed abnormal blood vessel development and lethality in heterozygous VEGF-deficient embryos [40].

Having in mind those findings one can attribute the particular significance to HO-1, since it is known as an important collaborator of VEGF, acting both upstream and downstream of the latter (Fig. 10.1). First link between HO-1 and VEGF was suggested by Dulak et al. by showing that HO-1 activity modulates

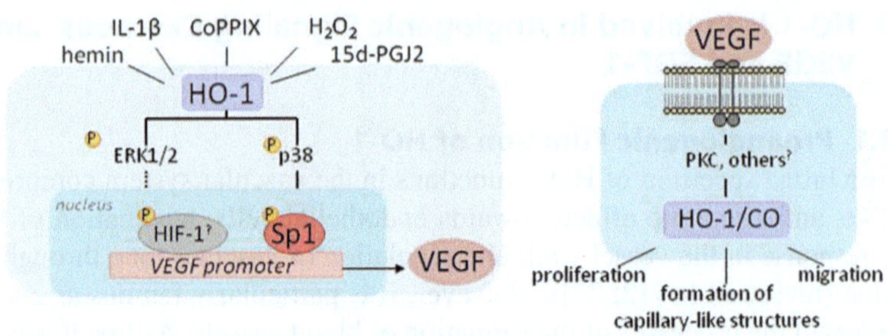

Fig. 10.1 HO-1 functions upstream and downstream of VEGF signaling (description in the text)

VEGF synthesis in vascular smooth muscle cells (VSMC) [42]. The inhibition of HO-1 activity by tin protoporphyrin IX (SnPPIX) abolished hemin- and IL-1β-induced VEGF production in normoxic conditions and resulted in a profound reduction of hypoxia-induced VEGF synthesis. Accordingly, VSMC transfection with HO-1 expression plasmid enhanced VEGF level. Interestingly, the data indicated that HO-1 by-products may exert divergent effects: biliverdin showed no influence, iron ions inhibited, and CO (1 %) augmented VEGF production [42]. However, since the stimulatory effect of HO-1 was noticed, one can assume that CO effect prevails the action of iron.

Further studies revealed also the HO-1-dependent regulation of VEGF expression in other cell types. In HMEC-1 the stimulatory effect of 15d-PGJ2 as well as cobalt protoporphyrin (CoPPIX) on VEGF production was reversed by SnPPIX [38, 43, 44]. As a mechanism of HO-1-mediated upregulation of VEGF synthesis in response to 15d-PGJ2, the phosphorylation of extracellular signal-regulated kinase (ERK1/2) was proposed [45]. Namely, transient transfection of human breast cancer (MCF-7) cells with dominant-negative ERK abrogated 15d-PGJ2-induced VEGF expression, while HO-1 inhibitor, zinc protoporphyrin (ZnPP), abolished ERK1/2 phosphorylation in response to 15d-PGJ2 [45]. Moreover, an involvement of HIF-1 or Sp1 transcription factors in VEGF upregulation could be considered [46].

Moreover, HO-1 was shown to be crucial for H_2O_2-induced VEGF synthesis in NIH 3T3 fibroblasts and in HaCaT keratinocytes [47]. In the latter silencing of HO-1 expression by specific siRNA as well as SnPPIX treatment attenuated also the hemin-induced effects [48]. Furthermore, studies using other HO-1 inhibitor, stannic mesoporphyrin (SnMP), showed also the involvement of HO-1 in VEGF production in human macrophages stimulated with prolactin, the pituitary hormone [49].

Importantly, the genetic evidence confirming the regulation of VEGF expression by HO-1 was provided by the use of aortic endothelial cells obtained from HO-1 knockout (HO-$1^{-/-}$) mice. Data indicated that either basal or H_2O_2–, hemin-, lysophosphatidylcholine-, and 15d-PGJ2-induced VEGF synthesis can be attenuated by the lack of HO-1 [47]. On the other hand, retrovirus-mediated human HO-1 gene transfer into rat lung microvessel endothelial cells resulted in enhancement of VEGF level [50].

The involvement of HO-1 in VEGF regulation was also demonstrated using in vivo models. For example, in portal hypertension model, intraperitoneal injection of SnPPIX for 7 days starting immediately after partial portal vein ligation markedly decreased (by 74 %) VEGF protein expression in the mesentery of rats [51]. On the other hand, systemic induction of HO-1 in mice injected intravenously with adenovirus-bearing HO-1 gene led to elevated serum levels of VEGF (and SDF-1; see below) [52]. The significance of HO-1 actions upstream VEGF was underlined by Lin et al. working on murine model of myocardial infarction [53]. They demonstrated that HO-1 overexpression obtained by hypoxia-regulated adeno-associated viral (AAV) vectors promotes neovascularization in ischemic heart by co-induction of VEGF and SDF-1 (see below) [53]. As a mechanism of increased expression of VEGF in myocytes, CO-mediated phosphorylation of p38 kinase and the following induction of Sp1 transcription factor were suggested based on in vitro studies [54].

The complexity of the interactions between HO-1 and VEGF arises from the fact that HO-1 may function both upstream and downstream of VEGF. Owing to the latter HO-1 serves as a mediator of VEGF-dependent proangiogenic, proliferative, and pro-migratory response of endothelial cells and progenitors. In 2003 our group for the first time reported that HO-1 is required for proangiogenic action of VEGF in endothelial cells [38]. We showed that HO-1 inhibition by SnPPIX diminishes the VEGF-elicited angiogenic activities of human umbilical vein endothelial cells (HUVEC) by decreasing their proliferation, migration, as well as the formation of capillary-like structures on Matrigel and in spheroidal culture in collagen gel. Accordingly, augmentation of capillary sprouting was observed in endothelial spheroids overexpressing HO-1. Such an effect was mimicked by incubation of cells with CO-releasing molecules (CO-RM) suggesting crucial importance of CO in HO-1-dependent angiogenesis [38].

In addition, Fernandez et al. revealed that HO-1 expression may be upregulated by VEGF in vivo in the chick embryo chorioallantoic membrane by a mechanism dependent on the increase in cytosolic calcium levels and activation of protein kinase C [55]. In this model, VEGF-stimulated angiogenesis was significantly attenuated by the HO-1 inhibitor, zinc mesoporphyrin (ZnMP) [55]. VEGF-dependent HO-1 regulation was confirmed by Bussolati et al. in human endothelial cells, both HMEC-1 and HUVEC [56]. Interestingly they showed that although blockade of HO-1 inhibits VEGF-induced in vitro angiogenesis, it increases the angiogenic effect of VEGF in murine Matrigel model in vivo. Such discrepancy may be related to leukocyte chemoattractant activity of VEGF, which, in the absence of HO-1, may lead to the induction of inflammatory blood vessel formation. Thus, it seems that HO-1 may modulate proinflammatory and proangiogenic actions of VEGF and may be of crucial importance in the treatment of chronic inflammatory diseases [56].

10.3.2.3 Interactions with SDF-1 Signaling

SDF-1, a member of the CXC group of chemokines, seems to be another player closely related to HO-1 functioning. It has been shown that mice lacking SDF-1 or its primary physiological receptor, CXC-chemokine receptor type 4 (CXCR4), die perinatally or in utero, respectively, and are defective in vascular development, hematopoiesis, and cardiogenesis [57, 58]. Particularly, SDF-1 is known to play a role in cell trafficking. SDF-1 expression at the site of injury/ischemia enables homing of circulating CXCR4-positive and other stem/progenitor cells for organ regeneration and tissue repair [59, 60]. For example, in the murine hind limb ischemia model overexpression of SDF-1α (a predominant splicing form of SDF-1) stimulated endothelial progenitor cells' (EPC) mobilization and increased blood flow and capillary density in the ischemic limb [61].

Similarly as in the case of VEGF, the regulation of SDF-1 by HO-1 was reported. Recently we showed that genetic or pharmacological activation of HO-1 in C2C12 myoblasts augments production of SDF-1 [37]. As an in vivo example, as mentioned above, in ischemic hearts of mice with permanent coronary artery ligation AAV-mediated HO-1 overexpression highly induced SDF-1 [53]. Accordingly, treatment of Wistar rats with CoPPIX to induce HO-1 prior to coronary artery ligation resulted in increased expression of SDF-1α and promoted myocardial neovascularization [32]. On the other hand, systemic induction of HO-1 obtained by intravenous administration of adenoviral vectors led to elevated serum levels of SDF-1, the effect mimicked by animal exposure to CO [52]. Recently, HO-1 was also reported to be involved in AMP-activated kinase (AMPK)-mediated upregulation of SDF-1α production in EPCs [62]. Nonetheless, the mechanism of SDF-1 control by HO-1 needs to be examined.

In contrast, the mechanism of HO-1 role in SDF-1-elicited blood vessel formation was well understood. In their studies Deshane et al. demonstrated that SDF-1 through the CXCR-4 receptor transcriptionally activates HO-1 in endothelial cells via the atypical protein kinase Cζ isoform-dependent and VEGF-independent mechanism [39]. Further it was proposed that as a result of HO-1 enzyme activity CO is produced, which phosphorylates vasodilator-stimulated phosphoprotein (VASP), a cytoskeletal-associated protein involved in cell migration. In vitro, HO-1 was shown to mediate SDF-1-induced angiogenesis in human aortic endothelial cells (HAECs) and mouse aortic endothelial cells (MAECs). The blockade of HO-1 by ZnPP or siRNA strategy resulted in inhibition of SDF-1-induced tube formation in HAECs. Similar effect was observed in case of MAECs isolated from HO-1$^{-/-}$ mice. Moreover, SDF-1 stimulated capillary-like sprouting in aortic rings from wild type (HO-1$^{+/+}$) mice but not HO-1$^{-/-}$ mice, a defect reversed by CO. Strikingly, the functional relevance of HO-1 in the SDF-1-elicited proangiogenic actions was confirmed in Matrigel plug, wound healing, and retinal ischemia models in vivo [39].

10.3.3 HO-1 Function in Therapeutic Angiogenesis and Progenitor Cells

10.3.3.1 HO-1 in Regeneration of Damaged Tissue

Broadly defined cytoprotective, antioxidant, and anti-inflammatory capacity of HO-1 was shown many times to protect against injuries in different tissues. For example, cardiac-specific expression of HO-1 protected against ischemia and reperfusion injury in transgenic mice, thus improving the recovery of cardiac function [63]. Accordingly, in a porcine model of myocardial infarction intracoronary delivery of allogenic HO-1-transduced bone marrow cells led to faster recovery of left ventricular ejection fraction in comparison to control cells [64]. On the other hand, the absence of HO-1 significantly increased myocardial infarct size, while its overexpression protected against myocardial ischemic injury in diabetic mice [19].

On the other hand, a growing body of evidence points at particular significance of proangiogenic function of HO-1 in the regenerative neovascularization of ischemic/damaged tissues. As mentioned above the first evidence suggesting in vivo role of HO-1 in therapeutic neovascularization was reported by Suzuki et al. working on the rat model of hind limb ischemia [31]. After intra-arterial bolus injection of adenoviral vectors encoding HO-1, an improvement in blood flow and capillary density was observed in the ischemic muscles, the effect reversed by ZnPP treatment [31]. Accordingly, HO-1 inhibition by SnPPIX in mice subjected to femoral artery ligation resulted in the attenuation of both neovascularization and functional recovery of injured tissue [35]. As a complement and extension of those results our group demonstrated similar effects in HO-1$^{-/-}$ mice [30]. In contrast we observed that in wild-type mice transient hypoxia/ischemia-induced HO-1 overexpression, obtained by intramuscular administration of plasmid vector carrying HO-1 gene driven by hypoxia response elements (HREs), significantly improved the post-ischemic foot blood flow. In such conditions reduced levels of pro-inflammatory cytokines (IL-6 and CXCL1) and decreased number of apoptotic cells were observed. Moreover, overexpression of HO-1 in ischemic skeletal muscles influenced also the levels of molecules controlling myocyte regeneration such as myogenin, Pax3, Pax7, as well as small noncoding RNAs (microRNAs): miR-146a and miR-206 [30]. In addition, other studies provided evidence that HO-1 may promote also neovascularization after myocardial infarction [32]. Thus it seems that HO-1 plays crucial therapeutic role in counteracting detrimental effects of tissue ischemia.

On the other hand, as mentioned above, Deshane et al. observed impaired wound healing and a diminished angiogenic response after full-thickness excisional skin wound injury in HO-1$^{-/-}$ mice [39]. The absence of HO-1 decreased the percentage of reepithelialized surface and the number of CD31-positive vessels in the wounds and diminished also the neovascularization associated with topical application of SDF-1 to the wound surface [39]. In accordance with those results SnPPIX-mediated inhibition of HO-1 in wild-type mice resulted in retardation of wound closure [33]. Moreover, keratinocyte-specific overexpression of HO-1 was enough to improve the neovascularization and accelerate wound healing in transgenic mice. Strikingly, in diabetic mice HO-1 upregulation after skin injury was delayed as compared to wild-type mice. Since chronic nonhealing wounds are one of the major consequences of diabetes, HO-1 importance was suggested and indeed confirmed when HO-1 transgene delivery into the injured tissue was able to improve the healing process in such animals [33].

Interestingly, Bhang et al. pointed at beneficial effects of combined delivery of HO-1 with other proangiogenic molecules, such as fibroblast growth factor-2 (FGF-2) or HIF-1α for therapeutic angiogenesis [65, 66]. In the murine hind limb ischemia model FGF-2 (in fibrin gels) could rescue muscle necrosis prior to the exogenous expression of HO-1 and enhanced HO-1 gene transfection in ischemic limbs. As an effect angiogenesis was greater in mice treated with both HO-1 and FGF-2 gene therapy in comparison to HO-1 delivered alone [65]. On the other side, also the combined gene therapy of HO-1 and HIF-1α turned out to be superior (enhanced anti-apoptosis and neovascularization) to both single-gene therapies, resulting in rapid expression of HIF-1α gene (which protein degrades quickly) and long-term maintenance of expressed HO-1 protein (which gene expression occurs a couple of days after transfection) [66].

10.3.3.2 HO-1 in Proangiogenic Cells

In 1997 the studies of Asahara et al. revealed that blood vessel formation in ischemic/injured tissues depends not only on resident endothelial cells but may be supported by EPCs, which are believed to be mobilized from bone marrow and contribute to postnatal vascular regeneration by forming a structural component of capillaries and/or by secreting angiogenic factors [67, 68]. Since among the major inducers of EPC homing one can find VEGF and SDF-1 [69, 70], it is reasonable to expect HO-1 involvement in those processes.

The first finding linking HO-1 and functioning of EPCs was described in 2003 by Kong et al. working on balloon injury model of carotid artery [71]. However, EPCs isolated from rabbit peripheral blood, expanded in culture, transduced with retroviral vectors expressing HO-1, and finally transplanted into the injured vessel showed no effect over control, green fluorescent protein (GFP)-expressing EPCs. Nonetheless, re-endothelialization was enhanced and neointimal thickening was reduced in the EPC-transplanted vessels independently of gene transfer (HO-1 or GFP) [71].

In contrast, further studies revealed a significant effect of HO-1 on EPC mobilization and recruitment to the damaged tissue. Using murine hind limb ischemia model Tongers et al. showed that HO-1 inhibition by SnPPIX significantly reduces the number of circulating progenitor cells (Sca-1+/Kdr+) [35]. Moreover, in transgenic mice expressing β-galactosidase under the control of the endothelial cell-selective Tie2 promoter, in response to femoral artery ligation fewer β-galactosidase-positive cells were detected in the ischemic muscle when HO-1 was inhibited. Such observations could be explained by SnPPIX-dependent disruption of the establishment of SDF-1 gradient between the ischemic limb and bone marrow [35]. Accordingly, in HO-1$^{-/-}$ mice significant attenuation of EPC mobilization and recruitment as well as re-endothelialization of denuded vessels in the wire-induced carotid artery injury model was observed [52].

On the other side, an approach of HO-1 overexpression confirmed its function in EPC trafficking. Systemic HO-1 induction after intravenous injection of the HO-1-encoding adenoviral vectors in addition to mentioned elevated serum levels of VEGF and SDF-1 led to an increase in circulating EPCs (CD34+/Flk-1+) and facilitated re-endothelialization after vascular injury [52]. The increase in EPC mobilization and re-endothelialization of injured vessels was also visible in mice exposed to CO, which was accompanied by SDF-1 (but not VEGF) enhancement [52]. In the other studies AAV-mediated HO-1 gene transfer increased the number of stem cells (c-kit+) recruited to the infarcted area and enhanced vascularization after coronary artery ligation [53]. As mentioned before VEGF and SDF-1 were highly induced in HO-1-transduced myocardium and colocalized with mononuclear cell infiltration. Importantly, neutralizing antibodies against VEGF and SDF-1 attenuated HO-1-mediated angiogenesis and cardiac function recovery. Thus, HO-1 overexpression was shown to protect ischemic myocardium by promoting neovascularization through induction of proangiogenic factors and the recruitment of progenitor/stem cells [53].

A confirmation of HO-1 contribution to vascular repair by increasing circulating EPCs was reported also by Wu et al. [72]. In a rabbit model of aortic balloon injury, pharmacological inducer of HO-1, probucol, promoted re-endothelialization and increased the number of circulating progenitor cells (Sca-1+/Flk1+) as well as Sca-1+ and Flk1+ cells on the luminal side of injured vessels, the actions reversed by SnPPIX co-treatment. Similar effects were observed for a derivative of probucol, succinobucol. In addition, in noninjured wild-type mice succinobucol also increased bone marrow (CD34+/Flk1+/CD45−) and circulating (CD34+/CD45−/lin−) EPCs. Moreover, in vitro cultured bone marrow cells from HO-1$^{-/-}$ mice generated fewer colony-forming unit endothelial cells (CFU-EC or CFU-Hill) (early outgrowth EPCs) as well as endothelial colony-forming cells (late outgrowth EPCs) (for review, see [73]). Confirming the role of HO-1 in the formation of progenitor cells, succinobucol promoted CFU-Hill colony formation in wild-type mice, but had no effect in HO-1$^{-/-}$ littermates [72].

Finally, very recent reports point at increased utility of HO-1-overexpressing EPCs, combining gene and cell therapy approach, for neovascularization and recovery of injured tissues. Since EPCs are suggested to be dysfunctional as a result of cardiovascular risk factors, their modification by gene transfer of cytoprotective and proangiogenic HO-1, alone or in combination with other pro-survival factors, may be a reasonable therapeutic approach to treat vascular diseases. Indeed, retroviral transduction with both HO-1 and protein kinase B (Akt) improved survival, adhesion, and migration of late outgrowth EPCs in a simulated infarct-like environment [74]. Importantly, transplantation of Akt/HO-1-expressing EPCs after myocardial infarction improved neovascularization and cardiac recovery in mice [74]. In the other approach, transplantation of VEGF- and HO-1-modified EPCs into the ischemic limbs of rats significantly increased the microvessel density, the recovery of blood flow, and the rate of limb salvage compared with either VEGF- or HO-1-expressing EPCs [75].

In addition, several studies provided evidence of the advantage of combined gene and cell therapy over either strategy alone also in the case of the use of HO-1-overexpressing adipose tissue-derived stem cells [76] or particularly mesenchymal stem cells (MSCs) [77–80]. As an example of the latter, although MSCs provided a beneficial effect on swine cardiac function after ischemia/reperfusion by the induction of therapeutic angiogenesis, this effect was amplified by HO-1 overexpression [77].

10.3.4 HO-1 Function in Tumor Angiogenesis and Growth

It is well known that tumor growth is strictly dependent on angiogenesis and blood supply [4]. Proangiogenic capacity of HO-1, in addition to its cytoprotective, antioxidant function conferring survival support and protection against anticancer therapies, is considered to promote tumor progression by stimulation of blood vessel formation. Although elevated HO-1 levels have been detected in many types of tumors, including renal adenocarcinoma, hepatoma, glioblastoma, melanoma, prostate cancers, squamous carcinoma, pancreatic cancer, and others [81], its effect on angiogenesis-based tumor growth seems to be tissue-dependent.

The first report showing a positive correlation between HO-1 expression and vascular density was shown in 1999 for human gliomas [82]. However, a direct evidence that HO-1 may indeed accelerate tumor angiogenesis was demonstrated by Sunamura et al. 4 years later [34]. Marked and rapid development of capillary-like network in vitro in coculture of human HO-1-transduced pancreatic cancer cells and endothelial cells was observed. In vivo, using severe combined immune-deficient mice they showed that HO-1 stimulates angiogenesis of pancreatic carcinoma affecting tumor vascular density and proliferation rate. Accordingly, inhibition of HO-1 activity by SnMP was able to transiently delay tumor growth [34].

Similarly, overexpression of HO-1 increased cell proliferation and angiogenic potential of murine melanoma B16(F10) cells in vitro [83]. In accordance with that, tumors developing after subcutaneous injection of melanoma cells stably transfected with HO-1 into the syngenic mice displayed augmented vascularization and stronger production of VEGF [83]. Further studies suggested also positive effect of HO-1 on angiogenesis in urothelial carcinoma of the urinary bladder (T24 cell line) and hepatocellular carcinoma (Hepa129 cell line) [36, 84].

As a possible mechanism of HO-1-elicited effects in tumor angiogenesis, its interaction with VEGF signaling was proposed. For example, HO-1 involvement in 15d-PGJ2-induced upregulation of VEGF in breast cancer cell line (MCF-7) was revealed (see above) [45]. Accordingly, nuclear HO-1 activated the transcriptional activity of VEGF and promoted VEGF secretion in prostate cancer cells [85], while in urothelial carcinoma inhibition of HO-1 decreased tumor growth and microvessel density by suppressing VEGF and HIF-1α [36]. On the other side HO-1 was also suggested to mediate the angiogenic effect of an enzyme thymidine phosphorylase (TP) in bladder carcinoma cells [86] (see also Chap. 17).

Despite reports advocating proangiogenic function of HO-1 in some tumors there are also a few suggesting the opposite action of the enzyme in other cancer types. Indeed, latest findings by Skrzypek et al. revealed anti-angiogenic effect of HO-1 in human non-small cell lung cancer (NSCLC) [87]. Stable HO-1 overexpression significantly downregulated angiopoietin-1 expression and reduced proliferation, migration, as well as angiogenic potential of NSCLC NCI-H292 cell line. Particularly, elevated HO-1 level decreased expression of a group of microRNAs capable of promoting angiogenesis, so-called angiomirs: miR-17–92, miR-210, miR-378, whereas increased that of anti-angiomirs, such as miR-424. In vivo, HO-1-overexpressing tumors were both less vascularized and oxygenated. Strikingly, overexpression of miR-378 exerted opposite effects [87].

In some cases controversial data exist showing different effects of HO-1 even in the same cancer cell line as for prostate cancer. Namely, as mentioned above HO-1 stimulated VEGF in human prostate cancer cell line, PC3 [85]. In contrast, Ferrando et al. noticed that in response to HO-1 overexpression in the same prostate cancer PC3 cells, several proangiogenic genes are downregulated, in particular VEGF, VEGF-C, HIF-1α, and $\alpha 5\beta 1$ integrin [88]. Moreover, intradermal inoculation of PC3 cells stably transfected with HO-1 generated less vascularized tumors with decreased microvessel density [88]. Nonetheless, it seems that in majority of cases the effect of HO-1 on tumor angiogenesis seems to be tissue/cell type-dependent.

10.4 Novel Proangiogenic Function of Nrf2

10.4.1 The Effect of Nrf2 Knockout

In contrast to HO-1$^{-/-}$ mice, homozygous Nrf2 mutant mice (Nrf2$^{-/-}$) are not anemic and develop normally. Since they reach adulthood with any abnormality of major organs and are able to reproduce, it seems that Nrf2 is dispensable for mouse development [89]. Nonetheless, in the harmful conditions arising from oxidative stress and chemical injuries Nrf2 turned out to be crucial for switching on cellular protection mechanisms. For example, Nrf2 ablation rendered mice highly sensitive to the toxic effects of food preservative butylated hydroxytoluene (BHT) [90]. The same doses of BHT were tolerated by wild-type mice, which led to the development of acute respiratory distress syndrome in Nrf2$^{-/-}$ mice, the effect accompanied by decreased expression of Nrf2-dependent cytoprotective genes, such as HO-1 or NQO1 [90]. Similarly, Nrf2-deficient mice were more vulnerable to acetaminophen hepatotoxicity [91] and many other toxicities.

Moreover, the lack of Nrf2 is associated with increased sensitivity to carcinogenesis. The strong induction of phase 2 enzymes in wild-type mice by a chemoprotective agent oltipraz was shown to be abrogated in the Nrf2$^{-/-}$ mice. Accordingly, oltipraz was not able to rescue Nrf2$^{-/-}$ mice from significantly higher burden of gastric neoplasia after treatment with benzo[a]pyrene [92]. Interestingly, it was shown also that the aged Nrf2$^{-/-}$ female mice developed lupus-like autoimmune nephritis [93]. Overall, it seems that Nrf2-deficient mice are more susceptible to disease development [93].

Interestingly, based on the recent data, Nrf2 deficiency has been correlated also with the processes of blood vessel formation. Therefore, Nrf2$^{-/-}$ mice as well as cells lacking Nrf2 expression are increasingly being used to study such a new angiogenic face of this transcription factor, which will be discussed below.

10.4.2 Nrf2 Place in Angiogenic Signaling Pathways

Very recently several papers have suggested an interplay between Nrf2 and some angiogenic signaling pathways (summarized in Table 10.1). In 2010 Afonyushkin et al. first published that VEGF may be regulated downstream of Nrf2 [94]. In endothelial cells oxidized phospholipid (OxPL)-induced ATF4 transcription factor and VEGF expression are Nrf2-dependent. Nrf2 was able to bind to ARE in ATF4 gene promoter. Knockdown of Nrf2 inhibited OxPL-stimulated elevation of both ATF4 and VEGF expression as well as sprouts formation by endothelial cells [94]. Interesting interplay between VEGF and Nrf2 was shown in the human placental choriocarcinoma cell line, BeWo, by Kweider et al. studying preeclampsia [95]. A positive feedback loop was described in which without oxidative stress stimulus VEGF activated Nrf2 in an ERK1/2-dependent manner resulting in the upregulation of HO-1 expression and subsequent CO production, which in turn upregulated VEGF expression. Thus, VEGF through Nrf2 pathway may protect against oxidative stress and upregulate the expression of itself [95].

In addition, other reports suggested the upregulation of VEGF in response to the Nrf2-dependent HO-1 induction (compare with Sect. 10.3.2.2) [96–98]. Shibuya et al. demonstrated that sofalcone, an antiulcer and gastric mucosa protective agent, increased HO-1 expression in rat gastric epithelial cells (RGM-1) via Nrf2 and in turn leads to the elevation of VEGF. Treatment of RGM-1 cells with SnPPIX or transfection with siRNA against HO-1 attenuated sofalcone-induced VEGF production [96]. Accordingly, Meng et al. gently suggested the contribution of Nrf2/HO-1 pathway in arsenic-mediated angiogenesis in vitro and observed that knockdown of HO-1 expression decreased arsenite-induced VEGF expression, cell migration, and tube formation [98]. In the other approach, Chao et al. reported that diesel exhaust particles (DEPs) through induction of ROS in capillary-like endothelial tube cells can sequentially lead to Nrf2 nuclear translocation, HO-1 upregulation, increased VEGF production, and endothelial permeability [97].

In addition, Nrf2 cross talk with HIF-1, a major hypoxia-induced transcription factor exerting proangiogenic functions mostly through stimulation of VEGF production [99], was also reported by several groups in different experimental settings. Malec et al. were first to demonstrate that HIF-1 signaling may be augmented by induction of the Nrf2 pathway [100]. They showed that under conditions of intermittent hypoxia in lung adenocarcinoma A549 cells high levels of ROS are generated by NADPH oxidase subunit NOX1. In such conditions the induction of Nrf2 and its target gene, thioredoxin 1, turned out to enhance HIF-1 signaling. Noteworthy, continuous hypoxia did not exert the same effects, inducing HIF-1 without an involvement of Nrf2 [100]. Actually, our group reported that hypoxia mimic, dimethyloxaloylglycine (DMOG), as well as HIF-1α overexpression may diminish Nrf2 and subsequent IL-8 expression in endothelial cells [101]. Thus one could hypothesize that in case of HIF-1 inhibition, Nrf2 could exert alternative proangiogenic effects.

Nevertheless, further studies confirmed and underlined the role of Nrf2 as an upstream regulator of HIF-1. Recently, Zhang et al. using human epithelial ovarian

Table 10.1 The summary of the reported Nrf2 actions in blood vessel formation and its cross talk with proangiogenic signaling pathways

Experimental approach	Observed activities	Cell type/animal model	References
• Arsenite treatment – HO-1 siRNA transfection	Increased Nrf2 binding to HO-1 gene; knockdown of HO-1 expression decreases arsenite-induced VEGF expression, cell migration, and tube formation	Human microvascular endothelial cells (HMVECs)	Meng et al. (2010) [98]
• OxPLs stimulation – Nrf2 siRNA transfection	Knockdown of Nrf2 inhibits ATF4 and VEGF expression and sprouts formation in vitro	Human umbilical vein endothelial cells (HUVECs)	Afonyushkin et al. (2010) [94]
• Sofalcone treatment – HO-1 siRNA transfection – SnPPIX	Nrf2-dependent increase of HO-1 and VEGF expression; HO-1 silencing/inhibition attenuates sofalcone-induced VEGF production	Rat gastric epithelial cells (RGM-1)	Shibuya et al. (2010) [96]
Air or hyperoxia (75% oxygen)	Decreased secondary capillary network formation in the retina at P9 in air-reared $Nrf2^{-/-}$; Hyperoxia-exposed $Nrf2^{-/-}$ mice are more susceptible to increased vaso-obliteration at P9	$Nrf2^{-/-}$ mice	Uno et al. (2010) [110]
VEGF treatment	Activated Nrf2/HO-1/CO pathway upregulates VEGF expression (positive feedback loop of VEGF)	Human placental choriocarcinoma cells (BeWo)	Kweider et al. (2011) [95]
• VEGF and IGF-1 stimulation – Nrf2 siRNA transfection – Keap-1 plasmid transfection	Nrf2 deficiency impairs proliferation, migration, adhesion to vitronectin and collagen, ability to form capillary-like structures in vitro	Human coronary arterial endothelial cells (CAECs)	Valcarcel-Ares et al. (2012) [104]
• Diesel exhaust particles treatment	ROS-activated Nrf2 through HO-1 stimulates VEGF expression and endothelial permeability	Capillary-like endothelial tube cells	Chao et al. (2012) [97]
• Hypoxia mimic (DMOG) treatment – HIF-1α adenoviral transduction	Diminished Nrf2 decreases IL-8 expression	Human microvascular endothelial cells (HMEC-1)	Loboda et al. (2009) [101]

(continued)

Table 10.1 (continued)

Experimental approach	Observed activities	Cell type/animal model	References
• Intermittent hypoxia	ROS-activated Nrf2 increases thioredoxin 1 expression enhancing HIF-1 signaling	Lung adenocarcinoma A549	Malec et al. (2010) [100]
• Hypoxia – Nrf2 shRNA lentiviral transduction – Nude mouse xenograft model – CAM model	Under hypoxia Nrf2-deficient cells failed to accumulate HIF-1α protein attenuating VEGF expression; in vivo, knockdown of Nrf2 suppresses tumor growth and reduces blood vessel formation and VEGF expression as well as exerts antiangiogenic effects in CAM and endothelial tube formation assays	Human colon cancer cells (HCT116/HT29)	Kim et al. (2011) [103]
• Follicle-stimulating hormone treatment – Nrf2 siRNA transfection	ROS-activated Nrf2 by upregulating HIF-1 stimulates VEGF production; knockdown of Nrf2 inhibits FSH-induced VEGF expression as well as impairs HIF1α signaling activation	Human epithelial ovarian cancer	Zhang et al. (2013) [102]
– Nrf2 knockdown – Nude mouse xenograft model	Decreased microvessel density and tumor growth	Human glioma cells (U251MG)	Ji et al. (2013) [119]

cancer convincingly presented that in response to follicle-stimulating hormone (FSH), the ROS-dependent activation of Nrf2 occurs, which by upregulating HIF-1 may stimulate VEGF production. Knockdown of Nrf2 inhibited FSH-induced VEGF expression as well as impaired HIF-1α signaling activation [102].

Accordingly, as an in vivo example, Kim et al. reported that Nrf2 blockade may inhibit hypoxia-induced activation of HIF-1α in turn suppressing colon tumor angiogenesis (see below) [103]. Antiangiogenic effects of Nrf2 knockdown in cancer cells were noticed also in chicken embryo chorioallantoic membrane (CAM) assay. On the other hand, conditioned media from hypoxia-exposed Nrf2-deficient cancer cells attenuated the ability of HUVEC to form capillary-like structures on Matrigel as compared to media collected from control cells cultured under the same conditions. Specifically, under hypoxic conditions Nrf2-deficient cancer cells failed to accumulate HIF-1α protein attenuating VEGF expression, the mechanism associated with reduced mitochondrial O_2 consumption in cells lacking Nrf2 [103].

Importantly, genetic evidence of the hypothesis that Nrf2 may be crucial for angiogenic capacity of endothelial cells was elegantly shown by Valcarcel-Ares et al. [104]. In their studies an approach of the silencing of Nrf2 signaling by siRNA was tested in cultured human coronary arterial endothelial cells (CAECs) stimulated with VEGF and insulin-like growth factor-1 (IGF-1). It was convincingly demonstrated that Nrf2-deficient cells exhibited impaired proliferation, migration (measured by a wound healing assay), and adhesion to vitronectin and collagen and suppressed ability to form capillary-like structures in vitro. Noteworthy, similar effects were evoked by overexpression of Keap1, the cytosolic repressor of Nrf2 [104].

As an extension of all those results, our very recent data suggest that in endothelial cells Nrf2 may function as a mediator linking different proangiogenic signal transduction pathways (Florczyk et al., in press). Using HMEC-1 we showed that treatment with growth factors such as VEGF, SDF-1, and IL-8 may result in the activation of Nrf2 as well as enhanced HO-1 expression. Moreover, activation of Nrf2 by sulforaphane stimulated capillary-like network formation on Matrigel, whereas its siRNA-mediated inhibition decreased angiogenic response of HMEC-1, the latter effect reversed by HO-1 overexpression (Florczyk et al. in press).

10.4.3 Nrf2 in Regenerative Neovascularization of Damaged Tissue

A great spectrum of antioxidant and anti-inflammatory actions of Nrf2 has been reported using in vivo models of several drug-induced toxicities as well as ischemia/reperfusion injury, wire vascular injury, hyperoxia-induced acute lung injury and inflammation, and many others [10, 105–108]. Nonetheless, in contrast to HO-1, hitherto there are just a few studies linking Nrf2 and regenerative neovascularization of ischemic/damaged tissues.

Interestingly, although HO-1 was documented as crucial for wound healing-related reepithelialization and neovascularization [33, 39], it seems that its upstream regulator,

Nrf2, could be not required in those processes. Namely, auf dem Keller et al. reported that full-thickness skin wounds can heal normally in transgenic mice expressing a dominant-negative mutant of Nrf2 in basal keratinocytes of the epidermis [109]. Thus, one can assume that Nrf2-mediated gene expression in keratinocytes is dispensable for wound healing [109]. Nonetheless, still no data are available concerning Nrf2 function in blood vessel formation in the wounds.

On the other hand, Uno et al. being the first who investigated Nrf2 in normal murine vascular development pointed at important role of Nrf2 in the secondary capillary network formation in the retina [110]. The numbers of deep or secondary capillaries in air-reared Nrf2$^{-/-}$ mice were decreased in comparison to wild-type mice at postnatal day 9 (P9). However, at P9 the absence of Nrf2 made animals more vulnerable to hyperoxia-stimulated vaso-obliteration, associated with oxidative stress. Interestingly, at P12 retinal vasculature was able to recover from the initial hyperoxic insult and normal development of the deep capillary system approached normal by P12 in Nrf2$^{-/-}$ mice [110].

In contrast to expectations, further experiments revealed that Nrf2 deficiency in vivo may actually exert positive effect on neovascularization accompanying damaged/ischemic tissue regeneration. Nijmeh et al. demonstrated that Nrf2$^{-/-}$ mice subjected to left pulmonary artery ligation (LPAL) show increased blood flow in response to lung ischemia as compared to wild-type counterparts [111]. As a mechanism an elevation of ROS production was suggested and the ratio of antioxidant-reduced glutathione (GSH) and its oxidized form (GSSG) was shown to be inversely associated with the magnitude of neovascularization [111]. Similar data were obtained by Ichihara et al. using murine model of hind limb ischemia [112]. As an explanation of increased ischemia-induced neovascularization in the absence of Nrf2, impaired antioxidant defense and increased ROS accumulation in endothelial cells, associated with enhanced inflammatory response, were proposed. Specifically, Nrf2 target genes, HO-1 and thioredoxin 1, and the concentration of total glutathione were reduced, whereas the infiltration of inflammatory cells, the level of adhesion molecules, MCP-1, tumor necrosis factor-α, and finally angiogenic factors were increased in the ischemic hindlimb of Nrf2$^{-/-}$ mice [112]. Very recently also our group confirmed the advantage of inflammatory neovascularization in Nrf2-deficient ischemic environment (Florczyk et al. in press). In addition we showed that Nrf2 deficiency may influence the functioning of PACs (formerly defined as EPCs) attenuating their survival, proliferation, migration, and angiogenic potential in vitro, while in vivo suppressing PAC mobilization after hind limb ischemia in mice. Thus, since the revascularization of ischemic limb was shown to be better in Nrf2$^{-/-}$ mice, one could question the significance of PACs in such conditions (Florczyk et al. in press). Actually, it is the first report linking Nrf2 and PACs/EPCs. In addition, there are data proving that Nrf2 may regulate function of their close relatives, hematopoietic stem progenitor cells, affecting survival, differentiation, migration, and retention in their niche, and engraftment after bone marrow transplantation [113, 114].

10.4.4 Nrf2 in Tumor Angiogenesis

Many studies point at preventive function of Nrf2 during tumor initiation. Indeed, $Nrf2^{-/-}$ mice demonstrated higher susceptibility toward cancer development in a variety of chemical carcinogenesis models [115]. In contrast, Nrf2 involvement in tumor progression was also reported [116, 117]. Considering Nrf2 contribution to tumor growth and taking into account suggested proangiogenic function of Nrf2, recently its role in tumor angiogenesis is being tested [118].

As mentioned above, elegant data showing an involvement of Nrf2 in the formation of blood vessels nourishing the tumor were published by Kim et al. [103]. In nude mouse xenograft model, stable RNAi-mediated knockdown of Nrf2 in human colon cancer cells was shown to suppress tumor growth with a concomitant reduction in blood vessel formation and VEGF expression [103].

A proangiogenic effect of Nrf2 in tumors could be also concluded from the studies of Zhang et al. (see above), who showed Nrf2 involvement in FSH-induced expression of VEGF in human epithelial ovarian cancer [102]. As such Nrf2 might act as a mediator of well-defined FSH-induced tumor angiogenesis.

In the other approach glioma cells (U251MG) with stable Nrf2 knockdown were established [119]. Using such cells in mouse xenograft model, decreased microvessel density and tumor growth were observed [119].

Conclusions

A wide spectrum of cytoprotective functions, including the induction of HO-1 expression, make Nrf2 a key player guarding the healthy phenotype of vascular cells when exposed to oxidative/ischemic conditions and maintaining the functional integrity of the vasculature [104]. Recently, a novel function of Nrf2 acting as a mediator of blood vessel formation was suggested, and its cross talk with known angiogenic signaling pathways, comprising HO-1, HIF-1, and VEGF, was documented.

A growing body of evidence points at particular significance of proangiogenic function of HO-1 in the regeneration after ischemia/injury by influencing progenitor cell mobilization and neovascularization and stimulating functional recovery of the tissue. Although little is known about such actions of Nrf2, recent data suggest its involvement in normal vascular development. Strikingly, verification of Nrf2 proangiogenic function in vivo using common models of tissue ischemia turned out to be tricky, since Nrf2 knockout may actually exert positive effect on oxidative stress-related inflammatory neovascularization. Thus, further studies are definitely needed for a thorough assessment of Nrf2 place in the processes of blood vessel formation.

On the other hand, since tumor growth is tightly dependent on blood supply, mediators governing angiogenesis are of obvious therapeutic importance. Both Nrf2 and HO-1 are well known as cancer preventive agents, protecting healthy tissues against carcinogen-induced insults. In contrast, Nrf2/HO-1 overexpression reported in many cancer types may benefit tumor survival and growth. Additionally, taking into account the potential function of Nrf2 in the processes of blood vessel formation and well-known proangiogenic function of HO-1, one can assume their crucial role in cancer progression where efficient neovascularization would enable the growth of stress-resistant tumors. However, the tissue-specific HO-1 actions in tumors should be remembered.

Acknowledgments The Faculty of Biochemistry, Biophysics and Biotechnology of the Jagiellonian University is a beneficiary of the structural funds from the European Union and the Polish Ministry of Science and Higher Education (grants No: POIG.02.01.00-12 064/08, POIG 01.01.02-00-109/09, POIG.02.02.00-014/08, and 01.01.02-00-069/09).

References

1. Baird L, Dinkova-Kostova AT (2011) The cytoprotective role of the Keap1-Nrf2 pathway. Arch Toxicol 85(4):241–272
2. Zhang X, Chen X, Song H, Chen HZ, Rovin BH (2005) Activation of the Nrf2/antioxidant response pathway increases IL-8 expression. Eur J Immunol 35(11):3258–3267
3. Alam J, Stewart D, Touchard C, Boinapally S, Choi AM, Cook JL (1999) Nrf2, a Cap'n'Collar transcription factor, regulates induction of the heme oxygenase-1 gene. J Biol Chem 274(37):26071–26078
4. Folkman J (1971) Tumor angiogenesis: therapeutic implications. N Engl J Med 285(21): 1182–1186
5. Misra MK, Sarwat M, Bhakuni P, Tuteja R, Tuteja N (2009) Oxidative stress and ischemic myocardial syndromes. Med Sci Monit 15(10):RA209–RA219
6. Bemeur C, Ste-Marie L, Montgomery J (2007) Increased oxidative stress during hyperglycemic cerebral ischemia. Neurochem Int 50(7–8):890–904
7. Mitra S, Deshmukh A, Sachdeva R, Lu J, Mehta JL (2011) Oxidized low-density lipoprotein and atherosclerosis implications in antioxidant therapy. Am J Med Sci 342(2):135–142
8. Donovan EL, McCord JM, Reuland DJ, Miller BF, Hamilton KL (2012) Phytochemical activation of Nrf2 protects human coronary artery endothelial cells against an oxidative challenge. Oxid Med Cell Longev 2012:132931
9. Kim M, Kim S, Lim JH, Lee C, Choi HC, Woo CH (2012) Laminar flow activation of ERK5 protein in vascular endothelium leads to atheroprotective effect via NF-E2-related factor 2 (Nrf2) activation. J Biol Chem 287(48):40722–40731
10. Ashino T, Yamamoto M, Yoshida T, Numazawa S (2013) Redox-sensitive transcription factor Nrf2 regulates vascular smooth muscle cell migration and neointimal hyperplasia. Arterioscler Thromb Vasc Biol 33(4):760–768
11. Bitar MS, Al-Mulla F (2011) A defect in Nrf2 signaling constitutes a mechanism for cellular stress hypersensitivity in a genetic rat model of type 2 diabetes. Am J Physiol Endocrinol Metab 301(6):E1119–E1129
12. Loboda A, Jazwa A, Grochot-Przeczek A, Rutkowski AJ, Cisowski J, Agarwal A, Jozkowicz A, Dulak J (2008) Heme oxygenase-1 and the vascular bed: from molecular mechanisms to therapeutic opportunities. Antioxid Redox Signal 10(10):1767–1812
13. Castilho A, Aveleira CA, Leal EC, Simoes NF, Fernandes CR, Meirinhos RI, Baptista FI, Ambrosio AF (2012) Heme oxygenase-1 protects retinal endothelial cells against high glucose- and oxidative/nitrosative stress-induced toxicity. PLoS One 7(8):e42428
14. Ishikawa K, Sugawara D, Wang X, Suzuki K, Itabe H, Maruyama Y, Lusis AJ (2001) Heme oxygenase-1 inhibits atherosclerotic lesion formation in ldl-receptor knockout mice. Circ Res 88(5):506–512
15. Yet SF, Layne MD, Liu X, Chen YH, Ith B, Sibinga NE, Perrella MA (2003) Absence of heme oxygenase-1 exacerbates atherosclerotic lesion formation and vascular remodeling. FASEB J 17(12):1759–1761

16. Kim JY, Cho HJ, Sir JJ, Kim BK, Hur J, Youn SW, Yang HM, Jun SI, Park KW, Hwang SJ, Kwon YW, Lee HY, Kang HJ, Oh BH, Park YB, Kim HS (2009) Sulfasalazine induces haem oxygenase-1 via ROS-dependent Nrf2 signalling, leading to control of neointimal hyperplasia. Cardiovasc Res 82(3):550–560
17. Sambuceti G, Morbelli S, Vanella L, Kusmic C, Marini C, Massollo M, Augeri C, Corselli M, Ghersi C, Chiavarina B, Rodella LF, L'Abbate A, Drummond G, Abraham NG, Frassoni F (2009) Diabetes impairs the vascular recruitment of normal stem cells by oxidant damage, reversed by increases in pAMPK, heme oxygenase-1, and adiponectin. Stem Cells 27(2): 399–407
18. Fan J, Xu G, Jiang T, Qin Y (2012) Pharmacologic induction of heme oxygenase-1 plays a protective role in diabetic retinopathy in rats. Invest Ophthalmol Vis Sci 53(10):6541–6556
19. Liu X, Wei J, Peng DH, Layne MD, Yet SF (2005) Absence of heme oxygenase-1 exacerbates myocardial ischemia/reperfusion injury in diabetic mice. Diabetes 54(3): 778–784
20. Poss KD, Tonegawa S (1997) Heme oxygenase 1 is required for mammalian iron reutilization. Proc Natl Acad Sci U S A 94(20):10919–10924
21. Poss KD, Tonegawa S (1997) Reduced stress defense in heme oxygenase 1-deficient cells. Proc Natl Acad Sci U S A 94(20):10925–10930
22. True AL, Olive M, Boehm M, San H, Westrick RJ, Raghavachari N, Xu X, Lynn EG, Sack MN, Munson PJ, Gladwin MT, Nabel EG (2007) Heme oxygenase-1 deficiency accelerates formation of arterial thrombosis through oxidative damage to the endothelium, which is rescued by inhaled carbon monoxide. Circ Res 101(9):893–901
23. Yachie A, Niida Y, Wada T, Igarashi N, Kaneda H, Toma T, Ohta K, Kasahara Y, Koizumi S (1999) Oxidative stress causes enhanced endothelial cell injury in human heme oxygenase-1 deficiency. J Clin Invest 103(1):129–135
24. Radhakrishnan N, Yadav SP, Sachdeva A, Pruthi PK, Sawhney S, Piplani T, Wada T, Yachie A (2011) Human heme oxygenase-1 deficiency presenting with hemolysis, nephritis, and asplenia. J Pediatr Hematol Oncol 33(1):74–78
25. Taha H, Skrzypek K, Guevara I, Nigisch A, Mustafa S, Grochot-Przeczek A, Ferdek P, Was H, Kotlinowski J, Kozakowska M, Balcerczyk A, Muchova L, Vitek L, Weigel G, Dulak J, Jozkowicz A (2010) Role of heme oxygenase-1 in human endothelial cells: lesson from the promoter allelic variants. Arterioscler Thromb Vasc Biol 30(8):1634–1641
26. Koizumi S (2007) Human heme oxygenase-1 deficiency: a lesson on serendipity in the discovery of the novel disease. Pediatr Int 49(2):125–132
27. Kim YM, Pae HO, Park JE, Lee YC, Woo JM, Kim NH, Choi YK, Lee BS, Kim SR, Chung HT (2011) Heme oxygenase in the regulation of vascular biology: from molecular mechanisms to therapeutic opportunities. Antioxid Redox Signal 14(1):137–167
28. Deramaudt BM, Braunstein S, Remy P, Abraham NG (1998) Gene transfer of human heme oxygenase into coronary endothelial cells potentially promotes angiogenesis. J Cell Biochem 68(1):121–127
29. Malaguarnera L, Pilastro MR, Quan S, Ghattas MH, Yang L, Mezentsev AV, Kushida T, Abraham NG, Kappas A (2002) Significance of heme oxygenase in prolactin-mediated cell proliferation and angiogenesis in human endothelial cells. Int J Mol Med 10(4):433–440
30. Jazwa A, Stepniewski J, Zamykal M, Jagodzinska J, Meloni M, Emanueli C, Jozkowicz A, Dulak J (2013) Pre-emptive hypoxia-regulated HO-1 gene therapy improves post-ischaemic limb perfusion and tissue regeneration in mice. Cardiovasc Res 97(1):115–124
31. Suzuki M, Iso-o N, Takeshita S, Tsukamoto K, Mori I, Sato T, Ohno M, Nagai R, Ishizaka N (2003) Facilitated angiogenesis induced by heme oxygenase-1 gene transfer in a rat model of hindlimb ischemia. Biochem Biophys Res Commun 302(1):138–143
32. Lakkisto P, Kyto V, Forsten H, Siren JM, Segersvard H, Voipio-Pulkki LM, Laine M, Pulkki K, Tikkanen I (2010) Heme oxygenase-1 and carbon monoxide promote neovascularization after myocardial infarction by modulating the expression of HIF-1alpha, SDF-1alpha and VEGF-B. Eur J Pharmacol 635(1–3):156–164

33. Grochot-Przeczek A, Lach R, Mis J, Skrzypek K, Gozdecka M, Sroczynska P, Dubiel M, Rutkowski A, Kozakowska M, Zagorska A, Walczynski J, Was H, Kotlinowski J, Drukala J, Kurowski K, Kieda C, Herault Y, Dulak J, Jozkowicz A (2009) Heme oxygenase-1 accelerates cutaneous wound healing in mice. PLoS One 4(6):e5803
34. Sunamura M, Duda DG, Ghattas MH, Lozonschi L, Motoi F, Yamauchi J, Matsuno S, Shibahara S, Abraham NG (2003) Heme oxygenase-1 accelerates tumor angiogenesis of human pancreatic cancer. Angiogenesis 6(1):15–24
35. Tongers J, Knapp JM, Korf M, Kempf T, Limbourg A, Limbourg FP, Li Z, Fraccarollo D, Bauersachs J, Han X, Drexler H, Fiedler B, Wollert KC (2008) Haeme oxygenase promotes progenitor cell mobilization, neovascularization, and functional recovery after critical hindlimb ischaemia in mice. Cardiovasc Res 78(2):294–300
36. Miyake M, Fujimoto K, Anai S, Ohnishi S, Kuwada M, Nakai Y, Inoue T, Matsumura Y, Tomioka A, Ikeda T, Tanaka N, Hirao Y (2011) Heme oxygenase-1 promotes angiogenesis in urothelial carcinoma of the urinary bladder. Oncol Rep 25(3):653–660
37. Kozakowska M, Ciesla M, Stefanska A, Skrzypek K, Was H, Jazwa A, Grochot-Przeczek A, Kotlinowski J, Szymula A, Bartelik A, Mazan M, Yagensky O, Florczyk U, Lemke K, Zebzda A, Dyduch G, Nowak W, Szade K, Stepniewski J, Majka M, Derlacz R, Loboda A, Dulak J, Jozkowicz A (2012) Heme oxygenase-1 inhibits myoblast differentiation by targeting myomirs. Antioxid Redox Signal 16(2):113–127
38. Jozkowicz A, Huk I, Nigisch A, Weigel G, Dietrich W, Motterlini R, Dulak J (2003) Heme oxygenase and angiogenic activity of endothelial cells: stimulation by carbon monoxide and inhibition by tin protoporphyrin-IX. Antioxid Redox Signal 5(2):155–162
39. Deshane J, Chen S, Caballero S, Grochot-Przeczek A, Was H, Li Calzi S, Lach R, Hock TD, Chen B, Hill-Kapturczak N, Siegal GP, Dulak J, Jozkowicz A, Grant MB, Agarwal A (2007) Stromal cell-derived factor 1 promotes angiogenesis via a heme oxygenase 1-dependent mechanism. J Exp Med 204(3):605–618
40. Carmeliet P, Ferreira V, Breier G, Pollefeyt S, Kieckens L, Gertsenstein M, Fahrig M, Vandenhoeck A, Harpal K, Eberhardt C, Declercq C, Pawling J, Moons L, Collen D, Risau W, Nagy A (1996) Abnormal blood vessel development and lethality in embryos lacking a single VEGF allele. Nature 380(6573):435–439
41. Ferrara N, Carver-Moore K, Chen H, Dowd M, Lu L, O'Shea KS, Powell-Braxton L, Hillan KJ, Moore MW (1996) Heterozygous embryonic lethality induced by targeted inactivation of the VEGF gene. Nature 380(6573):439–442
42. Dulak J, Jozkowicz A, Foresti R, Kasza A, Frick M, Huk I, Green CJ, Pachinger O, Weidinger F, Motterlini R (2002) Heme oxygenase activity modulates vascular endothelial growth factor synthesis in vascular smooth muscle cells. Antioxid Redox Signal 4(2):229–240
43. Jozkowicz A, Huk I, Nigisch A, Weigel G, Weidinger F, Dulak J (2002) Effect of prostaglandin-J(2) on VEGF synthesis depends on the induction of heme oxygenase-1. Antioxid Redox Signal 4(4):577–585
44. Loboda A, Jazwa A, Wegiel B, Jozkowicz A, Dulak J (2005) Heme oxygenase-1-dependent and -independent regulation of angiogenic genes expression: effect of cobalt protoporphyrin and cobalt chloride on VEGF and IL-8 synthesis in human microvascular endothelial cells. Cell Mol Biol (Noisy-le-grand) 51(4):347–355
45. Kim EH, Na HK, Surh YJ (2006) Upregulation of VEGF by 15-deoxy-Delta12,14-prostaglandin J2 via heme oxygenase-1 and ERK1/2 signaling in MCF-7 cells. Ann N Y Acad Sci 1090:375–384
46. Choi YK, Kim CK, Lee H, Jeoung D, Ha KS, Kwon YG, Kim KW, Kim YM (2010) Carbon monoxide promotes VEGF expression by increasing HIF-1alpha protein level via two distinct mechanisms, translational activation and stabilization of HIF-1alpha protein. J Biol Chem 285(42):32116–32125
47. Cisowski J, Loboda A, Jozkowicz A, Chen S, Agarwal A, Dulak J (2005) Role of heme oxygenase-1 in hydrogen peroxide-induced VEGF synthesis: effect of HO-1 knockout. Biochem Biophys Res Commun 326(3):670–676

48. Jazwa A, Loboda A, Golda S, Cisowski J, Szelag M, Zagorska A, Sroczynska P, Drukala J, Jozkowicz A, Dulak J (2006) Effect of heme and heme oxygenase-1 on vascular endothelial growth factor synthesis and angiogenic potency of human keratinocytes. Free Radic Biol Med 40(7):1250–1263
49. Malaguarnera L, Imbesi RM, Scuto A, D'Amico F, Licata F, Messina A, Sanfilippo S (2004) Prolactin increases HO-1 expression and induces VEGF production in human macrophages. J Cell Biochem 93(1):197–206
50. Abdel-Aziz MT, el-Asmar MF, el-Miligy D, Atta H, Shaker O, Ghattas MH, Hosni H, Kamal N (2003) Retrovirus-mediated human heme oxygenase-1 (HO-1) gene transfer into rat endothelial cells: the effect of HO-1 inducers on the expression of cytokines. Int J Biochem Cell Biol 35(3):324–332
51. Angermayr B, Mejias M, Gracia-Sancho J, Garcia-Pagan JC, Bosch J, Fernandez M (2006) Heme oxygenase attenuates oxidative stress and inflammation, and increases VEGF expression in portal hypertensive rats. J Hepatol 44(6):1033–1039
52. Lin HH, Chen YH, Yet SF, Chau LY (2009) After vascular injury, heme oxygenase-1/carbon monoxide enhances re-endothelialization via promoting mobilization of circulating endothelial progenitor cells. J Thromb Haemost 7(8):1401–1408
53. Lin HH, Chen YH, Chang PF, Lee YT, Yet SF, Chau LY (2008) Heme oxygenase-1 promotes neovascularization in ischemic heart by coinduction of VEGF and SDF-1. J Mol Cell Cardiol 45(1):44–55
54. Lin HH, Lai SC, Chau LY (2011) Heme oxygenase-1/carbon monoxide induces vascular endothelial growth factor expression via p38 kinase-dependent activation of Sp1. J Biol Chem 286(5):3829–3838
55. Fernandez M, Bonkovsky HL (2003) Vascular endothelial growth factor increases heme oxygenase-1 protein expression in the chick embryo chorioallantoic membrane. Br J Pharmacol 139(3):634–640
56. Bussolati B, Ahmed A, Pemberton H, Landis RC, Di Carlo F, Haskard DO, Mason JC (2004) Bifunctional role for VEGF-induced heme oxygenase-1 in vivo: induction of angiogenesis and inhibition of leukocytic infiltration. Blood 103(3):761–766
57. Tachibana K, Hirota S, Iizasa H, Yoshida H, Kawabata K, Kataoka Y, Kitamura Y, Matsushima K, Yoshida N, Nishikawa S, Kishimoto T, Nagasawa T (1998) The chemokine receptor CXCR4 is essential for vascularization of the gastrointestinal tract. Nature 393 (6685):591–594
58. Nagasawa T, Hirota S, Tachibana K, Takakura N, Nishikawa S, Kitamura Y, Yoshida N, Kikutani H, Kishimoto T (1996) Defects of B-cell lymphopoiesis and bone-marrow myelopoiesis in mice lacking the CXC chemokine PBSF/SDF-1. Nature 382(6592):635–638
59. Lau TT, Wang DA (2011) Stromal cell-derived factor-1 (SDF-1): homing factor for engineered regenerative medicine. Expert Opin Biol Ther 11(2):189–197
60. Wang Y, Luther K (2012) Genetically manipulated progenitor/stem cells restore function to the infarcted heart via the SDF-1alpha/CXCR4 signaling pathway. Prog Mol Biol Transl Sci 111:265–284
61. Hiasa K, Ishibashi M, Ohtani K, Inoue S, Zhao Q, Kitamoto S, Sata M, Ichiki T, Takeshita A, Egashira K (2004) Gene transfer of stromal cell-derived factor-1alpha enhances ischemic vasculogenesis and angiogenesis via vascular endothelial growth factor/endothelial nitric oxide synthase-related pathway: next-generation chemokine therapy for therapeutic neovascularization. Circulation 109(20):2454–2461
62. Li FY, Lam KS, Tse HF, Chen C, Wang Y, Vanhoutte PM, Xu A (2012) Endothelium-selective activation of AMP-activated protein kinase prevents diabetes mellitus-induced impairment in vascular function and reendothelialization via induction of heme oxygenase-1 in mice. Circulation 126(10):1267–1277
63. Yet SF, Tian R, Layne MD, Wang ZY, Maemura K, Solovyeva M, Ith B, Melo LG, Zhang L, Ingwall JS, Dzau VJ, Lee ME, Perrella MA (2001) Cardiac-specific expression of heme

oxygenase-1 protects against ischemia and reperfusion injury in transgenic mice. Circ Res 89 (2):168–173
64. Wojakowski W, Tendera M, Cybulski W, Zuba-Surma EK, Szade K, Florczyk U, Kozakowska M, Szymula A, Krzych L, Paslawska U, Paslawski R, Milewski K, Buszman PP, Nabialek E, Kuczmik W, Janiszewski A, Dziegiel P, Buszman PE, Jozkowicz A, Dulak J (2012) Effects of intracoronary delivery of allogenic bone marrow-derived stem cells expressing heme oxygenase-1 on myocardial reperfusion injury. Thromb Haemost 108(3): 464–475
65. Bhang SH, Kim JH, Yang HS, La WG, Lee TJ, Sun AY, Kim GH, Lee M, Kim BS (2009) Combined delivery of heme oxygenase-1 gene and fibroblast growth factor-2 protein for therapeutic angiogenesis. Biomaterials 30(31):6247–6256
66. Bhang SH, Kim JH, Yang HS, La WG, Lee TJ, Kim GH, Kim HA, Lee M, Kim BS (2011) Combined gene therapy with hypoxia-inducible factor-1alpha and heme oxygenase-1 for therapeutic angiogenesis. Tissue Eng Part A 17(7–8):915–926
67. Jujo K, Ii M, Losordo DW (2008) Endothelial progenitor cells in neovascularization of infarcted myocardium. J Mol Cell Cardiol 45(4):530–544
68. Asahara T, Murohara T, Sullivan A, Silver M, van der Zee R, Li T, Witzenbichler B, Schatteman G, Isner JM (1997) Isolation of putative progenitor endothelial cells for angiogenesis. Science 275(5302):964–967
69. Asahara T, Takahashi T, Masuda H, Kalka C, Chen D, Iwaguro H, Inai Y, Silver M, Isner JM (1999) VEGF contributes to postnatal neovascularization by mobilizing bone marrow-derived endothelial progenitor cells. EMBO J 18(14):3964–3972
70. Ceradini DJ, Kulkarni AR, Callaghan MJ, Tepper OM, Bastidas N, Kleinman ME, Capla JM, Galiano RD, Levine JP, Gurtner GC (2004) Progenitor cell trafficking is regulated by hypoxic gradients through HIF-1 induction of SDF-1. Nat Med 10(8):858–864
71. Kong D, Melo LG, Mangi AA, Zhang L, Lopez-Ilasaca M, Perrella MA, Liew CC, Pratt RE, Dzau VJ (2004) Enhanced inhibition of neointimal hyperplasia by genetically engineered endothelial progenitor cells. Circulation 109(14):1769–1775
72. Wu BJ, Midwinter RG, Cassano C, Beck K, Wang Y, Changsiri D, Gamble JR, Stocker R (2009) Heme oxygenase-1 increases endothelial progenitor cells. Arterioscler Thromb Vasc Biol 29(10):1537–1542
73. Krenning G, van Luyn MJ, Harmsen MC (2009) Endothelial progenitor cell-based neovascularization: implications for therapy. Trends Mol Med 15(4):180–189
74. Brunt KR, Wu J, Chen Z, Poeckel D, Dercho RA, Melo LG, Funk CD, Ward CA, Li RK (2012) Ex vivo Akt/HO-1 gene therapy to human endothelial progenitor cells enhances myocardial infarction recovery. Cell Transplant 21(7):1443–1461
75. Long J, Wang S, Zhang Y, Liu X, Zhang H (2013) The therapeutic effect of vascular endothelial growth factor gene- or heme oxygenase-1 gene-modified endothelial progenitor cells on neovascularization of rat hindlimb ischemia model. J Vasc Surg 58(3):756–765, e2
76. Yang JJ, Yang X, Liu ZQ, Hu SY, Du ZY, Feng LL, Liu JF, Chen YD (2012) Transplantation of adipose tissue-derived stem cells overexpressing heme oxygenase-1 improves functions and remodeling of infarcted myocardium in rabbits. Tohoku J Exp Med 226(3):231–241
77. Jiang YB, Zhang XL, Tang YL, Ma GS, Shen CX, Wei Q, Zhu Q, Yao YY, Liu NF (2011) Effects of heme oxygenase-1 gene modulated mesenchymal stem cells on vasculogenesis in ischemic swine hearts. Chin Med J (Engl) 124(3):401–407
78. Hou C, Shen L, Huang Q, Mi J, Wu Y, Yang M, Zeng W, Li L, Chen W, Zhu C (2013) The effect of heme oxygenase-1 complexed with collagen on MSC performance in the treatment of diabetic ischemic ulcer. Biomaterials 34(1):112–120
79. Zeng B, Lin G, Ren X, Zhang Y, Chen H (2010) Over-expression of HO-1 on mesenchymal stem cells promotes angiogenesis and improves myocardial function in infarcted myocardium. J Biomed Sci 17:80
80. Kearns-Jonker M, Dai W, Gunthart M, Fuentes T, Yeh HY, Gerczuk P, Pera M, Mummery C, Kloner RA (2012) Genetically engineered mesenchymal stem cells influence gene expression in donor cardiomyocytes and the recipient heart. J Stem Cell Res Ther. S1.

81. Jozkowicz A, Was H, Dulak J (2007) Heme oxygenase-1 in tumors: is it a false friend? Antioxid Redox Signal 9(12):2099–2117
82. Nishie A, Ono M, Shono T, Fukushi J, Otsubo M, Onoue H, Ito Y, Inamura T, Ikezaki K, Fukui M, Iwaki T, Kuwano M (1999) Macrophage infiltration and heme oxygenase-1 expression correlate with angiogenesis in human gliomas. Clin Cancer Res 5(5):1107–1113
83. Was H, Cichon T, Smolarczyk R, Rudnicka D, Stopa M, Chevalier C, Leger JJ, Lackowska B, Grochot A, Bojkowska K, Ratajska A, Kieda C, Szala S, Dulak J, Jozkowicz A (2006) Overexpression of heme oxygenase-1 in murine melanoma: increased proliferation and viability of tumor cells, decreased survival of mice. Am J Pathol 169(6):2181–2198
84. Sass G, Leukel P, Schmitz V, Raskopf E, Ocker M, Neureiter D, Meissnitzer M, Tasika E, Tannapfel A, Tiegs G (2008) Inhibition of heme oxygenase 1 expression by small interfering RNA decreases orthotopic tumor growth in livers of mice. Int J Cancer 123(6):1269–1277
85. Birrane G, Li H, Yang S, Tachado SD, Seng S (2013) Cigarette smoke induces nuclear translocation of heme oxygenase 1 (HO-1) in prostate cancer cells: nuclear HO-1 promotes vascular endothelial growth factor secretion. Int J Oncol 42(6):1919–1928
86. Sengupta S, Sellers LA, Matheson HB, Fan TP (2003) Thymidine phosphorylase induces angiogenesis in vivo and in vitro: an evaluation of possible mechanisms. Br J Pharmacol 139 (2):219–231
87. Skrzypek K, Tertil M, Golda S, Ciesla M, Weglarczyk K, Collet G, Guichard A, Kozakowska M, Boczkowski J, Was H, Gil T, Kuzdzal J, Muchova L, Vitek L, Loboda A, Jozkowicz A, Kieda C, Dulak J (2013) Interplay between heme oxygenase-1 and miR-378 affects non-small cell lung carcinoma growth, vascularization and metastasis. Antioxid Redox Signal 19(7):644–660
88. Ferrando M, Gueron G, Elguero B, Giudice J, Salles A, Leskow FC, Jares-Erijman EA, Colombo L, Meiss R, Navone N, De Siervi A, Vazquez E (2011) Heme oxygenase 1 (HO-1) challenges the angiogenic switch in prostate cancer. Angiogenesis 14(4):467–479
89. Chan K, Lu R, Chang JC, Kan YW (1996) NRF2, a member of the NFE2 family of transcription factors, is not essential for murine erythropoiesis, growth, and development. Proc Natl Acad Sci USA 93(24):13943–13948
90. Chan K, Kan Y (1999) Nrf2 is essential for protection against acute pulmonary injury in mice. Proc Natl Acad Sci USA 96(22):12731–12736
91. Enomoto A, Itoh K, Nagayoshi E, Haruta J, Kimura T, O'Connor T, Harada T, Yamamoto M (2001) High sensitivity of Nrf2 knockout mice to acetaminophen hepatotoxicity associated with decreased expression of ARE-regulated drug metabolizing enzymes and antioxidant genes. Toxicol Sci 59(1):169–177
92. Ramos-Gomez M, Kwak M, Dolan PM, Itoh K, Yamamoto M, Talalay P, Kensler TW (2001) Sensitivity to carcinogenesis is increased and chemoprotective efficacy of enzyme inducers is lost in nrf2 transcription factor-deficient mice. Proc Natl Acad Sci USA 98(6):3410–3415
93. Yoh K, Itoh K, Enomoto A, Hirayama A, Yamaguchi N, Kobayashi M, Morito N, Koyama A, Yamamoto M, Takahashi S (2001) Nrf2-deficient female mice develop lupus-like autoimmune nephritis. Kidney Int 60(4):1343–1353
94. Afonyushkin T, Oskolkova OV, Philippova M, Resink TJ, Erne P, Binder BR, Bochkov VN (2010) Oxidized phospholipids regulate expression of ATF4 and VEGF in endothelial cells via NRF2-dependent mechanism: novel point of convergence between electrophilic and unfolded protein stress pathways. Arterioscler Thromb Vasc Biol 30(5):1007–1013
95. Kweider N, Fraqoulis A, Rosen C, Pecks U, Rath W, Pufe T, Wruck CJ (2011) Interplay between vascular endothelial growth factor (VEGF) and nuclear factor erythroid 2-related factor-2 (Nrf2): implications for preeclampsia. J Biol Chem 286(50):42863–42872
96. Shibuya A, Onda K, Kawahara H, Uchiyama Y, Nakayama H, Omi T, Nagaoka M, Matsui H, Hirano T (2010) Sofalcone, a gastric mucosa protective agent, increases vascular endothelial growth factor via the Nrf2-heme-oxygenase-1 dependent pathway in gastric epithelial cells. Biochem Biophys Res Commun 398(3):581–584

97. Chao MW, Po IP, Laumbach RJ, Koslosky J, Cooper K, Gordon MK (2012) DEP induction of ROS in capillary-like endothelial tubes leads to VEGF-A expression. Toxicology 297(1–3): 34–46
98. Meng D, Wang X, Chang Q, Hitron A, Zhang Z, Xu M, Chen G, Luo J, Jiang B, Fang J, Shi X (2010) Arsenic promotes angiogenesis in vitro via a heme oxygenase-1-dependent mechanism. Toxicol Appl Pharmacol 244(3):291–299
99. Forsythe JA, Jiang BH, Iyer NV, Agani F, Leung SW, Koos RD, Semenza GL (1996) Activation of vascular endothelial growth factor gene transcription by hypoxia-inducible factor 1. Mol Cell Biol 16(9):4604–4613
100. Malec V, Gottschald OR, Li S, Rose F, Seeger W, Hanze J (2010) HIF-1 alpha signaling is augmented during intermittent hypoxia by induction of the Nrf2 pathway in NOX1-expressing adenocarcinoma A549 cells. Free Radic Biol Med 48:1626–1635
101. Loboda A, Stachurska A, Florczyk U, Rudnicka D, Jazwa A, Wegrzyn J, Kozakowska M, Stalinska K, Poellinger L, Levonen AL, Yla-Herttuala S, Jozkowicz A, Dulak J (2009) HIF-1 induction attenuates Nrf2-dependent IL-8 expression in human endothelial cells. Antioxid Redox Signal 11(7):1501–1517
102. Zhang Z, Wang Q, Ma J, Yi X, Zhu Y, Xi X, Feng Y, Jin Z (2013) Reactive oxygen species regulate FSH-induced expression of vascular endothelial growth factor via Nrf2 and HIF1alpha signaling in human epithelial ovarian cancer. Oncol Rep 29(4):1429–1434
103. Kim TH, Hur EG, Kang SJ, Kim JA, Thapa D, Lee YM, Ku SK, Jung Y, Kwak MK (2011) NRF2 blockade suppresses colon tumor angiogenesis by inhibiting hypoxia-induced activation of HIF-1alpha. Cancer Res 71(6):2260–2275
104. Valcarcel-Ares MN, Gautam T, Warrington JP, Bailey-Downs L, Sosnowska D, de Cabo R, Losonczy G, Sonntag WE, Ungvari Z, Csiszar A (2012) Disruption of Nrf2 signaling impairs angiogenic capacity of endothelial cells: implications for microvascular aging. J Gerontol A Biol Sci Med Sci 67(8):821–829
105. Wei Y, Gong J, Yoshida T, Eberhart CG, Xu Z, Kombairaju P, Sporn MB, Handa JT, Duh EJ (2011) Nrf2 has a protective role against neuronal and capillary degeneration in retinal ischemia-reperfusion injury. Free Radic Biol Med 51(1):216–224
106. Reddy NM, Kleeberger SR, Kensler TW, Yamamoto M, Hassoun PM, Reddy SP (2009) Disruption of Nrf2 impairs the resolution of hyperoxia-induced acute lung injury and inflammation in mice. J Immunol 182(11):7264–7271
107. Copple IM, Goldring CE, Kitteringham NR, Park BK (2008) The Nrf2-Keap1 defence pathway: role in protection against drug-induced toxicity. Toxicology 246(1):24–33
108. Cho HY, Jedlicka AE, Reddy SP, Kensler TW, Yamamoto M, Zhang LY, Kleeberger SR (2002) Role of NRF2 in protection against hyperoxic lung injury in mice. Am J Respir Cell Mol Biol 26(2):175–182
109. Auf dem Keller U, Huber M, Beyer TA, Kumin A, Siemes C, Braun S, Bugnon P, Mitropoulos V, Johnson DA, Johnson JA, Hohl D, Werner S (2006) Nrf transcription factors in keratinocytes are essential for skin tumor prevention but not for wound healing. Mol Cell Biol 26(10):3773–3784
110. Uno K, Prow TW, Bhutto IA, Yerrapureddy A, McLeod DS, Yamamoto M, Reddy SP, Lutty GA (2010) Role of Nrf2 in retinal vascular development and the vaso-obliterative phase of oxygen-induced retinopathy. Exp Eye Res 90(4):493–500
111. Nijmeh J, Moldobaeva A, Wagner EM (2010) Role of ROS in ischemia-induced lung angiogenesis. Am J Physiol Lung Cell Mol Physiol 299(4):L535–L541
112. Ichihara S, Yamada Y, Liu F, Murohara T, Itoh K, Yamamoto M, Ichihara G (2010) Ablation of the transcription factor Nrf2 promotes ischemia-induced neovascularization by enhancing the inflammatory response. Arterioscler Thromb Vasc Biol 30(8):1553–1561
113. Merchant AA, Singh A, Matsui W, Biswal S (2011) The redox-sensitive transcription factor Nrf2 regulates murine hematopoietic stem cell survival independently of ROS levels. Blood 118(25):6572–6579

114. Tsai JJ, Dudakov JA, Takahashi K, Shieh JH, Velardi E, Holland AM, Singer NV, West ML, Smith OM, Young LF, Shono Y, Ghosh A, Hanash AM, Tran HT, Moore MA, van den Brink MR (2013) Nrf2 regulates haematopoietic stem cell function. Nat Cell Biol 15(3):309–316
115. Yu X, Kensler T (2005) Nrf2 as a target for cancer chemoprevention. Mutat Res 591(1–2): 93–102
116. Satoh H, Moriguchi T, Takai J, Ebina M, Yamamoto M (2013) Nrf2 prevents initiation but accelerates progression through the Kras signaling pathway during lung carcinogenesis. Cancer Res 73(13):4158–4168
117. Bauer AK, Cho HY, Miller-Degraff L, Walker C, Helms K, Fostel J, Yamamoto M, Kleeberger SR (2011) Targeted deletion of Nrf2 reduces urethane-induced lung tumor development in mice. PLoS One 6(10):e26590
118. Zhou S, Ye W, Zhang M, Liang J (2012) The effects of nrf2 on tumor angiogenesis: a review of the possible mechanisms of action. Crit Rev Eukaryot Gene Expr 22(2):149–160
119. Ji XJ, Chen SH, Zhu L, Pan H, Zhou Y, Li W, You WC, Gao CC, Zhu JH, Jiang K, Wang HD (2013) Knockdown of NF-E2-related factor 2 inhibits the proliferation and growth of U251MG human glioma cells in a mouse xenograft model. Oncol Rep 30(1):157–164

Metabolism of Methylarginines and Angiogenesis

11

Hilda Tsang, Lucio Iannone, and Beata Wojciak-Stothard

Contents

11.1 Introduction 242
11.2 Methylarginines: Definition, Synthesis and Metabolism 242
11.3 Regulatory Role of Methylarginines in Angiogenesis: Experimental and Clinical Evidence 244
11.3.1 Angiogenesis in Cultured Cells and Knockout Animals 244
11.3.2 Dysregulation of ADMA Metabolism in Diseases Associated with Defective Angiogenesis 246
11.3.3 The Effects of ADMA/DDAH on Endothelial Progenitor Cell Function 248
11.4 The ADMA/DDAH Pathway in Angiogenesis 248
11.4.1 ADMA Metabolism Regulates Angiogenic Responses in Endothelial Cells 249
11.4.2 Signalling Mediators Involved in the Regulation of Angiogenesis by Methylarginines 250
11.4.3 Regulation of Actin Dynamics by ADMA: Rho GTPases and Cell Adhesion and Migration 251
References 255

Abstract

Endogenously produced analogues of the amino acid arginine, monomethylarginine (L-NMMA), asymmetric dimethylarginine (ADMA) and symmetric dimethylarginine (SDMA) play an important regulatory role in angiogenesis.

Hilda Tsang and Lucio Iannone are joint first authorship

H. Tsang • L. Iannone
Department of Experimental Medicine and Toxicology, Hammersmith Campus, Imperial College London, London, UK

B. Wojciak-Stothard (✉)
Department of Experimental Medicine and Toxicology, Hammersmith Campus, Imperial College London, Office 535, Laboratory 504, 5th floor ICTEM Building, Du Cane Road, London W12 0NN, UK
e-mail: b.wojciak-stothard@imperial.ac.uk

J. Dulak et al. (eds.), *Angiogenesis and Vascularisation*,
DOI 10.1007/978-3-7091-1428-5_11, © Springer-Verlag Wien 2013

This chapter provides an overview of the role of methylarginines and the methylarginine metabolising enzymes dimethylaminohydrolases (DDAH) on endothelial angiogenic responses in vitro and in vivo, in animal models, and in human disease. We also discuss molecular and cellular mechanisms involved, including nitric oxide (NO) pathway, transforming growth factor β (TGF-β) and Ras and Rho protein signalling.

Keywords
Angiogenesis • Nitric oxide • Methylarginines • DDAH • Endothelial

11.1 Introduction

Angiogenesis requires coordinated changes in endothelial cell–cell adhesion, proliferation, apoptosis, migration and differentiation. Each stage of this complex process is controlled by a range of growth factors, chemokines, endothelial-specific receptors, angiogenic enzymes, extracellular matrix proteins and adhesion molecules.

NO, a short-lived, cell-permeable gas produced by nitric oxide synthases (NOS), is one of the key regulators of angiogenesis. Apart from having anti-apoptotic, pro-proliferative and pro-migratory effects on endothelial cells, NO enhances matrix–endothelial cell interaction and increases degradation of the extracellular matrix [1–4].

Angiogenic growth factors, such as vascular endothelial growth factor (VEGF) or basic fibroblast growth factor (bFGF), upregulate NO production or require NO to mediate their effects on angiogenesis [5, 6]. In a positive-feedback mechanism, NO can also induce the synthesis and release of VEGF from vascular cells and regulate VEGF-induced eNOS expression ([7]). Vascular repair and angiogenesis are attenuated in eNOS-knockout mice, whereas they are enhanced in eNOS-transgenic mice or in rats undergoing local gene transfer of the human eNOS gene [8, 9].

Mounting evidence indicates that methylated analogues of arginine, L-NMMA, ADMA and, to a lesser extent, SDMA play a regulatory role in angiogenesis. ADMA and L-NMMA are endogenous inhibitors of all three isoforms of NOS-endothelial NOS (eNOS), inducible NOS (iNOS) and neuronal NOS (nNOS) [10].

11.2 Methylarginines: Definition, Synthesis and Metabolism

Methylarginines are generated via protein methylation by protein arginine *N*-methyltransferase enzymes (PRMTs) (Fig. 11.1). Methylation of arginine side chain guanidino groups is the most prevalent form of post-translational modification found in nuclear and cytoplasmic proteins in mammalian cells [11]. Arginines can be irreversibly methylated by PRMTs in three different ways on their terminal guanidino nitrogen group (ω-NG). All PRMTs can add a single methyl group to the

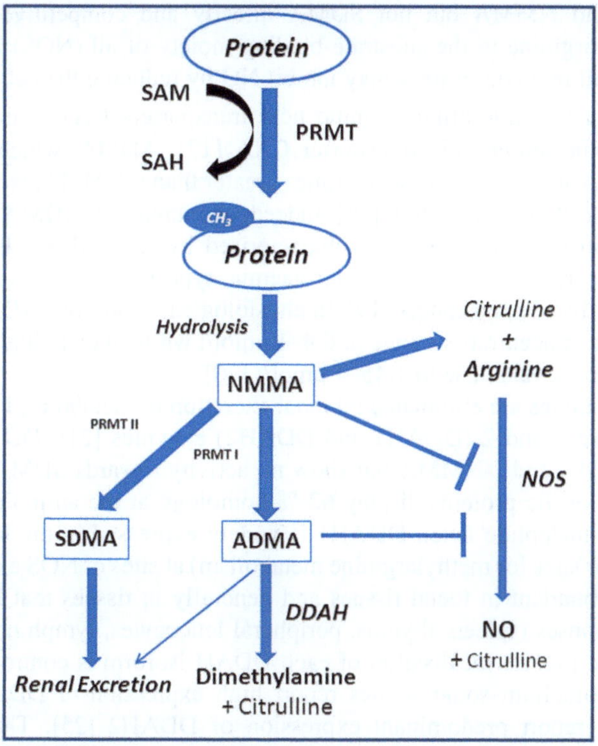

Fig. 11.1 Generation and metabolism of methylarginines. Protein methylation is carried out by arginine *N*-methyltransferases I and II (PMRTs). PMRTs transfer methyl group from S-adenosyl-L-methionine (SAM) to the arginine residues in proteins, providing methylated proteins and S-adenosyl-L-homocysteine (SAH). Methylarginines, ω-NG mono-methyl arginine (NMMA), symmetric dimethylarginine (SDMA) or asymmetric dimethylarginine (ADMA), are produced as a result of proteolytic degradation of methylated proteins. ADMA and L-NMMA but not SDMA act as competitive inhibitors of nitric oxide synthases (NOS). ADMA and L-NMMA are predominantly metabolised by dimethylaminohydrolases (DDAH), while SDMA is removed by renal excretion

arginine guanidino group forming ω-NG mono-methyl arginine (L-NMMA). Subsequently, a second methyl group is added by type I PRMTs (PRMT 1, 2, 3, 4, 6, 8) to generate ω-*NG*,*NG* asymmetric-di-methyl arginine (ADMA) or by type II PRMTs (PRMT 5, 7) generating ω-NG,NG symmetric di-methyl arginine (SDMA). PRMTs use S-adenosyl-methionine (SAM) as a universal methyl donor which is converted to S-adenosyl homocysteine (SAH). SAH in turn acts as a potent feedback inhibitor of PRMTs activity [12]. The methylation of arginine residues is thought to be involved in the regulation of RNA export, control of transcription, DNA repair, protein localisation, protein–protein interaction, signal transduction and recycling of receptors [13–15]. Subsequent proteolysis of methylated proteins releases free methylarginines into the cytosol [10].

ADMA and NMMA but not SDMA directly and competitively inhibit the binding of L-arginine to the substrate-binding moiety of all (NOS) isoforms [16]. In addition, all methylarginines may inhibit NO by reducing the cellular uptake of L-arginine via the high-affinity amino acid transporters CAT-1 and/or the low-affinity cationic amino acid transporter CAA [17]. ADMA, which circulates in human plasma at a concentration ten times greater than L-NMMA, is considered the main inhibitor of NO production [18]. Indeed, an increase in ADMA plasma levels is observed in several pathologies characterised by reduced NO bioavailability, including hypertension, hypercholesterolaemia, type II diabetes, congestive heart failure, renal disease and cancer [19]. In physiological conditions ADMA circulates in plasma in a concentration range of 0.4–1 μmol/l while in pathological conditions ADMA plasma levels raise to 1.45–4 μmol/l [20].

Methylarginines are eliminated by renal excretion or metabolised by dimethylaminohydrolase 1 and 2 (DDAH1 and DDAH2) enzymes [21]. DDAH hydrolyse 90 % of ADMA and L-NMMA but show no activity towards SDMA. Structurally these two cytosolic proteins display 62 % homology at the amino acid level and 63 % at the nucleotide level. DDAH1 is widely expressed in the lung, liver and kidney (major sites for methylarginine metabolism) at sites of NOS expression [22]. DDAH2 is abundant in foetal tissues and generally in tissues that have a role in immune responses (spleen, thymus, peripheral leucocytes, lymph nodes and bone marrow). The precise localisation of each DDAH isoform is controversial. In the vascular endothelium some studies report high expression of DDAH1 [23, 24], while others report predominant expression of DDAH2 [25]. The key role of endothelial DDAH1 in the regulation of vascular function was confirmed in a mice model with endothelium-specific deletion of DDAH1. The endo-DDAH1−/− mice showed no apparent defects in growth or development, but ADMA levels were significantly elevated and DDAH1 expression was greatly reduced in kidney, lung, brain and liver, indicating that in these organs DDAH1 is mainly distributed in vascular endothelial cells. The endo-DDAH1−/− mice also exhibited significantly attenuated acetylcholine-induced NO production and vessel relaxation in isolated aortic rings [24]. The effects of genetic manipulation of DDAH expression on angiogenesis in vitro and in vivo are described below and summarised in Table 11.1.

11.3 Regulatory Role of Methylarginines in Angiogenesis: Experimental and Clinical Evidence

11.3.1 Angiogenesis in Cultured Cells and Knockout Animals

Exogenous ADMA inhibits tubulogenesis in cultured human endothelial cells and endothelial progenitor cells [26]. These effects of ADMA can be prevented by overexpression of DDAH1 [26, 38] or DDAH2 [36]. DDAH1 heterozygous knockout mice, characterised by elevated plasma and tissue ADMA, show significantly

Table 11.1 The effects of genetic manipulation of DDAH expression on angiogenesis

Organism/cell type	DDAH	Effect	References
Mouse	DDAH1 heterozygous knockout	Reduced angiogenesis in Matrigel plugs in vivo	[26]
Mouse/pulmonary microvascular endothelial cells	DDAH1 heterozygous knockout	Decrease in endothelial cell migration, decrease in aortic capillary sprouting	[27]
Mouse	Global DDAH1 knockout	Inhibition of capillary sprouting from aortic rings Increased ADMA, L-NMMA, Reduction in DDAH activity, DDAH2 expression unchanged	[28, 29]
Mouse	Endothelial DDAH1 knockout	Attenuated acetylcholine-induced NO production and vessel relaxation	[24]
Human/human umbilical vein endothelial cells (HUVECs)	DDAH1 overexpression	Increased VEGF-induced tubulogenesis	[26]
Mouse	Transgenic DDAH1 overexpression	Resistance to the inhibitory effects of ADMA on angioadaptation after hindlimb ischaemia Improved endothelial regeneration after femoral artery injury, reduced systemic vascular resistance	[30, 31]
Mouse	Transgenic DDAH2 overexpression	Inhibition of ADMA/angiotensin II-induced medial thickening and fibrosis in coronary microvessels	[32]
Human/HUVEC cell line SGHEC-7	DDAH1 overexpression	Increased endothelial cell migration and VEGF production	[33]
Rat	DDAH1 overexpression	Prevention of ADMA-induced loss of renal peritubular capillaries	[34]
Rat	DDAH1 overexpression	Amelioration of tubulointerstitial ischaemia	[35]
Human/endothelial cell line ECV304 Mouse/sEnd.1 cell line	DDAH2 overexpression	Increased levels of VEGF mRNA and tube formation	[36]
Human/HUVEC	DDAH1 overexpression	Increased endothelial cell proliferation, migration and tube formation	[28]
Human/HUVEC Bovine/bovine aortic endothelial cells	DDAH2 overexpression	Increased VEGF production, endothelial cell proliferation and migration	[37]
Porcine/human pulmonary artery endothelial cells	DDAH1 and DDAH2 overexpression	Increased endothelial cell migration	[27]

reduced angiogenesis in aortic ring explants and reduced neovascularisation of subcutaneously implanted Matrigel plugs compared to wild-type littermates [26]. Conversely, transgenic mice overexpressing DDAH1 are more resistant to the inhibitory effect of ADMA on angioadaptation (angiogenesis and arteriogenesis) after hindlimb ischaemia, show increased blood vessel formation in the fibrovascular disc system and show improved endothelial regeneration after femoral artery injury [30, 31]. These mice exhibit greater tissue DDAH activity, reduced plasma ADMA levels, increased NOS activity and reduced systemic vascular resistance [31]. Global deletion of DDAH1 significantly elevates ADMA and L-NMMA levels and reduces DDAH activity while DDAH2 expression levels remain unchanged [29]. This further reinforces the view that DDAH1 and not DDAH2 is the main enzyme responsible for ADMA metabolism.

Currently, there is no published data describing the effects of global or endothelium-specific deletion of DDAH2 in mice. DDAH2 overexpressing, transgenic mice show a reduction in plasma ADMA and an elevation in cardiac NO levels but no changes in systemic blood pressure (SBP) compared with wild-type mice. DDAH2 overexpression prevents ADMA/angiotensin II-induced medial thickening and perivascular fibrosis in coronary microvessels [32], but the effects on angiogenesis have not been elucidated.

11.3.2 Dysregulation of ADMA Metabolism in Diseases Associated with Defective Angiogenesis

Ischaemia. Hindlimb ischaemia, induced by surgical ligation of the femoral artery, leads to decreased DDAH1 expression, increased tissue ADMA level and reduced NOS expression and activity [39]. Genetic overexpression of DDAH1 enhances neovascularisation and limb perfusion after surgery, by reducing plasma and tissue ADMA levels and enhancing tissue NOS activity. Exogenous ADMA or L-nitroarginine (L-NAME), a NOS antagonist that cannot be degraded by DDAH, significantly inhibits neovascularisation in response to hindlimb ischaemia [31]. Conversely, DDAH1 transgenic animals are resistant to the effects of exogenous ADMA but not to L-NAME [31, 39].

Hypercholesterolaemia. Elevated plasma levels of ADMA reduce NO synthesis in hypercholesterolaemic animals [40] and patients [41, 42]. The angiogenic response to a persistent elevation of ADMA was studied in the hyperlipidaemic apolipoprotein E-knockout (apoE-KO) mouse. Unlike most mouse strains, which are highly resistant to atherosclerosis, apoE-KO mice develop atherosclerotic lesions in a distribution closely resembling human disease [43]. Hypercholesterolaemia in apoE-KO mice led to a sustained increase in plasma ADMA levels, reduced NO biogenesis and reduced vascularisation of implanted polyvinyl alcohol sponge discs [40]. Further, reduction of DDAH I activity in hypercholesterolaemic mice inhibited basal and FGF-induced angiogenesis. This effect can be mimicked in normocholesterolaemic animals by administration of ADMA and reversed by oral administration of L-arginine [40].

Cancer. Angiogenesis is a prerequisite for tumour growth and is highly regulated by growth factors and cytokines, some of which are modulated by NO signalling. NOS activity correlates with lymph node metastases and tumour vascularisation [44] and is elevated in tissue specimens from cancer patients. Consistent with this, increased DDAH activity has been detected in a series of human tumours [33, 45]. Animal studies show that DDAH promotes tumour growth by modulating the development of tumour vasculature without altering the growth properties of cancer cells [33, 46]. Increased DDAH expression results in decreased tumour ADMA, increased NO production, tumour growth and angiogenesis and that these effects are mediated through changes in the expression of VEGF [33]. Interestingly, different types of cancer tissue also show increased expression of PRMTs, possibly reflecting the key role of protein methylation in the control of tumour growth [47]. The same study also identified increased levels of ADMA in serum from cancer patients [47], in contrast to other reports showing low ADMA levels in cancer tissues. It is conceivable that ADMA may promote or inhibit cell growth, depending on the cell origin, reflecting differential effects of NO on cell proliferation [48, 49].

Renal disease. Deprivation of NO impairs angiogenesis and contributes to tubulointerstitial fibrosis and renal damage in progressive renal disease [50]. Kidney resection or chronic administration of exogenous ADMA in rats increases plasma ADMA levels followed by kidney fibrosis and reduction in the number of renal peritubular capillaries [51, 52]. Adenoviral overexpression of DDAH1 reduces plasma levels of ADMA and prevents the progression of renal dysfunction [34].

Diabetes mellitus. Endothelial dysfunction contributes to cardiovascular complications in patients with diabetes mellitus, including increased vascular resistance and atherogenesis [53]. Endothelial dysfunction in diabetic patients is believed to be caused by NO deficiency, likely to result from increased oxidative degradation of NO [54] or increased levels of ADMA [55]. Increase in ADMA is detectable early in the course of the kidney disease in patients with type 1 diabetes mellitus, even before the development of vascular complications [56, 57]. In type 2 diabetes mellitus, several common polymorphisms in the DDAH1 and DDAH2 genes associate strongly with elevated serum ADMA levels. SNPs in DDAH1 and DDAH2 have an additive effect on ADMA levels [58]. Experiments on cultured vascular smooth muscle and endothelial cells show that high glucose impairs DDAH activity, increases ADMA levels and inhibits NO signalling. This mechanism is likely to contribute to the endothelial dysfunction characteristic of diabetic nephropathy [59].

Progressive diabetic nephropathy is one of the major complications in diabetic patients. While neo-angiogenesis of the glomerular capillaries may be seen at an early stage of diabetes, loss of capillaries in the glomerulus and in the interstitium are key events in advanced disease. Loss of microvessels leads to tissue hypoxia which induces glomerulosclerosis and tubulointerstitial fibrosis, both of which feature in advanced diabetic kidney disease [60]. High plasma ADMA levels predict the progression of nephropathy in adult type 2 diabetic patients [61]. The rs805304 DDAH2-1151 polymorphism, which significantly increases ADMA levels in diabetic patients, is strongly associated with diabetic renal impairment [62]. The mechanism of how DDAH gene variations influence ADMA levels and what role this plays in pathogenesis of the disease have yet to be elucidated.

Additional evidence of the role of ADMA/DDAH in vascular dysfunction in diabetes is provided by animal models of the disease. Rats with streptozotocin (STZ)-induced type 2 diabetes mellitus exhibit reduced DDAH activity associated with markedly elevated plasma ADMA levels compared to wild-type animals [59]. Overexpression of DDAH1 in these rats ameliorates tubulointerstitial ischaemia [35].

11.3.3 The Effects of ADMA/DDAH on Endothelial Progenitor Cell Function

Endothelial progenitor cells (EPCs) play an important role in angiogenesis and vasculogenesis [63]. EPCs, which can be cultured in vitro from mononuclear cells found in peripheral blood or bone marrow, express both hematopoietic stem cell and endothelial cell markers on their surface. Their ability to differentiate into endothelial cells and to secrete protective cytokines and growth factors indicates that they participate in endothelial repair and postnatal angiogenesis. Mounting evidence suggests that circulating EPCs are reduced and dysfunctional in various diseases associated with increased ADMA levels, including hypertension, diabetes, coronary heart disease and ischaemic stroke. ADMA is thought to inhibit mobilisation, differentiation and function of endothelial progenitor cells [38], thus impairing endogenous regeneration of diseased blood vessels and contributing to the progressive development of vascular lesions. In humans, plasma ADMA levels correlate with the severity of coronary artery disease and inversely correlate with the number of circulating CD34+/CD133+ progenitor cells and colony-forming units [38]. Accordingly, a reduction in ADMA levels might accelerate endothelial regeneration by promoting mobilisation of bone-marrow-derived endothelial cells [64]. Exercise significantly reduces systemic ADMA levels and increases numbers of EPCs in patients with advanced atherosclerosis, suggesting enhanced angiogenesis and improved endothelial function [65].

11.4 The ADMA/DDAH Pathway in Angiogenesis

ADMA interferes with basic biological processes required for effective angiogenesis to occur, such as cell proliferation, cell migration, intercellular adhesion and cell-to-cell communication. In the following section we discuss the regulatory role of ADMA in angiogenesis and identify potential signalling mediators involved in this process. A summary of the associated pathways is presented in Fig. 11.2.

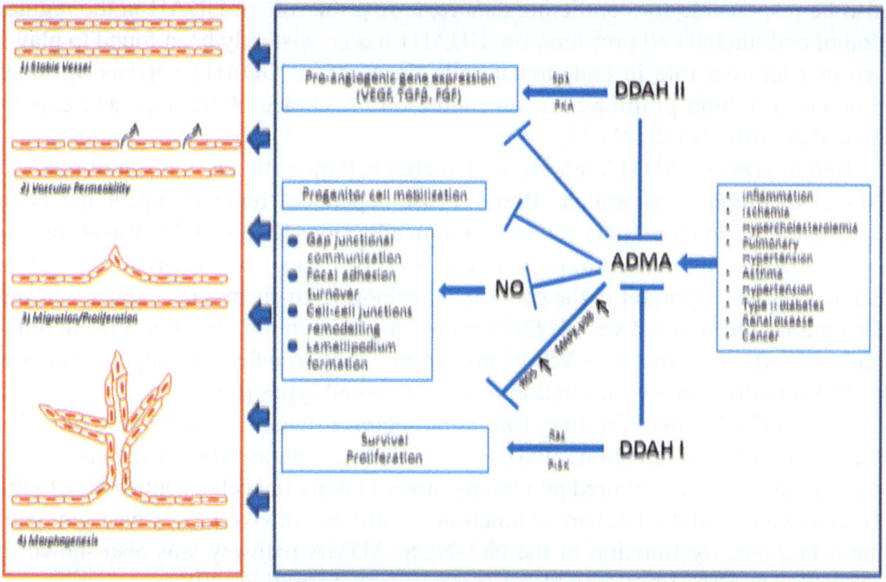

Fig. 11.2 The effects of ADMA on angiogenic responses in endothelial cells

11.4.1 ADMA Metabolism Regulates Angiogenic Responses in Endothelial Cells

Proliferation and apoptosis. Conflicting evidence exists regarding the role of ADMA in proliferation and apoptosis. Whereas one study claims that ADMA has pro-apoptotic and anti-proliferative effects in certain endothelial cell types [66], other reports show no such correlation [23, 36]. ADMA was shown to induce apoptosis in cultured human umbilical vein endothelial cells (HUVECs) via an elevation of intracellular ROS and signalling through the p38 MAPK/caspase-3-dependent pathway [66]. The phosphodiesterase 3/4 inhibitor tolafentrine significantly decreased ADMA-induced apoptosis in HUVECs via adenosine 3′,5′-monophosphate (cAMP)/protein kinase A (PKA)-dependent pathway by induction of DDAH2 [67]. Chronic nebulisation of tolafentrine in rats with monocrotaline-induced pulmonary hypertension suppressed apoptosis in endothelial cells and increased lung vascularisation [67]. Using the same experimental approach employing the tetrazolium compound, 3-(4,5-dimethylthiazol-2-yl)-5-(3-carboxymethoxyphenyl)-2-(4-sulfophenyl)-2H-tetrazolium (MTS), ADMA was found to significantly inhibit cell proliferation in bovine retinal capillary endothelial cells [68], but had no effect on cell proliferation or apoptosis in HUVECs [26]. It is worth noting that the pro-apoptotic effects of ADMA were particularly prominent in cells cultured in media deprived of growth factors and with substantially reduced serum levels [67], which is likely to account for the increased sensitivity to ADMA treatment. Lack of a unified experimental approach may also be responsible for conflicting data regarding the role of DDAH in the regulation of endothelial cell proliferation. DDAH1 has consistently been found to play a pro-proliferative role in endothelial cells. In contrast, DDAH2 appears to either promote or inhibit proliferation, depending on the choice of cell type and experimental conditions [28, 36, 37].

Cell migration. ADMA inhibits endothelial cell migration as a result of reduced NO bioavailability as well as Rho GTPase-dependent remodelling of the actin cytoskeleton and increase in cell–substratum adhesions [27]. ADMA also abrogates polarity of endothelial cell movement and prevents proper alignment of endothelial cells, which is important in the endothelial responses to chemotactic signals during formation of new blood vessels [26]. In contrast to its inhibitory effects in endothelial cells, ADMA stimulates VSMC migration, likely to reflect the opposing effects of NO signalling on cell migration in these two cell types [69].

Cell–cell adhesion. Capillary formation requires dynamic remodelling of endothelial intercellular junctions. Exogenous and endogenous ADMA disrupts intercellular adhesions in cultured pulmonary artery endothelial cells, contributing to the breakdown of endothelial barrier function in vitro and in vivo [23]. Consistent with these findings, dysfunction of the NO–NOS–ADMA pathway was also shown to contribute to retinal [68] and blood–brain barrier disruption [70].

Gap junctional communication. In addition to its inhibitory effect on the endothelial junctional integrity, ADMA inhibits gap junctional communication in endothelial cells [71]. Gap junctions connect the cytoplasm of adjacent cells and allow exchange of small signalling molecules such as cyclic nucleotides, Ca^{2+}, ATP and inositol 1,4,5-trisphosphate between cells. Knockdown of endothelial gap junctional proteins connexins abrogates angiogenesis in vitro and induces several cardiovascular abnormalities in vivo [72–74]. ADMA was shown to decrease connexin 43 (Cx43) expression and inhibit gap junctional communication in human umbilical vein endothelial cells [71]. Cx43 augments several processes important in angiogenesis such as cell–cell adhesion, cell migration and expression of genes in VEGF and TGF-β signalling pathways [75, 76]. The interaction of Cx43 with submembrane actin cytoskeleton helps coordinate formation of protein scaffolds containing proteins such as ZO-1, cadherins and integrins and therefore is important for the formation and maintenance of cell–substratum and cell–cell adhesions. Consistent with the postulated role of Cx43, the carboxyl tail of Cx43 was shown to augment Hela cell migration in a gap junction-independent manner [77].

11.4.2 Signalling Mediators Involved in the Regulation of Angiogenesis by Methylarginines

The most intuitive explanation of the effects of ADMA is reduction in NO bioavailability. Downstream mediators of NO, including cAMP- and cGMP-dependent protein kinases, PKA and PKG (protein kinase G), Rho GTPases and ROS, have been implicated in the anti-angiogenic effects of ADMA. ADMA may reduce signalling of pro-angiogenic growth factors such as VEGF, TGF-β and FGF and inhibit gene expression of factors that promote angiogenesis, such as αvβ3 integrin [78]. Consistent with the role of NO in the ADMA/DDAH pathway, DDAH I gene deletion in mice leads to inhibition of angiogenic responses, similar to that seen in eNOS-knockout mice [26]. In contrast to DDAH1, DDAH2 appears to have a limited effect on NO levels and signal through a NO-independent pathway [37]. While both DDAH1 and 2 are capable of affecting NO signalling in an overexpression system [28, 79], their roles in vivo may differ depending on their interactions with regulatory proteins and cellular localisation. Here we provide a brief account of the role NO-dependent and NO-independent signalling in ADMA/ DDAH pathway.

11.4.3 Regulation of Actin Dynamics by ADMA: Rho GTPases and Cell Adhesion and Migration

Changes in NO levels affect the activity of Rho (Ras homologous) GTPases, key regulators of actin dynamics [80, 81]. Rho GTPases play a major role in the control of endothelial permeability, motility, angiogenesis, proliferation, differentiation and apoptosis [82]. Coordinated activation of Rho GTPases Rac1, RhoA and Cdc42 is required for cell motility [83]. Rac1 and Cdc42 mediate focal contact/ complex formation and polymerisation of the actin cytoskeleton at the leading edge of the cell, whereas RhoA controls stability of focal adhesions, stress fibre formation in the main body of the cell and retraction at the rear. Rho GTPase activity regulates remodelling of inter-endothelial junctions: Rac1 and Cdc42 enhance intercellular tethering forces whereas RhoA primarily acts antagonistically to impair intercellular adhesions.

ADMA activates RhoA and inhibits Rac1 and Cdc42 in pulmonary endothelial cells, leading to inhibition of spontaneous endothelial cell motility and vessel sprouting from aortic rings [27]. These effects are mediated by NO and its downstream effector, PKG. PKG-mediated phosphorylation of RhoA at Ser188 prevents RhoA activation at the cell membrane [84]. Consistent with this model, ADMA was shown to decrease Ser188 phosphorylation of RhoA in cultured human pulmonary endothelial cells and cells from DDAH I HT-knockout mice, resulting in activation of RhoA, an increase in cell–substratum adhesion and inhibition of spontaneous endothelial cell motility in vitro [27].

We have shown that ADMA also inhibits VEGF-induced endothelial chemotaxis and angiogenesis in vitro and in vivo [26]. This effect is predominantly associated with the inhibition of Rac1 as a result of decreased phosphorylation of a scaffolding protein VASP (vasodilator-stimulated phosphoprotein). VASP links adherens junction proteins and integrins to the actin cytoskeleton and regulates polymerisation of cortical actin cytoskeleton [26]. Rac1 activation requires NO/PKG-induced phosphorylation of VASP, but the mechanisms are not fully understood. It is likely that the interactions of VASP with p120Ras GAP (GTPase-activating protein), GTP exchange factors for Rac1 or the tight junction component ZO-1 (zonula occludens 1) protein play a role in the regulation of Rac1 activity [85, 86]. Rac1-induced lamellipodia formation is crucial for establishing directional endothelial cell motility, matrix invasion and alignment into capillary networks [87]. Consistent with the role of ADMA in the regulation of Rac1 activity, endothelial-specific Rac1 haploinsufficient mice (EC-Rac1+/−) show mild hypertension, impaired angiogenesis and decreased NO signalling [88], reminiscent of the phenotype of the DDAH I HT- and eNOS-knockout mice [27, 89–91]. Dysregulation of Rho GTPases activity by ADMA may also affect other processes important in angiogenesis. ADMA destabilises endothelial junctions as a result of Rac1 inhibition and RhoA activation in vitro and in vivo [23]. Destabilisation of endothelial adherens and tight junctions precedes endothelial cell migration from pre-existing blood vessels, whereas reassembly is required during capillary formation. In addition, activation of RhoA and inhibition of Rac1 by ADMA may further exacerbate endothelial dysfunction by reducing NO production in cells. RhoA and Rho kinase reduce eNOS activity and expression, whereas Rac1 has a stimulatory effect [88, 92, 93].

Interestingly, DDAH1 has been shown to associate with Ras, a protein related to Rho GTPases [28]. The interaction of DDAH1 with Ras is necessary for the subsequent activation of PI3 kinase and its downstream effector, kinase Akt and stimulation of endothelial cell proliferation, migration and angiogenesis in vitro. While the effects of DDAH1 on PI3 kinase/Akt activity are independent of NO signalling, DDAH1 overexpression increased NO production in cells. This increase was caused in part by degradation of ADMA and in part by Akt-mediated phosphorylation of eNOS on Ser1177 [28]. DDAH-1 was also shown to bind directly to the C-terminal domain and to the cysteine/serine-rich domain of the Ras pathway regulator, neurofibromin 1 (NF-1) [94]. Binding of DDAH-1 to NF-1 increases NF-1 phosphorylation by PKA. While the physiological significance of this phosphorylation is not clear, it may have a regulatory effect on angiogenesis, as NF-1 haploinsufficiency augments angiogenesis [95].

Future experiments will have to determine the role of NO signalling and delineate specific functions for Ras and Rho GTPases in ADMA/DDAH-induced effects. Ras and Rac have overlapping functions and regulate cell behaviour through a common set of regulatory proteins [96]. A summary of ADMA/DDAH-induced effects on Rho and Ras-regulated actin dynamics is shown in Fig. 11.3.

11.4.3.1 Modulation of Gene Expression by Methylarginines

ADMA affects the expression of several genes involved in the regulation of endothelial cell metabolism, growth, transcription and organisation of the cell cytoskeleton [97], but the relevance of these changes in the anti-angiogenic effects of ADMA remains to be established. One of the genes affected, bone morphogenetic protein 2-inducible kinase (BMP2K), is a signalling mediator in the bone morphogenetic protein (BMP) pathway. Expression of BMP2K increases by 2.3-fold and 2.7-fold in response to 2 μM and 100 μM ADMA, respectively, and is increased in several organs of DDAH1 heterozygous knockout mice [97]. Bone morphogenetic proteins are members of TGF-β superfamily, involved in control of cell proliferation, differentiation and angiogenesis [98, 99]. Recombinant BMP2

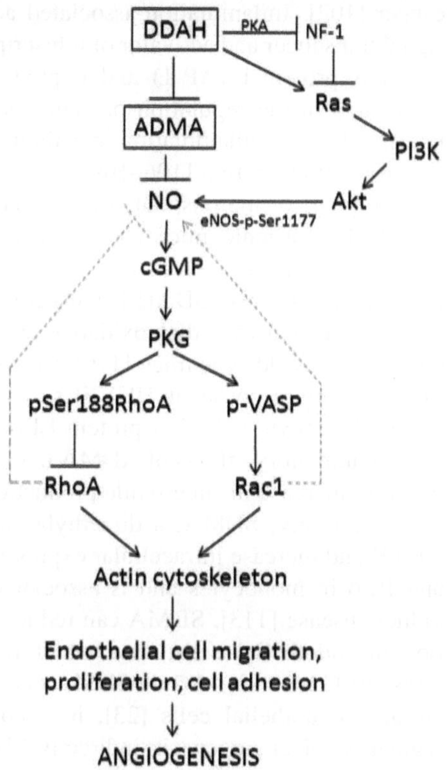

Fig. 11.3 Proposed mechanisms involved in the regulation of actin dynamics by ADMA/DDAH

induces endothelial cell migration, invasion and tube formation in vitro [100]. Interestingly, defective BMP signalling is involved in the pathogenesis of pulmonary hypertension, characterised by endothelial dysfunction and deregulated angiogenesis [101].

While DDAH1 appears to act on angiogenesis predominantly by increasing NO bioavailability, DDAH2 can promote angiogenesis by enhancing the VEGF mRNA expression in endothelial cells. Overexpression of DDAH2, but not DDAH1, stimulates VEGF expression and secretion in a range of endothelial cells, and this effect is associated with enhanced cell proliferation and migration [36, 37]. These effects are accomplished in the absence of a change in NO generation. DDAH2 binds to PKA, leading to the phosphorylation of the transcription factor specificity protein 1 (Sp1), which translocates to the nucleus and binds to the promoter region of the VEGF gene, leading to its transcription [37]. DDAH2 localises in the nucleus of vascular endothelial and smooth muscle cells, but the significance of this finding in the regulation of gene expression will require further studies [25].

11.4.3.2 Oxidative Stress and Inflammation

Inflammation and angiogenesis are coupled together in many chronic inflammatory disorders including psoriasis, rheumatoid arthritis, Crohn's disease, diabetes and cancer [102]. Inflammation-associated activation of transcription factors such as signal transducer and activator of transcription (STAT), nuclear factor-κB (NF-κB), activator protein-1 (AP-1) and hypoxia-inducible factor-1α (HIF-1α) has been implicated in the regulation of angiogenesis [103–105]. ADMA levels correlate with markers of inflammation in asthma, rheumatoid arthritis, atherosclerosis and systemic inflammation [106–109].

In support of the postulated anti-inflammatory role of DDAH1/NO signalling, DDAH I transgenic mice show reduced inflammatory response and enhanced vascular repair and increased NOS activity following femoral artery injury [64]. Consistent with this, DDAH I transgenic mice undergoing cardiac transplantation show reduced myocardial oxidative stress, inflammation and tissue damage, as compared to wild-type mice [110]. Conversely, ADMA was shown to induce pro-inflammatory responses in HUVECs by increasing expression of ICAM-1, a ligand for the leucocyte adhesion protein LFA-1, in a NF-kB-dependent manner [111]. Pro-inflammatory effects of ADMA may result from inhibition of NO bioavailability or an increase in superoxide production as a result of eNOS uncoupling [112].

Interestingly, SDMA, a dimethylarginine considered to be inert, can activate NF-κB and increase intracellular expression of pro-inflammatory cytokines TNF-α and IL-6 in monocytes and is associated with inflammatory markers in chronic kidney disease [113]. SDMA can reduce NO synthesis and enhance ROS production in endothelial cells, as a result from competitive inhibition of L-arginine transport [114, 115]). While SDMA does not affect motility or barrier function in cultured endothelial cells [23], it is conceivable that it may contribute to the regulation of angiogenesis indirectly via modulation of inflammatory responses in vivo.

Conclusions

Clinical and experimental evidence shows that ADMA impairs endothelial repair and angiogenesis in vitro and in vivo. Modulation of dimethyl-arginine dimethylaminohydrolases, the key enzymes metabolising ADMA, may offer a potential therapeutic strategy for the prevention and treatment of angiogenic disorders. Potential ways of targeting DDAH, reviewed by Cooke [116], involve activation of DDAH promoters by farnesoid X receptor agonists or treatment with statins. The cellular localisation of DDAH, the role of its different isoforms and the role of NO as a mediator of ADMA/DDAH signalling remain unclear. Efforts should be made to adopt a unified experimental approach to address the controversies and inconsistencies present in the current literature.

Acknowledgements The authors wish to thank Lisa Storck for her kind help in preparation of this chapter.

References

1. Dimmeler S, Hermann C, Galle J, Zeiher AM (1999) Upregulation of superoxide dismutase and nitric oxide synthase mediates the apoptosis-suppressive effects of shear stress on endothelial cells. Arterioscler Thromb Vasc Biol 19(3):656–664
2. Morbidelli L, Chang CH, Douglas JG, Granger HJ, Ledda F, Ziche M (1996) Nitric oxide mediates mitogenic effect of VEGF on coronary venular endothelium. Am J Physiol 270(1 Pt 2):H411–H415
3. Murohara T, Witzenbichler B, Spyridopoulos I, Asahara T, Ding B, Sullivan A et al (1999) Role of endothelial nitric oxide synthase in endothelial cell migration. Arterioscler Thromb Vasc Biol 19(5):1156–1161
4. Ziche M, Parenti A, Ledda F, Dell'Era P, Granger HJ, Maggi CA et al (1997) Nitric oxide promotes proliferation and plasminogen activator production by coronary venular endothelium through endogenous bFGF. Circ Res 80(6):845–852
5. Babaei S, Teichert-Kuliszewska K, Monge JC, Mohamed F, Bendeck MP, Stewart DJ (1998) Role of nitric oxide in the angiogenic response in vitro to basic fibroblast growth factor. Circ Res 82(9):1007–1015
6. Papapetropoulos A, Garcia-Cardena G, Madri JA, Sessa WC (1997) Nitric oxide production contributes to the angiogenic properties of vascular endothelial growth factor in human endothelial cells. J Clin Invest 100(12):3131–3139
7. Dulak J, Jozkowicz A, Dembinska-Kiec A, Guevara I, Zdzienicka A, Zmudzinska-Grochot D et al (2000) Nitric oxide induces the synthesis of vascular endothelial growth factor by rat vascular smooth muscle cells. Arterioscler Thromb Vasc Biol 20(3):659–666
8. Amano K, Matsubara H, Iba O, Okigaki M, Fujiyama S, Imada T et al (2003) Enhancement of ischemia-induced angiogenesis by eNOS overexpression. Hypertension 41(1):156–162
9. Smith RS Jr, Lin KF, Agata J, Chao L, Chao J (2002) Human endothelial nitric oxide synthase gene delivery promotes angiogenesis in a rat model of hindlimb ischemia. Arterioscler Thromb Vasc Biol 22(8):1279–1285
10. Leiper J, Nandi M (2011) The therapeutic potential of targeting endogenous inhibitors of nitric oxide synthesis. Nat Rev Drug Discov 10(4):277–291
11. Najbauer J, Johnson BA, Young AL, Aswad DW (1993) Peptides with sequences similar to glycine, arginine-rich motifs in proteins interacting with RNA are efficiently recognized by methyltransferase(s) modifying arginine in numerous proteins. J Biol Chem 268 (14):10501–10509
12. Cheng D, Yadav N, King RW, Swanson MS, Weinstein EJ, Bedford MT (2004) Small molecule regulators of protein arginine methyltransferases. J Biol Chem 279 (23):23892–23899
13. Bedford MT, Richard S (2005) Arginine methylation an emerging regulator of protein function. Mol Cell 18(3):263–272
14. Boisvert FM, Cote J, Boulanger MC, Cleroux P, Bachand F, Autexier C et al (2002) Symmetrical dimethylarginine methylation is required for the localization of SMN in Cajal bodies and pre-mRNA splicing. J Cell Biol 159(6):957–969
15. El-Andaloussi N, Valovka T, Toueille M, Hassa PO, Gehrig P, Covic M et al (2007) Methylation of DNA polymerase beta by protein arginine methyltransferase 1 regulates its binding to proliferating cell nuclear antigen. FASEB J 21(1):26–34
16. Pope AJ, Karuppiah K, Cardounel AJ (2009) Role of the PRMT-DDAH-ADMA axis in the regulation of endothelial nitric oxide production. Pharmacol Res 60(6):461–465
17. Closs EI, Simon A, Vekony N, Rotmann A (2004) Plasma membrane transporters for arginine. J Nutr 134(10 Suppl):2752S–2759S, discussion 65S–67S
18. Vallance P, Leone A, Calver A, Collier J, Moncada S (1992) Endogenous dimethylarginine as an inhibitor of nitric oxide synthesis. J Cardiovasc Pharmacol 20(Suppl 12):S60–S62
19. Cooke JP (2004) Asymmetrical dimethylarginine: the Uber marker? Circulation 109 (15):1813–1818

20. Cardounel AJ, Cui H, Samouilov A, Johnson W, Kearns P, Tsai AL et al (2007) Evidence for the pathophysiological role of endogenous methylarginines in regulation of endothelial NO production and vascular function. J Biol Chem 282(2):879–887
21. Achan V, Broadhead M, Malaki M, Whitley G, Leiper J, MacAllister R et al (2003) Asymmetric dimethylarginine causes hypertension and cardiac dysfunction in humans and is actively metabolized by dimethylarginine dimethylaminohydrolase. Arterioscler Thromb Vasc Biol 23(8):1455–1459
22. Nijveldt RJ, Teerlink T, Siroen MP, van Lambalgen AA, Rauwerda JA, van Leeuwen PA (2003) The liver is an important organ in the metabolism of asymmetrical dimethylarginine (ADMA). Clin Nutr 22(1):17–22
23. Wojciak-Stothard B, Torondel B, Zhao L, Renné T, Leiper JM (2009) Modulation of Rac1 activity by ADMA/DDAH regulates pulmonary endothelial barrier function. Mol Biol Cell 20(1):33–42. doi:10.1091/mbc.E08-04-0395
24. Hu X, Xu X, Zhu G, Atzler D, Kimoto M, Chen J, Schwedhelm E, Lüneburg N, Böger RH, Zhang P, Chen Y (2009) Vascular endothelial-specific dimethylarginine dimethylaminohydrolase-1-deficient mice reveal that vascular endothelium plays an important role in removing asymmetric dimethylarginine. Circulation 120(22):2222–2229. doi:10.1161/CIRCULATIONAHA.108.819912
25. Palm F, Onozato ML, Luo Z, Wilcox CS (2007) Dimethylarginine dimethylaminohydrolase (DDAH): expression, regulation, and function in the cardiovascular and renal systems. Am J Physiol Heart Circ Physiol 293(6):H3227–H3245
26. Fiedler LR, Bachetti T, Leiper J, Zachary I, Chen L, Renne T et al (2009) The ADMA/DDAH pathway regulates VEGF-mediated angiogenesis. Arterioscler Thromb Vasc Biol 29 (12):2117–2124
27. Wojciak-Stothard B, Torondel B, Tsang LY, Fleming I, Fisslthaler B, Leiper JM et al (2007) The ADMA/DDAH pathway is a critical regulator of endothelial cell motility. J Cell Sci 120 (Pt 6):929–942
28. Zhang P, Hu X, Xu X, Chen Y, Bache RJ (2011) Dimethylarginine dimethylaminohydrolase 1 modulates endothelial cell growth through nitric oxide and Akt. Arterioscler Thromb Vasc Biol 31(4):890–897
29. Hu X, Atzler D, Xu X, Zhang P, Guo H, Lu Z et al (2011) Dimethylarginine dimethylaminohydrolase-1 is the critical enzyme for degrading the cardiovascular risk factor asymmetrical dimethylarginine. Arterioscler Thromb Vasc Biol 31(7):1540–1546
30. Cooke JP (2003) NO and angiogenesis. Atheroscler Suppl 4(4):53–60
31. Jacobi J, Sydow K, von Degenfeld G, Zhang Y, Dayoub H, Wang B et al (2005) Overexpression of dimethylarginine dimethylaminohydrolase reduces tissue asymmetric dimethylarginine levels and enhances angiogenesis. Circulation 111(11):1431–1438
32. Hasegawa K, Wakino S, Tatematsu S, Yoshioka K, Homma K, Sugano N et al (2007) Role of asymmetric dimethylarginine in vascular injury in transgenic mice overexpressing dimethylarginie dimethylaminohydrolase 2. Circ Res 101(2):e2–e10
33. Kostourou V, Robinson SP, Cartwright JE, Whitley GS (2002) Dimethylarginine dimethylaminohydrolase I enhances tumour growth and angiogenesis. Br J Cancer 87(6):673–680
34. Matsumoto Y, Ueda S, Yamagishi S, Matsuguma K, Shibata R, Fukami K et al (2007) Dimethylarginine dimethylaminohydrolase prevents progression of renal dysfunction by inhibiting loss of peritubular capillaries and tubulointerstitial fibrosis in a rat model of chronic kidney disease. J Am Soc Nephrol 18(5):1525–1533
35. Shibata R, Ueda S, Yamagishi S, Kaida Y, Matsumoto Y, Fukami K et al (2009) Involvement of asymmetric dimethylarginine (ADMA) in tubulointerstitial ischemia in the early phase of diabetic nephropathy. Nephrol Dial Transplant 24(4):1162–1169
36. Smith CL, Birdsey GM, Anthony S, Arrigoni FI, Leiper JM, Vallance P (2003) Dimethylarginine dimethylaminohydrolase activity modulates ADMA levels, VEGF expression, and cell phenotype. Biochem Biophys Res Commun 308(4):984–989

37. Hasegawa K, Wakino S, Tanaka T, Kimoto M, Tatematsu S, Kanda T et al (2006) Dimethylarginine dimethylaminohydrolase 2 increases vascular endothelial growth factor expression through Sp1 transcription factor in endothelial cells. Arterioscler Thromb Vasc Biol 26(7):1488–1494
38. Thum T, Tsikas D, Stein S, Schultheiss M, Eigenthaler M, Anker SD et al (2005) Suppression of endothelial progenitor cells in human coronary artery disease by the endogenous nitric oxide synthase inhibitor asymmetric dimethylarginine. J Am Coll Cardiol 46(9):1693–1701
39. Achan V, Ho HK, Heeschen C, Stuehlinger M, Jang JJ, Kimoto M et al (2005) ADMA regulates angiogenesis: genetic and metabolic evidence. Vasc Med 10(1):7–14
40. Jang JJ, Ho HK, Kwan HH, Fajardo LF, Cooke JP (2000) Angiogenesis is impaired by hypercholesterolemia: role of asymmetric dimethylarginine. Circulation 102(12):1414–1419
41. Böger RH, Bode-Boger SM, Szuba A, Tsao PS, Chan JR, Tangphao O et al (1998) Asymmetric dimethylarginine (ADMA): a novel risk factor for endothelial dysfunction: its role in hypercholesterolemia. Circulation 98(18):1842–1847
42. Chobanyan-Jürgens K, Fuchs AJ, Tsikas D, Kanzelmeyer N, Das AM, Illsinger S et al (2012) Increased asymmetric dimethylarginine (ADMA) dimethylaminohydrolase (DDAH) activity in childhood hypercholesterolemia type II. Amino Acids 43(2):805–811
43. Nakashima Y, Plump AS, Raines EW, Breslow JL, Ross R (1994) ApoE-deficient mice develop lesions of all phases of atherosclerosis throughout the arterial tree. Arterioscler Thromb 14(1):133–140
44. Gallo O, Masini E, Morbidelli L, Franchi A, Fini-Storchi I, Vergari WA et al (1998) Role of nitric oxide in angiogenesis and tumor progression in head and neck cancer. J Natl Cancer Inst 90(8):587–596
45. Ronquist KG, Ronquist G, Larsson A, Carlsson L (2010) Proteomic analysis of prostate cancer metastasis-derived prostasomes. Anticancer Res 30(2):285–290
46. Kostourou V, Robinson SP, Whitley GS, Griffiths JR (2003) Effects of overexpression of dimethylarginine dimethylaminohydrolase on tumor angiogenesis assessed by susceptibility magnetic resonance imaging. Cancer Res 63(16):4960–4966
47. Yoshimatsu M, Toyokawa G, Hayami S, Unoki M, Tsunoda T, Field HI et al (2011) Dysregulation of PRMT1 and PRMT6, Type I arginine methyltransferases, is involved in various types of human cancers. Int J Cancer 128(3):562–573
48. Jeremy JY, Rowe D, Emsley AM, Newby AC (1999) Nitric oxide and the proliferation of vascular smooth muscle cells. Cardiovasc Res 43(3):580–594
49. Moncada S, Higgs EA (2006) Nitric oxide and the vascular endothelium. Handb Exp Pharmacol (176 Pt 1):213–254
50. Kang DH, Hughes J, Mazzali M, Schreiner GF, Johnson RJ (2001) Impaired angiogenesis in the remnant kidney model: II. Vascular endothelial growth factor administration reduces renal fibrosis and stabilizes renal function. J Am Soc Nephrol 12(7):1448–1457
51. Matsuguma K, Ueda S, Yamagishi S, Matsumoto Y, Kaneyuki U, Shibata R et al (2006) Molecular mechanism for elevation of asymmetric dimethylarginine and its role for hypertension in chronic kidney disease. J Am Soc Nephrol 17(8):2176–2183
52. Mihout F, Shweke N, Bige N, Jouanneau C, Dussaule JC, Ronco P et al (2011) Asymmetric dimethylarginine (ADMA) induces chronic kidney disease through a mechanism involving collagen and TGF-beta1 synthesis. J Pathol 223(1):37–45
53. Hadi HA, Suwaidi JA (2007) Endothelial dysfunction in diabetes mellitus. Vasc Health Risk Manag 3(6):853–876
54. Tesfamariam B, Cohen RA (1992) Free radicals mediate endothelial cell dysfunction caused by elevated glucose. Am J Physiol 263(2 Pt 2):H321–H326
55. Valkonen VP, Tuomainen TP, Laaksonen R (2005) DDAH gene and cardiovascular risk. Vasc Med 10(Suppl 1):S45–S48
56. Altinova AE, Arslan M, Sepici-Dincel A, Akturk M, Altan N, Toruner FB (2007) Uncomplicated type 1 diabetes is associated with increased asymmetric dimethylarginine concentrations. J Clin Endocrinol Metab 92(5):1881–1885

57. Tarnow L, Hovind P, Teerlink T, Stehouwer CD, Parving HH (2004) Elevated plasma asymmetric dimethylarginine as a marker of cardiovascular morbidity in early diabetic nephropathy in type 1 diabetes. Diabetes Care 27(3):765–769
58. Abhary S, Burdon KP, Kuot A, Javadiyan S, Whiting MJ, Kasmeridis N et al (2010) Sequence variation in DDAH1 and DDAH2 genes is strongly and additively associated with serum ADMA concentrations in individuals with type 2 diabetes. PLoS One 5(3):e9462
59. Lin KY, Ito A, Asagami T, Tsao PS, Adimoolam S, Kimoto M et al (2002) Impaired nitric oxide synthase pathway in diabetes mellitus: role of asymmetric dimethylarginine and dimethylarginine dimethylaminohydrolase. Circulation 106(8):987–992
60. Advani A, Gilbert RE (2012) The endothelium in diabetic nephropathy. Semin Nephrol 32 (2):199–207
61. Hanai K, Babazono T, Nyumura I, Toya K, Tanaka N, Tanaka M et al (2009) Asymmetric dimethylarginine is closely associated with the development and progression of nephropathy in patients with type 2 diabetes. Nephrol Dial Transplant 24(6):1884–1888
62. Marra M, Marchegiani F, Ceriello A, Sirolla C, Boemi M, Franceschi C et al (2013) Chronic renal impairment and DDAH2-1151 A/C polymorphism determine ADMA levels in type 2 diabetic subjects. Nephrol Dial Transplant 28(4):964–971
63. Ahn GO, Brown JM (2009) Role of endothelial progenitors and other bone marrow-derived cells in the development of the tumor vasculature. Angiogenesis 12(2):159–164
64. Konishi H, Sydow K, Cooke JP (2007) Dimethylarginine dimethylaminohydrolase promotes endothelial repair after vascular injury. J Am Coll Cardiol 49(10):1099–1105
65. Schlager O, Giurgea A, Schuhfried O, Seidinger D, Hammer A, Groger M et al (2011) Exercise training increases endothelial progenitor cells and decreases asymmetric dimethylarginine in peripheral arterial disease: a randomized controlled trial. Atherosclerosis 217(1):240–248
66. Jiang DJ, Jia SJ, Dai Z, Li YJ (2006) Asymmetric dimethylarginine induces apoptosis via p38 MAPK/caspase-3-dependent signaling pathway in endothelial cells. J Mol Cell Cardiol 40 (4):529–539
67. Pullamsetti SS, Savai R, Schaefer MB, Wilhelm J, Ghofrani HA, Weissmann N et al (2011) cAMP phosphodiesterase inhibitors increases nitric oxide production by modulating dimethylarginine dimethylaminohydrolases. Circulation 123(11):1194–1204
68. Chen YH, Xu X, Sheng MJ, Zheng Z, Gu Q (2011) Effects of asymmetric dimethylarginine on bovine retinal capillary endothelial cell proliferation, reactive oxygen species production, permeability, intercellular adhesion molecule-1, and occludin expression. Mol Vis 17:332–340
69. Sun Y, Zhou Q, Zhang W, Fu Y, Huang H (2002) ASYMMETRIC LEAVES1, an Arabidopsis gene that is involved in the control of cell differentiation in leaves. Planta 214 (5):694–702
70. Li N, Worthmann H, Deb M, Chen S, Weissenborn K (2011) Nitric oxide (NO) and asymmetric dimethylarginine (ADMA): their pathophysiological role and involvement in intracerebral hemorrhage. Neurol Res 33(5):541–548
71. Jia SJ, Zhou Z, Zhang BK, Hu ZW, Deng HW, Li YJ (2009) Asymmetric dimethylarginine damages connexin43-mediated endothelial gap junction intercellular communication. Biochem Cell Biol 87(6):867–874
72. Gärtner C, Ziegelhoffer B, Kostelka M, Stepan H, Mohr FW, Dhein S (2012) Knock-down of endothelial connexins impairs angiogenesis. Pharmacol Res 65(3):347–357
73. Reaume AG, de Sousa PA, Kulkarni S, Langille BL, Zhu D, Davies TC et al (1995) Cardiac malformation in neonatal mice lacking connexin43. Science 267(5205):1831–1834
74. Simon AM, McWhorter AR (2002) Vascular abnormalities in mice lacking the endothelial gap junction proteins connexin37 and connexin40. Dev Biol 251(2):206–220
75. Giepmans BN (2004) Gap junctions and connexin-interacting proteins. Cardiovasc Res 62 (2):233–245

76. Walker DL, Vacha SJ, Kirby ML, Lo CW (2005) Connexin43 deficiency causes dysregulation of coronary vasculogenesis. Dev Biol 284(2):479–498
77. Behrens J, Kameritsch P, Wallner S, Pohl U, Pogoda K (2010) The carboxyl tail of Cx43 augments p38 mediated cell migration in a gap junction-independent manner. Eur J Cell Biol 89(11):828–838
78. Fiedler LR, Wojciak-Stothard B (2009) The DDAH/ADMA pathway in the control of endothelial cell migration and angiogenesis. Biochem Soc Trans 37(Pt 6):1243–1247
79. Torondel B, Nandi M, Kelly P, Wojciak-Stothard B, Fleming I, Leiper J (2010) Adenoviral-mediated overexpression of DDAH improves vascular tone regulation. Vasc Med 15 (3):205–213
80. Hall A (1998) Rho GTPases and the actin cytoskeleton. Science 279(5350):509–514
81. Loirand G, Guilluy C, Pacaud P (2006) Regulation of Rho proteins by phosphorylation in the cardiovascular system. Trends Cardiovasc Med 16(6):199–204
82. Storck EM, Wojciak-Stothard B (2013) Rho GTPases in pulmonary vascular dysfunction. Vascul Pharmacol 58(3):202–210
83. Ridley AJ (2001) Rho GTPases and cell migration. J Cell Sci 114(Pt 15):2713–2722
84. Sauzeau V, Rolli-Derkinderen M, Marionneau C, Loirand G, Pacaud P (2003) RhoA expression is controlled by nitric oxide through cGMP-dependent protein kinase activation. J Biol Chem 278(11):9472–9480
85. Schlegel N, Burger S, Golenhofen N, Walter U, Drenckhahn D, Waschke J (2008) The role of VASP in regulation of cAMP- and Rac 1-mediated endothelial barrier stabilization. Am J Physiol Cell Physiol 294(1):C178–C188
86. Comerford KM, Lawrence DW, Synnestvedt K, Levi BP, Colgan SP (2002) Role of vasodilator-stimulated phosphoprotein in PKA-induced changes in endothelial junctional permeability. FASEB J 16(6):583–585
87. Connolly JO, Simpson N, Hewlett L, Hall A (2002) Rac regulates endothelial morphogenesis and capillary assembly. Mol Biol Cell 13(7):2474–2485
88. Sawada N, Salomone S, Kim HH, Kwiatkowski DJ, Liao JK (2008) Regulation of endothelial nitric oxide synthase and postnatal angiogenesis by Rac1. Circ Res 103(4):360–368
89. Murohara T, Asahara T (2002) Nitric oxide and angiogenesis in cardiovascular disease. Antioxid Redox Signal 4(5):825–831
90. Huang PL, Huang Z, Mashimo H, Bloch KD, Moskowitz MA, Bevan JA et al (1995) Hypertension in mice lacking the gene for endothelial nitric oxide synthase. Nature 377 (6546):239–242
91. Leiper J, Nandi M, Torondel B, Murray-Rust J, Malaki M, O'Hara B, Rossiter S, Anthony S, Madhani M, Selwood D, Smith C, Wojciak-Stothard B, Rudiger A, Stidwill R, McDonald NQ, Vallance P (2007) Disruption of methylarginine metabolism impairs vascular homeostasis. Nat Med 13(2):198–203
92. Rikitake Y, Liao JK (2005) Rho GTPases, statins, and nitric oxide. Circ Res 97 (12):1232–1235
93. Rikitake Y, Liao JK (2005) Rho-kinase mediates hyperglycemia-induced plasminogen activator inhibitor-1 expression in vascular endothelial cells. Circulation 111(24):3261–3268
94. Tokuo H, Yunoue S, Feng L, Kimoto M, Tsuji H, Ono T et al (2001) Phosphorylation of neurofibromin by cAMP-dependent protein kinase is regulated via a cellular association of N (G), N(G)-dimethylarginine dimethylaminohydrolase. FEBS Lett 494(1–2):48–53
95. Wu M, Wallace MR, Muir D (2006) Nf1 haploinsufficiency augments angiogenesis. Oncogene 25(16):2297–2303
96. Crowe DL (2004) Overlapping functions of Ras and Rac GTPases in regulating cancer cell proliferation and invasion. Anticancer Res 24(2B):593–597
97. Smith CL, Anthony S, Hubank M, Leiper JM, Vallance P (2005) Effects of ADMA upon gene expression: an insight into the pathophysiological significance of raised plasma ADMA. PLoS Med 2(10):e264

98. Kingsley DM (1994) The TGF-beta superfamily: new members, new receptors, and new genetic tests of function in different organisms. Genes Dev 8(2):133–146
99. Mahmoud M, Upton PD, Arthur HM (2011) Angiogenesis regulation by TGFbeta signalling: clues from an inherited vascular disease. Biochem Soc Trans 39(6):1659–1666
100. Finkenzeller G, Hager S, Stark GB (2012) Effects of bone morphogenetic protein 2 on human umbilical vein endothelial cells. Microvasc Res 84(1):81–85
101. Morrell NW (2010) Role of bone morphogenetic protein receptors in the development of pulmonary arterial hypertension. Adv Exp Med Biol 661:251–264
102. Costa C, Incio J, Soares R (2007) Angiogenesis and chronic inflammation: cause or consequence? Angiogenesis 10(3):149–166
103. Subramaniam A, Shanmugam MK, Perumal E, Li F, Nachiyappan A, Dai X et al (2013) Potential role of signal transducer and activator of transcription (STAT)3 signaling pathway in inflammation, survival, proliferation and invasion of hepatocellular carcinoma. Biochim Biophys Acta 1835(1):46–60
104. Mantovani A, Garlanda C, Allavena P (2010) Molecular pathways and targets in cancer-related inflammation. Ann Med 42(3):161–170
105. Kim YW, West XZ, Byzova TV (2013) Inflammation and oxidative stress in angiogenesis and vascular disease. J Mol Med (Berl) 91(3):323–328
106. Scott JA, North ML, Rafii M, Huang H, Pencharz P, Subbarao P et al (2011) Asymmetric dimethylarginine is increased in asthma. Am J Respir Crit Care Med 184(7):779–785
107. Kwaśny-Krochin B, Gluszko P, Undas A (2012) Plasma asymmetric dimethylarginine in active rheumatoid arthritis: links with oxidative stress and inflammation. Pol Arch Med Wewn 122(6):270–276
108. Tripepi G, Mattace Raso F, Sijbrands E, Seck MS, Maas R, Boger R et al (2011) Inflammation and asymmetric dimethylarginine for predicting death and cardiovascular events in ESRD patients. Clin J Am Soc Nephrol 6(7):1714–1721
109. van der Zwan LP, Scheffer PG, Dekker JM, Stehouwer CD, Heine RJ, Teerlink T (2011) Systemic inflammation is linked to low arginine and high ADMA plasma levels resulting in an unfavourable NOS substrate-to-inhibitor ratio: the Hoorn Study. Clin Sci (Lond) 121 (2):71–78
110. Tanaka M, Sydow K, Gunawan F, Jacobi J, Tsao PS, Robbins RC, Cooke JP (2005) Dimethylargininedimethylaminohydrolase overexpression suppresses graft coronary artery disease. Circulation 112(11):1549–1556
111. Wei-Kang G, Dong-Liang Z, Xin-Xin W, Wei K, Zhang Y, Qi-Dong Z et al (2011) Actin cytoskeleton modulates ADMA-induced NF-kappaB nuclear translocation and ICAM-1 expression in endothelial cells. Med Sci Monit 17(9):BR242–BR247
112. Antoniades C, Shirodaria C, Leeson P, Antonopoulos A, Warrick N, Van-Assche T et al (2009) Association of plasma asymmetrical dimethylarginine (ADMA) with elevated vascular superoxide production and endothelial nitric oxide synthase uncoupling: implications for endothelial function in human atherosclerosis. Eur Heart J 30(9):1142–1150
113. Schepers E, Barreto DV, Liabeuf S, Glorieux G, Eloot S, Barreto FC et al (2011) Symmetric dimethylarginine as a proinflammatory agent in chronic kidney disease. Clin J Am Soc Nephrol 6(10):2374–2383
114. Closs EI, Basha FZ, Habermeier A, Forstermann U (1997) Interference of L-arginine analogues with L-arginine transport mediated by the y + carrier hCAT-2B. Nitric Oxide 1 (1):65–73
115. Bode-Böger SM, Scalera F, Kielstein JT, Martens-Lobenhoffer J, Breithardt G, Fobker M, Reinecke H (2006) Symmetrical dimethylarginine: a new combined parameter for renal function and extent of coronary artery disease. J Am Soc Nephrol 17(4):1128–1134
116. Cooke JP (2010) DDAH: a target for vascular therapy? Vasc Med 15(3):235–238

The Vasohibin Family: Novel Regulators of Angiogenesis 12

Yasufumi Sato

Contents

12.1 Introduction 262
12.2 Vasohibin-1 262
  12.2.1 Isolation of VASH1 262
  12.2.2 VASH1 in Pathophysiological Conditions 262
  12.2.3 Role of VASH1 in Survival or Death of ECs 264
12.3 Vasohibin-2 264
  12.3.1 Isolation of VASH2 264
  12.3.2 Opposing Role of VASH2 in Angiogenesis 265
  12.3.3 VASH2 in Pathophysiological Conditions 265
References 266

Abstract

A number of endogenous regulators of angiogenesis have been found in the body. We isolated vasohibin-1 (VASH1) as a negative feedback regulator of angiogenesis produced by endothelial cells, and then VASH2 as a homologue of VASH1. We further explored that VASH1 is expressed in endothelial cells to terminate angiogenesis, whereas VASH2 is expressed in non-endothelial cells in the sprouting front to promote angiogenesis. This chapter will focus on this vasohibin family members, novel regulators of angiogenesis.

Keywords

Vasohibin-1 • Vasohibin-2 • Angiogenesis inhibitors • VEGF • Cancer • Tumor angiogenesis

Y. Sato (✉)
Department of Vascular Biology, Institute of Development, Aging and Cancer, Tohoku University, 4-1 Seiryo-machi, Aoba-ku, Sendai 980-8575, Japan
e-mail: y-sato@idac.tohoku.ac.jp

J. Dulak et al. (eds.), *Angiogenesis and Vascularisation*,
DOI 10.1007/978-3-7091-1428-5_12, © Springer-Verlag Wien 2013

12.1 Introduction

The balance between angiogenesis stimulators and inhibitors determines angiogenesis, and the body contains a number of endogenous angiogenesis stimulators and inhibitors. Most of angiogenesis stimulators are growth factors and cytokines, whereas angiogenesis inhibitors are variable, including hormones, chemokines, proteins accumulated in the extracellular matrix, proteolytic fragments of various proteins, and so forth. Those angiogenesis inhibitors are mainly extrinsic to the vasculature; some are constitutively expressed and act as barriers to prevent the invasion of neo-vessels, while the others are generated in response to stimuli and prevent this process. In addition, vascular cells may produce certain angiogenesis inhibitors during the course of angiogenesis [1].

12.2 Vasohibin-1

12.2.1 Isolation of VASH1

We performed microarray analysis to characterize genes whose expressions in endothelial cells (ECs) were modulated by the vascular endothelial growth factor (VEGF) stimulation [2]. Ninety-seven genes were induced more than twofold in ECs after 24 h incubation with VEGF. Among those VEGF-inducible 92 genes, we focused our attention on nine genes whose functions were uncharacterized. We applied functional assays for angiogenesis, isolated one gene having anti-angiogenic activity, and designated it as vasohibin-1 (VASH1) [3].

The gene for human *VASH1* gene is located on chromosome 14q24.3 and consists of seven exons. There are two isoforms of human VASH1: full length VASH1A and the spliced variant VASH1B (Fig. 12.1). Human VASH1A protein is composed of 365 amino acid residues, while human VASH1B protein is composed of 204 amino acid residues.

The VEGF-induced expression of VASH1 in ECs is mediated via VEGF receptor 2 and its downstream protein kinase C delta (PKCδ) [4]. VASH1 can also be induced by fibroblast growth factor 2 (FGF-2), another potent angiogenic factor [3, 4], and this induction is again mediated by PKCδ [4]. Accordingly, the principal signaling pathways for the induction of VASH1 by these two representative angiogenic growth factors considerably overlap. Importantly, this induction of VASH1 in ECs is absent under a hypoxic condition or in the presence of inflammatory cytokines, tumor necrosis factor α (TNFα), and interleukin 1 (IL-1) [4].

12.2.2 VASH1 in Pathophysiological Conditions

Immunohistochemical analysis has revealed that VASH1 protein is present in ECs in the developing human or mouse embryo but is reduced in expression in the post-neonate [5]. Nimmagadda et al. independently showed by in situ hybridization that

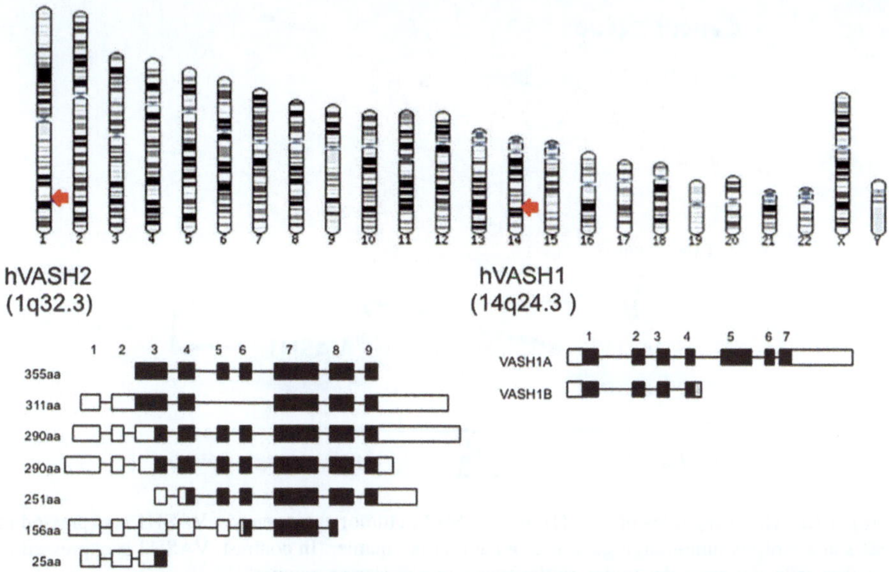

Fig. 12.1 *VASH1* and *VASH2* genes and their transcripts. Human *VASH1* gene is encoded in 14q24.3, while human *VASH2* gene is encoded in 1q32.3. There are multiple transcripts in both human VASH1 and VASH2. *Black squares* encode proteins

VASH1 mRNA is expressed in a wide range of tissues and organs in the chicken embryo and suggested that the expression of VASH1 might not be limited to ECs [6]. Nevertheless, our immunohistochemical analysis preferentially detects VASH1 protein in ECs at the site of angiogenesis [3, 5].

We further investigated the expression of VASH1 under various conditions accompanied by pathological angiogenesis and a related condition. The presence of VASH1 in ECs is evident in various cancers, atherosclerotic lesions, age-dependent macular degeneration (AMD), diabetic retinopathy, rheumatoid arthritis, and arterial re-endothelialization after denudation [7–19], and thus the expression of VASH1 in ECs can be a biomarker of angiogenesis. Alternatively, with regard to its function, increased tumor growth and tumor angiogenesis were evident when Lewis lung carcinoma (LCC) cells were inoculated into *VASH1*(−/−) mice [12]. Indeed, decreased expression of VASH1 correlated with poor prognosis of certain human cancers [19, 20]. These observations suggest that endogenous VASH1 negatively regulates the course of pathological conditions associated with angiogenesis (Fig. 12.2).

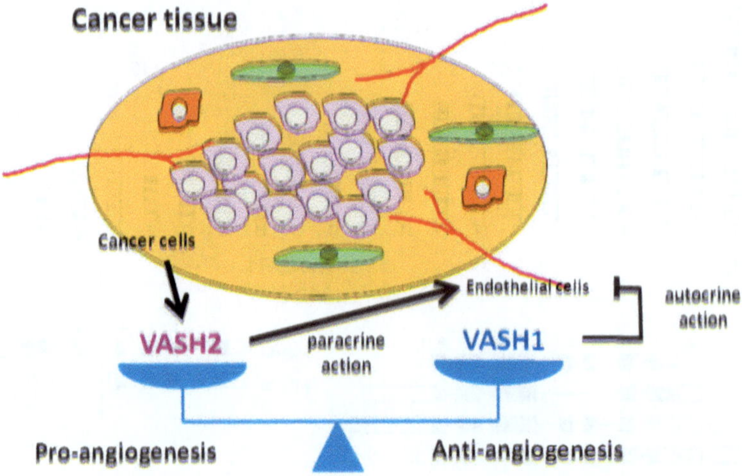

Fig. 12.2 Opposing roles of VASH1 and VASH2 in tumor angiogenesis. VASH1 is expressed in ECs and inhibits tumor angiogenesis in an autocrine manner. In contrast, VASH2 is expressed in cancer cells and promotes tumor angiogenesis in a paracrine manner

12.2.3 Role of VASH1 in Survival or Death of ECs

In general, angiogenesis stimulators promote EC survival, while angiogenesis inhibitors induce EC death [21]. As VASH1 is produced by ECs, it may act as a suicidal molecule for ECs. We examined the role of VASH1 in EC survival/death [22]. Interestingly, the knockdown of VASH1 induced premature senescence of ECs, and those ECs were easily killed by various cellular stresses. Alternatively, over-expression of VASH1 made ECs resistant to premature senescence and cell death caused by cellular stresses. Thus, as an angiogenesis inhibitor, VASH1 exhibits unique characteristics that increase stress tolerance and enhance survival of ECs. We then defined its underlying mechanism. What we have noticed is that VASH1 increases the expression of superoxide dismutase 2 (SOD2), an enzyme known to quench reactive oxygen species (ROS). Simultaneously, VASH1 augments the synthesis of sirtuin 1 (SIRT1), which improved stress tolerance of ECs [22]. This close relationship among VASH1, SOD2, and SIRT1 may suggest the protective role of VASH1 in the vascular system.

12.3 Vasohibin-2

12.3.1 Isolation of VASH2

By searching in the database, we found one gene homologous to VASH1, and designated it as vasohibin-2 (VASH2) [5]. The gene for human *VASH2* is located on chromosome 1q32.3. So far, nine exons for the *VASH2* gene have been shown in the database to form multiple transcripts owing to alternative splicing (Fig. 12.1). The overall homology between full length human VASH1 and VASH2 is 52.5 % at the amino acid level.

The recent database search reveals that sea squirt posses one common ancestry vasohibin gene whereas vertebrates have VASH1 and VASH2, and homology between this common ancestry gene and human VASH1 or human VASH2 is about 40 %, respectively. Moreover, amino acid sequences of vertebrate VASH1 and VASH2 are well conserved between species. Thus, we estimate that a common ancestry gene is divided into VASH1 and VASH2 during the evolution [23].

12.3.2 Opposing Role of VASH2 in Angiogenesis

To disclose the expression and function of VASH2, we examined its spatiotemporal expression and function during angiogenesis [24]. Our analysis using the mouse subcutaneous angiogenesis model revealed that while VASH1 is expressed not in ECs at the sprouting front but in newly formed blood vessels behind the sprouting front where angiogenesis terminates, VASH2 is expressed preferentially in mononuclear cells (MNCs) that are mobilized from the bone marrow and infiltrate the sprouting front. Furthermore, while *VASH1 KO* mice contain numerous immature microvessels in the area where angiogenesis should be terminated, *VASH2 KO* mice contain less neo-vessels in the sprouting front of angiogenesis [24]. These results indicate that while VASH1 is expressed in ECs in the termination zone of angiogenesis to terminate angiogenesis, VASH2 is mainly expressed in MNCs in the sprouting front and promotes angiogenesis (Fig. 12.2).

12.3.3 VASH2 in Pathophysiological Conditions

Our initial analysis revealed that VASH2 protein was also present in ECs in the developing human or mouse embryo and faded in the post-neonate [5]. However, when the expression of VASH2 was examined in postnatal angiogenesis, it was shown not in ECs but in infiltrating MNCs [24]. We extended our analysis to cancers thereafter, and showed the expression of VASH2 in cancer cells of ovarian serous adenocarcinoma [25] and hepatocellular carcinoma (HCC) [26]. This increased expression of VASH2 in cancer cells can be mediated by the decrease of mir-200b [25] or modulating the methylation of its promoter region [26]. Importantly, the knockdown of VASH2 expression in cancer cells prominently inhibited both tumor growth and tumor angiogenesis [25–27]. These results indicate that VASH2 is expressed in cancer cells and stimulates tumor angiogenesis in a paracrine manner (Fig. 12.2).

Concluding Remarks

This chapter has focused on the role of newly discovered VASH family members, VASH1 and VASH2, in angiogenesis. VASH1 and VASH2 are highly conserved in vertebrates, but their roles in the regulation of angiogenesis are distinct and may be contradictory. It seems that VASH1 is to terminate angiogenesis whereas VASH2 is to promote angiogenesis. Further studies are required to disclose the entire function of these unique family proteins in terms of their receptors and downstream signaling pathways.

References

1. Sato Y (2006) Update on endogenous inhibitors of angiogenesis. Endothelium 13:147–155
2. Abe M, Sato Y (2001) cDNA microarray analysis of the gene expression profile of VEGF-induced human umbilical vein endothelial cells. Angiogenesis 4:289–298
3. Watanabe K, Hasegawa Y, Yamashita H, Shimizu K, Ding Y, Abe M, Ohta H, Imagawa K, Hojo K, Maki H, Sonoda H, Sato Y (2004) Vasohibin as an endothelium-derived negative feedback regulator of angiogenesis. J Clin Invest 114:898–907
4. Shimizu K, Watanabe K, Yamashita H, Abe M, Yoshimatsu H, Ohta H, Sonoda H, Sato Y (2005) Gene regulation of a novel angiogenesis inhibitor, vasohibin, in endothelial cells. Biochem Biophys Res Commun 327:700–706
5. Shibuya T, Watanabe K, Yamashita H, Shimizu K, Miyashita H, Abe M, Moriya T, Ohta H, Sonoda H, Shimosegawa T, Tabayashi K, Sato Y (2006) Isolation and characterization of vasohibin-2 as a homologue of VEGF-inducible endothelium-derived angiogenesis inhibitor vasohibin. Arterioscler Thromb Vasc Biol 26:1051–1057
6. Nimmagadda S, Geetha-Loganathan P, Pröls F, Scaal M, Christ B, Huang R (2007) Expression pattern of vasohibin during chick development. Dev Dyn 236:1358–1362
7. Yamashita H, Abe M, Watanabe K, Shimizu K, Moriya T, Sato A, Satomi S, Ohta H, Sonoda H, Sato Y (2006) Vasohibin prevents arterial neointimal formation through angiogenesis inhibition. Biochem Biophys Res Commun 345:919–925
8. Yoshinaga K, Ito K, Moriya T, Nagase S, Takano T, Niikura H, Yaegashi N, Sato Y (2008) Expression of vasohibin as a novel endothelium-derived angiogenesis inhibitor in endometrial cancer. Cancer Sci 99:914–919
9. Wakusawa R, Abe T, Sato H, Yoshida M, Kunikata H, Sato Y, Nishida K (2008) Expression of vasohibin, an antiangiogenic factor, in human choroidal neovascular membranes. Am J Ophthalmol 146:235–243
10. Tamaki K, Moriya T, Sato Y, Ishida T, Maruo Y, Yoshinaga K, Ohuchi N, Sasano H (2008) Vasohibin-1 in human breast carcinoma: a potential negative feedback regulator of angiogenesis. Cancer Sci 100:88–94
11. Sato H, Abe T, Wakusawa R, Asai N, Kunikata H, Ohta H, Sonoda H, Sato Y, Nishida K (2009) Vitreous levels of vasohibin-1 and vascular endothelial growth factor in patients with proliferative diabetic retinopathy. Diabetologia 52:359–361
12. Hosaka T, Kimura H, Heishi T, Suzuki Y, Miyashita H, Ohta H, Sonoda H, Moriya T, Suzuki S, Kondo T, Sato Y (2009) Vasohibin-1 expressed in endothelium of tumor vessels regulates angiogenesis. Am J Pathol 175:430–439
13. Miyake K, Nishida K, Kadota Y, Yamasaki H, Nasu T, Saitou D, Tanabe K, Sonoda H, Sato Y, Maeshima Y, Makino H (2009) Inflammatory cytokine-induced expression of vasohibin-1 by rheumatoid synovial fibroblasts. Acta Med Okayama 63:349–358
14. Tamaki K, Sasano H, Maruo Y, Takahashi Y, Miyashita M, Moriya T, Sato Y, Hirakawa H, Tamaki N, Watanabe M, Ishida T, Ohuchi N (2010) Vasohibin-1 as a potential predictor of aggressive behavior of ductal carcinoma in situ of the breast. Cancer Sci 101:1051–1058

15. Yoshinaga K, Ito K, Moriya T, Nagase S, Takano T, Niikura H, Sasano H, Yaegashi N, Sato Y (2011) Roles of intrinsic angiogenesis inhibitor, vasohibin, in cervical carcinomas. Cancer Sci 102:446–451
16. Miyazaki Y, Kosaka T, Mikami S, Kikuchi E, Tanaka N, Maeda T, Ishida M, Miyajima A, Nakagawa K, Okada Y, Sato Y, Oya M (2012) The prognostic significance of vasohibin-1 expression in patients with upper urinary tract urothelial carcinoma. Clin Cancer Res 18:4145–4153
17. Bai X, Margariti A, Hu Y, Sato Y, Zeng L, Ivetic A, Habi O, Mason JC, Wang X, Xu Q (2010) PKCδ-deficiency accelerates neointimal lesion of mouse injured artery involving delayed reendothelialization and vasohibin-1 accumulation. Arterioscler Thromb Vasc Biol 30:2467–2474
18. Yazdani S, Miki Y, Tamaki K, Ono K, Iwabuchi E, Abe K, Suzuki T, Sato Y, Kondo T, Sasano H (2013) Proliferation and maturation of intratumoral blood vessels in non-small cell lung cancer. Hum Pathol 44:1586–1596
19. Kanomata N, Sato Y, Miyaji Y, Nagai A, Moriya T (2013) Vasohibin-1 is a new predictor of disease-free survival in operated renal cell carcinoma patients. J Clin Pathol 66:613–619
20. Lu C, Han HD, Mangala LS, Ali-Fehmi R, Newton CS, Ozbun L, Armaiz-Pena GN, Hu W, Stone RL, Munkarah A, Ravoori MK, Shahzad MM, Lee JW, Mora E, Langley RR, Carroll AR, Matsuo K, Spannuth WA, Schmandt R, Jennings NB, Goodman BW, Jaffe RB, Nick AM, Kim HS, Guven EO, Chen YH, Li LY, Hsu MC, Coleman RL, Calin GA, Denkbas EB, Lim JY, Lee JS, Kundra V, Birrer MJ, Hung MC, Lopez-Berestein G, Sood AK (2010) Regulation of tumor angiogenesis by EZH2. Cancer Cell 18:185–197
21. Chavakis E, Dimmeler S (2002) Regulation of endothelial cell survival and apoptosis during angiogenesis. Arterioscler Thromb Vasc Biol 22:887–893
22. Miyashita H, Watanabe T, Hayashi H, Suzuki Y, Nakamura T, Ito S, Ono M, Hoshikawa Y, Okada Y, Kondo T, Sato Y (2012) Angiogenesis inhibitor vasohibin-1 enhances stress resistance of endothelial cells via induction of SOD2 and SIRT1. PLoS One 7:e46459
23. Sato Y (2012) The vasohibin family: novel regulators of angiogenesis. Vascul Pharmacol 56:262–266
24. Kimura H, Miyashita H, Suzuki Y, Kobayashi M, Watanabe K, Sonoda H, Ohta H, Fujiwara T, Shimosegawa T, Sato Y (2010) Distinctive localization and opposed roles of vasohibin-1 and vasohibin-2 in the regulation of angiogenesis. Blood 113:4810–4818
25. Takahashi Y, Koyanagi T, Suzuki Y, Saga Y, Kanomata N, Moriya T, Suzuki M, Sato Y (2012) Vasohibin-2 expressed in human serous ovarian adenocarcinoma accelerates tumor growth by promoting angiogenesis. Mol Cancer Res 10:1135–1146
26. Xue X, Gao W, Sun B, Xu Y, Han B, Wang F, Zhang Y, Sun J, Wei J, Lu Z, Zhu Y, Sato Y, Sekido Y, Miao Y, Kondo Y (2013) Vasohibin 2 is transcriptionally activated and promotes angiogenesis in hepatocellular carcinoma. Oncogene 32:1724–1734
27. Koyanagi T, Saga Y, Takahashi Y, Suzuki Y, Suzuki M, Sato Y (2013) Downregulation of vasohibin-2, a novel angiogenesis regulator, suppresses tumor growth by inhibiting angiogenesis in endometrial cancer cells. Oncol Lett 5:1058–1062

Part III

Diseases and Therapy

Hypoxia-Induced Pathological Angiogenesis in Zebrafish

13

Lasse D. Jensen, Pegah Rouhi, and Yihai Cao

Contents

13.1 Introduction 272
13.2 Role of Hypoxia in Tumor Progression and Metastasis 275
13.3 Embryo Zebrafish Metastasis and Tumor Model 277
13.4 Hypoxia-Induced Angiogenic Retinopathy Model in Adult Zebrafish 280
13.5 Acute Hypoxia Models in Zebrafish and Glass Catfish 285
References 290

Abstract

Deregulated angiogenesis is a major underlying cause of many severe diseases including cancer, retinopathy, diabetes, myocardial infarction, and stroke. In these diseases, tissue hypoxia is the main cause of the pathological vascular phenotypes. While the mechanisms behind hypoxia-induced changes in cellular signaling have been extensively studied in vitro, much less is known regarding the effects of hypoxia in physiological or pathological settings in vivo. The highly hypoxia-tolerant zebrafish and glass catfish provide excellent systems for studying the effects of hypoxia on angiogenesis and vascular pathology in vertebrate disease models. Here we present and discuss the benefits and drawbacks in using zebrafish to study basic mechanisms of hypoxia in disease, with special emphasis on the role of angiogenesis and vascular function. Specifically, we will in detail discuss zebrafish models of hypoxia-induced

L.D. Jensen (✉) • Y. Cao (✉)
Department of Microbiology, Tumor and Cell Biology, Karolinska Institutet, 17177 Stockholm, Sweden

Department of Medicine and Health Sciences, Linköping University, 581 83 Linköping, Sweden
e-mail: lasse.jensen@liu.se; Yihai.cao@ki.se

P. Rouhi
Department of Microbiology, Tumor and Cell Biology, Karolinska Institutet, 17177 Stockholm, Sweden

J. Dulak et al. (eds.), *Angiogenesis and Vascularisation*,
DOI 10.1007/978-3-7091-1428-5_13, © Springer-Verlag Wien 2013

angiogenesis in the retina and tumor, as well as acute hypoxia models using glass catfish, and discuss the usefulness of these models to elucidate key mechanisms behind pathological vascular disruption in retinopathy and cancer. At the end of the chapter we contextualize the hypoxia-induced angiogenesis-mediated zebrafish disease models and discuss the perspectives in using zebrafish for medical research on hypoxia and angiogenesis.

Keywords
Angiogenesis • Zebrafish • Hypoxia • Vasculature • Cancer • Retinopathy

13.1 Introduction

Common and detrimental human diseases such as cardiovascular diseases—including arteriosclerosis, myocardial infarction and stroke, retinopathies—including retinopathy of prematurity, diabetic retinopathy (DR), and age-related macular degeneration (AMD), metabolic disorders—including obesity and diabetes, and malignancies—including solid and soft tissue tumor growth and metastasis are all highly dependent on pathological changes in blood vessels [1–3]. As such, angiogenesis, characterized as the growth of new blood vessels from an existing vasculature, is a key process for the progression of these diseases from an early, presymptomatic stage into clinical disease.

In cancer, for example, the now late Dr. Judah Folkman more than 40 years ago discovered that tumor growth is dependent on the ability of the tumor cells to switch on angiogenesis in order to ensure sufficient supply of oxygen and nutrients and thereby to sustain growth [4]. He suggested in a seminal paper in the New England Journal of Medicine that if tumor angiogenesis could be blocked therapeutically, tumors would remain dormant as small nodules and would not lead to disease [4]. Today 40 years later, this concept is now appreciated as a major strategy for development of the next generation of anticancer drugs and has spurred rigorous research in the science, now known as tumor angiogenesis.

Similarly, in retinopathies such as DR and AMD, angiogenesis marks a switch in the disease progression in patients. Diabetic or elderly patients with DR or AMD, respectively, do not suffer great vision impairments at so-called pre-proliferative, dry, or non-angiogenic states of the disease. However, once angiogenesis in the retina is switched on, the disease quickly progresses and the vision rapidly deteriorates. Again, it has been shown that blocking angiogenesis in these diseases is a highly effective course of treatment [5, 6]—at least in a transient setting, as the drugs used today may cause side effects or resistance during prolonged use.

In these and other angiogenesis-dependent diseases, tissue hypoxia such as in the tumor and the eye is the main inducer of pathological angiogenesis. Hypoxia arise either from an expansion of the tissue such as in tumor growth or obesity which render the existing vasculature insufficient for providing enough oxygen and nutrients. In other diseases such as in retinopathy or cardiovascular diseases, the impairment of blood flow in one or several blood vessels lead to reduced perfusion of oxygen and nutrients to a part of the tissue, which renders it hypoxic. Hypoxia does not only cause angiogenesis but also lead to a destabilization and inhibit maturation of the vasculature. This means that hypoxia-induced angiogenesis leads to formation of low quality blood vessels which have a fragile and under-developed vascular wall, are highly leaky and prone to rupture and thus often cause hemorrhages, and are not efficient in supporting circulation.

In recognition of the importance of tissue hypoxia—in combination with lack of blood flow also known as ischemia—in the induction of pathological angiogenesis, it has become an important task to identify hypoxia-induced factors which lead to angiogenesis and a pathological phenotype of the blood vessels. One such angiogenic factor, which is highly up-regulated by hypoxia, is vascular endothelial growth factor or VEGF [6, 7]. VEGF is directly and potently induced by the oxygen-sensitive transcription factor hypoxia-inducible factor (HIF)-1. HIF-1 consists of two subunits, -alfa and -beta, and especially the -alfa subunit is rapidly removed from cells under sufficient oxygenation. In response to hypoxia, however, the -alfa subunit is stabilized and HIF-1 is activated leading to a rapid increase in the production of VEGF [8]. This pathway has been a major focus for targeting hypoxia-induced angiogenesis in cancer and retinopathy in the past and still to this day. Currently, there are several FDA approved drugs available which neutralize VEGF, in clinical use to prevent progression of retinopathy and cancer [9]. However, these drugs are associated with severe problems and drawbacks. While anti-VEGF drugs are quite effective against retinopathy, they have to be administrated via invasive intravitreal injections on a regular basis. These injections are associated with severe administration-related side effects such as cataracts [10]. Also these drugs are so expensive that only a minority of patients with these diseases is able to afford long-term treatment. Systemic treatment with anti-VEGF drugs in cancer patients are associated with severe side effects including bone-marrow suppression, hypertension, and hypothyroidism [11, 12]. Due to these drawbacks, treatment is usually given in intervals, with a resting period of no treatment between periods of treatment. However, during these resting periods, tumor vessels rapidly regrow, and the effect of anti-VEGF treatment is therefore short-lasting. Many retinopathy and cancer patients furthermore develop resistance to anti-VEGF treatment, or are from the very beginning intrinsically resistant, which furthermore limit the use of these drugs [13].

Today it is of great importance to identify other, important pathways of hypoxia-induced angiogenesis which may be targeted for disease treatment/prevention. Common disease models that are used to generate knowledge of such pathways are however difficult to manipulate in the sense of regulating the level, location, and duration of hypoxia in the tissue, and it is therefore difficult to determine causal effects of hypoxia-induced pathways in disease. For example, in the classical tumor models, in which tumors are implanted under the skin in mice, alternatively in an organ, or develop spontaneously, hypoxia within the tumor cannot be regulated. Researchers are only able to correlate the existence of hypoxia with that of certain factors, but it is not possible to determine whether the particular factor is related to hypoxia or regulated by an independent pathway. These problems have meant that sophisticated studies, which are needed to delineate mechanistic consequences of hypoxia on the vasculature, have been very difficult to do in rodents—and other methods are therefore required.

Zebrafish have emerged as a highly versatile model organism for human diseases including diseases of the vasculature. The zebrafish embryos are transparent, develop very rapidly, are produced in large numbers, develop outside the mother in regular tap water or defined media, and are highly amenable to both genetic and pharmacologic manipulation of protein functions [14]. These characteristics have prompted the substantial interest in this animal as a system to study developmental aspects of blood vessel growth and to delineate mechanisms of vascular disease in molecular detail, although in a living organism [15]. Recently, its use has been further extended as several highly relevant disease models have been generated which progress in a highly similar way as what is observed in the clinic. In particular, zebrafish are highly tolerant to hypoxia, and they do well in water with as little as only 10 % of the oxygen levels found in air-equilibrated water [3, 6]. This tolerability enables studies on how tissues respond to hypoxia in vivo. For example, water oxygen levels can quickly and accurately be modified, and as the oxygen levels in the water determine the oxygen level in the tissue within the zebrafish, tissue hypoxia can be quickly and accurately manipulated in this model. Tissue hypoxia in this respect occurs while blood flow is uninterrupted, leading to no changes in the mechanical stimulation of the endothelium. This is potentially important as it is known that the lack of flow itself lead to pronounced changes in the factors being generated by the endothelium, leading to its destabilization and eventual degeneration—which may not be hypoxia-dependent processes per se [16]. Thus, this model can be used to answer questions regarding what happens at intermediate and severe hypoxia in the tissue, respectively, as well as the dynamic aspects of the vascular and tissue changes in response to hypoxia and reperfusion.

As the transparency of the zebrafish embryos enables researchers to study changes within the living embryo in real time as they happen, it is of great value to also have a model system which remains transparent as adults. Such transparent adult fish would potentially be very important for studying cell movements and functions in diseases which are known to develop later in life including immunological disorders, metabolic disorders, cardiovascular disorders, and cancer. Zebrafish unfortunately lose transparency as adults, and although non-pigmented zebrafish strains exist, where the tissue is more light permeable than in wild-type zebrafish, even these strains are not truly transparent and cannot be used for high-resolution imaging of single-cell behavior in the adult fish. However, there are a different fish species, *Kryptopterus bicirrhis* or the glass catfish, which remain completely transparent throughout their lifetime—at least in the majority of the tissues. These fish are therefore very well suited for studying dynamic aspects of hypoxia in an adult setting.

There are, however, also several drawbacks associated with zebrafish and glass catfish models of angiogenesis that researchers have to keep in mind when designing experiments or evaluating results based on this system. First, due to a genome duplication, which took place after the segregation of the teleost-lineage from the mammalian lineage, zebrafish often have duplicate genes for each gene found in mammals. As such, zebrafish, for example, have two genes for VEGF-A, two genes for VEGFR-2, and two genes for HIF1α, just to mention a few relevant examples. Sometimes only one of these genes is associated with the effects we can observe from the mammalian gene product, but sometimes the genes exhibit redundant functions and both have to be silenced in order to achieve a similar phenotype as in a knockout mouse, for example. Second some genetic and molecular tools are less developed for zebrafish compared to mice, including (1) specific antibodies for staining or neutralization of various proteins and (2) genetic recombination with engineered DNA fragments to produce, for example, LoxP-flanked genes used to achieve inducible or cell-type-specific knockout animals. However, the lack of antibodies, which may be used for staining or neutralization of zebrafish proteins involved in angiogenesis, has to a significant extent been overcome by the generation of transgenic lines which express fluorescent proteins under promoters of vascular-specific genes including *fli1a* or *vegfr2*. Also the problem of generating inducible or tissue-specific knockout strains in zebrafish has recently been solved with the development of the CRISPR-Cas technology. This technology is, however, new and it still remains to be seen if it will live up to the high expectations of the zebrafish community in the future.

In this chapter we will discuss the use of zebrafish and glass catfish in the study of hypoxia on angiogenesis and vascular biology by highlighting three conceptual examples of how such studies can be done. In the first part of the chapter, we will discuss how hypoxia influence the zebrafish embryo and in particular how hypoxia modulates tumor angiogenesis and metastatic dissemination in a recently developed embryonic zebrafish tumor xenograft model. In the second part we will show how hypoxia induces angiogenesis in the retina of adult zebrafish and discuss applications of this technique in the study of retinopathy. In the third part of the chapter we will discuss acute hypoxia models in adult zebrafish and glass catfish and how such models may give valuable insights into early effects of hypoxia on the blood vessels which may be important for early pathological progression in, for example, ischemic diseases. We will conclude the chapter with an outlook on how we envision zebrafish and glass catfish models could be used to continue to generate novel and highly clinically relevant data which may lead to an increased understanding of how hypoxia modulate disease processes in humans.

13.2 Role of Hypoxia in Tumor Progression and Metastasis

Tumor progression from benign to metastatic disease is a multistep process that consists of tumor growth and recruitment of tumor blood vessels, local tissue invasion, intravasation into peri- and intra-tumoral blood vessels, survival in circulation, extravasation, and regrowth at the metastatic site. Growth of the solid tumor at the primary site or at the metastatic sites to a macroscopic size leads to reduction of available oxygen for the highly metabolically active tumor cells inside, and also those surrounding, the tumor [17]. Consumption of oxygen through accelerated and irregular cell proliferation in the tumor causes hypoxia that leads to activation of hypoxia-inducible transcription factors (HIFs) and mTOR and UPR pathways [18]. Reduction of oxygen levels in the cellular environment in combination with genetic or other environmental factors drives tumor cells to adopt aerobic fermentation, to express a multitude of angiogenic factors, and to elevated production of intracellular reactive oxygen species. These hypoxia-induced changes within the tumor cells lead to further accelerated growth and destabilization of tumor blood vessels, a more mutagenic environment inside the tumor cells and to switch on a more motile and less differentiated phenotype known as the epithelial-to-mesenchymal transition (EMT), which is prerequisite for the invasive and metastatic behavior of the cells [19].

Tumor cells in general, even in presence of oxygen, prefer aerobic glycolysis as a means of energy production. Many tumor types exhibit such sustained glycolysis, presumably due to the loss of tumor suppressor genes and activation of oncogenes which lead to such a pathologic metabolic profile in the tumor cells [20]. This phenomenon is also known as the Warburg effect after Otto Warburg who was one of the first to describe this phenomenon more than half a century ago [21]. As the Warburg effect render tumor cell metabolism highly inefficient, tumor cells require much higher levels of sugar for their survival. Also, this leads to production of large amounts of toxic metabolic bi-products including reactive oxygen species (ROS) and acidosis. Moderate increase in level of ROS is beneficial for cells and induces cell proliferation and differentiation, but excessive ROS can damage membranes, lipids, proteins, DNA, and also can be mutagenic. It is believed that cancer cells handle the oxidative stress in a different way than normal cells and therefore ROS will be accumulated in the cancer cell environment. ROS also stabilizes HIF-1, which mean that the Warburg effect leads to a pseudo-hypoxic environment in the tumor, where HIF-1, even in normoxia, activates a plethora of genes that are involved in anaerobic metabolism, survival, angiogenesis, and invasion [19].

Intratumoral endothelial cells also react strongly to hypoxia. Hypoxia-induced HIF-1 leads to increased levels of VEGF which destabilize blood vessels by peeling off vascular mural cells (pericytes), inducing fenestrations (holes) in the endothelium, activates the endothelial cells to assume a migratory phenotype, and induces degradation of the basement membrane. These pathological vascular characteristics aid tumor invasion and dissemination of tumor cells and is crucial for successful metastatic colonization in distal organs. This hypoxia-induced angiogenesis is mounted by the body as a defense against hypoxia, but in tumors, the induction of angiogenesis is excessive and uncontrolled, leading to formation of a very high density of irregular and nonfunctional blood vessels of poor structural integrity, which give rise to tumor edema and hemorrhages, high intratumoral fluid pressure, and reduced perfusion. Thus, as it is also the case in ischemic disorders and in retinopathies, the hypoxia-induced tumor angiogenesis paradoxically lead to further elevated tumor hypoxia, thereby locking the tumors in a futile circle, which drives disease progression.

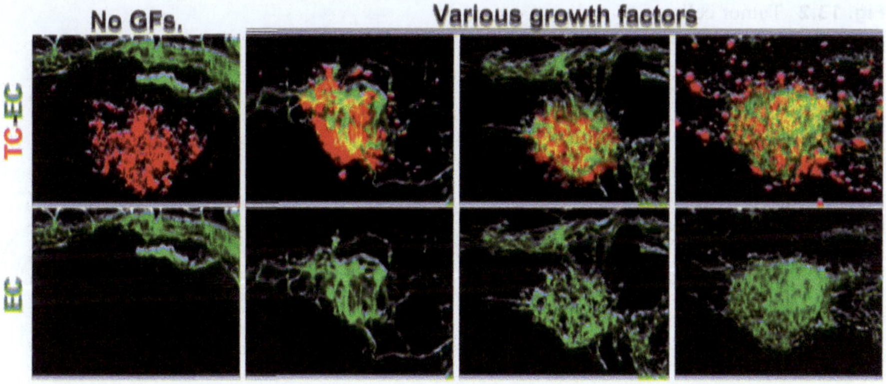

Fig. 13.1 Mammalian tumor cells expressing different angiogenic factors are labeled in vitro with DiI (*red color*) and transplanted into the perivitelline space of 2-day-old fli1:EGFP transgenic zebrafish embryos in which the blood vessels are labeled with EGFP (*green color*). Three days after implantation, it is clear that the tumor-produced angiogenic factors induce different morphologies of the tumor vessels, depending on the factor that is over-expressed by the tumor cells. *TC* tumor cells, *EC* endothelial cells

13.3 Embryo Zebrafish Metastasis and Tumor Model

Tumor cell dissemination, intravasation, extravasation, formation of metastatic niches, and regrowth to clinically detectable metastatic lesions are different steps of the metastatic cascade. Tumor cell invasion into their neighboring tissues and into circulation is considered the most critical step in the metastatic cascade, as this is the foundation for systemic spread in the organism. By inhibiting this process, it would be possible to block metastasis all together. Most studies on tumor metastasis, however, has been and are still done in mice, in which either metastatic seeds from a primary tumor mass or tumor cells injected intravenously colonize the liver or the lung and are evaluated only after they have grown to a relatively large size that can be detected visually, i.e., several million cells in one mass. Small micrometastatic nodules cannot be investigated in these models, as available imaging techniques for rodents are not sufficiently sensitive to detect small amount of cells [1, 2, 7]. Therefore these models are not useful to study what happens during early metastatic progression or how to possibly interfere with these processes. Zebrafish, however, represent a good candidate organism to study initial tumor cell dissemination and invasion [1, 2, 7]. Transparency of the embryos during early days of development allows us to look through the body and trace any fluorescently labeled implanted tumor cells through a fluorescent or confocal microscope in living embryos in a noninvasive manner (Fig. 13.1). Zebrafish embryos lack an adaptive immune system and therefore cells from any origin can be implanted without risk of rejection [7]. In early days of development, embryos absorb oxygen by penetration through their skin and also in a similar way, they will take up water-soluble drugs and compounds, e.g., Sunitinib [1]. Actually, because cutaneous oxygen absorption is thought to be the most important means of tissue

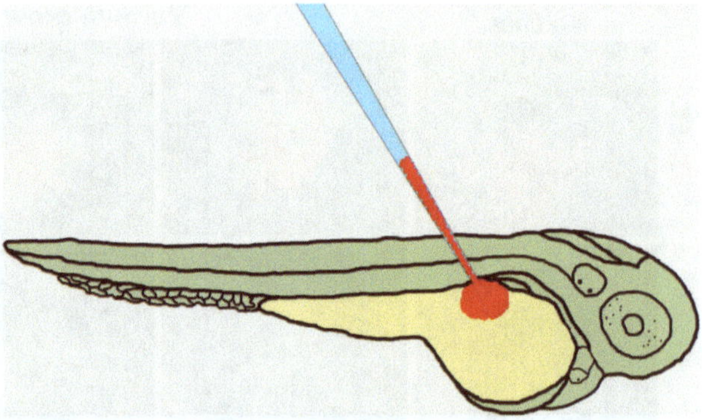

Fig. 13.2 Tumor cells labeled with the red fluorescent dye DiI in vitro are carefully injected into the perivitteline space of 2-day-old zebrafish embryos. Usually 100–1,000 tumor cells are injected in 1–5 nL of medium per zebrafish embryo

oxygenation in zebrafish embryos, regulation of tissue hypoxia in this model is as simple as regulating the oxygen level of the water. Also the passive uptake of drugs over the skin or gills facilitate equal delivery of drugs to all the zebrafish embryos independent of size, intake, or blood flow simply by adding such drugs to the water [1, 2]. Thus, zebrafish embryos are highly amenable to studies on hypoxia as well as pharmacological intervention. As the embryos are small, and develop without problems inside a 96-well plate, this system has previously been used for medium throughput screens of compound libraries which has led to the identification of new drugs that potentially correct a pathological phenotype such as congenital vascular malformations [22]. Such screens could also be performed in the future to identify novel compounds that inhibit tumor angiogenesis and -metastasis. This is a unique benefit of zebrafish embryos, which undoubtedly will be further developed and exploited in the future.

The zebrafish embryo model allows us to study tumor cell dissemination and invasion in a completely noninvasive fashion. Fluorescently labeled tumor cells are implanted into the perivitelline space of 48-h-old zebrafish embryos (Fig. 13.2). Once cells are implanted in this cavity, they form a primary tumor and subsequently start to invade the nearby tissues and metastasize to more distal parts of the fish body, e.g., trunk, head, and tail. Invading tumor cells intravasate into the blood vessels which lie in the vicinity of the primary tumor and move away from the site and stay in circulation until they find a new location to extravasate and form a metastatic lesion [2, 7].

The existing vasculature of the host as well as the tumor-induced angiogenesis can be affected by different growth factors and cytokines expressed and secreted by tumor cells (Fig. 13.1). Leaky and tortuous vessels are convenient routes for intravasation, since tumor cells can easily cross the endothelium of such vessels to gain access to the circulation.

In order to study the effects of different genes on invasiveness and metastatic abilities of cells, different kinds of tumor cell manipulation, e.g., knock down or knock in of the specific gene can be done using different tools, e.g., siRNA, shRNA, mRNA, or expression vectors. These manipulated cells can be then implanted into the embryo, and the effect of the gene on the dissemination and invasion pattern of the cells can be easily studied. On the other hand, blocking gene expression in the host can be easily done using "morpholinos" which are oligomers used to block base pairing regions of RNA and subsequently impairing protein expression. Also, gene expression in the zebrafish can be driven by injection of mRNA or expression vectors [15], which therefore allow both positive and negative regulation of any factor in either the tumor cells or the host. This makes molecular studies on tumor–host interactions much more feasible in this model.

In our attempt to study the angiogenic effects of various cell lines and also to trace disseminating single cells and foci, we use the Tg(fli1:EGFP) zebrafish line, in which the zebrafish blood vessels are labeled with green fluorescent protein. Transparency of zebrafish embryos allows us to easily detect tumor cells disseminating through the vasculature with the help of a fluorescent microscope. To visualize tumor cells, they should therefore also transmit a fluorescent signal. Cells can be either labeled permanently to express a fluorescent protein or they can be labeled with fluorescent dye, such as DiI, which emit red light, shortly prior to implantation (Fig. 13.1).

It is preferred that tumors are implanted into a non-vascularized region in the perivitelline space. In this way tumor-induced angiogenesis and the morphology of tumor blood vessels in different types of tumor models can be studied (Fig. 13.1). Since tumor cells need blood vessels to metastasize to distal parts of body, the hematologous dissemination of tumor cells can be evaluated by counting the number of cells in, for example, the posterior, ventral venous plexus.

Zebrafish embryos will be dechorionated shortly before implantation. They will be anesthetized and transferred onto a modified gel to simplify the implantation process. Tumor cells will be implanted into embryos with a micromanipulator which is connected to a microinjector that supplies pressure for injections (Fig. 13.2). Embryos may subsequently be treated differently according to the purpose of the study. In case of drug treatment under hypoxic and/or normoxic conditions, water-soluble drugs will be added to the house-keeping water to which the tumor-bearing embryos will be exposed during the experiment.

If acute effects of hypoxia on migratory properties of a specific cell line are the particular purpose of the study, then transplanted embryos will be transferred into hypoxic water right after postimplantation screening (Fig. 13.3). Oxygen levels in the hypoxic chamber/aquarium are tightly regulated with a sensor located inside the chamber at the bottom. The sensor is connected to a nitrogen capsule which blows in nitrogen according to the appropriate oxygen level (Fig. 13.3). But if tumor cell dissemination and invasion of is going to be studied under normoxic conditions, embryos implanted with tumor cells will be transferred into the aquarium/chamber perfused with air instead of nitrogen. In all conditions above fish embryos will be kept at 28 °C and will be investigated according to the experimental design.

In order to investigate each embryo thoroughly and individually, they should be kept in a separate well of a multi-well plate throughout the experiment. Labeling cells with fluorescent dye allow us to follow and investigate each single tumor cell in vivo in a completely noninvasive fashion using a fluorescent microscope. Green fluorescent color endothelial cells in the fli1:EGFP transgenic strain allow us to see

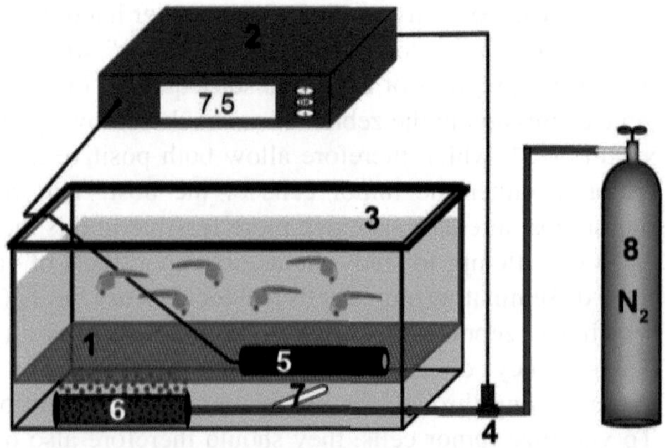

Fig. 13.3 Representation of the hypoxia setup used with both embryonic and adult zebrafish, but shown for zebrafish embryos. (1) Net, (2) Oxygen controller, (3) Hypoxia aquarium, (4) Valve, (5) Oxygen electrode, (6) Air-diffuser, (7) Magnetic stirrer, (8) Nitrogen capsule. Figure adapted from [2]

the effect of various growth factors expressed from tumor cells on the native vasculature and also investigate the consequence of various treatments on the vasculature (Fig. 13.1).

In order to get more detailed, three dimensional information on the structure of the tumor vasculature and the tumor cell-endothelial cell interactions during intravasation and extravasation, confocal imaging can be performed either during the experiment or on euthanized embryos mounted in so called viewing chambers, which allow imaging with high-power magnification (Fig. 13.1). By the use of two photon confocal microscopy on tumor baring embryos which develop in a soft agarose gel with anesthetic so they are sleeping and cannot move away during the investigations, it is possible to take time-lapse videos of how tumor cells crawl over the endothelium and into circulation, and how this process is modulated by tissue hypoxia and/or certain drugs added to the medium. This method thus enables the capture of highly dynamic images or movies in single-cell resolution, which is impossible in most other metastasis models. Such studies may potentially provide a much more detailed investigation of the crucial early metastatic process using this zebrafish model compared to the mouse models; the latter, however, may be more suitable for studying late phases of the metastatic process such as regrowth into clinical detectable metastatic lesions.

13.4 Hypoxia-Induced Angiogenic Retinopathy Model in Adult Zebrafish

Retinopathies including retinopathy of pre-maturity (ROP), diabetic retinopathy (DR) and age-related macular degeneration (AMD) are the leading causes of vision loss in infants, long-term diabetics and the elderly respectively [10, 23]. The sudden loss of vision is a major debilitating factor which greatly impacts on quality of life in the patients as well as on socio-economic stress in the community. Collectively there are millions of people world-wide affected by these diseases, and they are therefore considered to be a severe public health issue [10, 23]. Furthermore, due to the change in lifestyle in the western countries over the last 50–100 years, as well as the ongoing change in life-style coupled to improved economic status in many Asian countries, the incidence of diabetic people are expected to increase, and as a consequence also the incidence of patients with DR are expected to go up. Also, due to recent and projected advances in medicine, the average lifetime and the amount of people reaching old age is expected to increase in the future, and thereby also the incidence of people with AMD. Thus, from a public health perspective, there is an urgent need to understand the development of and find better treatment options for these diseases.

The disease-progression in ROP, DR and AMD employs a presymptomatic phase in which the disease slowly progresses in silence without causing any major discomfort to the patient [10, 23]. Therefore, it is rare that patients are admitted to ophthalmology clinics at presymptomatic stages. In the case of ROP, very premature infants are sometimes put in hyperoxia in order to aid the oxygen uptake of the infant, which otherwise would be limited due to an underdeveloped heart and lungs. This is important to secure proper development. However, it may also lead to a regression of infant hyaloid vessels and inhibited growth of the retinal vasculature [24]. Upon transfer to a normoxic environment, the under-developed retinal vessels are not sufficient to supply the retina with the oxygen needed, and retinal hypoxia ensues which induce a rapid and VEGF-dependent angiogenic response leading to the formation of a high density immature retinal vasculature that is highly leaky and prone to rupture [24]. In later stages, multiple microhemorrhages, pronounced retinal edema and scar formation lead to impaired vision, blood–retinal barrier breakdown, neuronal loss, retinal detachment from the retinal pigment epithelium and ultimately blindness [24].

In diabetic retinopathy, prolonged high blood glucose levels lead to irritation of the retinal microvessels which become obstructed and/or perforated and leaky which give rise to retinal exudates, hemorrhages and edema. As retinal edema presses back onto the blood vessels, and as obstructed flow and hemorrhages disrupt the vasculature, this lead to poor perfusion of blood in the retina and retinal hypoxia. Eventually the disease goes through the same steps as described above for ROP, with the end result that patients completely lose vision unless measures are taken to alleviate the symptoms of the disease before it goes that far [23].

In age-related macular degeneration, fat and macromolecular debris accumulate as small depots between the retina and Brucks membrane, structures known as drusen. Drusen are present in practically everyone above the age of 60, but if the number and size is high, it will cause obstruction of choroidal circulation and create hypoxic areas in the back of the retina. Again, hypoxia will lead to angiogenesis as the disease progress from a “dry” to a “wet” stage with the formation of poorly organized and low-quality blood vessels, which are highly leaky and causing damage to the retina. Ensuing edema, microhemorrhages etc. will eventually lead to a disease progression similar to that described for DR and ROP, and ultimately to blindness if the disease remain untreated [10].

Thus, in all three cases of retinopathy described above, the main switch from a presymptomatic to a rapidly progressing disease is the event of retinal hypoxia and

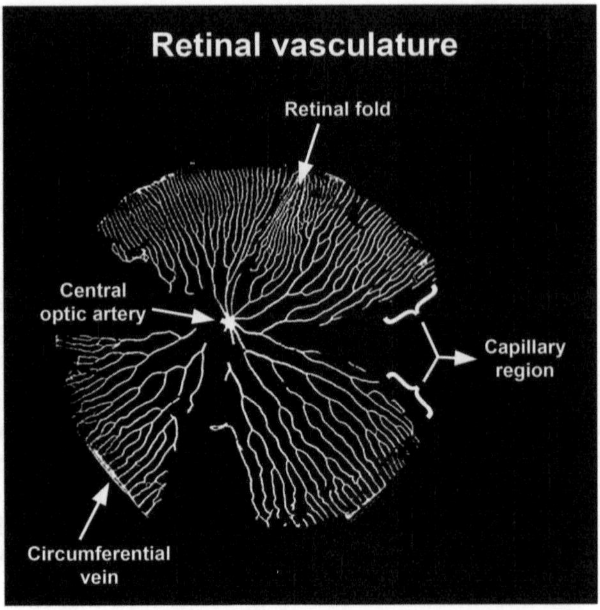

Fig. 13.4 Retinal vasculature in flat-mounted retinas of adult fli1:EGFP transgenic fish shown in *white*. The position of the central optic artery, retinal capillaries, retinal fold, and circumferential vein has been indicated by *white arrows*. Image adapted from [6]

hypoxia-induced angiogenesis. Hypoxia is a complicated physiological phenomenon and therefore most researchers studying retinal hypoxia employ models in which blood vessels are ligated or burned by chemicals or lasers. This creates an environment of severe inflammation in combination with hypoxia, lack of blood perfusion, acidosis, neuronal cell death, and multiple other physiological phenomena which in their own right give rise to complex signaling mechanisms. Therefore, the commonly used mammalian models are not able to uncouple the effects of hypoxia in the retina from the spectrum of other responses that are also induced by burning or blocking retinal blood flow. In order to create a model in which the effects of retinal hypoxia can be isolated and studied in detail, we have taken advantage of the hypoxia-tolerant adult zebrafish, which can be exposed to as little as 10 % of their normal oxygen level in the water for extended periods of time [6]. We constructed a hypoxia setup, which highly resembles the setup for exposing zebrafish embryos to hypoxia (Fig. 13.3), in which nitrogen gas was perfused into the fish aquarium under regulation of a hypoxia-controller. When the water oxygen levels dropped below a preset value, the nitrogen gas flow was shut off, and the oxygen levels would very slowly increase as an effect of atmospheric air leaking into the aquarium. As the water oxygen levels increased above a slightly higher preset value, the nitrogen gas flow would be re-enabled and thus the oxygen level in the water would be kept between the two preset levels, in our experiments between 9.5 and 10 % relative air saturation (Fig. 13.3) [6]. The fish would stay under these conditions for 6–14 days which was well tolerated by the fish.

Upon first examination, we were amazed to find that the retinal vasculature of adult zebrafish is exceptionally simple in structure (Fig. 13.4) and therefore very well suited for studies on angiogenesis. The retinal vasculature exhibits a very clear organization in which a central optic artery, emerging from the back of the eye where it is running alongside the optic nerve, branches out at the center of the optic disc. These arterial branches run from the center towards the periphery of the retina, branching a few times along the way to create a tree-shaped retinal vasculature that is highly organized, with smooth vessels without sprouts, only few branches, and increasing density as the vasculature approaches the circumferential vein (Fig. 13.4). Approximately 250 μm from the vein, the vessels become thinner, more sensitive to hypoxia-stimulation, and more leaky than in other parts of the retina. We have therefore characterized these vessels as retinal capillaries, as they respond in very much the same way as human and mouse retinal capillaries to vascular stressors including hypoxia. These capillaries empty into the circumferential vein, which is the only vein in the retinal vasculature and which carries the blood away from the retina and back into circulation (Fig. 13.4). Thus, blood flow in the adult zebrafish retina is unidirectional coming from the center and running to the periphery, and the mere location of the vessels is sufficient to predict whether they are arterial, capillary, or venous in nature—a level of organization that is not shared in mammals, in which arteries, capillaries, and veins are intermingled to a much higher extent.

The highly organized nature of these blood vessels makes it very simple to identify changes in retinal vascular parameters such as sprouting, branching, leakage, and regression. We found that upon exposure to hypoxia, the retinal vasculature responds by sprouting specifically in the capillary region but not in the venous or arterial regions (Fig. 13.5). Sprouting and other capillary phenomenon occur early—already after 3 days in hypoxia. However, if the fish are maintained in hypoxic water for prolonged periods, we found that these sprouts progresses to form branches, which support blood flow and greatly increase the vascular density in the retina over the course of 7–14 days, after which time the angiogenic response plateaus [3, 6]. These findings are highly interesting as it is the first time hypoxia-induced retinal angiogenesis has been studied as an isolated event, uncoupled from inflammation, tissue damage, or lack of blood flow.

We further characterized the level of hypoxia, which was required to induce retinal angiogenesis in this model. As fish are exceptionally tolerant to hypoxia, it was no surprise that the water oxygen level had to be decreased greatly—to only 10 % of that in fully oxygenated water—in order to induce branching in the retina. This level of hypoxia is however well tolerated in healthy and well-fed zebrafish and poses no threat to the well-being of the fish. However, if also a drastic drop in tissue oxygenation is required for retinal angiogenesis in mammals or whether it would also occur in the moderately hypoxic retina is not known. Inferring from knowledge about the disease progression of retinopathies, it is likely, however, that highly hypoxic areas are needed for inducing angiogenesis. In AMD, for example, a significant amount of patients progress to clinical disease states in absence of angiogenesis [10]. This type of AMD is classified as dry AMD, and the disease is mainly characterized by many, large drusen in the macula, which induce widespread ischemia (global ischemia) but for some reason fail to induce angiogenesis.

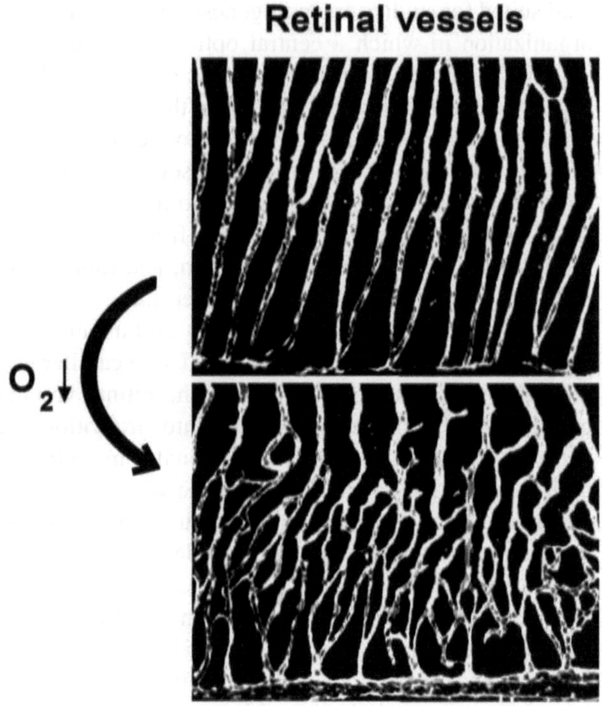

Fig. 13.5 Retinal vessels observed in flat-mounted retinas of adult fli:EGFP zebrafish after exposure to fully air-saturated water (*upper panel*) or water with 10 % relative air saturation (*lower panel*) for 6 days. Vessels are shown in *white*. It is clear that dramatic angiogenic expansion is observed in the hypoxia-exposed fish in the capillary region specifically. Image adapted from [6]

Thus, it is likely that even though blood perfusion in these dry-AMD cases are severely disrupted by the drusen, and this lead to hypoxia-induced apoptosis in retinal neurons and therefore vision loss, the level of hypoxia may not be sufficient to induce angiogenesis. In most AMD and DR patients, however, the disease progresses from an ischemic state to an angiogenic/proliferative/wet state in which the very low levels of oxygen in the retina evoke a massive angiogenic response—similar to what is found in zebrafish after approximately 1–2 weeks of hypoxia-treatment.

We have found that blockade of the Notch signaling pathway achieved by adding the chemical inhibitor of gamma-secretase, DAPT, to the hypoxic water during incubation of the fish led to extensive sprouting of arterial vessels. This finding is interesting as neither hypoxia nor DAPT treatment under normoxia was sufficient to induce sprouting in this vascular bed by itself [6]. Hypoxia signaling therefore exhibits a cross talk with the Notch pathway to induce destabilization of the arterial vessels in the retina which then become sensitive to react to hypoxia and start sprouting. As previously described in other models, DAPT treatment under normoxia also induce marked sprouting, but only in the capillary bed, in which the vessels are presumably more sensitive to stimulation [6]. This inherent stimulant in the retina is, however, not sufficient to induce angiogenesis in the arteries when the fish are not also exposed to hypoxia. These findings extend into considerations on how to take advantage of such signaling pathways to promote arteriogenesis—the specific growth of arteries in tissues—to treat ischemic diseases of the heart and brain, such as myocardial infarction and stroke. In these diseases the tissue is rendered severely hypoxic by a block in arterial blood perfusion. This hypoxic tissue does not, however, induce potent arteriogenesis by itself to alleviate the tissue hypoxia, and the tissue starts dying. Many attempts have been made to try and speed up or induce the arteriogenic process in these diseases, but without much luck. Most examinations have focused on evaluating the use of factors that are already induced by hypoxia—the most studied of which being VEGF. However, and perhaps not surprisingly, such interventions do not help, at least not in the clinical setting. We envision that perhaps an initial destabilization of the surrounding (also known as collateral) arterial network around the ischemic infarct would render arteries in particular more sensitive to respond to the hypoxia-induced VEGF and other factors and start growing into the tissue to alleviate the hypoxia. This somewhat provocative notion needs further preclinical testing before it can be tried in human patients.

In conclusion, the retinal-angiogenesis model in zebrafish has proved highly valuable to elucidate specific mechanisms involved in hypoxia-induced angiogenesis in the retina, which are important for progression of DR, AMD, and ROP to clinical stages. We believe that also in the future, this model is going to have great potential for delineating novel pathways involved in this pathologic response, as well as a tool for discovering general principles governing vessel growth, regression, and function, which may be used in other diseases in which hypoxia and angiogenesis (or the lack thereof) play a major role.

13.5 Acute Hypoxia Models in Zebrafish and Glass Catfish

The species *K. bicirrhis* (Glass or Ghost catfish), which is endemic to south East Asia, has a naturally transparent body in which all tissues can be readily observed (Fig. 13.6). This fish species has therefore been used previously in studies on peripheral nerve activity, on the dynamics of neuromuscular junctions, and on how hypoxia affects blood and lymphatic vessels, areas of research in which the transparency of the peripheral tissues has been crucial. Whereas embryos and early stage larvae from zebrafish and other fish species are highly exploited due to their transparent nature; these fish start losing transparency already at 2–3 days of age unless pigmentation is inhibited with chemicals, which themselves may affect the development of the fish. On the other hand, the glass catfish remain transparent throughout their lifespan, in a nongenetically manipulated setting, which is a unique quality of this species (Fig. 13.6). Furthermore, the fish has adopted a play dead mechanism of defense which means that when taken from its natural environment and into for example a petri dish for observation under the microscope, the fish will lie completely still on its side for hours, without the need for using anesthesia. As many drugs used for anesthesia—which often in other fish species as well as in rodents are only used to keep the animal still during investigations—also affect neuronal activity and vascular tone, it is expected that non-anesthetized animals

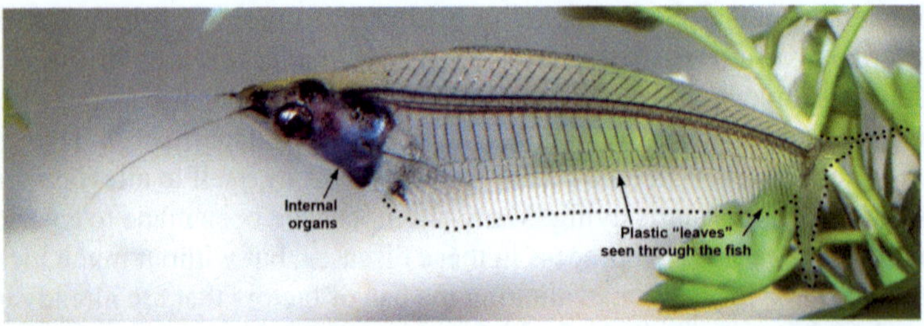

Fig. 13.6 A photograph of a glass catfish (*Kryptopterus bicirrhis*) located in front of a *green* plastic "plant." Observe that the "leaves" can be visualized just as clearly through the fish as next to the fish. *Black dots* have been placed around the otherwise nearly invisible anal and caudal fins of the fish in order to denote their location. The position of the internal organs within the *silvery* pouch just posterior to the head has been marked with a *black arrow*. Image adapted from [25]

would provide a more clean system in which the experimental conditions are not confounded by possible side effects of the anesthetics.

Acute hypoxia is a major pathologic factor in vascular diseases including myocardial infarction and stroke. As a main initial response to hypoxia, the organism tries to increase blood flow and oxygen carrying ability to the tissue in several ways. While most studies on hypoxia in disease has investigated the long-term effects of hypoxia such as how it affects angiogenesis and red blood cell formation, the main pathological consequence in cardiovascular diseases are much more acute and arise from the immediate effects of hypoxia on the vasculature as well as other cells in the affected tissues (i.e., cardiomyocytes in myocardial infarction and neurons in stroke). Due to technical difficulties in current state-of-the-art imaging techniques to visualize deep tissues at high resolution in rodents, it is practically impossible to study the very early effects of hypoxia in the currently used mouse and rat models of arterial-ligation-induced hypoxia in the heart and brain. Also, as previously mentioned these methods are not clean and result in induction of signaling pathways related to blocked blood flow, which may not arise as a consequence of hypoxia per se. Just as it was the case for zebrafish, glass catfish are also highly tolerant to low levels of oxygen in the water, which mean that this fish would provide a magnificent model system for studying immediate effects of hypoxia on the peripheral vasculature [25].

A major pathologic hypoxia-induced vascular event is increased vascular permeability in the hypoxic tissue. This lead to extravasation of plasma and thereby edema, which in many cases have detrimental effect for example in retinopathy, stroke, myocardial infarction and cancer. As edema is physiologically eliminated by lymphatic vessels, it would be of great importance to understand how lymphatic function is impaired or affected by hypoxia. However, to date very little is known regarding the effects of hypoxia on lymphatic vessels. It was recently rediscovered that fish, including zebrafish and glass catfish, have lymphatic or at least lymphatic-like vessels [25–27]. These vessels share critical anatomical, physiological, developmental, and functional traits with their mammalian counterparts as they are lacking a basement membrane and mural cells coverage, are important for draining extracellular fluids and develop from venous endothelial cells in response to VEGF-C—VEGFR3 signaling [25–27]. Furthermore, they are located in the same anatomical position in the fish as in mice and humans, and genes implicated in lymphatic disorders in humans have similar functions in the zebrafish. These findings have spurred the interest in using zebrafish to study lymphatic pathologies. However, there are also key differences in fish lymphatics that researchers need to take into account when using these models. First of all, fish lack some secondary lymphatic structures such as lymph nodes, peyers patchcs, and tonsils, and the key lymphatic organ is the head kidney and not the spleen. Also the lymph vessels themselves seem to have retained to some extent the direct connections with blood vessels, such that they do not solely arise from blind-ended lymphatic sacs in the periphery [25, 28]. This means that fish lymphatic vessels are not exclusively afferent in nature, they may also carry lymph-fluid efferently from the heart or arteries to the tissues. The connections to the blood vessels have led some researchers to classify these vessels as a secondary circulation [28], rather than as a true lymphatic vasculature, as the definition of lymphatic vessels have been thought of strictly as vessels that exclusively start in blind ended sacks in the periphery. However, evidence is emerging that in the rat mesentery, there are also direct connections between lymph and blood vessels [29], the importance of which, however, has not been defined. In fish, the connections between the lymphatic and blood vasculatures have been suggested to act as a filtration barrier and allow specifically white blood cells to enter into the lymphatics, thereby creating an immunologic function of these vessels [28]. This makes sense as the fish lymphatics are primarily present in tissues exposed to an exogenous milieu such as the intestine and the skin and would therefore be important for mobilizing an adaptive immune defense upon injury and exposure to pathogens. Also these vessels are highly plastic and can change diameter up to fivefold thus leading to a capacity to hold and carry tremendous amounts of liquid [25]. This is possibly a main reason why fish that have been genetically manipulated to lack these lymphatic vessels present with massive edema already at relatively early stages of development [30]. However, as transarterial blood pressure in fish is very low, there is hardly any leakage across the arterial wall, which means that lymphatics are not required to pick up and drain away extracellular liquid in fish. It seems that these vessels are more involved in buffering the amount of liquid in the fish when the capacity of the blood vasculature has been reached.

We have recently taken advantage of the transparency and hypoxia tolerance of the glass catfish to study immediate hypoxia-induced changes in the blood and lymphatic vasculatures in these fish as well as in zebrafish. We found that, as expected, the fins of these fish contain predominantly lymphatic vessels and not many blood vessels [25]. Due to the ability of fish to absorb oxygen over the skin directly from the water, and because the fins are so thin that all cells in this tissue may receive sufficient amounts of oxygen by this passive diffusion, it makes sense that there is no physiological reason for having blood vessels in these parts of the fish. However, as the fins are usually quite exposed in fish, and often sites of injury, it is potentially important to have vessels which would transport immune cells to fight off invading pathogens. In agreement with this hypothesis, we saw that there are vessels in the fin, but that these do not contain cells (or only very few) under resting conditions. Furthermore, these vessels expressed markers of lymphatic vessels in mammals including LYVE1 and PROX1, but did not express blood vessel markers such as VEGFR2, and we therefore characterized them as lymphatics [25]. In the glass catfish we found that these lymphatic vessels connected directly with segmental arteries [25], just as it has been described previously also for other fish species [28]. However, we saw that the lymphatic vasculature in the fins of both glass catfish and zebrafish became heavily perfused with blood when the fish were exposed to hypoxia [25]. Mechanistically, the connections between the lymphatic vasculature and segmental arteries, termed arteriolymphatic conduits (ALCs) were, in resting fish, highly tangled and constricted, and thus did not allow transport of cells through to the lymphatics. However, these ALCs opened up and straightened out in response to hypoxia to allow blood flow into the lymphatic vessels and thus immediately and efficiently increase perfusion to the tissues [25]. Furthermore, as these connections are located in the smooth muscle enriched arterial wall, this hypoxia-induced opening of the fish lymphatics was shown to be dependent on nitric oxide, produced by endothelial cell nitric oxide synthase and acting on smooth muscle cell guanylyl cyclase [25], which has been shown also in the mouse and humans to be an important pathway for smooth muscle relaxation and arterial dilation. Interestingly, while there were no smooth muscle cells surrounding the fish lymphatics, these vessels were able to increase their diameter several fold in response to the increased amount of liquid and cells travelling through these vessels in hypoxia-stimulated fish [25]. These studies have established zebrafish and glass catfish as excellent models for studying immediate responses to hypoxia, and how lymphatic vessel plasticity may be involved in such responses. These finding could further extend to the notion that such arterio-lymphatic conduits, if present also in peritumoral tissues, could be important for lymphatic dissemination of tumor cells. It has previously been a mystery how tumors, which do not have detectable amounts of intratumoral lymphatic vessels, could primarily disseminate to lymph nodes. The finding that peritumoral hypoxia may establish a route whereby tumor cells could spread via blood vessels to the lymphatic circulation could provide an explanation for this puzzling phenomenon.

Conclusions and Perspectives

Tissue hypoxia is perhaps the most important pathological driving force of angiogenesis and progression of diseases such as retinopathies, stroke, myocardial infarction, and cancer. Sophisticated studies on mechanisms behind hypoxia-induced pathways that may be important for disease progression have, however, been hampered by a lack of suitable animal models which would allow precise and dynamic alterations in tissue oxygenation. Zebrafish and glass catfish have recently entered the arena of hypoxia research due to the high hypoxia tolerance and transparency of these animals. Models based on adult glass catfish or zebrafish, as well as zebrafish embryos, hold great promise as a mirror to the mechanisms involved in hypoxia-induced pathological angiogenesis in living, intact animals and thus could potentially be used to discover important pathways involved in disease progression. For example, tumor cell dissemination and metastasis has been described to be induced by hypoxia, but the mechanism behind such regulation in vivo has been lacking, until it was shown that hypoxia-induced VEGF, expressed by the tumor cells, could lead to VEGFR2-dependent tumor angiogenesis and formation of a chaotic tumor vasculature which enable systemic dissemination of tumor cells via blood vessels to the entire organism [7]. Similarly, hypoxia is known to be a driving force of retinopathy, via induction of VEGF, but it has not been possible to precisely study retinal hypoxia in the mice without concomitantly inducing severe inflammation and lack of blood flow. The adult zebrafish hypoxia-induced retinal angiogenesis model [6] not only allow for the study of angiogenesis in the adult retina but also offer a means to uncouple hypoxia from other pathways induced in the highly invasive rodent models. Finally, acute hypoxia is important for ischemic disorders such as stroke and myocardial infarction, but as the rodent models are suffering from problems in visualizing the hypoxic tissue while it is hypoxic and the animal is still alive, it is very difficult to examine the immediate effects of hypoxia, and how this may lead to changes in blood flow through different vascular beds. Zebrafish and glass catfish offer a way to not only study what happens immediately following a hypoxic insult but also what level of hypoxia is required to elicit a certain response and how tissues may respond differently and dynamically to changes in oxygen availability.

We envision that the zebrafish and glass catfish hypoxia models described in this chapter would be a highly valuable tool in the future. As genetically modified zebrafish strains become more and more abundant, and as imaging technologies become more and more sophisticated, these hypoxia models will give researchers the opportunity to study the effects of hypoxia on cells, tissues, and even signaling pathways in real time and thus generate information on exactly how different cells respond to varying oxygen levels over time, and how such responses can be modulated pharmacologically. In essence, this would allow the generation of specific drugs, which target hypoxia-induced pathological pathways such as hypoxia-induced angiogenesis or pathological changes in blood vessels, which would be a novel strategy for treating multiple diseases with the same drug. These models described here could therefore be used to develop therapeutic regimens which potentially could help millions of people world wide suffering from cardiovascular, malignant, or other types of ischemic diseases.

References

1. Jensen LD, Rouhi P, Cao Z, Lanne T, Wahlberg E, Cao Y (2011) Zebrafish models to study hypoxia-induced pathological angiogenesis in malignant and nonmalignant diseases. Birth Defects Res Part C Embryo Today 93:182–193
2. Rouhi P, Jensen LD, Cao Z, Hosaka K, Lanne T, Wahlberg E, Steffensen JF, Cao Y (2010) Hypoxia-induced metastasis model in embryonic zebrafish. Nat Protoc 5:1911–1918
3. Cao Z, Jensen LD, Rouhi P, Hosaka K, Lanne T, Steffensen JF, Wahlberg E, Cao Y (2010) Hypoxia-induced retinopathy model in adult zebrafish. Nat Protoc 5:1903–1910
4. Folkman J (1971) Tumor angiogenesis: therapeutic implications. N Engl J Med 285:1182–1186
5. Ferrara N (2010) Vascular endothelial growth factor and age-related macular degeneration: from basic science to therapy. Nat Med 16:1107–1111
6. Cao R, Jensen LD, Soll I, Hauptmann G, Cao Y (2008) Hypoxia-induced retinal angiogenesis in zebrafish as a model to study retinopathy. PLoS One 3:e2748
7. Lee SL, Rouhi P, Dahl Jensen L, Zhang D, Ji H, Hauptmann G, Ingham P, Cao Y (2009) Hypoxia-induced pathological angiogenesis mediates tumor cell dissemination, invasion, and metastasis in a zebrafish tumor model. Proc Natl Acad Sci USA 106:19485–19490
8. Makino Y, Cao R, Svensson K, Bertilsson G, Asman M, Tanaka H, Cao Y, Berkenstam A, Poellinger L (2001) Inhibitory PAS domain protein is a negative regulator of hypoxia-inducible gene expression. Nature 414:550–554
9. Cao Y (2011) Antiangiogenic cancer therapy: why do mouse and human patients respond in a different way to the same drug? Int J Dev Biol 55:557–562
10. Folk JC, Stone EM (2010) Ranibizumab therapy for neovascular age-related macular degeneration. N Engl J Med 363:1648–1655
11. Zhang D, Hedlund EM, Lim S, Chen F, Zhang Y, Sun B, Cao Y (2011) Antiangiogenic agents significantly improve survival in tumor-bearing mice by increasing tolerance to chemotherapy-induced toxicity. Proc Natl Acad Sci USA 108:4117–4122
12. Kamba T, McDonald DM (2007) Mechanisms of adverse effects of anti-VEGF therapy for cancer. Br J Cancer 96:1788–1795
13. Bergers G, Hanahan D (2008) Modes of resistance to anti-angiogenic therapy. Nat Rev Cancer 8:592–603
14. Jensen LD, Cao Y (2013) Clock controls angiogenesis. Cell Cycle 12:405–408
15. Jensen LD, Cao Z, Nakamura M, Yang Y, Brautigam L, Andersson P, Zhang Y, Wahlberg E, Lanne T, Hosaka K, Cao Y (2012) Opposing effects of circadian clock genes bmal1 and period2 in regulation of VEGF-dependent angiogenesis in developing zebrafish. Cell Rep 2:231–241
16. Nicoli S, Standley C, Walker P, Hurlstone A, Fogarty KE, Lawson ND (2010) MicroRNA-mediated integration of haemodynamics and Vegf signalling during angiogenesis. Nature 464:1196–1200
17. Semenza GL (2012) Hypoxia-inducible factors: mediators of cancer progression and targets for cancer therapy. Trends Pharmacol Sci 33:207–214
18. Bailey KM, Wojtkowiak JW, Hashim AI, Gillies RJ (2012) Targeting the metabolic microenvironment of tumors. Adv Pharmacol 65:63–107
19. Keith B, Simon MC (2007) Hypoxia-inducible factors, stem cells, and cancer. Cell 129:465–472
20. Dang CV (2012) Links between metabolism and cancer. Genes Dev 26:877–890
21. Warburg O (1956) On the origin of cancer cells. Science 123:309–314
22. Peterson RT, Shaw SY, Peterson TA, Milan DJ, Zhong TP, Schreiber SL, MacRae CA, Fishman MC (2004) Chemical suppression of a genetic mutation in a zebrafish model of aortic coarctation. Nat Biotechnol 22:595–599
23. Antonetti DA, Klein R, Gardner TW (2012) Diabetic retinopathy. N Engl J Med 366:1227–1239

24. Chen J, Smith LE (2007) Retinopathy of prematurity. Angiogenesis 10:133–140
25. Dahl Ejby Jensen L, Cao R, Hedlund EM, Soll I, Lundberg JO, Hauptmann G, Steffensen JF, Cao Y (2009) Nitric oxide permits hypoxia-induced lymphatic perfusion by controlling arterial-lymphatic conduits in zebrafish and glass catfish. Proc Natl Acad Sci USA 106:18408–18413
26. Yaniv K, Isogai S, Castranova D, Dye L, Hitomi J, Weinstein BM (2006) Live imaging of lymphatic development in the zebrafish. Nat Med 12:711–716
27. Kuchler AM, Gjini E, Peterson-Maduro J, Cancilla B, Wolburg H, Schulte-Merker S (2006) Development of the zebrafish lymphatic system requires VEGFC signaling. Curr Biol 16:1244–1248
28. Rasmussen KJ, Steffensen JF, Buchmann K (2013) Differential occurrence of immune cells in the primary and secondary vascular systems in rainbow trout, *Oncorhynchus mykiss* (Walbaum). J Fish Dis 36:675–679
29. Robichaux JL, Tanno E, Rappleye JW, Ceballos M, Stallcup WB, Schmid-Schonbein GW, Murfee WL (2010) Lymphatic/Blood endothelial cell connections at the capillary level in adult rat mesentery. Anat Rec (Hoboken) 293:1629–1638
30. Hogan BM, Bos FL, Bussmann J, Witte M, Chi NC, Duckers HJ, Schulte-Merker S (2009) Ccbe1 is required for embryonic lymphangiogenesis and venous sprouting. Nat Genet 41:396–398

Tumor Angiogenesis: *Fishing* for Screening Models

14

Dafne Gays, Vera Mugoni, and Massimo M. Santoro

Contents

14.1 Tumor Angiogenesis 294
14.1.1 Understanding Tumor Angiogenesis 294
14.1.2 Overview of Angiogenic Mechanisms Associated to Tumor Angiogenesis in Endothelial Cells 295
14.2 The Zebrafish Model 296
14.2.1 *Danio rerio* as a Powerful Tool to Study Vessel Formation and Homeostasis 296
14.2.2 Zebrafish and microRNA 297
14.2.3 Genetic Engineering in Zebrafish 299
14.3 Zebrafish as a Model to Study Tumor Angiogenesis 300
14.3.1 Zebrafish as a Model System for Modeling Cancer in Lower Vertebrates . . 300
14.3.2 Xenotransplants Models in Zebrafish for Anti-angiogenic Therapies 301
14.4 Antitumor Angiogenic Drug Screening in Zebrafish 303
14.4.1 Drugs Targeting Tumor Angiogenesis In Vivo 303
14.4.2 Performing a Drug Screening in Zebrafish 304
14.4.3 Toxicity Assay in Zebrafish 306
14.4.4 Drug Target Validation 306
14.5 Perspectives and Conclusions 308
14.5.1 Zebrafish in Pharmaceutical Companies 308
14.5.2 Can We Use the Zebrafish Cancer Models to Identify New Therapies in Clinics? 308
References 308

Abstract

The process of de novo vessel formation, called angiogenesis, is essential for tumor progression and metastasis. The identification and targeting of the molecular pathways involved in this process are becoming critical issues for

D. Gays • V. Mugoni • M.M. Santoro (✉)
Department of Molecular Biotechnology and Health Science, Molecular Biotechnology Center, University of Torino, Torino, Italy
e-mail: massimo.santoro@unito.it

J. Dulak et al. (eds.), *Angiogenesis and Vascularisation*,
DOI 10.1007/978-3-7091-1428-5_14, © Springer-Verlag Wien 2013

anti-angiogenic cancer therapies. To pursue these molecular and pharmacological approaches, researchers need to develop better preclinical models to study tumor angiogenesis and then test anti-angiogenic therapies. As a vertebrate, the zebrafish model system is equipped with easy and powerful transgenesis and imaging tools to investigate not only angiogenesis but also tumor development and its progression. In this chapter we will illustrate how a small tropical fish can help to better understand the tumor angiogenesis process and identify new pharmacological therapies for tumor angiogenesis. Lastly, we describe in what way this model can act as a preclinical model for screening new chemical compounds which are able to selectively block tumors but not the normal healthy angiogenesis in vivo.

Keywords
Tumor angiogenesis • Zebrafish • Xenograft • Drug screening

14.1 Tumor Angiogenesis

14.1.1 Understanding Tumor Angiogenesis

De novo formation of blood vessels is essential for supplying growing tissues with nutrients and in the removal of cellular waste. Impairment in this functional vascular network causes tissue to collapse and die [1]. Tumors initially are avascular masses of a few transformed cells. Later, when they increase in size, new blood vessels are required to sustain the proliferating mass. This induction of vessels around the tumor mass is a critical step for tumor survival [2]. Such a process has also been reported as a rate-limiting step in progression of solid cancers [3]. For several decades until now, tumor angiogenesis has been considered as one of the factors that influence tumor outcome in humans. In fact, it has been established that the recruitment of blood vessels is essential for tumor invasion and metastasization [4]. The phase of tumor progression, when the tumor mass recruits blood vessels, is called the "angiogenic switch." This phase defines the transition of tumors from an avascular, quiescent mass to a vascularized, growing mass that often shows a more aggressive phenotype [5].

Anti-angiogenic studies are important for establishing new therapeutic strategies to cure cancer. Initially, the anti-angiogenic therapy was established with the purpose of "starving" a tumor mass by impairing blood vessel recruitment. According to this strategy, anti-angiogenic drugs were designed to block vessel formation by targeting angiogenic molecules such as the Vascular Endothelial Growth Factor (VEGF). Different clinical trials have shown that better results are obtained when the anti-angiogenic therapy is combined with chemotherapic drugs [6, 7].

Modulation of tumor vessel architecture is another important aspect in the study of tumor angiogenesis. Tumor vessels show abnormal vascular architecture irregularity compared to that of normal vessels [8]. Because of this, tumor vessels are less functional than normal vessels in delivering nutrients and oxygen. As a consequence, tumor mass perfusion is heterogeneous and blood flow is also discontinuous. It has been reported that such abnormal vascular patterns favor tumor cell metastasis and constitute an obstacle for the efficient delivery of drugs to cancer cells. Thus, targeting tumor vascular abnormalities and restoring protection from excessive pro-angiogenic stimuli represent a promising new strategy in blocking cancer progression and metastasis [9, 10].

In summary, the combination of these innovative strategies will provide a higher chance in blocking tumor vascularization and progression in the near future.

14.1.2 Overview of Angiogenic Mechanisms Associated to Tumor Angiogenesis in Endothelial Cells

In healthy adult tissues endothelial cells are quiescent and do not normally respond to pro-angiogenic factors. Thus, the tumor angiogenesis process requires that a normal quiescent endothelial cell reactivates itself and spreads over the newly formed cancer mass. Specific signaling events occur at this stage.

To sustain angiogenesis the tumor environment also releases angiogenic factors. VEGFs (VEGF-A, VEGF-B, and VEGF-C), VEGF receptors (VEGFR-1, VEGFR-2, VEGFR-3), and co-receptors neuropilins (NRP1, NRP2) have been established as a critical players for the developing vasculature [11]. During tumor angiogenesis, the gradient of VEGF released by stromal and myeloid cells increases vessel branching and guides vessel sprouting. VEGF factors directly assist and sustain tip cells guiding vessel sprouting. Analogous to VEGF, FGFs family (Fibroblast Growth Factor) plays a critical role in normal and tumor angiogenesis [12]. FGFs are frequently upregulated in tumors that escape anti-VEGF treatments. The signaling pathway represented by angiopoietin 1 (ANG-1), angiopoietin 2 (ANG-2), and TIE-2 receptor has great relevance in tumor angiogenesis. ANG-1 is expressed by mural and tumor cells and promotes vessels stability. ANG-2 is released by sprouting endothelial cells and enhances vessel dynamics by limiting mural cell adhesion. ANG-1 and ANG-2 mediate antagonistic signaling, and the overall effect is regulated by the tumor context.

The regulation of angiogenesis by hypoxia is an important component of homeostatic mechanisms that links vascular oxygen supply to metabolic demand. Low-oxygen levels present in the tumor mass promote angiogenesis by HIF1α activation. New insights into this process were the identification of hypoxia-inducible factor (HIF) as a key regulator of pro-angiogenic molecules such as VEGF and HIF regulation by the family of HIF hydroxylase in accordance with oxygen availability [13].

Besides growth factor receptor signaling, chemokines may also promote tumor angiogenesis. Chemokines are a large group of low molecular weight cytokines that are known to selectively attract and activate different cell types. It is well recognized that the primary function of chemokines is leukocyte attractants. Nevertheless recent evidences indicate that they also play a role in a number of tumor-related processes such as growth, angiogenesis, and metastasis [14]. Chemokines activate their seventh trans-membrane, G-protein-coupled receptors (GPCR) on the cell surfaces of target cells. The role played by chemokines and their receptors in tumor pathophysiology is complex, as some chemokines favor tumor growth and metastasis, while others may enhance antitumor immunity. Among the most known chemokines there is SDF-1α that is upregulated during hypoxia. The SDF-1α promotes vascularization by binding to CXCR4 receptor on endothelial tip cells. Another known chemokine factor is S1P (sphingosine-1-phosphate) that modulates vessel stability by binding to its G-protein-coupled receptors (S1PRs) in the tumor environment [6]. These diverse functions of chemokines establish themselves as key mediators between the tumor cells and their microenvironment, and play a critical role in tumor progression and metastasis.

In addition to gene products, there are novel ways of research that are oriented towards regulatory mechanisms of gene expression in tumor angiogenesis. The posttranscriptional regulation orchestrated by microRNAs has been investigated with significant in/to tumor angiogenesis. MicroRNAs (miRNAs) are noncoding single stranded RNAs that regulate the level of genes expression by targeting mRNAs and repressing their translation (see also Sect. 14.2.2). MiRNAs play a critical role in the regulation of physiological vasculogenesis and angiogenesis [15, 16]. Pro-angiogenic miRNAs promote angiogenesis by targeting negative regulators of angiogenesis, while anti-angiogenic miRNAs target positive regulators of angiogenesis. Different reports demonstrate that miRNAs expressions are modulated by tumor environment and pro-angiogenic factors released by cancer-associated cells [17]. Analogously, also hypoxia, a condition associated to tumor formation and progression, modulates expression of miRNAs [18]. In endothelial cells, miR-210 regulates the cellular response to hypoxia, and it has been reported as differentially modulated by cancer [19]. Recent data identify a miR-503 that targets both FGF2 and VEGFA in cancers, demonstrating the anti-angiogenic role of miR-503 in tumorigenesis [20]. Also, miR-145 reveals an inhibitory role in tumor angiogenesis acting by the posttranscriptional regulation of known oncogenic targets such as N-RAS and VEGF-A [21].

14.2 The Zebrafish Model

14.2.1 *Danio rerio* as a Powerful Tool to Study Vessel Formation and Homeostasis

Danio rerio, commonly known as zebrafish, has emerged as an optimal vertebrate model system to study vascular development and homeostasis. The zebrafish embryo and adult models have a relatively simple vascular system, and the molecular mechanisms underlying its vessel formation and morphogenesis are very similar to those of higher vertebrate. In comparison to the blood vessels of zebrafish with other vertebrates, there is a remarkable degree of anatomical and functional conservation [22]. These observations indicate that vascular development is genetically determined through evolutionarily conserved regulatory mechanisms. In vertebrates, the vascular system arises in a process called vasculogenesis, during which free angioblast progenitors aggregate and generate a primary vascular plexus. These vessels later acquire arterial and venous identities [23]. Therefore, new vessels sprout from arteries and veins through a process called angiogenesis [3, 24]. Consequently, the vascular plexus progressively expands and remodels into a highly organized and stereotyped network. A complete development and maturation of the vascular system requires smooth muscle or mural cell recruitment that also ultimately stabilize the vessels and regulate the blood flow [25, 26]. Angiogenesis contributes to organ growth during development but, during adulthood, most endothelial cells within blood vessels remain quiescent until they receive instructions to start proliferation and migration.

There are many model systems to study vasculogenesis and angiogenesis. However, most of them lack some of the features that have led to become the established model for studying vascular biology [27]. Zebrafish embryos develop externally, making them easily accessible to experimental manipulation and imaging. After fertilizations, the optically clear embryos rapidly develop and acquire functional cardiovascular system, already at 24 hpf (hour post fertilization). The availability of vascular-specific transgenic lines and advanced microscopy techniques (i.e., SPIM and confocal microangiography) also allow a high-resolution time-lapse imaging of blood vessels in developing embryo [22, 28–30]. Such dynamic imaging can also allow the dissection of mechanisms underlying vessel patterning, branching, and the regulation of vascular tip cell specification [31, 32].

Zebrafish are also suitable for forward and reverse genetic analysis thus providing an opportunity to identify novel genes involved in vascular development and homeostasis [33, 34]. Zebrafish embryos can develop and survive for several days without a functional cardiovascular system, since tissues are oxygenated by passive diffusion. Thus, the effects of genetic mutations affecting the cardiovascular system can be easily studied in zebrafish, especially when compared to mammalian systems where an abnormal heart or vascular development is fatal at early developmental stage. Recent studies have also established zebrafish as a useful model to study developmental lymphangiogenesis and lymphatic network formation [35].

14.2.2 Zebrafish and microRNA

MicroRNAs (miRNAs) are small noncoding RNAs that fine-tune gene expression acting at a posttranscriptional level. They are then synthesized as longer precursors called pri-miRNAs; they are cleaved by *Drosha* into 60–100-nucleotides hairpins named pre-miRNAs. In the cytoplasm, *Dicer* then processes pre-miRNAs into a duplex containing the mature miRNAs sequence. This duplex is recognized by the protein complex Argonaute to form the RNA-induced silencing complex (RISC). Thus, the "seed" region of mature miRNA (nucleotides 2–8) guides RISC on those mRNAs that contain the miRNA binding site(s) in their 3′UTR. This association leads to a block of protein synthesis through degradation of the target mRNAs and translation repression. Since miRNAs have numerous high and low affinity targets, and the mRNAs contain binding site for multiple miRNAs, it is possible to generate a complex network of gene regulation that fine-tune biological processes [36, 37]. Since their discovery miRNAs have been shown to play a role in a wide range of physio-pathologic processes, including development, differentiation, and cell proliferation [38, 39]. The complete loss of Dicer function leads to severe developmental defects in both zebrafish and mouse. Zebrafish maternal-zygotic Dicer mutants (that lack both maternal and zygotic Dicer) show morphogenetic defects during gastrulation, abnormal brain formation, somitogenesis, and heart development. However, early gastrulation defects could be partially rescued through the injection of miRNA-430, making it possible to analyze the loss of miRNAs in tissues that otherwise would not develop [40, 41]. The fundamental role of miRNAs in the regulation of cardiovascular development and function has been demonstrated by analyzing mutants of Dicer. In mouse, endothelial-specific deletion of Dicer demonstrated the importance of miRNAs in postnatal angiogenic responses to a variety of stimuli, including VEGF, tumors, limb ischemia, and wound healing [17]. MiRNAs are also indispensable for the proper development and functionality of vascular smooth muscle cells since Dicer silencing in this cell type lead to a dilated, thin-walled blood vessels that also exhibit impaired contraction [42]. Recent studies on zebrafish *Dicer* mutant embryos demonstrated excessive endocardial cell formation, suggesting a role for miRNAs in cardiac development [43].

Zebrafish have also proven to be a helpful system in studying miRNA functions during developmental processes [44]. Identification and characterization of miRNAs in zebrafish have provided important information regarding the conservation of miRNA genes in vertebrates. So far more than 400 different miRNAs has been isolated in zebrafish through prediction algorithms and large-scale sequencing of small-RNA cDNA libraries [45–47]. It is likely that this number will increase in the near future once the complete annotation of the zebrafish genome is known. Several techniques could be used to study miRNAs function in zebrafish. Modulation of miRNAs expression could be achieved in several ways. Morpholino injection at one single cell stage is widely used to analyze the effect of the downregulation of a specific miRNA during development. Transient overexpression of a miRNAs can be obtained through the injection of miRNA mimics. The role of a miRNA in larvae and adult animals can be investigated by using Cre-loxP or TALEN technologies that allow stable and conditional miRNA down or overexpression in a tissue-specific manner [48]. Several vascular specific miRNAs have been identified and studied in zebrafish inter alia miR-24, miR-92, miR-126, and miR-221 [44, 48]. Some of them act as anti-angiogenic factors. A representative example of such anti-angiogenic activity is given by the miR-24.

MiR-24 overexpression in zebrafish embryos impaired intersegmental vessel formation due to the downregulation of two different targets, gata2 and pak4 [49]. A similar effect is obtained through the overexpression of miR-92, a miRNA part of the cancer-associated miRNA cluster 17–92 [50]. A very interesting endothelial-specific miRNA is miR-126. It has been reported by several studies that miR-126 is a master regulator of angiogenic signaling and vascular integrity in vitro and in vivo. In zebrafish, the downregulation of miR-126 induces collapsed blood vessels and cranial hemorrhages, as a consequence of SPREAD1 (Sprouty-Related EVH1 domain-containing protcin 1) and PIK3R2 (a regulatory subunit of PI3K) deregulation [51, 52]. Another vascular miRNA identified in zebrafish is miR-221. MiR-221 regulates endothelial tip cell behavior during vascular development in zebrafish. Tip cells are present at the leading edge of intersomitic vascular sprouts and guide new vessels in their final destination. MiR-221 regulates the proliferation and migration of these cells through the modulation of cdkn1b (cyclin-dependent kinase inhibitor 1b) and pik3r1 (phosphoinositide-3-kinase regulatory subunit 1) levels [53].

Although it is known that several miRNAs are involved in angiogenesis (pro- and anti- angiogenesis) it mostly remain undetermined their vascular cell autonomous target [48]. Identification of miRNAs signature of different pathological vascular disorders such as tumor-induced angiogenesis could help to develop innovative anti-angiogenic cancer therapies. Zebrafish represent a very useful model to identify and characterize the function of novel miRNAs involved in vascular development and homeostasis.

14.2.3 Genetic Engineering in Zebrafish

Remarkable repertoires of genetic tools are available to modify the zebrafish genome. The whole genome of zebrafish has now been sequenced by the Sanger Center, and extensive genome annotation is available through the trans-National Institutes of Health Zebrafish Genome Initiative (Sanger, zfin.org, ENSEMBL). The zebrafish genome can be functionally accessed using both forward and reverse genetics-based approaches. Forward genetic screens (from phenotype to genotype) have generated thousands of mutations that affect organogenesis, physiology, and behavior. Many of these mutations have already been mapped, mainly through a laborious and time-consuming polymorphism analysis. Recently deep sequencing technology has greatly facilitated the identification of the mutated genes [54]. Mutant generation and characterization have provided a strategy to study the relationship between gene and their function in normal and diseased conditions. A complementary approach consists in reverse genetics analysis (from genotype to phenotype). In zebrafish reverse genetics can be achieved in a fast and cost-effective way using the morpholino (MO) knockdown technology [55]. MOs are chemically modified oligonucleotides that are injected in eggs at one cell stage to inhibit the expression of the target gene. MOs exert their effect throughout embryogenesis. Morpholino technology is then a rapid and cost-effective method to address gene function during development [55]. Another reverse genetic tool is represented by the TILLING (Targeting Induced Local Lesions in Genomes) technology. This method combines a standard and efficient technique of mutagenesis with a chemical mutagen such as Ethyl methanesulfonate (EMS) with a sensitive DNA screening technique that identifies single base mutations (also called point mutations) in a target gene. TILLING is a different reverse genetic tool that allows directed identification of mutations in a specific gene and in a chemically mutagenized population [56]. Finally, gene-specific knock out models can be obtained using the Zinc Finger Nuclease (ZFN) or TALEN or CRISPR technology [57–59].

Beside the "loss of function" tools (MOs, ZFN), other techniques can be used in zebrafish to analyze "gain of function" effects. The consequences of a transient gene overexpression can be assessed easily during development through the injection of messenger RNA. On the other hand, stable and conditional overexpression systems can be obtained by using transgenesis. Transgenic technology is one of the most powerful tools in zebrafish. Usually, a tissue-specific promoter is used to drive expression of a fluorescent reporter protein. However, more sophisticated genetic techniques based on the CreERT2/loxP system are now available for zebrafish [60, 61]. These techniques can be further engineered to allow temporal and tissue-specific gene expression in the zebrafish model system [62].

14.3 Zebrafish as a Model to Study Tumor Angiogenesis

14.3.1 Zebrafish as a Model System for Modeling Cancer in Lower Vertebrates

Due to its peculiar features the zebrafish can be used as a model for human disease, especially to study cancer [63]. In fact, the conventional tools for genome manipulations offer the opportunity to use forward and reverse genetic approaches to generate cancer mutants or to directly manipulate endogenous gene expression. In forward genetic zebrafish embryos are exposed to mutagenic agents to introduce random mutations in their genome. Then, mutations associated to a specific phenotype are selected, mapped, and characterized [64]. Using this procedure, zebrafish mutant lines have been selected for susceptibility to tumor development. Several zebrafish mutant lines were generated with this genetic approach and most of them showed increased incidence of different types of tumors including lymphomas, sarcomas, pancreatic, and gut cancers [65]. In addition, reverse genetics offer the possibility to manipulate a gene of interest such as an oncogene and to facilitate the characterization of its function [66]. Such an approach has been used to create zebrafish cancer models for some of the most widespread human carcinomas. The zebrafish line that expresses the inactivating mutation M214K in the p53 gene represents one example of zebrafish cancer model. The p53 gene is a known tumor suppressor gene and has been widely reported to be frequently mutated in several human tumors. As for mammals, loss of wild-type p53 in zebrafish is associated with the lack of cell cycle control. This zebrafish mutant line not only offered a new opportunity to model cancer in vivo but also suggested that molecular pathways associated to cancer are maintained in zebrafish as well as in higher vertebrates [67].

In alternative to genetic approaches, zebrafish cancer models can be obtained by treating animals with mutagenic chemicals [68]. As a matter of fact, different studies have shown that a large number of tumors could be modeled in zebrafish by chemical treatments [69, 70]. For example, the exposure of zebrafish to nitrocompounds (i.e., nitrosomorpholine, diethylnitrosamine, diethylnitrosamine) causes the formation of hepatocellular carcinoma and liver tumors. Chemically induced tumors are comparable to genetically caused tumors and resemble human cancer features. Analysis of micro array gene profiles demonstrated that most of the genes deregulated in zebrafish liver tumors induced by chemical treatment are comparable to those deregulated in human cancers [71].

Altogether chemical and genetic approaches allow the modeling of several kinds of human cancer in the zebrafish. Moreover, modeling tumorigenesis in zebrafish has been improved by application of technologies for controlling the expression of specific oncogenes in defined spaces and times. Promising strategies are represented by adapting zebrafish strategies that are well established for the overexpression of oncogenes in mice. An experimental example of such approach is the creation of zebrafish Cre/Lox and hsp70 inducible systems that can be put under control of temperature.

To fit cancer research in the zebrafish system much work still needs to be done. Increasing experimental evidences have indicated that zebrafish model is as innovative system to study cancer and to investigate related biological processes such as the tumor angiogenesis. Even though genetic models for tumor angiogenesis are not yet well established, a good opportunity to characterize the cross talk between tumor cells and blood vessels dynamic is represented by zebrafish xenotransplant models.

14.3.2 Xenotransplants Models in Zebrafish for Anti-angiogenic Therapies

The significance of the zebrafish model has been implemented by setting up xenotransplants of human cancer cells. As for mice, xenotransplants conjugates high genetic complexity of human cancer with the chance to manipulate in vivo cancer cells. In the beginning there were only a few melanoma cells transplanted in zebrafish [71]. Since then many human tumor cell lines have been used in this assay, including breast adenocarcinoma [72] and melanoma cells [73]. More recently, cells dissociated from human solid tumors such as colon, pancreas, and stomach carcinoma [74] or for primary leukemia cells derived from bone marrow of patients have also been used for xenotranplants in zebrafish [75, 76].

Performing xenotransplants in zebrafish allows tumor cells to interact with an especially hospitable and handling system. Zebrafish embryos do not develop a competent immune system until 48 hpf, and cells can be introduced in the host at

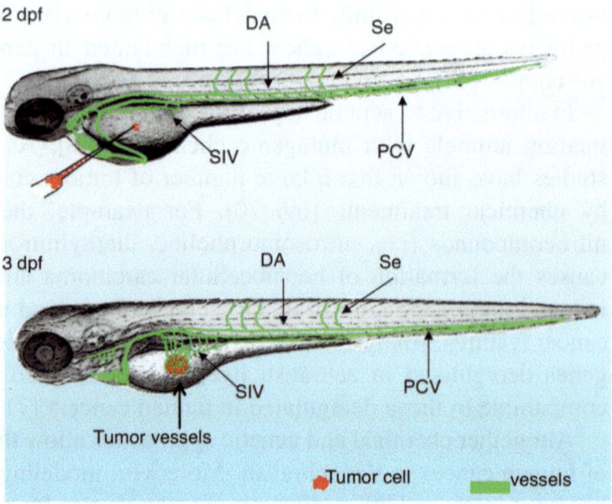

Fig. 14.1 Performing zebrafish tumor xenograft experiments. Tumor cells (*red*) are microinjected in the yolk sac of 2 or 3 dpf zebrafish embryo. 24 h after injection new vessels (*green*) sprouted from the subintestinal plexus (SIV) and vascularize the tumor mass. *DA* dorsal aorta, *PCV* posterior cardinal vein, *Se* intersegmental vessel, *SIV* subintestinal veins

earlier embryonic stages without any immune suppression. Otherwise, cells can be transplanted in adult fish immunosuppressed by irradiation or dexamethasone treatments. Tumor cells are usually transplanted at an early embryonic stage in the yolk, close to the Cuvier duct. Once tumor engraftments are performed the readout is available within a few days (Fig. 14.1).

Even though it is possible to transfer xenotransplant assays from mice to fish, there are some limitations that need to be considered. Zebrafish are tropical fish that normally live between 28 °C and 31 °C, while human cells are usually cultured at 37 °C; thus it is necessary to maintain host zebrafish at higher temperatures (34–35 °C) to avoid human cell death. Another problem that may arise is the dimensions of cells, given that human cells are larger than zebrafish cells. However, experimental evidence has shown that human cells can adapt to the anatomical structure of the zebrafish and invade tissues. As in mice, during xenotransplant experiments in zebrafish, it is required to support tumor cells with missing human growth factors such us specific cytokines [77].

Another advantage of using the zebrafish for tumor angiogenesis studies is the possibility to use vascular-specific fluorescent trangenic lines coupled to advanced imaging tools like light-sheet and dual-photon confocal microscopy. For example, performing a xenotransplant in transgenic lines with fluorescent vasculature may highlight new mechanisms of blood vessels recruitment and metastasis initiation [78]. Before transplantation, tumor cells can be modified for the expression of specific endothelial targets in order to validate their different ability in vessels recruitment. Correct modification of tumor cells before their introduction into hosts for expression of active growth factors, like FGF and VEGF, will regulate new blood vessel recruitment [79, 80].

As for now, xenotransplants in zebrafish are actively exploited for the investigation of anti-angiogenic pathways. Among the most applied techniques in combination with xenotransplants, there have been genetic and chemical approaches. In the first case molecular pathways sustaining blood vessel sprouting can be down regulated by morpholinos for endothelial genes and the effect can be followed over the time using confocal or light sheet microscopy. Also, treatment of xenotransplanted zebrafish with new generation drugs may make evident the innovative molecules for anti-angiogenic therapies. Joining genetic modified zebrafish lines with xenotransplant approach, important molecular pathways such as endothelial cell migration and drug resistance in tumor angiogenesis could be uncovered.

14.4 Antitumor Angiogenic Drug Screening in Zebrafish

14.4.1 Drugs Targeting Tumor Angiogenesis In Vivo

Inhibition of angiogenesis is a promising strategy, not only for the treatment of cancer, but also for limiting pathological conditions associated to vascular disorders. In spite of several efforts made so far, further understanding of vascular angiogenesis and identification of specific blockers of this process must be comprehended. Many different in vitro and in vivo angiogenesis assays are currently available for drug screening. Due to their straightforwardness in vitro models are extensively performed. These assays are mainly established on endothelial cell cultures and are mainly used to measure endothelial cell migration, proliferation, apoptosis, and tube formation. To copy the complexity of an in vivo system more complex assays are needed. Different in vivo assays have been developed in vertebrate animal models such as rodents, xenopus, and chick. These assays include the chick chorioallantoic membrane assay (CAM), corneal neovascularization assay, and matrigel plug assay. However, they are expensive and require specific experimental setups [27]. An alternative model is represented by the zebrafish that provides an optimal system for drug screening since it combines the biologic complexity of an in vivo system with the possibility to carry out high-throughput screening. Additionally, studies have demonstrated that anti-angiogenic compounds used clinically also inhibit the growth of blood vessels in zebrafish, suggesting that this small fish could help to discover new potential therapies [81]. Various studies have shown that different small-molecule inhibitors of VEGF signaling, known for anti-angiogenetic activity in cell culture or mouse model, have the same effect in zebrafish embryos [82].

The anti-angiogenic activity of molecules is commonly validated in zebrafish by the analysis of developing vasculature. In zebrafish embryos the primarily sprouting vessels, via angiogenesis, are the intersegmental vessels (ISVs) of the trunk that originate at 20 hpf from the dorsal aorta (DA) the main axial artery of the zebrafish. Live embryo microangiography and transgenic lines with fluorescent vasculature can allow a very simple and direct observation of normal and defective development of ISVs. For this reason, in different studies the specific development of ISVs is considered a parameter to analyze the anti-angiogenic potential of assayed compounds [83]. Otherwise, the development of the subintestinal plexus (SIV) that occurs later, at 48 hpf is currently used as a readout [81]. A further improvement in the anti-angiogenic drugs screening was obtained through the automated quantitative screening assay. This method takes advantage of transgenic lines that express fluorescence in endothelial cells and of an algorithm to quantify the growth of angiogenic vessels in zebrafish embryos. In this way, Tran and colleagues screened the LOPAC1280 compound library for anti-angiogenic activity and identified three hit compounds, one of which is the indirubin-3′-monoxime (IRO), not previously reported for anti-angiogenic activity [84]. Demonstrating that IRO also inhibits human endothelial cell tube formation and proliferation, these authors were the first researchers to discover a new anti-angiogenetic compound using the zebrafish model system [84]. Recently, another group identified new anti-angiogentic compounds developing a similar automated system for zebrafish screening and opening innovative methods in this field [85].

14.4.2 Performing a Drug Screening in Zebrafish

In several studies the zebrafish system has been successfully used to identify new drugs affecting different biological processes [86]. In those experiments a library of chemical compounds is screened for their effects, simply by adding them to fish water. After mating embryos are placed in 96- or 384-well plates. Small molecules of a synthesized or acquired library are added to the fish water (single or in combination) and at the end of the treatment phenotypic effects are analyzed visually or through an automated read-out [87]. Zebrafish embryos are small in size (less than 1 mm in diameter) allowing several embryos to fit in a single 384-well plate and using small quantities of drugs.

Zebrafish embryos permit a short assay time due to their rapid development. Furthermore, the dissection of the phenotype of interest and at the same time the evaluation of possible side effects and toxicity can be done simply. All these features offer the zebrafish as a system suitable for high-throughput screening (Fig. 14.2).

Peterson and colleagues published one of the first chemical screenings that was performed in zebrafish [88]. In this work, more than one thousand compounds were tested for their effects on the development of the central nervous system, cardiovascular system, pigmentation, and the ear. Different chemicals affecting those structures were identified by a visual examination of embryos. Interestingly, an automated high-throughput platform for in vivo chemical screenings on zebrafish embryos has been developed aiming for the highest possible throughput and minimization of human error. It includes automated methods for embryo collection and preparation, compound delivery, incubation, imaging, and analysis of the results. This method was applied to develop two different bioassays to evaluate a wide range of molecules for cardio-toxicity and angiogenesis properties. In this way, 15 compounds were selected as inhibitors of known regulators of angiogenesis [85].

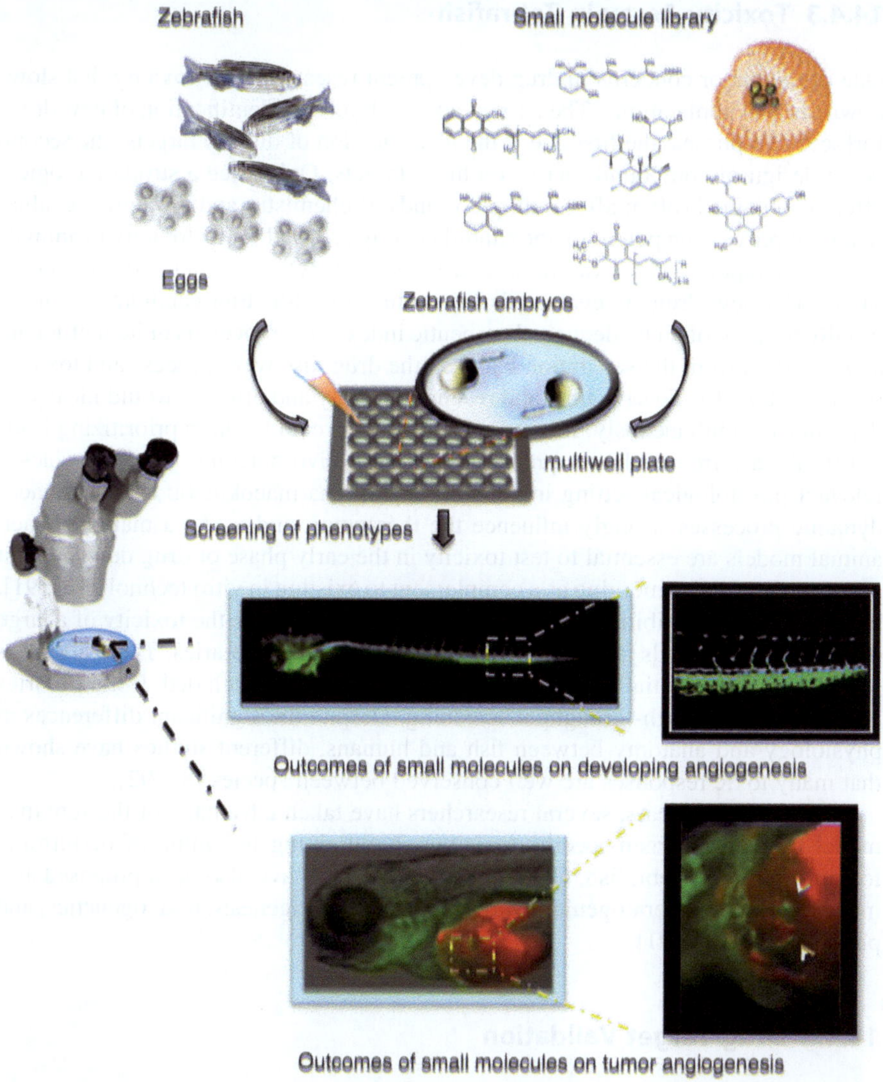

Fig. 14.2 Schematic representation of a typical small-scale drug screening in zebrafish embryos. Developing zebrafish embryos are prepared and maintained in multiwell plates (96 or 384 wells) and treated with drugs from earliest stages. Chemical compounds can be easily dissolved in water or delivered inside by diffusion or specific carriers such as liposomes. Specific anti-angiogenic drugs can be easily screened by analysis of developing vasculature inside tumor mass (*arrows head*) at stereomicroscope. Small molecules libraries can be in vivo selected for their effects on developmental angiogiogenesis or tumor angiogenesis in xenograft model assays

Identifying and validating novel drug targets remains a problem in drug discovery. Several strategies have been developed for target identification in zebrafish and a detailed analysis of them is discussed below.

14.4.3 Toxicity Assay in Zebrafish

One of the major concerns in drug development research is the toxicity that slows down clinical applications. The current approach for the identification of new drugs foresees two phases: the first one is the identification of disease targets; the second is the design of compounds active on those targets. Only once a strong biological effect is obtained (often after multiple rounds of chemistry and biological evaluation to improve drug potency), the candidate drug is tested for its toxicity in animals and the therapeutic index (a measure of both the beneficial effect and the relative safety of a new drug) is evaluated. The failure rate for drug candidates is high, usually because of an inadequate therapeutic index. This process is quite inefficient, and it suffers from the separation between the drug discovery process and toxicology tests [89, 90]. To save time and resources, safety and efficacy would ideally be determined simultaneously. In vitro, toxicity assays could help in prioritizing leads but results are frequently not predictive of the in vivo outcome. They also lack a relevant physiological setting in which complex pharmacokinetic and pharmacodynamic processes strongly influence the therapeutic index. As a matter of fact, animal models are essential to test toxicity in the early phase of drug development and the zebrafish is emerging as a complement to existing in vitro technologies [91]. In zebrafish the possibility to rapidly and efficiently evaluate the toxicity of a large number of compounds allows a prescreening of chemical libraries. Then, after this screening, all potentially unsafe compounds could be excluded from libraries before functional high-throughput screening. Despite the significant differences in physiology and anatomy between fish and humans, different studies have shown that many toxic responses are well conserved between species [87, 92].

In the past few years, several researchers have taken advantage of the zebrafish model to analyze organ-specific toxicities, highlighting the value of performing toxicity studies in zebrafish. Consequently, zebrafish have also been proposed as a model for studying teratogenicity, drug-induced tumorigenesis, toxicogenetics, and pharmacogenetics [91].

14.4.4 Drug Target Validation

Target validation is still a challenging phase of drug discovery. The tagging of the compounds may facilitate identification of their molecular targets, although tagging may negatively affect the potential of the compound. In vivo target validation using gene knockouts is a powerful method of predicting drug action. As previously explained, the zebrafish is easily accessible to genetic technologies that can be applied to assess the identification and validation of novel drug targets. Different approaches are based on phenotype-based screen and their strength derives from the availability of several zebrafish mutant lines and model disease. Functional correlations could be hypothesized when a mutated gene and a drug treatment cause a similar phenotype. At the same time off-target effects could be easily identified through the comparison of a mutant/morphant phenotype with the effects

Table 14.1 Advantages and limitations of the zebrafish model in modeling human diseases

Advantages	Limitations
Zebrafish husbandry	Experimental limits
• It is relatively inexpensive to obtain and maintain large number of adult and embryo zebrafish	• Few antibodies against zebrafish proteins are available so far. For this reason it is difficult to perform immunofluorescence analyses in zebrafish samples
Genetic manipulation	Embryo manipulation
• Generation of larvae with deletion or overexpression of specific genes can be easily accomplished using available tools (e.g. morpholinos, RNA, mimics) • Considerable availability of genetically manipulated zebrafish strains that are defective or have acquired functions for specific gene products (e.g. zinc-fingers, ENU, tilling). And also Transgenic zebrafish lines that express reporter genes in particular cell types are also available in the scientific community	• Due relative small size of zebrafish larvae, you need skillful and careful trainees to perform experiments • As most mammalian tumors grow at 37 °C, it is difficult to study the process of xenograft tumor growth at the optimal temperature
Chemical screening	Technical skills
• Addition of active chemical stimulators or inhibitors to the water enables analysis of intervention of these compounds on physiological and pathological processes	• Microinjection of tumor cells into the perivitelline space of a large number of zebrafish embryos is a tedious procedure and requires highly skillful micro-operations
Timing:	
• Turnover time for experiments is relatively short	
Anatomy of the zebrafish	
• Optical clarity of zebrafish embryos allows visualization of vascular and hematopoietic cells as well as tumor cell dissemination in living animals • Zebrafish embryos allows implantation of mammalian tumor cells, including human and mouse tumor cells, due to the absence of a functional immune system at this stage	

of a specific drug treatment. In this way, for example, Davidson and colleagues deduct that pifithrin-α (PFT-α), a pharmacological inhibitor of p53, was also active on p73. Infact PTF-α treated embryos showed developmental abnormalities strongly similar to those caused by p73 knockdown and not related to p53 downregulation [93]. Off-target evaluation is not secondary in the process of drug discovery in view of treatments for human health.

Moreover, drug screening could be performed in order to rescue a diseased phenotype and facilitate the identification of the drug target (if the mutated gene is known) or of the mutated gene (if the drug target is known).

14.5 Perspectives and Conclusions

Zebrafish have entered the stage of becoming a promising new model system to study human cancer. This has been largely due to the development of transgenic and xenograft models of cancer and their amenability to genetic and pharmacological testing. Cancer progression in these animals recapitulates many aspects of human disease and opens the door for further studies into identifying genetic and chemical modifiers of cancer. There are general advantages and limitations of using zebrafish as an in vivo model to study tumor angiogenesis (see Table 14.1).

14.5.1 Zebrafish in Pharmaceutical Companies

These models are already being utilized by academia and industry to search for genetic and chemical modifiers of cancer with success. The attention has been further stimulated by the amenability of zebrafish to pharmacological testing and the superior imaging properties of fish tissues that allow visualization of cancer progression and angiogenesis in live animals.

14.5.2 Can We Use the Zebrafish Cancer Models to Identify New Therapies in Clinics?

In the past several years, the zebrafish shown great promise as a powerful animal model to study cancer progression, making way for several laboratories to perform research using this model system. However, despite a significant progress, more work is required to fully explore how closely the processes in zebrafish may parallel mammalian cancer mechanisms and how well they might be translated to human diseases. These improvements will require more specific comparative studies among transgenic and xenograft zebrafish and mouse/mammalian models of cancer/tumor angiogenesis and their relevance to the disease. When all these studies and analyses will be performed, the zebrafish model will have the full potential to be used and embraced by the cancer research community. In particular, the zebrafish model represents an emerging vertebrate system in studying the angiogenic process and to better understand the modification of tumor microenvironment by anti-angiogenesis therapies.

Acknowledgments Support in MMS lab come from HFSP, Marie Curie Action, Telethon, and AIRC. We thank Ellen Jane Corcoran for editorial assistance.

References

1. Folkman J (2006) Angiogenesis. Annu Rev Med 57:1–18
2. Naumov GN, Akslen LA, Folkman J (2006) Role of angiogenesis in human tumor dormancy: animal models of the angiogenic switch. Cell Cycle 5:1779–1787

3. Carmeliet P (2005) Angiogenesis in life, disease and medicine. Nature 438:932–936
4. Naumov GN, Bender E, Zurakowski D, Kang SY, Sampson D, Flynn E, Watnick RS, Straume O, Akslen LA, Folkman J et al (2006) A model of human tumor dormancy: an angiogenic switch from the nonangiogenic phenotype. J Natl Cancer Inst 98:316–325
5. Baeriswyl V, Christofori G (2009) The angiogenic switch in carcinogenesis. Semin Cancer Biol 19:329–337
6. Carmeliet P, Jain RK (2011) Molecular mechanisms and clinical applications of angiogenesis. Nature 473:298–307
7. Jain RK, Carmeliet P (2012) SnapShot: tumor angiogenesis. Cell 149:1408–1408e1401
8. Jain RK (2005) Normalization of tumor vasculature: an emerging concept in antiangiogenic therapy. Science 307:58–62
9. Carmeliet P, Jain RK (2011) Principles and mechanisms of vessel normalization for cancer and other angiogenic diseases. Nat Rev Drug Discov 10:417–427
10. Abadie J, Abbott BP, Abbott R, Abernathy M, Accadia T, Acernese F, Adams C, Adhikari R, Ajith P, Allen B et al (2011) Directional limits on persistent gravitational waves using LIGO S5 science data. Phys Rev Lett 107:271102
11. Ferrara N, Gerber HP, LeCouter J (2003) The biology of VEGF and its receptors. Nat Med 9:669–676
12. Wesche J, Haglund K, Haugsten EM (2011) Fibroblast growth factors and their receptors in cancer. Biochem J 437:199–213
13. Rey S, Semenza GL (2010) Hypoxia-inducible factor-1-dependent mechanisms of vascularization and vascular remodelling. Cardiovasc Res 86:236–242
14. Singh S, Sadanandam A, Singh RK (2007) Chemokines in tumor angiogenesis and metastasis. Cancer Metastasis Rev 26:453–467
15. Wang S, Olson EN (2009) AngiomiRs–key regulators of angiogenesis. Curr Opin Genet Dev 19:205–211
16. Fish JE, Srivastava D (2009) MicroRNAs: opening a new vein in angiogenesis research. Sci Signal 2(52):1
17. Suarez Y, Fernandez-Hernando C, Yu J, Gerber SA, Harrison KD, Pober JS, Iruela-Arispe ML, Merkenschlager M, Sessa WC (2008) Dicer-dependent endothelial microRNAs are necessary for postnatal angiogenesis. Proc Natl Acad Sci U S A 105:14082–14087
18. Chen Z, Lai TC, Jan YH, Lin FM, Wang WC, Xiao H, Wang YT, Sun W, Cui X, Li YS et al (2013) Hypoxia-responsive miRNAs target argonaute 1 to promote angiogenesis. J Clin Invest 123:1057–1067
19. Pulkkinen K, Malm T, Turunen M, Koistinaho J, Yla-Herttuala S (2008) Hypoxia induces microRNA miR-210 in vitro and in vivo ephrin-A3 and neuronal pentraxin 1 are potentially regulated by miR-210. FEBS Lett 582:2397–2401
20. Zhou B, Ma R, Si W, Li S, Xu Y, Tu X, Wang Q (2013) MicroRNA-503 targets FGF2 and VEGFA and inhibits tumor angiogenesis and growth. Cancer Lett 333:159–169
21. Zou C, Xu Q, Mao F, Li D, Bian C, Liu LZ, Jiang Y, Chen X, Qi Y, Zhang X et al (2012) MiR-145 inhibits tumor angiogenesis and growth by N-RAS and VEGF. Cell Cycle 11:2137–2145
22. Isogai S, Horiguchi M, Weinstein BM (2001) The vascular anatomy of the developing zebrafish: an atlas of embryonic and early larval development. Dev Biol 230:278–301
23. Lawson ND, Weinstein BM (2002) Arteries and veins: making a difference with zebrafish. Nat Rev Genet 3:674–682
24. Herbert SP, Stainier DY (2011) Molecular control of endothelial cell behaviour during blood vessel morphogenesis. Nat Rev Mol Cell Biol 12:551–564
25. Santoro MM, Pesce G, Stainier DY (2009) Characterization of vascular mural cells during zebrafish development. Mech Dev 126:638–649
26. Alexander MR, Owens GK (2012) Epigenetic control of smooth muscle cell differentiation and phenotypic switching in vascular development and disease. Annu Rev Physiol 74:13–40
27. Hasan J, Shnyder SD, Bibby M, Double JA, Bicknel R, Jayson GC (2004) Quantitative angiogenesis assays in vivo–a review. Angiogenesis 7:1–16
28. Weinstein BM, Stemple DL, Driever W, Fishman MC (1995) Gridlock, a localized heritable vascular patterning defect in the zebrafish. Nat Med 1:1143–1147

29. Lawson ND, Weinstein BM (2002) In vivo imaging of embryonic vascular development using transgenic zebrafish. Dev Biol 248:307–318
30. Kamei M, Isogai S, Weinstein BM (2004) Imaging blood vessels in the zebrafish. Methods Cell Biol 76:51–74
31. Torres-Vazquez J, Gitler AD, Fraser SD, Berk JD, Van NP, Fishman MC, Childs S, Epstein JA, Weinstein BM (2004) Semaphorin-plexin signaling guides patterning of the developing vasculature. Dev Cell 7:117–123
32. Siekmann AF, Lawson ND (2007) Notch signalling limits angiogenic cell behaviour in developing zebrafish arteries. Nature 445:781–784
33. Stainier DY, Fouquet B, Chen JN, Warren KS, Weinstein BM, Meiler SE, Mohideen MA, Neuhauss SC, Solnica-Krezel L, Schier AF et al (1996) Mutations affecting the formation and function of the cardiovascular system in the zebrafish embryo. Development 123:285–292
34. Jin SW, Herzog W, Santoro MM, Mitchell TS, Frantsve J, Jungblut B, Beis D, Scott IC, D'Amico LA, Ober EA et al (2007) A transgene-assisted genetic screen identifies essential regulators of vascular development in vertebrate embryos. Dev Biol 307:29–42
35. Gore AV, Monzo K, Cha YR, Pan W, Weinstein BM (2012) Vascular development in the zebrafish. Cold Spring Harb Perspect Med 2:a006684
36. Seitz H (2009) Redefining microRNA targets. Curr Biol 19:870–873
37. Didiano D, Hobert O (2008) Molecular architecture of a miRNA-regulated 3' UTR. RNA 14:1297–1317
38. Plasterk RH (2006) Micro RNAs in animal development. Cell 124:877–881
39. Bushati N, Cohen SM (2008) MicroRNAs in neurodegeneration. Curr Opin Neurobiol 18:292–296
40. Giraldez AJ, Cinalli RM, Glasner ME, Enright AJ, Thomson JM, Baskerville S, Hammond SM, Bartel DP, Schier AF (2005) MicroRNAs regulate brain morphogenesis in zebrafish. Science 308:833–838
41. Giraldez AJ, Mishima Y, Rihel J, Grocock RJ, Van Dongen S, Inoue K, Enright AJ, Schier AF (2006) Zebrafish MiR-430 promotes deadenylation and clearance of maternal mRNAs. Science 312:75–79
42. Albinsson S, Suarez Y, Skoura A, Offermanns S, Miano JM, Sessa WC (2010) MicroRNAs are necessary for vascular smooth muscle growth, differentiation, and function. Arterioscler Thromb Vasc Biol 30:1118–1126
43. Lagendijk AK, Goumans MJ, Burkhard SB, Bakkers J (2011) MicroRNA-23 restricts cardiac valve formation by inhibiting Has2 and extracellular hyaluronic acid production. Circ Res 109:649–657
44. Follo C, Ozzano M, Mugoni V, Castino R, Santoro M, Isidoro C (2011) Knock-down of cathepsin D affects the retinal pigment epithelium, impairs swim-bladder ontogenesis and causes premature death in zebrafish. PLoS One 6:e21908
45. Chen PY, Manninga H, Slanchev K, Chien M, Russo JJ, Ju J, Sheridan R, John B, Marks DS, Gaidatzis D et al (2005) The developmental miRNA profiles of zebrafish as determined by small RNA cloning. Genes Dev 19:1288–1293
46. Kloosterman WP, Steiner FA, Berezikov E, de Bruijn E, van de Belt J, Verheul M, Cuppen E, Plasterk RH (2006) Cloning and expression of new microRNAs from zebrafish. Nucleic Acids Res 34:2558–2569
47. Thatcher EJ, Bond J, Paydar I, Patton JG (2008) Genomic organization of zebrafish microRNAs. BMC Genomics 9:253
48. Gays D, Santoro MM (2013) The admiR-able advances in cardiovascular biology through the zebrafish model system. Cell Mol Life Sci 70:2489–2503
49. Fiedler J, Jazbutyte V, Kirchmaier BC, Gupta SK, Lorenzen J, Hartmann D, Galuppo P, Kneitz S, Pena JT, Sohn-Lee C et al (2011) MicroRNA-24 regulates vascularity after myocardial infarction. Circulation 124:720–730
50. Bonauer A, Carmona G, Iwasaki M, Mione M, Koyanagi M, Fischer A, Burchfield J, Fox H, Doebele C, Ohtani K et al (2009) MicroRNA-92a controls angiogenesis and functional recovery of ischemic tissues in mice. Science 324:1710–1713

51. Fish JE, Santoro MM, Morton SU, Yu S, Yeh RF, Wythe JD, Ivey KN, Bruneau BG, Stainier DY, Srivastava D (2008) miR-126 regulates angiogenic signaling and vascular integrity. Dev Cell 15:272–284
52. Nicoli S, Standley C, Walker P, Hurlstone A, Fogarty KE, Lawson ND (2010) MicroRNA-mediated integration of haemodynamics and Vegf signalling during angiogenesis. Nature 464:1196–1200
53. Nicoli S, Knyphausen CP, Zhu LJ, Lakshmanan A, Lawson ND (2012) miR-221 is required for endothelial tip cell behaviors during vascular development. Dev Cell 22:418–429
54. Voz ML, Coppieters W, Manfroid I, Baudhuin A, Von Berg V, Charlier C, Meyer D, Driever W, Martial JA, Peers B (2012) Fast homozygosity mapping and identification of a zebrafish ENU-induced mutation by whole-genome sequencing. PLoS One 7:e34671
55. Nasevicius A, Ekker SC (2000) Effective targeted gene 'knockdown' in zebrafish. Nat Genet 26:216–220
56. Wienholds E, Schulte-Merker S, Walderich B, Plasterk RH (2002) Target-selected inactivation of the zebrafish rag1 gene. Science 297:99–102
57. Zhu C, Smith T, McNulty J, Rayla AL, Lakshmanan A, Siekmann AF, Buffardi M, Meng X, Shin J, Padmanabhan A et al (2011) Evaluation and application of modularly assembled zinc-finger nucleases in zebrafish. Development 138:4555–4564
58. Bedell VM, Wang Y, Campbell JM, Poshusta TL, Starker CG, Krug RG 2nd, Tan W, Penheiter SG, Ma AC, Leung AY et al (2012) In vivo genome editing using a high-efficiency TALEN system. Nature 491:114–118
59. Chang N, Sun C, Gao L, Zhu D, Xu X, Zhu X, Xiong JW, Xi JJ (2013) Genome editing with RNA-guided Cas9 nuclease in zebrafish embryos. Cell Res 23:465–472
60. Hans S, Freudenreich D, Geffarth M, Kaslin J, Machate A, Brand M (2011) Generation of a non-leaky heat shock-inducible Cre line for conditional Cre/lox strategies in zebrafish. Dev Dyn 240:108–115
61. Mosimann C, Zon LI (2011) Advanced zebrafish transgenesis with Tol2 and application for Cre/lox recombination experiments. Methods Cell Biol 104:173–194
62. Ramachandran R, Reifler A, Wan J, Goldman D (2012) Application of Cre-loxP recombination for lineage tracing of adult zebrafish retinal stem cells. Methods Mol Biol 884:129–140
63. Goessling W, North TE, Zon LI (2007) New waves of discovery: modeling cancer in zebrafish. J Clin Oncol 25:2473–2479
64. Patton EE, Zon LI (2001) The art and design of genetic screens: zebrafish. Nat Rev Genet 2:956–966
65. Amsterdam A, Hopkins N (2006) Mutagenesis strategies in zebrafish for identifying genes involved in development and disease. Trends Genet 22:473–478
66. Amatruda JF, Patton EE (2008) Genetic models of cancer in zebrafish. Int Rev Cell Mol Biol 271:1–34
67. Berghmans S, Murphey RD, Wienholds E, Neuberg D, Kutok JL, Fletcher CD, Morris JP, Liu TX, Schulte-Merker S, Kanki JP et al (2005) tp53 mutant zebrafish develop malignant peripheral nerve sheath tumors. Proc Natl Acad Sci U S A 102:407–412
68. Pliss GB, Khudoley VV (1975) Tumor induction by carcinogenic agents in aquarium fish. J Natl Cancer Inst 55:129–136
69. Spitsbergen JM, Tsai HW, Reddy A, Miller T, Arbogast D, Hendricks JD, Bailey GS (2000) Neoplasia in zebrafish (Danio rerio) treated with N-methyl-N′-nitro-N-nitrosoguanidine by three exposure routes at different developmental stages. Toxicol Pathol 28:716–725
70. Topczewska JM, Postovit LM, Margaryan NV, Sam A, Hess AR, Wheaton WW, Nickoloff BJ, Topczewski J, Hendrix MJ (2006) Embryonic and tumorigenic pathways converge via Nodal signaling: role in melanoma aggressiveness. Nat Med 12:925–932
71. Lee LM, Seftor EA, Bonde G, Cornell RA, Hendrix MJ (2005) The fate of human malignant melanoma cells transplanted into zebrafish embryos: assessment of migration and cell division in the absence of tumor formation. Dev Dyn 233:1560–1570
72. Nicoli S, Presta M (2007) The zebrafish/tumor xenograft angiogenesis assay. Nat Protoc 2:2918–2923

73. Haldi M, Ton C, Seng WL, McGrath P (2006) Human melanoma cells transplanted into zebrafish proliferate, migrate, produce melanin, form masses and stimulate angiogenesis in zebrafish. Angiogenesis 9:139–151
74. Marques IJ, Weiss FU, Vlecken DH, Nitsche C, Bakkers J, Lagendijk AK, Partecke LI, Heidecke CD, Lerch MM, Bagowski CP (2009) Metastatic behaviour of primary human tumours in a zebrafish xenotransplantation model. BMC Cancer 9:128
75. Pruvot B, Jacquel A, Droin N, Auberger P, Bouscary D, Tamburini J, Muller M, Fontenay M, Chluba J, Solary E (2011) Leukemic cell xenograft in zebrafish embryo for investigating drug efficacy. Haematologica 96:612–616
76. Corkery DP, Dellaire G, Berman JN (2011) Leukaemia xenotransplantation in zebrafish—chemotherapy response assay in vivo. Br J Haematol 153:786–789
77. Konantz M, Balci TB, Hartwig UF, Dellaire G, Andre MC, Berman JN, Lengerke C (2012) Zebrafish xenografts as a tool for in vivo studies on human cancer. Ann N Y Acad Sci 1266:124–137
78. Stoletov K, Montel V, Lester RD, Gonias SL, Klemke R (2007) High-resolution imaging of the dynamic tumor cell vascular interface in transparent zebrafish. Proc Natl Acad Sci U S A 104:17406–17411
79. Nicoli S, Ribatti D, Cotelli F, Presta M (2007) Mammalian tumor xenografts induce neovascularization in zebrafish embryos. Cancer Res 67:2927–2931
80. Nicoli S, De Sena G, Presta M (2009) Fibroblast growth factor 2-induced angiogenesis in zebrafish: the zebrafish yolk membrane (ZFYM) angiogenesis assay. J Cell Mol Med 13:2061–2068
81. Serbedzija GN, Flynn E, Willett CE (1999) Zebrafish angiogenesis: a new model for drug screening. Angiogenesis 3:353–359
82. Bayliss PE, Bellavance KL, Whitehead GG, Abrams JM, Aegerter S, Robbins HS, Cowan DB, Keating MT, O'Reilly T, Wood JM et al (2006) Chemical modulation of receptor signaling inhibits regenerative angiogenesis in adult zebrafish. Nat Chem Biol 2:265–273
83. Cross LM, Cook MA, Lin S, Chen JN, Rubinstein AL (2003) Rapid analysis of angiogenesis drugs in a live fluorescent zebrafish assay. Arterioscler Thromb Vasc Biol 23:911–912
84. Tran TC, Sneed B, Haider J, Blavo D, White A, Aiyejorun T, Baranowski TC, Rubinstein AL, Doan TN, Dingledine R et al (2007) Automated, quantitative screening assay for antiangiogenic compounds using transgenic zebrafish. Cancer Res 67:11386–11392
85. Letamendia A, Quevedo C, Ibarbia I, Virto JM, Holgado O, Diez M, Izpisua Belmonte JC, Callol-Massot C (2012) Development and validation of an automated high-throughput system for zebrafish in vivo screenings. PLoS One 7:e36690
86. Lessman CA (2011) The developing zebrafish (Danio rerio): a vertebrate model for high-throughput screening of chemical libraries. Birth Defects Res C Embryo Today 93:268–280
87. Zon LI, Peterson RT (2005) In vivo drug discovery in the zebrafish. Nat Rev Drug Discov 4:35–44
88. Peterson RT, Link BA, Dowling JE, Schreiber SL (2000) Small molecule developmental screens reveal the logic and timing of vertebrate development. Proc Natl Acad Sci U S A 97:12965–12969
89. Ulrich R, Friend SH (2002) Toxicogenomics and drug discovery: will new technologies help us produce better drugs? Nat Rev Drug Discov 1:84–88
90. Borchers C, Boer R, Klemm K, Figala V, Denzinger T, Ulrich WR, Haas S, Ise W, Gekeler V, Przybylski M (2002) Characterization of the dexniguldipine binding site in the multidrug resistance-related transport protein P-glycoprotein by photoaffinity labeling and mass spectrometry. Mol Pharmacol 61:1366–1376
91. Peterson RT, Macrae CA (2012) Systematic approaches to toxicology in the zebrafish. Annu Rev Pharmacol Toxicol 52:433–453
92. McGrath P, Li CQ (2008) Zebrafish: a predictive model for assessing drug-induced toxicity. Drug Discov Today 13:394–401
93. Davidson W, Ren Q, Kari G, Kashi O, Dicker AP, Rodeck U (2008) Inhibition of p73 function by Pifithrin-alpha as revealed by studies in zebrafish embryos. Cell Cycle 7:1224–1230

Vascular Complications in Diabetes

15

Louise L. Dunn, Kim Hoe Chan, Martin K.C. Ng, and Roland Stocker

Contents

15.1 Introduction 314
15.1.1 Diabetes Mellitus 315
15.2 Pathways Underpinning Diabetic Vascular Complications 316
15.3 Microangiopathy in Diabetes Mellitus 318
15.3.1 Retinopathy 319
15.3.2 Nephropathy 321
15.3.3 Neuropathy 323
15.3.4 Wound Healing 324
15.4 Macroangiopathy in Diabetes Mellitus 326
15.4.1 Coronary Artery Disease 328
15.4.2 Peripheral Arterial Disease 329
15.4.3 Cerebral Ischemia 329
15.5 Future Directions 330
References 332

Abstract

The vascular complications of diabetes mellitus are accountable for significant morbidity and mortality of the disease worldwide. A striking feature of diabetes is the heterogeneity in the dysregulation of angiogenesis. Excessive and disordered angiogenesis predominate in microvessels, leading to retinopathy and nephropathy. Insufficient neovascularization features heavily in the diabetic wound. This is aggravated by neuropathy and poor nutritive blood flow due to

L.L. Dunn (✉) • R. Stocker
Victor Chang Cardiac Research Institute, Lowy Packer Building, 405 Liverpool Street, Darlinghurst 2010, Australia

Faculty of Medicine, The University of New South Wales, Kensington 2052, Australia
e-mail: l.dunn@victorchang.edu.au

K.H. Chan • M.K.C. Ng
Department of Cardiology, Royal Prince Alfred Hospital, Camperdown 2050, Australia

J. Dulak et al. (eds.), *Angiogenesis and Vascularisation*,
DOI 10.1007/978-3-7091-1428-5_15,

peripheral vascular disease. These complications can converge leading to amputation. Failure of neovascularization or collateralization in atheroocclusive diseased macrovessels can precipitate myocardial infarction and stroke. In this chapter, the features and mechanisms underlying the various vascular complications associated with diabetes will be reviewed.

Keywords
Diabetes mellitus • Angiogenesis • Neovascularization • Diabetic vasculopathies

Abbreviations

AGE	Advanced glycation end products
CAD	Coronary artery disease
DAG	Diacylglycerol
eNOS	Endothelial nitric oxide synthase
EPC	Endothelial progenitor cell
MAPK	Mitogen activated protein kinase
PAD	Peripheral artery disease
PKC	Protein kinase C
RAGE	Receptor for advanced glycation end products
T1DM	Type 1 diabetes mellitus
T2DM	Type 2 diabetes mellitus
VEGF	Vascular endothelial growth factor

15.1 Introduction

The vascular complications of diabetes mellitus are responsible for much of the morbidity and mortality associated with this disease worldwide. In 2011, The World Health Organization reported that 347 million people suffered from diabetes worldwide and predicted that the number of deaths from diabetes will double between 2005 and 2030 [1]. While diabetes has been associated with the Western lifestyle, 80 % of those afflicted with the disease reside in low- and middle-income countries [1]. As the population will continue to grow and age, and developing countries become increasingly industrialized and affluent, diabetes and its ensuing vascular complications are expected to continue to rise. Currently, there is no silver bullet for the vascular complication of diabetes. The primary treatment approach is normalization of blood sugar levels combined with rigorous management of cardiovascular risk factors. Strict blood glucose control mitigates microvascular complications whereas it does not prevent macrovascular disease associated with diabetes. Aside glucose control, angiotensin converting enzyme inhibitors and fenofibrates also reduce diabetic microvascular complications. Thus, other factors such as disordered lipid metabolism (dyslipidemia) and chronic inflammation also must be considered in the context of vascular disease in the setting of diabetes.

15.1.1 Diabetes Mellitus

Diabetes mellitus is a group of metabolic diseases defined by high blood sugar concentration (hyperglycemia), which leads to symptoms of excessive thirst (polydipsia), excessive hunger (polyphagia), and excessive urination (polyuria). The three types of diabetes mellitus are (1) type 1 diabetes mellitus (T1DM), (2) type 2 diabetes mellitus (T2DM), and (3) gestational diabetes mellitus. T1DM (previously known as insulin-dependent diabetes mellitus) accounts for approximately 10 % of diabetes cases and usually presents early in life as ketoacidosis and rapid weight loss [1]. T1DM is caused by the autoimmune destruction of the insulin secreting pancreatic beta cells. Both genetics and environmental factors contribute to initiating the autoimmune destruction. T1DM patients have no endogenous insulin production and cannot regulate their blood glucose levels without a strict regime of administered insulin. T2DM accounts for the remaining 90 % of diabetes cases worldwide [1]. T2DM typically develops later in life and has strong associations with obesity, dyslipidemia, and lifestyle. T2DM is characterized by insulin resistance of tissues and an initial increase in production of insulin by the beta cells to compensate for the resistance. In the long term, insulin production may ultimately decrease as beta cells are exhausted, and thereby T2DM may also lead to a requirement for exogenous insulin. Treatment for T2DM includes weight loss, oral hypoglycemic drugs and, in some instances, insulin therapy. Gestational diabetes mellitus occurs during pregnancy in women who have no previous history of diabetes and affects up to 10 % of pregnancies [1]. Gestational diabetes is characterized by a failure of the insulin receptor to function properly, which leads to elevations in blood glucose concentration. Women who experience gestational diabetes have a higher risk of developing T2DM later in life.

The hyperglycemia of diabetes is a major contributor to vascular complications associated with the disease. High blood glucose levels can lead to glycation (nonenzymatic addition of sugars) of plasma proteins, such as hemoglobin. Glycated hemoglobin (also known as HbA1c) has been correlated with vascular complications in the major vessels, including the coronary, peripheral, and cerebral circuits, and also the vascular beds servicing the eye, neurons, and dermis [2–8]. HbA1c concentration per se is an indicator of prolonged, elevated glucose concentrations rather than a cause of diabetic vascular complications. High glucose concentrations can lead to glycation of plasma and tissue proteins, resulting in carbonyl and oxidative stress that may contribute to the pathophysiology underlying diabetic vascular complications.

The single-cell layer lining the lumen of blood vessels (endothelium), being the interface between the blood and tissues, is particularly sensitive to hyperglycemia. Dysfunction of the endothelium is therefore considered a hallmark of diabetic vascular complications and is highly prognostic of vascular events [9, 10].

Endothelial dysfunction is typically defined as a decrease in the ability of vessels to relax (vasodilation), with an overall decrease in nitric oxide availability. Endothelial dyfunction is commonly associated with increases in oxidative and/or carbonyl stress, the propensity to coagulation, and the expression of adhesion molecules, facilitating the development of atherosclerosis and vessel stiffening [11]. Other functions of the endothelium include volume and electrolyte control, and also the growth of new blood vessels (angiogenesis). Angiogenesis in response to atherosclerotic disease and tissue ischemia is particularly deregulated in diabetes.

15.2 Pathways Underpinning Diabetic Vascular Complications

In diabetes, high blood glucose levels contribute to defective angiogenesis in some tissues and disordered neovascularization in others. High blood glucose levels can exert their deleterious effects on the vasculature via four major yet highly interrelated pathways [12]. These include (1) the polyol pathway, (2) hexosamine flux, (3) the diacylglycerol/protein kinase C (DAG/PKC) pathway, and (4) advanced glycation end products (AGE), as indicated in Fig. 15.1. Imbalances in these pathways can lead to carbonyl and oxidative stress.

Carbonyl stress occurs when there is an increase in reactive carbonyl species resulting from elevated glucose concentrations. Carbonyl species may be formed by nonenzymatic and enzymatic reactions and are precursors to AGE. AGE can be formed nonenzymatically, e.g., excessive glucose can lead to Maillard, Schiff base, and Amadori reactions of carbohydrates with the amino acids on proteins. The products of these reactions may then degrade to carbonyl compounds such as methlyglyoxal, glyoxal, and deoxyglucosone [13]. It should be noted that AGE may also be formed by "autoxidative glycosylation" and "glycoxidation" [14]. AGE can exert their deleterious effects via receptor-independent or -dependent mechanisms. Receptor-independent mechanisms include, but are not limited to, the cross-linking of collagen that increases vascular stiffness and damage, and expansion of basement membrane and matrices [15]. Receptors of AGE (RAGE) are expressed on a number of cell types including endothelial cells, endothelial progenitor cells (EPC), vascular smooth muscle cells, mononuclear cell, neurons, and fibroblasts, making these cells vulnerable to damage mediated by high concentrations of glucose. AGE may bind to receptors (RAGE) to activate intracellular signaling pathways [16]. These include increases in NADPH oxidase mediated formation of reactive oxygen species and activation of mitogen activated protein kinases (MAPK) to increase the activity of the transcription factor, nuclear factor-κB. In turn, nuclear factor-κB initiates transcription of pro-inflammatory and growth factor genes. AGE binding to RAGE can also decrease endothelial nitric oxide synthase (eNOS) activity and nitric oxide bioavailability, further increasing vascular stiffness [17]. In addition, glycosylation of low-density lipoprotein has been

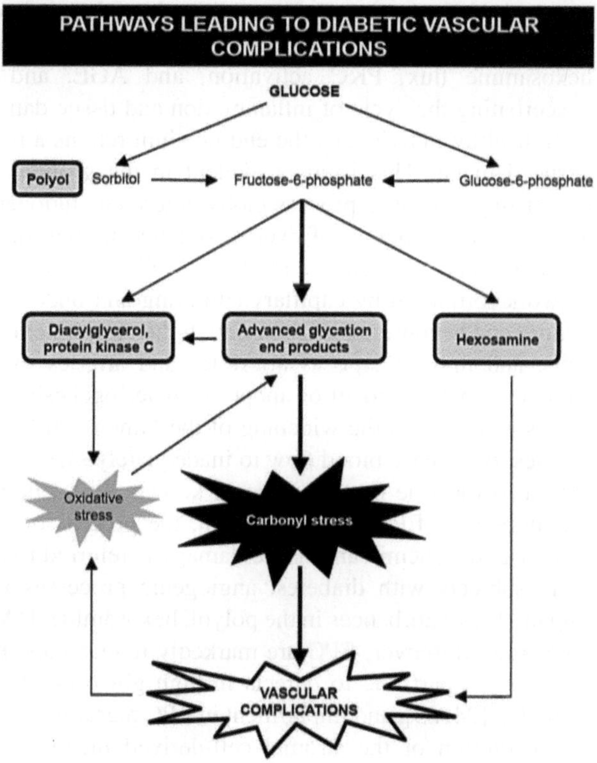

Fig. 15.1 Pathways leading to diabetic vascular complications. Four convergent pathways considered responsible for the toxic effects of glucose that lead to diabetic vascular complications. These include (1) the polyol pathway, (2) the hexosamine pathway, (3) the diacylglycerol/protein kinase C pathway, and (4) advanced glycation end products, which result in carbonyl and oxidative stress

suggested to increase low-density lipoprotein uptake in macrophages, increasing lipid deposition in atherosclerotic lesions [18].

Oxidative stress has long been associated with diabetic vascular complications and variously attributed as a key driver [12]. Indeed, oxidative stress can accelerate AGE formation. However, the Baynes and Thorpe model elegantly proposes that carbonyl stress is a causal driver in the development of complications, while oxidative stress is a consequence of tissue damage and carbonyl stress that may exacerbate disease progression [13]. Indeed, hydrogen peroxide, commonly derived from superoxide, is critical for cellular signaling under normal physiological conditions, and antioxidant enzymes, e.g., superoxide dismutase [19] and glutathione reductase [20], themselves may become inactivated as a result of glycosylation. Also, a recent study comparing the antioxidant content and extent of lipid oxidation in carotid lesions of subjects with T2DM and age-matched controls did not find significant differences in support of the notion that diabetes increases oxidative damage in arterial lesions [21]. Moreover, to date, antioxidant therapies have not been shown to conclusively protect or treat diabetic vascular complications whereas substrate reduction, i.e., glucose lowering, has had demonstrable effects [2–8]. Nevertheless, once present at supra-physiological levels, reactive oxygen species can cause tissue damage by a number of mechanisms including induction of apoptosis, inflammation, and modification of carbohydrates, lipids, and proteins. Excessive production of superoxide further increases polyol pathway flux, hexosamine flux, PKC activation, and AGE, and decreases eNOS activity, exacerbating the cycle of inflammation and tissue damage [12].

In healthy individuals, the endothelium retains a remarkable ability to adapt to tissue damage. This is most evident in the growth of new blood vessels from preexisting vessels, a process globally termed angiogenesis. This process involves a coordinated program of events, signaled by vascular endothelial growth factor (VEGF), that stimulates endothelial cell proliferation, migration, and vascular network formation by capillary sprouting and budding [22]. New capillaries may mature and become stabilized through the recruitment of pericytes and other mural cells, and may emerge as arterioles and arteries over time. Alternatively, new capillaries may remodel or apoptose. Arteriogenesis and collateralization are processes that refer to the widening of the lumen diameter and growth of preexisting arteries to increase blood flow to inadequately served vascular beds. Blood vessels may also form de novo, a process known as vasculogenesis, which involves the recruitment of EPC [22, 23]. Often, the process of new blood vessel growth in response to ischemia and tissue damage is referred to as neovascularization.

In subjects with diabetes, angiogenic processes are markedly disordered or impaired as disturbances in the polyol, hexosamine, DAG/PKC, and AGE pathways converge. Moreover, EPC are markedly reduced and exhibit impaired angiogenic function, in part due to defects in high glucose-induced VEGF and nitric oxide signaling [24, 25], and impairment in EPC recruitment to sites of neovascularization via disruption of the stromal cell-derived factor-1 signaling [26, 27]. Diabetic vascular complications may be grouped according to the vessels that they affect, e.g., the microvessels (microangiopathy) and macrovessels (macroangiopathy).

15.3 Microangiopathy in Diabetes Mellitus

The microvessels comprise arterioles, capillaries, and venules. Capillaries may be further subdivided into continuous, fenestrated, e.g., glomeruli in the kidney, or discontinuous, e.g., liver sinusoids. Continuous capillaries only allow small molecules to diffuse. Fenestrated capillaries have pores that allow some proteins to diffuse. Discontinuous capillaries have openings in the endothelium that allow transit of blood cells and plasma proteins. To enable efficient transfer of nutrients and waste, capillaries are only one cell thick. However, pericytes are embedded in the basement membrane around capillaries and venules, providing stability and assisting in the maintenance of capillary constriction and dilation. Nevertheless, this thinness makes capillaries, and their associated arterioles and venules, highly susceptible to damage. In diabetics, the predominantly affected microvascular beds include the eye (retinopathy), kidney (nephropathy), nerves (neuropathy), and skin (dermis; wound healing), all of which are nonresponsive to insulin-mediated glucose uptake.

15.3.1 Retinopathy

Diabetic retinopathy is attributed to ~5 % of blindness or severe visual impairment cases worldwide [1]. However, in developed countries (where other eye diseases such as cataracts are preventable or easily treated), diabetic retinopathy may account for ~15–20 % of blindness or visual impairment [28]. The prevalence of diabetic retinopathy increases with disease duration. After two decades of disease, ~100 % of T1DM patients and 50 % of T2DM patients have diabetic retinopathy; 10 % of all diabetics will have severe visual impairment and 2 % will become blind [1]. The severity of diabetic retinopathy is a continuum, ranging from nonproliferative to proliferative. Nonproliferative diabetic retinopathy may be mild to severe, and proliferative retinopathy is when abnormal neovascularization occurs. Disease development, progression, and severity are correlated strongly with poor glycemic control [2, 29]. Treatment of hypertension demonstrates the utility in mitigating diabetic retinopathy [30, 31]. Similarly, the lipid-lowering drug fenofibrate appears to have beneficial effects on diabetic retinopathy, albeit independent of its lipid lowering action [32].

The earliest feature of diabetic retinopathy is the loss of pericytes around the capillaries due to apoptosis and migration [33, 34]. The loss of pericytes impairs capillary tone, i.e., the ability of capillaries to dilate and constrict. This observation is associated with endothelial cell injury and basement membrane thickening. The latter paradoxically increases vascular permeability due to changes in the matrix that lead to increases in pore size between endothelial cells and also to changes in electrical charge at these pores [35]. The net effect is the transport across the endothelium of large molecules such as albumin that are usually excluded from passaging. Basement membrane thickening also decreases the diffusion of oxygen across the endothelium to the underlying cells and tissues. Furthermore, loss of pericytes and a cycle of tissue damage, basement membrane thickening, ischemia, and disordered neovascularization create an environment susceptible to the deposition of fibrous connective tissue. Concurrent increases in ocular pressure due to edema aggravate disease progression. Clinically, the earliest signs of nonproliferative diabetic retinopathy are micro-aneurysms and hemorrhages in the retina. As the disease progresses, damaged nerve fibers (resulting in swelling, referred to as cotton wool spots), venous hemorrhaging, and capillary proliferation are observed. Proliferative retinopathy is evident with the expansive growth of neovessels, vitreous hemorrhage, and fibrosis, which can compound in retinal detachment, leading to partial or total vision loss. Central vision loss may occur from retinal vasculature leakage and resultant edema.

The mechanisms underpinning diabetic retinopathy are driven, in a major part, by hyperglycemia. High glucose increases the substrate pool for the polyol, hexosamine, AGE, and DAG/PKC pathways (Fig. 15.2). In the polyol pathway, glucose is reduced by aldose reductase to sorbitol, which may accumulate in cells and possibly lead to osmotic stress and increased permeability [36, 37]. Metabolism of sorbitol to fructose and subsequent phosphorylation of fructose produces the glycosylating agents, fructose-3-phosphate, and 3-deoxyglucosone, which are precursors of AGE. The metabolism of

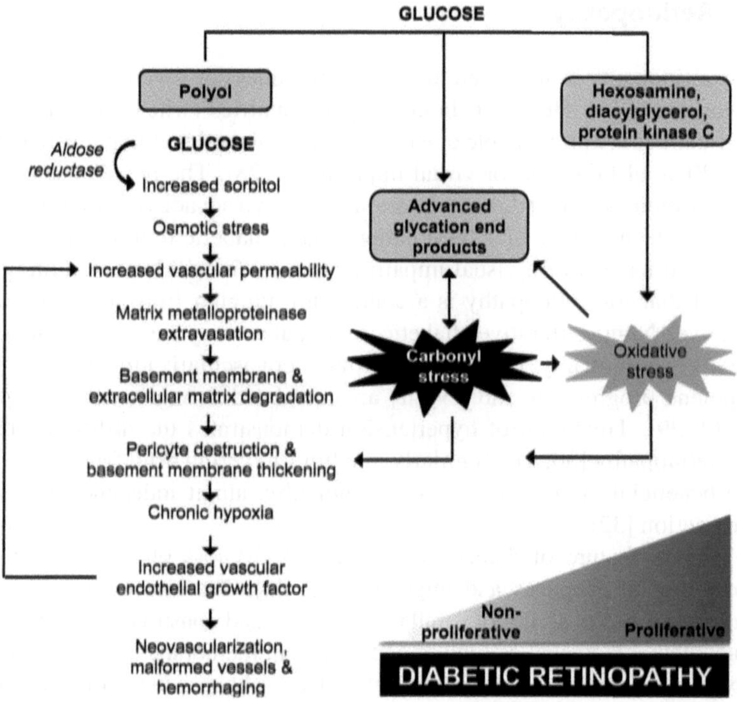

Fig. 15.2 Excessive neovascularization in diabetic retinopathy. High glucose concentrations increase the polyol pathway, and diacylglycerol and protein kinase C activation, and hexosamine pathway. Increased sorbitol may lead to osmotic stress resulting in expansion and degradation of the basement membrane and extracellular matrix, with increased vascular permeability. Degradation of the extracellular matrix is a vicious cycle that results in destruction of pericytes and ultimately leads to thickening of the basement membrane and chronic hypoxia. Low oxygen tension induces the production of vascular endothelial growth factor and other growth factors leading to neovascularization. However, neovessels are malformed, lacking pericytes and intact basement membranes, and vascular endothelial growth factor increases vascular leakage leading to hemorrhage and intensification of the cycle. Glucose also increases carbonyl stress, advanced glycation end products and their receptors, and oxidative stress to exacerbate the damage. Diabetic retinopathy may be categorized as nonproliferative, in which vessels rupture and leak, or proliferative, in which excessive abnormal neovascularization occurs

glucose to fructose also consumes NADPH and NAD^+ resulting in a decrease in reducing equivalents for glutathione reductase to maintain cellular glutathione, and an increased ratio of NADH/NAD^+, respectively. An increase in the NADH/NAD^+ ratio can lead to pseudo-hypoxia and a switch to glycolytic metabolism [38], which may assist in sustaining cell proliferation. Binding of AGE to RAGE increases oxidant production, decreases eNOS expression, and upregulates VEGF to drive neovascularization [39]. Increased inflammation and oxidative stress leads to apoptosis [40, 41]. Both VEGF and tumor necrosis factor-α increase the expression of adhesion molecules, such as intracellular adhesion molecule-1, that attracts white blood cells and can lead to leukostasis and contribute in part to retinal ischemia [42–44]. Furthermore, high glucose- and hypoxia-induced VEGF gene transcription is dependent upon PKC activation, while VEGF binding to its receptor, in turn, increases DAG to activate PKC [45].

15.3.2 Nephropathy

The World Health Organization estimates between 10 and 20 % of diabetes subjects die of renal failure [1]. Following the appearance of overt nephropathy ~75 % of T1DM and ~20 % of T2DM subjects will have progressed to end stage renal disease after 20 years [1]. Diabetic nephropathy progresses through four stages. High blood glucose concentrations initially cause the kidney to filter larger quantities of blood, a process called hyperfiltration. Microalbuminuria is the next stage when small amounts of protein leak into the urine. The presence of microalbuminuria is recognized clinically as overt nephropathy. Over time, larger amounts of protein may leak, a process termed macroalbuminuria that precedes end stage renal disease, where the kidneys have lost their ability to filter toxins leading to a buildup of waste in the blood. At the end stage dialysis or kidney transplantation is the only option. A common consequence of diabetic nephropathy is hypertension, and management of hypertension may prove useful in delaying the progression of nephropathy [31]. The risk of development and progression of diabetic nephropathy can also be significantly reduced by intensive blood glucose control [2, 5]. Blockade of the renin–angiotensin system with angiotensin converting enzyme inhibitors or angiotensin II receptor blockers can decrease efferent arteriolar pressure thereby reducing intra-glomerular pressure and helping to protect the glomerulus from further damage [30, 31]. More recently, fenofibrate has also been reported to slow progression of diabetic nephropathy [46].

The glomeruli are the functional units of the kidney. The glomerulus is a highly specialized network of capillaries that are the primary filter of blood. In both T1DM and T2DM there are abnormal and excessive blood vessels in the glomeruli. The abnormal vessels may be dilated with adjacent basement membrane thinning apparent. Extra abnormal vessels are also found with swollen endothelial cells and a thickening of the vascular wall. The thickening of the vascular wall and associated changes to the matrix can increase the pore size between endothelial cells, leading to vascular permeability and the passage of larger molecules across the endothelium. This neovascularization is associated with glomerular hypertrophy and hyperfiltration, as the kidneys work harder due to elevated glucose. A consequence of this hypertrophy is Kimmelstiel–Wilson nodules characterized by the thickening of the basement membrane, mesangial matrix expansion, tubulointerstitial fibrosis, and consequently albuminuria.

Pathways leading to diabetic nephropathy are outlined in Fig. 15.3. AGE play a key role in diabetic nephropathy, as the kidney is responsible for eliminating AGE from the circulation [47]. Studies in human and rodents are somewhat conflicting in regard to renal VEGF, although this could be due to the duration of experimental models and the lack of assessment of VEGF in the early stages of diabetic

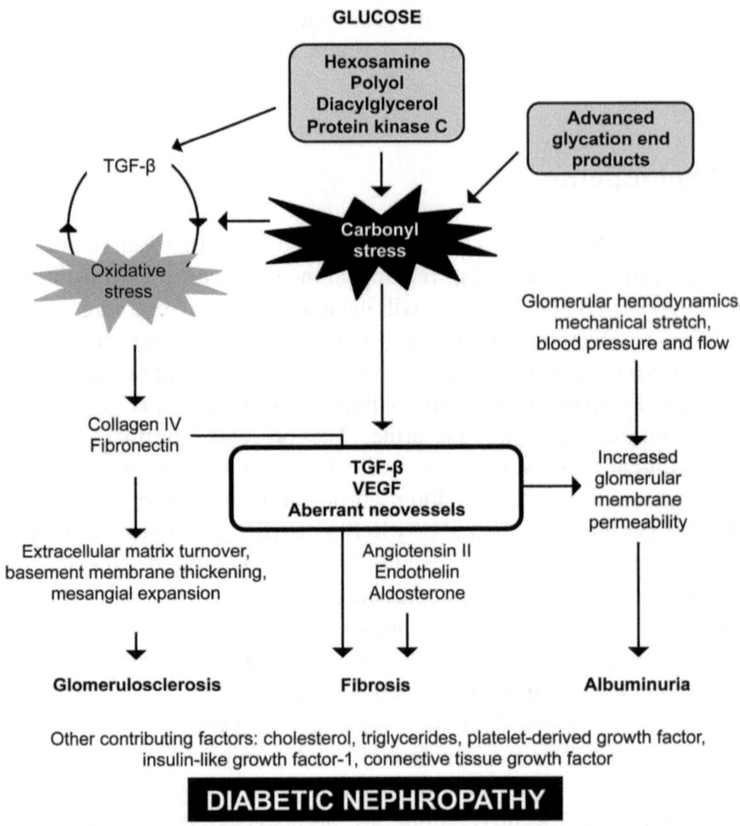

Fig. 15.3 Excessive neovascularization in diabetic nephropathy. High glucose increases hexosamine pathway flux, diacylglycerol, the polyol pathway, and protein kinase C activation resulting in carbonyl and oxidative stress, and increases in advanced glycation end products and transforming growth factor-β (TGF-β). Elevated TGF-β leads to increased production of collagen and fibronectin, resulting in mesangial expansion and extracellular matrix synthesis/degradation, which contributes to glomerulosclerosis and tubulointerstitial fibrosis. Increased TGF-β and vascular endothelial growth factor (VEGF) lead to aberrant angiogenesis. Increased vascularization around the glomeruli in turn leads to membrane permeability and hyperfiltration resulting in albuminuria

nephropathy in humans. Nevertheless, in animals AGE can increase VEGF in podocytes leading to basement membrane derangements [48]. AGE also increase the expression of collagen IV and transforming growth factor-β in podocytes and tubular cells to cause expansion of the mesangial matrix and tubulointerstitial fibrosis [49, 50]. The binding of AGE to RAGE leads to elevated reactive oxygen species, activation of nuclear factor-κB, and MAPK signaling, which increases intracellular adhesion molecules, inflammation, angiotensin II, transforming growth factor-β, connective tissue growth factor, and PKC signaling [51]. PKC signaling, in turn, amplifies this response. The inability of the kidney to filter and remove AGE increases their circulating levels, which can aggravate other diabetic vascular complications. It is likely that the polyol pathway also contributes to diabetic nephropathy in some degree [52].

15.3.3 Neuropathy

Neuropathy is observed in ~50 % of diabetes subjects 10–20 years after diagnosis [1]. The possible effects of diabetic neuropathy are myriad and dependent upon the tissues and organs affected. Autonomic neuropathy can affect the cardiovascular system, lungs, gastrointestinal tract, sweat glands, and eyes, and it can lead to "silent" symptoms such as myocardial infarction and hypoglycemia unawareness. Peripheral neuropathy affects the upper limbs, lower limbs, and extremities, leading to pain and/or loss of sensation/pain. The loss of sensation and pain in the lower extremities, coupled with impairment in wound healing (Sect. 3.4) and reduced blood flow due to peripheral artery disease (PAD; Sect. 4.2), may lead to the development of ulcers. Between 10 and 25 % of diabetics suffer a foot ulcer in their lifetime and ~85 % of all nontraumatic lower extremity amputations in diabetics are preceded by foot ulceration [53, 54]. People with a history of diabetic foot ulcers have ~40 % increased mortality after 10 years than diabetics without foot ulcers [55]. Moreover, ~50 % of those that undergo one amputation will have their other limb amputated after 5 years [56]. However, these outcomes may be avoided if neuropathy is detected early and blood glucose is controlled tightly.

In diabetic neuropathy, the blood vessels and nerves are inextricably linked (Fig. 15.4)—normal nerve function depends upon sufficient blood supply while blood vessel function is dependent upon normal nerve function. The microvessels that supply blood to the nerves are known as *vasa nervorum* and the first changes observed in diabetes include vasoconstriction, endothelial hyperplasia due to pericyte destruction, basement membrane thickening, and initial axonal thickening. These changes reduce the blood supply to the nerve resulting in hypoxia and neuronal ischemia. As the blood supply diminishes, demyelination can occur leading to impaired nerve conduction [57]. Neuronal ischemia may eventually lead to atrophy and axonal loss. High glucose concentrations elicit these changes in the *vasa nervorum* via mechanisms similar to those seen in retinopathy and nephropathy. Polyol pathway and hexosamine flux, and PKC activation increases AGE and oxidative stress to disrupt the endothelium, causing vascular leakage and disordered neovascularization. Thickening of the basement membrane surrounding capillaries can lead to a decrease in blood flow and hence nutrient and oxygen supply to the neuron, and this may result in neuronal ischemia. These localized changes in the endothelial cells are mirrored in the neighboring neuron, creating a milieu of tissue damage that can lead to impaired nerve conduction velocity (via polyol pathway decreases in the Na^+/K^+ ATPase required for conduction), demyelination, and neuronal loss [58].

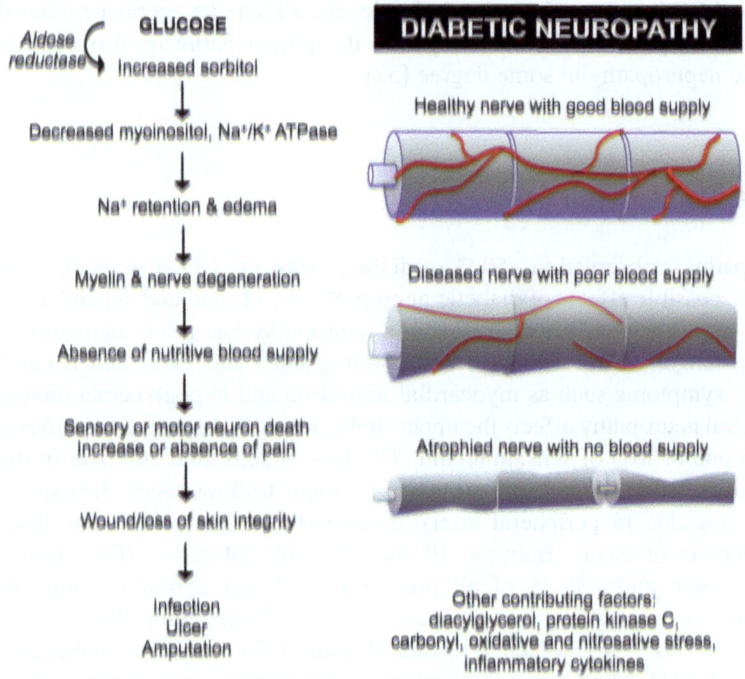

Fig. 15.4 Decreased blood supply in diabetic neuropathy. Chronic high glucose increases the polyol pathway and sorbitol levels that decrease the sodium-potassium adenosine triphosphatase (Na^+/K^+ ATPase), leading to sodium retention and subsequent edema. Edema at the myelin and axonal junctions results in neurodegeneration. The blood supply to the diseased nerve reduces leading to atrophy and cell death. Poor circulation, insufficient neovascularization, and the typically distal location of the wound can lead to infection and amputation

15.3.4 Wound Healing

Foot and lower extremity ulceration typify impaired wound healing in diabetes. Lower extremity wounds in diabetics are associated with neuropathy, lack of nutritive blood supply due to PAD, and increased infection due to reduced immune function, which may ultimately culminate in amputation [11]. Wound healing is a continuum of processes, in which there is significant overlapping [59]. Immediately following injury hemostatic and inflammatory processes are initiated. Inflammatory cells release a range of factors to encourage cell proliferation, migration, neovascularization, and re-epithelialization. Following re-epithelialization and new tissue formation, there is a phase of remodeling that involves apoptosis and the reorganization of matrix proteins such as collagen. In diabetes, defective recruitment of immune cell types and impaired signaling, e.g., VEGF, eNOS, and stromal cell-derived factor-1 between inflammatory cells, endothelial cells, EPC, fibroblasts, and keratinocytes, results in decreased neovascularization and wound healing [53] (Fig. 15.5). Of note, the expression of the cytroprotective enzyme heme oxygenase-1 was induced in normoglycemic mice

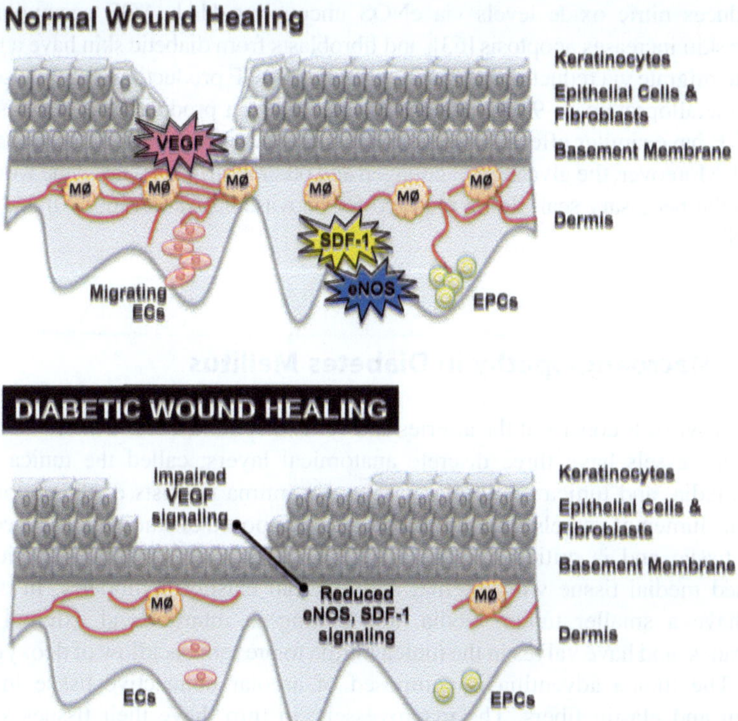

Fig. 15.5 Impaired neovascularization in diabetic wound healing. In mammals, wound healing comprises a number of over-lapping phases including, hemostasis, inflammation, re-epithelialization and remodeling that are compromised in diabetes patients. Neovascularization, including the recruitment of endothelial progenitor cells (EPC), underpins these processes and is impaired in diabetes patients due to high glucose. The secretion of vascular endothelial growth factor (VEGF) by keratinocytes, fibroblasts, endothelial cells (EC), and macrophages (Mø) is decreased. Similarly, a decrease in endothelial nitric oxide synthase (eNOS) and stromal cell-derived factor-1 (SDF-1) leads to a failure of the signaling axis that recruits EPC. The reduced blood supply to the wound leads to a deficit of oxygen and nutrients, further delaying the processes of cell proliferation, migration and neovascularization. Ultimately, the diabetic wound may ulcerate, leading to gangrene and amputation (not shown). Should sufficient wound healing occur in a diabetes patient, the remodeling of vessels and re-organization of matrix proteins such as collagen may be delayed and wound tensile strength become decreased

following wounding, while in *db/db* diabetic mice induction of heme oxygenase-1 was delayed [60]. Also, overexpression of heme oxygenase-1 in keratinocytes accelerated wound healing in *db/db* mice and was associated with increased neovascularization. Moreover, heme oxygenase-1-deficient mice have impaired wound healing due to a decrease in EPC numbers and capillary growth [61]. These observations are consistent with the known role of heme oxygenase-1 in modulating VEGF and stromal cell-derived factor-1 [62].

The high glucose-induced mechanisms underpinning impaired wound healing in diabetes include excessive superoxide from the polyol pathway, and also PKC activation that reduces nitric oxide levels via eNOS uncoupling [11]. AGE accumulation in diabetic skin increases apoptosis [63], and fibroblasts from diabetic skin have a reduced ability to migrate via reductions in nitric oxide and VEGF production [64, 65]. Levels of matrix metalloproteinase-9 are increased while collagen production is decreased [65, 66]. AGE have similar effects on nitric oxide in endothelial cells and their progenitors [67, 68]. Moreover, the glycation of collagen and other matrix proteins in the wound bed impairs the necessary scaffolding required for neovascularization [69, 70] and wound remodeling.

15.4 Macroangiopathy in Diabetes Mellitus

The macrovessels consist of the arteries and veins. In contrast to the microvasculature the macrovessels have three discrete anatomical layers, called the tunica intima, tunica media, and tunica adventitia. The tunica intima consists of the endothelium lining the lumen of vessels. The tunica media is composed of smooth muscle cells and elastic tissue and is critical for the maintenance of vascular tone. Arteries have increased medial tissue with internal and external elastic membranes. In contrast, veins have a smaller tunica media, are lacking in internal and external elastic membranes, and have valves in the tunica intima to prevent backflow of deoxygenated blood. The tunica adventitia is composed of areolar connective tissue including collagen and elastin fibers. The macrovessels, in turn, have their tissues supplied with blood via the *vasa vasorum*, a network of microvessels around the adventitia.

Cardiovascular disease is associated with a wide spectrum of clinical presentations, including asymptomatic disease, angina pectoris, heart failure, and cardiac arrhythmias, and remains a leading cause of death worldwide. While the mortality from cardiovascular disease has declined significantly in the USA since its peak in 1968, it is expected that this decline will be offset by an increasing prevalence of obesity and diabetes [1]. Indeed, approximately 50 % of diabetes subjects will die of heart disease and stroke [1]. Cardiovascular disease affects the macrovessels of the coronary, peripheral, and cerebral circulations; in each of these circuits the hyperglycemia of diabetes exerts common effects, which are hallmarked by endothelial dysfunction and impaired collateralization in response to tissue ischemia. However, while hyperglycemia contributes, in part, to the development of cardiovascular disease in diabetes by driving endothelial dysfunction, the deposition of oxidized lipids within arterial walls due to dyslipidemia, the resultant inflammatory milieu, and subsequent atheroocclusive and/or thrombotic disease are also considered to be key features. Indeed, while intensive blood glucose control has been shown to improve microvascular disease in diabetes, these beneficial effects have not fully extended to macrovascular disease [2, 5, 71]. Cardiovascular disease also has a microvessel component, e.g., in diabetes subjects, neovessels may permeate the vessel walls of the atherosclerotic plaque facilitating macrophage entry and also hemorrhage and plaque rupture. Key examples of neovascularization in cardiovascular disease are shown in Fig. 15.6.

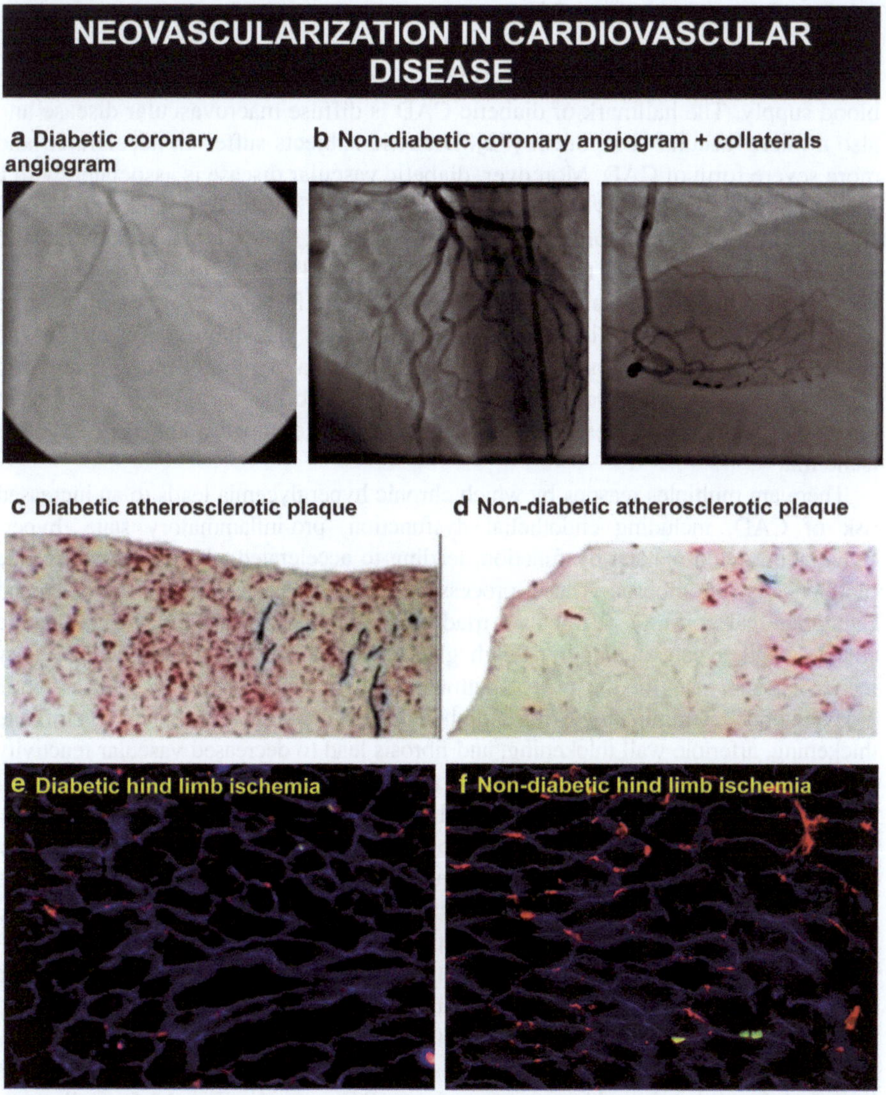

Fig. 15.6 Neovascularization in cardiovascular disease. Cardiovascular disease in diabetes subjects affects macro- and micro-vascular components. In ischemic heart disease the coronary arteries that supply oxygen and nutrients become blocked, e.g., with a thrombus, which can lead to heart attack. (**a**, **b**) Diabetes patients have more aggressive and diffuse atherosclerotic disease with marked impairment of coronary collateralization in contrast to non-diabetes subjects. (**c**, **d**) Neovessels (stained in *blue*) may be increased in plaques of diabetes patients compared with non-diabetes subjects, which can enhance macrophage infiltration (stained in *red*), accumulation of lipids, plaque rupture, hemorrhaging and thrombosis (not shown). (**e**, **f**) In a murine model of hind limb ischemia, neovessels (stained in *red*) and arterioles (stained in *green*) are dramatically decreased in diabetic mice compared with nondiabetic mice. Photomicrographs in **c** and **d**: Reprinted from *Cardiovascular Pharmacology*, 22, Purushothaman et al. Expression of angiotensin-converting enzyme 2 and its end product angiotensin 1–7 is increased in diabetic atheroma: implications for inflammation and neovascularization, Page No. 42–44, Copyright (2013) with Permission from Elsevier

15.4.1 Coronary Artery Disease

Coronary artery disease (CAD) affects the vessels providing the heart with nutritive blood supply. The hallmark of diabetic CAD is diffuse macrovascular disease and also a microvascular component [72]. Diabetic subjects suffer an accelerated and more severe form of CAD. Moreover, diabetic vascular disease is associated with a state of hypercoagulability that leads to thrombosis [73]. In large epidemiological studies, the incidence of CAD and myocardial infarction among diabetic individuals is two to four times compared with nondiabetic individuals, independent of other classical cardiovascular risk factors [74]. Unlike the microangiopathies, intensive blood glucose control has limited impact on the progression of CAD. Diabetes subjects with CAD have a much more severe and diffuse atherosclerotic disease, reducing the applicability and success of current revascularization strategies. Moreover, concurrent neuropathy can lead to silent ischemia.

There are multiples reasons by which chronic hyperglycemia leads to an increased risk of CAD, including endothelial dysfunction, pro-inflammatory state, hypercoagulability, and platelet dysfunction, leading to accelerated atherosclerosis, plaque rupture, and thrombosis. These processes may be grouped across three broad categories that constitute Virchow's triad of endothelial dysfunction, hemodynamic changes, and hypercoagulability. High glucose levels increase superoxide, e.g., by increasing substrate for the polyol pathway, which results in reduced nitric oxide bioavailability, culminating in endothelial dysfunction. Basement membrane thickening, arteriole wall thickening, and fibrosis lead to decreased vascular reactivity and arterial stiffness. Decreases in eNOS activity and methylglyoxal modification of hypoxia inducible factor-1α [75] can inhibit hypoxia-induced signaling pathways, leading to impaired angiogenesis and neovascularization. AGE can also glycosylate superoxide dismutase [76], which can lead to increased levels of superoxide and a reduction in the subsequent nitric oxide bioavailability and coronary dilation. AGE also increase cross-linking of collagen further contributing to vascular stiffness [77].

In a significant proportion of nondiabetic CAD subjects, the coronary vasculature adapts to atheroocclusive disease and tissue ischemia via collateralization and neovascularization. However, this adaptive and reparative capacity is markedly reduced in the coronary vessels of diabetes subjects [78], in part, via the disturbances in growth factor levels and impairment of signaling cascades (Fig. 15.6). Collateralization and arteriogenesis require the recruitment of monocytes and monocytes isolated from diabetics have reduced capacity to migrate towards a VEGF stimulus [25]. In myocardial biopsies of T2DM subjects, VEGF mRNA and protein were increased compared to nondiabetic controls [79]. However, while VEGF receptor-1 and VEGF receptor-2 mRNA were also increased in diabetes, protein levels of both receptors were decreased, as was VEGF receptor-2 protein phosphorylation. This, in turn, led to reduced phosphorylation of Akt and eNOS, which could underpin impaired neovascularization in the diabetic myocardium. These findings may extend to the collateral vessels.

15.4.2 Peripheral Arterial Disease

PAD is a condition characterized by atherosclerotic occlusive disease of blood vessels supplying the lower extremities. Although commonly presenting as intermittent claudication, severe PAD with critical limb ischemia results in gangrene and eventual lower limb amputation. In diabetics PAD is confounded by a significant microvascular component evidenced by neuropathy and impaired wound healing. While PAD is a major risk factor for lower extremity amputation, it is also accompanied by a high likelihood for symptomatic cardiovascular and cerebrovascular disease. The presence of diabetes is associated with an eightfold increased risk of PAD and amputation events [80]. Despite modern therapy, PAD remains a major burden on the healthcare system, with at least one amputation due to diabetes occurring every 30 s worldwide [81]. Relatively, little is known about the biology of PAD in individuals with diabetes. However, it is thought that changes observed with other manifestations of atherosclerotic disease, such as metabolic disturbances and impaired molecular signaling discussed for CAD (Sect. 3.1), also apply to patients with both PAD and diabetes. Attention to classic vascular risk factors such as hypertension and hypercholesterolemia and intensive blood glucose control has failed to substantially reduce the risk of amputation [2, 31, 82], thus highlighting the need to further delineate and understand the pathophysiology of PAD and to help to identify new therapeutic targets for its treatment.

15.4.3 Cerebral Ischemia

Stroke results in significant morbidity and disability and is the second leading cause of death in Western society [1]. Many epidemiological studies have shown convincingly that diabetes is one of the leading risk factors for ischemic stroke with approximately a two- to fivefold increase in risk compared with nondiabetic individuals [83]. The biology and impact of diabetes on the cerebrovascular circulation is similar to that in CAD, with atherosclerosis of large vessels as well as microvascular disease. However, diabetes is associated with a hypercoagulative state and an increased risk of atrial fibrillation, resulting in an increased risk of embolic stroke [84]. Diabetic retinopathy and nephropathy are also associated with an increased risk for stroke [85] that may result from tissue ischemia (ischemic stroke) due to thrombosis or embolism, or from hemorrhage (hemorrhagic stroke). In thrombotic stroke, a thrombus can form over an atherosclerotic lesion within the cerebral circulation leading to occlusion of the vessel. Should the thrombus break off from a cerebral or other atherosclerotic plaque, it becomes an embolus and may block an artery to cause an embolic stroke. Hemorrhagic stroke following rupture of cerebral vessels is more closely associated with hypertension and other angiopathies [86]. However, following an ischemic stroke, hemorrhagic transformation may occur. In a European cohort (>18,000 subjects), diabetes was found to be a strong risk factor for ischemic stroke but not hemorrhagic stroke [87]. Consistent with findings for CAD and PAD, aggressive blood glucose control has not reduced the risk of stroke in diabetes mellitus [2, 5, 71]. Further understanding of the pathogenesis of stroke in diabetes mellitus is necessary to facilitate development of new therapies that may ultimately reduce this risk.

Given the difficulties in assessing the vascular, metabolic, and neuronal changes following stroke in humans, a number of insights have necessarily been obtained from rodent models. Following ischemia, VEGF and matrix metalloproteinases have been found to be elevated [88–90]. This adaptive response could lead to neovascularization and adaptive arteriogenesis, characterized by increases in collateral vessels and lumen diameter, as well as vessel tortuosity leading to vessel leakage and rupture (hemorrhagic transformation). Chronic and acute hyperglycemia following injury and reperfusion worsens hemorrhagic transformation and tissue damage [91]. Edema and increased endothelial swelling lead to basement membrane thickening. High glucose further induces degenerative changes in the smooth muscle and pericytes leading to impaired vascular tone and abnormal vascular structures. These changes are driven, in part, by the generation of reactive oxygen species via the polyol pathway. Similar processes affect the neurons as they increase their anaerobic metabolism to adapt to the tissue ischemia, leading to intracellular acidosis that can result in neuron and glial damage. This acidosis can increase reactive oxygen species, contributing to the inflammatory milieu and tissue damage of adjacent vessels. High glucose concentrations can also increase excitory amino acids such as glutamate, which may lead to neuronal death [92]. PKC pathway modulation improves vascular tone [93] and increase thrombolysis [94], implicating PKC in the pathogenesis of cerebral vascular complications.

15.5 Future Directions

Unfortunately, once initiated, the vascular complications of diabetes are progressive due to a vicious feedback loop of tissue damage and maladaptive endothelial changes (Fig. 15.7). T1DM onset and progression cannot be prevented in a majority of subjects, necessitating the use of insulin to normalize blood glucose levels. While strict blood glucose control can reduce the development and progression of microvascular complications such as retinopathy, nephropathy, and neuropathy, these effects do not extend to the macrovessels. Moreover, intensive management of glucose with other comorbidities such as hypertension and dyslipidemia is not sufficient either in the prevention of disease progression in macrovessels. Therefore, an understanding of the interplay between nonglucose and glucose-related mechanisms in the macrovessels is imperative. While the majority of T2DM cases could be easily prevented through lifestyle choices and education, we must face the reality that the burden of this disease will continue to increase in the foreseeable future, necessitating the development of treatments to prevent the vascular complications of both T1DM and T2DM.

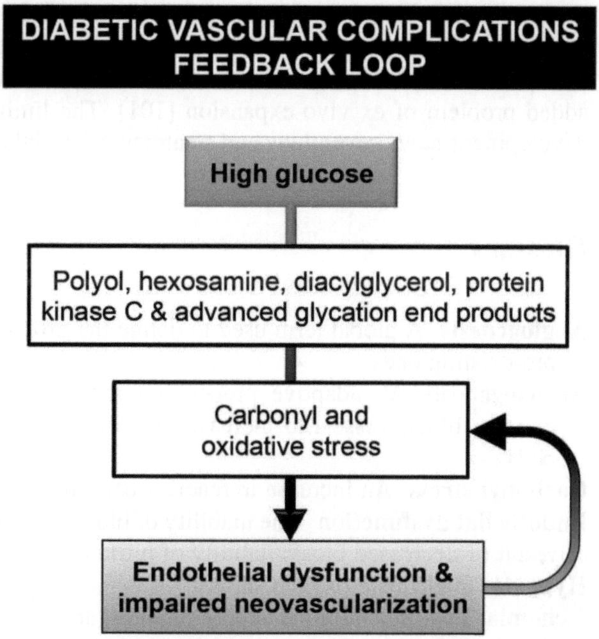

Fig. 15.7 Diabetic vascular complications feedback loop. High glucose leads to carbonyl and oxidative stress that contribute to endothelial dysfunction and impaired angiogenesis. The resulting tissue damage exacerbates disease by creating a positive feedback loop

Lipid-lowering drugs such as fenofibrate have shown some success in attenuating diabetic vascular complications [32, 95, 96]. Similarly, anti-hypertensives including angiotensin converting enzyme inhibitors and angiotensin-II receptor blockers can reduce diabetic microvascular disease [30, 31, 97]. However, to date, only a small number of other drugs have been trialed for the prevention and treatment of diabetic vascular complications. The PKC inhibitor ruboxistaurin has been shown to limit vision loss in diabetic retinopathy by reducing edema [98]. The anti-VEGF antibodies, bevacizumab, ranibizumab, and pegaptanib, have been administered intra-ocularly to delay the progression of diabetic retinopathy by a number of mechanisms, including decreased retinal neovascularization [99]. Alternative approaches to small molecules include gene and cell based therapies. While some results in initial studies show promise, the benefits may not extend to larger cohorts. The inconsistent data obtained from gene therapy trials of angiogenic factors have been variously attributed to the limitations of preclinical animal models, dosage and delivery and duration of therapy, endpoint selection, and the use of "no option" patients in the advanced stages of disease [100]. While these attributes should not be understated, the possibility remains that the problem may lie with the downstream signaling rather than the production of the angiogenic factors itself, e.g., increased VEGF levels in macrovascular disease may be irrelevant if VEGF-receptor signaling is impaired or eNOS activity is reduced. Autologous bone marrow, EPC, mesenchymal and inducible pluripotent stem cell therapies while promising, to date, face similar limitations in clinical trials with the added problem of ex vivo expansion [101]. The limitations create an impetus to development new preventions and treatments for diabetic vascular complications.

Glossary

Angiogenesis A global term used to define the growth of new blood vessels from pre-existing ones

Arteriogenesis An adaptive process referring to the growth and widening of existing blood vessels to increase blood flow to inadequately served vascular beds

Carbonyl stress An increase in reactive carbonyl species

Endothelial dysfunction The inability of blood vessels to appropriately dilate as a result of decreased bioavailability of nitric oxide

Hypoxia Inadequate oxygen supply

Ischemia Insufficient blood supply to an organ

Neovascularization The process of new blood vessel growth in response to stress, injury or disease such as ischemia, tissue damage and cancer

Nephropathy Disease of the nephrons within the kidney

Neuropathy Diseases of the nerves

Oxidative stress A net increased in reactive oxygen species

Retinopathy Disease of the retina within the eye

Type 1 diabetes mellitus A disease characterized by high blood glucose levels due to the destruction of insulin secreting pancreatic beta cells that is responsible for ~10 % diabetes cases

Type 2 diabetes mellitus A disease characterized by high blood glucose levels, elevated insulin levels, and tissue resistance to insulin, which is responsible for ~90 % of diabetes cases

Vascular tone The ability of blood vessels to dilate and constrict

Vasculogenesis De novo production of new endothelial cells and blood vessels, a process in which EPC are involved.

References

1. WHO (2011) The top 10 causes of death | Fact sheet No. 310. World Health Organization. http://www.who.int/mediacentre/factsheets/fs310/en/index.html. Accessed 12 Aug 2013
2. UKPDS (1998) Intensive blood-glucose control with sulphonylureas or insulin compared with conventional treatment and risk of complications in patients with type 2 diabetes (UKPDS 33). UK Prospective Diabetes Study (UKPDS) Group. Lancet 352:837–853
3. Holman RR, Paul SK, Bethel MA et al (2008) 10-Year follow-up of intensive glucose control in type 2 diabetes. N Engl J Med 359:1577–1589

4. Nathan DM, Cleary PA, Backlund JY et al (2005) Intensive diabetes treatment and cardiovascular disease in patients with type 1 diabetes. N Engl J Med 353:2643–2653
5. Patel A, MacMahon S, Chalmers J et al (2008) Intensive blood glucose control and vascular outcomes in patients with type 2 diabetes. N Engl J Med 358:2560–2572
6. Shichiri M, Kishikawa H, Ohkubo Y et al (2000) Long-term results of the Kumamoto Study on optimal diabetes control in type 2 diabetic patients. Diabetes Care 23(Suppl 2):B21–B29
7. Stratton IM, Adler AI, Neil HA et al (2000) Association of glycaemia with macrovascular and microvascular complications of type 2 diabetes (UKPDS 35): prospective observational study. BMJ 321:405–412
8. UKPDS (1998) Effect of intensive blood-glucose control with metformin on complications in overweight patients with type 2 diabetes (UKPDS 34). UK Prospective Diabetes Study (UKPDS) Group. Lancet 352:854–865
9. Schalkwijk CG, Stehouwer CD (2005) Vascular complications in diabetes mellitus: the role of endothelial dysfunction. Clin Sci (Lond) 109:143–159
10. Le Brocq M, Leslie SJ, Milliken P et al (2008) Endothelial dysfunction: from molecular mechanisms to measurement, clinical implications, and therapeutic opportunities. Antioxid Redox Signal 10:1631–1674
11. Kolluru GK, Bir SC, Kevil CG (2012) Endothelial dysfunction and diabetes: effects on angiogenesis, vascular remodeling, and wound healing. Int J Vasc Med 2012(918267):1–30
12. Brownlee M (2005) The pathobiology of diabetic complications: a unifying mechanism. Diabetes 54:1615–1625
13. Baynes JW, Thorpe SR (1999) Role of oxidative stress in diabetic complications: a new perspective on an old paradigm. Diabetes 48:1–9
14. Baynes JW (1991) Role of oxidative stress in development of complications in diabetes. Diabetes 40:405–412
15. Brownlee M, Cerami A, Vlassara H (1988) Advanced glycosylation end products in tissue and the biochemical basis of diabetic complications. N Engl J Med 318:1315–1321
16. Ramasamy R, Yan SF, Schmidt AM (2011) Receptor for AGE (RAGE): signaling mechanisms in the pathogenesis of diabetes and its complications. Ann N Y Acad Sci 1243:88–102
17. Farmer DG, Kennedy S (2009) RAGE, vascular tone and vascular disease. Pharmacol Ther 124:185–194
18. Klein RL, Laimins M, Lopes-Virella MF (1995) Isolation, characterization, and metabolism of the glycated and nonglycated subfractions of low-density lipoproteins isolated from type I diabetic patients and nondiabetic subjects. Diabetes 44:1093–1098
19. Kawamura N, Ookawara T, Suzuki K et al (1992) Increased glycated Cu, Zn-superoxide dismutase levels in erythrocytes of patients with insulin-dependent diabetis mellitus. J Clin Endocrinol Metab 74:1352–1354
20. Morgan PE, Dean RT, Davies MJ (2002) Inactivation of cellular enzymes by carbonyls and protein-bound glycation/glycoxidation products. Arch Biochem Biophys 403:259–269
21. Kathir K, Dennis JM, Croft KD et al (2010) Equivalent lipid oxidation profiles in advanced atherosclerotic lesions of carotid endarterectomy plaques obtained from symptomatic type 2 diabetic and nondiabetic subjects. Free Radic Biol Med 49:481–486
22. Carmeliet P (2005) Angiogenesis in life, disease and medicine. Nature 438:932–936
23. Asahara T, Murohara T, Sullivan A et al (1997) Isolation of putative progenitor endothelial cells for angiogenesis. Science 275:964–967
24. Chen YH, Lin SJ, Lin FY et al (2007) High glucose impairs early and late endothelial progenitor cells by modifying nitric oxide-related but not oxidative stress-mediated mechanisms. Diabetes 56:1559–1568
25. Waltenberger J (2001) Impaired collateral vessel development in diabetes: potential cellular mechanisms and therapeutic implications. Cardiovasc Res 49:554–560

26. Gallagher KA, Liu ZJ, Xiao M et al (2007) Diabetic impairments in NO-mediated endothelial progenitor cell mobilization and homing are reversed by hyperoxia and SDF-1 alpha. J Clin Invest 117:1249–1259
27. Liu ZJ, Velazquez OC (2008) Hyperoxia, endothelial progenitor cell mobilization, and diabetic wound healing. Antioxid Redox Signal 10:1869–1882
28. Resnikoff S, Pascolini D, Etya'ale D et al (2004) Global data on visual impairment in the year 2002. Bull World Health Organ 82:844–851
29. Henricsson M, Nilsson A, Janzon L et al (1997) The effect of glycaemic control and the introduction of insulin therapy on retinopathy in non-insulin-dependent diabetes mellitus. Diabet Med 14:123–131
30. Schrier RW, Estacio RO, Esler A et al (2002) Effects of aggressive blood pressure control in normotensive type 2 diabetic patients on albuminuria, retinopathy and strokes. Kidney Int 61:1086–1097
31. UKPDS (1998) Tight blood pressure control and risk of macrovascular and microvascular complications in type 2 diabetes: UKPDS 38. UK Prospective Diabetes Study Group. BMJ 317:703–713
32. Keech AC, Mitchell P, Summanen PA et al (2007) Effect of fenofibrate on the need for laser treatment for diabetic retinopathy (FIELD study): a randomised controlled trial. Lancet 370:1687–1697
33. Garner A (1993) Histopathology of diabetic retinopathy in man. Eye (Lond) 7:250–253
34. Cai J, Boulton M (2002) The pathogenesis of diabetic retinopathy: old concepts and new questions. Eye (Lond) 16:242–260
35. Dokken BB (2008) The pathophysiology of cardiovascular disease and diabetes: beyond blood pressure and lipids. Diabetes Spectr 2008:160–165
36. Lorenzi M (2007) The polyol pathway as a mechanism for diabetic retinopathy: attractive, elusive, and resilient. Exp Diabetes Res 2007(61038):1–10
37. Tang WH, Martin KA, Hwa J (2012) Aldose reductase, oxidative stress, and diabetic mellitus. Front Pharmacol 3:1–8
38. Williamson JR, Chang K, Frangos M et al (1993) Hyperglycemic pseudohypoxia and diabetic complications. Diabetes 42:801–813
39. Goldin A, Beckman JA, Schmidt AM et al (2006) Advanced glycation end products: sparking the development of diabetic vascular injury. Circulation 114:597–605
40. Behl Y, Krothapalli P, Desta T et al (2008) Diabetes-enhanced tumor necrosis factor-alpha production promotes apoptosis and the loss of retinal microvascular cells in type 1 and type 2 models of diabetic retinopathy. Am J Pathol 172:1411–1418
41. Antonetti DA, Barber AJ, Bronson SK et al (2006) Diabetic retinopathy: seeing beyond glucose-induced microvascular disease. Diabetes 55:2401–2411
42. Lu M, Perez VL, Ma N et al (1999) VEGF increases retinal vascular ICAM-1 expression in vivo. Invest Ophthalmol Vis Sci 40:1808–1812
43. Miyamoto K, Khosrof S, Bursell SE et al (1999) Prevention of leukostasis and vascular leakage in streptozotocin-induced diabetic retinopathy via intercellular adhesion molecule-1 inhibition. Proc Natl Acad Sci USA 96:10836–10841
44. Vinores SA, Xiao WH, Shen J et al (2007) TNF-alpha is critical for ischemia-induced leukostasis, but not retinal neovascularization nor VEGF-induced leakage. J Neuroimmunol 182:73–79
45. Xia P, Aiello LP, Ishii H et al (1996) Characterization of vascular endothelial growth factor's effect on the activation of protein kinase C, its isoforms, and endothelial cell growth. J Clin Invest 98:2018–2026
46. Ting RD, Keech AC, Drury PL et al (2012) Benefits and safety of long-term fenofibrate therapy in people with type 2 diabetes and renal impairment: the FIELD Study. Diabetes Care 35:218–225
47. Saito A, Takeda T, Sato K et al (2005) Significance of proximal tubular metabolism of advanced glycation end products in kidney diseases. Ann N Y Acad Sci 1043:637–643

48. Cooper ME, Vranes D, Youssef S et al (1999) Increased renal expression of vascular endothelial growth factor (VEGF) and its receptor VEGFR-2 in experimental diabetes. Diabetes 48:2229–2239
49. Iglesias-de la Cruz MC, Ziyadeh FN, Isono M et al (2002) Effects of high glucose and TGF-beta1 on the expression of collagen IV and vascular endothelial growth factor in mouse podocytes. Kidney Int 62:901–913
50. Wendt TM, Tanji N, Guo J et al (2003) RAGE drives the development of glomerulosclerosis and implicates podocyte activation in the pathogenesis of diabetic nephropathy. Am J Pathol 162:1123–1137
51. Wolf G (2004) New insights into the pathophysiology of diabetic nephropathy: from haemodynamics to molecular pathology. Eur J Clin Invest 34:785–796
52. Chung SS, Ho EC, Lam KS et al (2003) Contribution of polyol pathway to diabetes-induced oxidative stress. J Am Soc Nephrol 14:S233–S236
53. Brem H, Tomic-Canic M (2007) Cellular and molecular basis of wound healing in diabetes. J Clin Invest 117:1219–1222
54. Singh N, Armstrong DG, Lipsky BA (2005) Preventing foot ulcers in patients with diabetes. JAMA 293:217–228
55. Iversen MM, Tell GS, Riise T et al (2009) History of foot ulcer increases mortality among individuals with diabetes: ten-year follow-up of the Nord-Trondelag Health Study, Norway. Diabetes Care 32:2193–2199
56. Izumi Y, Satterfield K, Lee S et al (2006) Risk of reamputation in diabetic patients stratified by limb and level of amputation: a 10-year observation. Diabetes Care 29:566–570
57. Sharma KR, Cross J, Farronay O et al (2002) Demyelinating neuropathy in diabetes mellitus. Arch Neurol 59:758–765
58. Greene DA, Sima AA, Stevens MJ et al (1992) Complications: neuropathy, pathogenetic considerations. Diabetes Care 15:1902–1925
59. Gurtner GC, Werner S, Barrandon Y et al (2008) Wound repair and regeneration. Nature 453:314–321
60. Grochot-Przeczek A, Lach R, Mis J et al (2009) Heme oxygenase-1 accelerates cutaneous wound healing in mice. PLoS One 4:e5803
61. Deshane J, Chen S, Caballero S et al (2007) Stromal cell-derived factor 1 promotes angiogenesis via a heme oxygenase 1-dependent mechanism. J Exp Med 204:605–618
62. Dulak J, Deshane J, Jozkowicz A et al (2008) Heme oxygenase-1 and carbon monoxide in vascular pathobiology: focus on angiogenesis. Circulation 117:231–241
63. Niu Y, Xie T, Ge K et al (2008) Effects of extracellular matrix glycosylation on proliferation and apoptosis of human dermal fibroblasts via the receptor for advanced glycosylated end products. Am J Dermatopathol 30:344–351
64. Lerman OZ, Galiano RD, Armour M et al (2003) Cellular dysfunction in the diabetic fibroblast: impairment in migration, vascular endothelial growth factor production, and response to hypoxia. Am J Pathol 162:303–312
65. Burrow JW, Koch JA, Chuang HH et al (2007) Nitric oxide donors selectively reduce the expression of matrix metalloproteinases-8 and -9 by human diabetic skin fibroblasts. J Surg Res 140:90–98
66. Wall SJ, Sampson MJ, Levell N et al (2003) Elevated matrix metalloproteinase-2 and -3 production from human diabetic dermal fibroblasts. Br J Dermatol 149:13–16
67. Li H, Zhang X, Guan X et al (2012) Advanced glycation end products impair the migration, adhesion and secretion potentials of late endothelial progenitor cells. Cardiovasc Diabetol 11:137–146
68. Bucala R, Tracey KJ, Cerami A (1991) Advanced glycosylation products quench nitric oxide and mediate defective endothelium-dependent vasodilatation in experimental diabetes. J Clin Invest 87:432–438
69. Liao H, Zakhaleva J, Chen W (2009) Cells and tissue interactions with glycated collagen and their relevance to delayed diabetic wound healing. Biomaterials 30:1689–1696

70. Wear-Maggitti K, Lee J, Conejero A et al (2004) Use of topical sRAGE in diabetic wounds increases neovascularization and granulation tissue formation. Ann Plast Surg 52:519–522
71. Gerstein HC, Miller ME, Byington RP et al (2008) Effects of intensive glucose lowering in type 2 diabetes. N Engl J Med 358:2545–2559
72. Laakso M (2011) Heart in diabetes: a microvascular disease. Diabetes Care 34(Suppl 2): S145–S149
73. Carr ME (2001) Diabetes mellitus: a hypercoagulable state. J Diabetes Complications 15:44–54
74. Sniderman A, Michel C, Racine N (1992) Heart disease in patients with diabetes mellitus. J Clin Epidemiol 45:1357–1370
75. Ceradini DJ, Yao D, Grogan RH et al (2008) Decreasing intracellular superoxide corrects defective ischemia-induced new vessel formation in diabetic mice. J Biol Chem 283: 10930–10938
76. Taniguchi N, Arai K, Kinoshita N (1989) Glycation of copper/zinc superoxide dismutase and its inactivation: identification of glycated sites. Methods Enzymol 179:570–581
77. Wolffenbuttel BH, Boulanger CM, Crijns FR et al (1998) Breakers of advanced glycation end products restore large artery properties in experimental diabetes. Proc Natl Acad Sci USA 95:4630–4634
78. Abaci A, Oguzhan A, Kahraman S et al (1999) Effect of diabetes mellitus on formation of coronary collateral vessels. Circulation 99:2239–2242
79. Sasso FC, Torella D, Carbonara O et al (2005) Increased vascular endothelial growth factor expression but impaired vascular endothelial growth factor receptor signaling in the myocardium of type 2 diabetic patients with chronic coronary heart disease. J Am Coll Cardiol 46:827–834
80. Johannesson A, Larsson GU, Ramstrand N et al (2009) Incidence of lower-limb amputation in the diabetic and nondiabetic general population: a 10-year population-based cohort study of initial unilateral and contralateral amputations and reamputations. Diabetes Care 32: 275–280
81. Bakker K, van Houtum WH, Riley PC (2005) 2005: the International Diabetes Federation focuses on the diabetic foot. Curr Diab Rep 5:436–440
82. Tongers J, Roncalli JG, Losordo DW (2008) Therapeutic angiogenesis for critical limb ischemia: microvascular therapies coming of age. Circulation 118:9–16
83. Stamler J, Vaccaro O, Neaton JD et al (1993) Diabetes, other risk factors, and 12-yr cardiovascular mortality for men screened in the Multiple Risk Factor Intervention Trial. Diabetes Care 16:434–444
84. Tsang TS, Petty GW, Barnes ME et al (2003) The prevalence of atrial fibrillation in incident stroke cases and matched population controls in Rochester, Minnesota: changes over three decades. J Am Coll Cardiol 42:93–100
85. Rosenson RS, Fioretto P, Dodson PM (2011) Does microvascular disease predict macrovascular events in type 2 diabetes? Atherosclerosis 218:13–18
86. O'Donnell MJ, Xavier D, Liu L et al (2010) Risk factors for ischaemic and intracerebral haemorrhagic stroke in 22 countries (the INTERSTROKE study): a case-control study. Lancet 376:112–123
87. Hyvarinen M, Tuomilehto J, Mahonen M et al (2009) Hyperglycemia and incidence of ischemic and hemorrhagic stroke-comparison between fasting and 2-hour glucose criteria. Stroke 40:1633–1637
88. Plate KH, Beck H, Danner S et al (1999) Cell type specific upregulation of vascular endothelial growth factor in an MCA-occlusion model of cerebral infarct. J Neuropathol Exp Neurol 58:654–666
89. Issa R, Krupinski J, Bujny T et al (1999) Vascular endothelial growth factor and its receptor, KDR, in human brain tissue after ischemic stroke. Lab Invest 79:417–425
90. Romanic AM, White RF, Arleth AJ et al (1998) Matrix metalloproteinase expression increases after cerebral focal ischemia in rats: inhibition of matrix metalloproteinase-9 reduces infarct size. Stroke 29:1020–1030

91. de Courten-Myers GM, Kleinholz M, Holm P et al (1992) Hemorrhagic infarct conversion in experimental stroke. Ann Emerg Med 21:120–126
92. Li PA, Shuaib A, Miyashita H et al (2000) Hyperglycemia enhances extracellular glutamate accumulation in rats subjected to forebrain ischemia. Stroke 31:183–192
93. Bright R, Steinberg GK, Mochly-Rosen D (2007) DeltaPKC mediates microcerebrovascular dysfunction in acute ischemia and in chronic hypertensive stress in vivo. Brain Res 1144: 146–155
94. Liang W, Chuan-Zhen L, Qiang D et al (2004) Reductions in mRNA of the neuroprotective agent, neuroserpin, after cerebral ischemia/reperfusion in diabetic rats. Brain Res 1015: 175–180
95. Rajamani K, Colman PG, Li LP et al (2009) Effect of fenofibrate on amputation events in people with type 2 diabetes mellitus (FIELD study): a prespecified analysis of a randomised controlled trial. Lancet 373:1780–1788
96. Keech A, Simes RJ, Barter P et al (2005) Effects of long-term fenofibrate therapy on cardiovascular events in 9795 people with type 2 diabetes mellitus (the FIELD study): randomised controlled trial. Lancet 366:1849–1861
97. Gross JL, de Azevedo MJ, Silveiro SP et al (2005) Diabetic nephropathy: diagnosis, prevention, and treatment. Diabetes Care 28:164–176
98. Aiello LP, Vignati L, Sheetz MJ et al (2011) Oral protein kinase c beta inhibition using ruboxistaurin: efficacy, safety, and causes of vision loss among 813 patients (1,392 eyes) with diabetic retinopathy in the Protein kinase C beta inhibitor-diabetic retinopathy study and the Protein kinase C beta inhibitor-diabetic retinopathy study 2. Retina 31:2084–2094
99. Simo R, Hernandez C (2009) Advances in the medical treatment of diabetic retinopathy. Diabetes Care 32:1556–1562
100. Gupta R, Tongers J, Losordo DW (2009) Human studies of angiogenic gene therapy. Circ Res 105:724–736
101. Leeper NJ, Hunter AL, Cooke JP (2010) Stem cell therapy for vascular regeneration: adult, embryonic, and induced pluripotent stem cells. Circulation 122:517–526

Angiogenesis in Rheumatoid Arthritis

16

Sofia Konisti, Serafim Kiriakidis, and Ewa M. Paleolog

Contents

16.1 Introduction 341
16.2 Role of Angiogenesis in RA 344
16.2.1 Angiogenic Factors Expressed In RA: A Key Role for VEGF 345
16.2.2 Regulation of Angiogenesis by Hypoxia in RA 347
16.2.3 Interplay Between Hypoxia, Angiogenesis and Inflammation 348
16.3 Angiogenesis Inhibition as a Therapeutic Option in RA 351
References 355

Abstract

Rheumatoid arthritis (RA) is a chronic inflammatory disease, associated with inflammation of the synovial tissue lining joints and tendons, which leads to degradation of underlying cartilage and bone. Extra-articular manifestations of RA, including depression and anaemia, combine with inflammation and joint destruction to impact significantly on patients' quality of life. RA is also associated with co-morbidities such as increased cardiovascular disease. Successful development of therapies to treat RA requires an understanding of the cellular and molecular events underlying the disease. Angiogenesis is now

S. Konisti
Department of Medicine, Imperial College London, London, UK

S. Kiriakidis
Department of Medicine, Imperial College London, London, UK

Kennedy Institute of Rheumatology, University of Oxford, Oxford, UK

E.M. Paleolog (✉)
Department of Medicine, Imperial College London, London, UK

Kennedy Institute of Rheumatology, University of Oxford, Oxford, UK

Nuffield Department of Orthopaedics, Rheumatology and Musculoskeletal Sciences, Kennedy Institute of Rheumatology, University of Oxford, Oxford, UK
e-mail: ewa.paleolog@kennedy.ox.ac.uk

J. Dulak et al. (eds.), *Angiogenesis and Vascularisation*,
DOI 10.1007/978-3-7091-1428-5_16, © Springer-Verlag Wien 2013

understood to play an important role. Inadequate oxygenation (hypoxia) is believed to drive the increase in synovial angiogenesis which occurs in RA, through expression of hypoxia-inducible molecules, including vascular endothelial growth factor (VEGF). This allows further infiltration of inflammatory cells and production of inflammatory mediators, perpetuating synovitis. In parallel, inflammatory molecules and cells, particularly activated macrophages, are observed in synovial tissue and can also directly affect angiogenesis.

The current chapter describes the importance of angiogenesis in RA and discusses whether angiogenesis may be a potential therapeutic target in RA. Furthermore, we will review how angiogenesis and inflammation may interact to promote, maintain and resolve synovitis in RA, with a particular focus on the functional responses of macrophages in the context of RA. Successful treatment of RA is associated with reduced levels of pro-angiogenic factors such as VEGF, supporting the concept that modulation of blockade could be of therapeutic benefit in RA.

Keywords
Hypoxia • Inflammation • VEGF • TNFα • Anti-angiogenic therapy

Abbreviations

ACPA	Antibodies to citrullinated protein antigens
Ang	Angiopoietin
ANGPTL4	Angiopoietin-like 4
CCP	Cyclic citrullinated peptides
CIA	Collagen-induced arthritis
CXCL	CXC chemokine ligand
DMARD	Disease modifying anti-rheumatic drug
DMOG	Dimethyloxaloylglycine
EFNA3	Ephrin A3
FGF	Fibroblast growth factor
FIH-1	Factor inhibiting HIF-1
FLS	Fibroblast-like synoviocytes
HGF	Hepatocyte growth factor
HIF	Hypoxia-inducible factor
HLA	Human leukocyte antigen
HRE	Hypoxia-response elements
IKK	IκB kinases
IL	Interleukin
iNOS	Inducible nitric oxide synthase
IκB	Inhibitor of NF-κB
LPS	Lipopolysaccharide

MHC	Major histocompatibility complex
MMP	Matrix metalloproteinases
MTX	Methotrexate
NFκB	Nuclear factor κ B
NRP-1	Neuropilin-1
OA	Osteoarthritis
PDGF	Platelet-derived growth factor
PHD	Prolyl hydroxylase domain-containing enzyme
PI3K	Phosphoinositide 3-kinase
RA	Rheumatoid arthritis
SE	Shared epitope
TAM	Tumour associated macrophages
TH	T helper cell-type
Tie	Tyrosine kinase with immunoglobulin-like and epidermal growth factor-like domains
TNFα	Tumour necrosis factor α
VEGF	Vascular endothelial growth factor
VEGFR	VEGF receptor

16.1 Introduction

Rheumatoid arthritis (RA) is a chronic systemic inflammatory disease, which affects approximately 1 % of the population worldwide, and is primarily characterised by inflammation of the synovial membrane which lines the joint spaces, leading to the localised invasion and destruction of underlying cartilage and bone. RA may begin at any time from the first few weeks of life until the ninth decade, although the peak time of presentation is 35–45 years of age. Early diagnosis of RA is challenging, due the heterogeneity of its clinical presentation and lack of sufficiently specific and sensitive laboratory tests. Patients present with painful, stiff and swollen joints, predominantly the small joints of the hands and wrists, as well as the metatarsophalangeal joints, ankles, knees and cervical spine. In most patients, symptoms appear over weeks to months, and peri-articular structures such as tendon sheaths may be inflamed. RA is associated with a range of other symptoms such as fatigue, anaemia, weight loss and vasculitis. Depressive symptoms occur in many patients, and there is some evidence that depression may exacerbate pain and disease activity [1, 2]. Indeed more recent clinical trials of new therapeutics in RA include at least one generic patient-reported outcome (PRO) instrument, usually the Health Assessment Questionnaire-Disability Index, to assess function, and/or the so-called 'SF-36 survey', to assess Health Related Quality of Life (HRQoL). These instruments have shown validity and sensitivity for assessment of changes in clinical trials of disease modifying anti-rheumatic drugs (DMARD) and biological therapies. Further instruments have been designed to address problems specific to the RA population. The Arthritis Impact Measurement Scale (AIMS) was one of the first PRO measures developed to evaluate HRQoL in RA [3]. In particular, RA Impact of Disease (RAID) is a new PRO instrument recently developed by EULAR (European League Against Rheumatism) for use in clinical trials, as a measure of the impact of RA on HRQoL, and takes into account pain, functional capacity, fatigue and physical and emotional well-being, together with outcomes such as quality of sleep [4].

Up to 30 % of people with RA become permanently work-disabled within 3 years of diagnosis if they do not have medical treatment [5]. For example, it has been documented in a recent study that at the time of first symptoms of RA, 86 % of men and 64 % of women below 65 years of age were working. More than a third (37 %) of these patients reported subsequent work disability, and the probabilities of continuing to work were 80 % and 68 % at 2 and 5 years, respectively [6]. Furthermore, the standardised mortality ratio for patients with RA is more than 1.5–2.5-fold higher than the ratio for the general population [7]. Although RA patients are at increased risk of dying of urogenital, gastrointestinal, respiratory infections and cancer [8, 9], the major cause of mortality (more than 40 % of deaths) is cardiovascular disease, including ischemic heart disease and heart failure [10]. The odds ratio for RA patients for the risk of all-category stroke was 1.64, with an odds ratio of 2.66 for ischemic stroke [11]. Another study in the USA found that subjects with RA had higher odds ratio for congestive heart failure (3.59) [12]. In a prospective cohort study, which comprised more than 100,000 women free of RA and cardiovascular disease at baseline, the adjusted relative risks of myocardial infarction and stroke in women subsequently diagnosed with RA were 2.00 and 1.48, respectively, when compared to women without RA [13]. A high 10-year risk of cardiovascular disease in newly diagnosed RA patients has been reported, with the absolute cardiovascular risk in RA patients similar to that in non-RA subjects who were 5–10 years older [14]. RA patients also have an increased risk of fatality following myocardial infarction (assessed as the 30-day mortality rates following a first acute cardiovascular event) [15]. A recent study of cardiovascular autopsy findings found that RA patients had more frequent myocardial abnormalities (21 %) than those without RA (12 %) [16].

In terms of disease pathogenesis, interplay between environmental and genetic factors, sex hormones and perhaps an infectious agent or other immune-activating factor is thought to initiate an autoimmune response that culminates in a disease with inflammatory and destructive features [17]. The most common genetic risk factors associated with RA comprise the shared epitope (SE) alleles of the human leukocyte antigen (HLA) gene and a polymorphism of protein tyrosine phosphatase N22 (PTPN22) [18, 19]. Over 80 % of Caucasian RA patients have SE conserved across the HLA-DR1 and HLA-DR4 haplotypes (0101, 0401, 0404 and 1402) [20]. This genetic predisposition has been additionally linked to environmental factors such as smoking [21] and certain autoantibodies. Smoking history, when combined with the presence of HLA-DR SE, increases the risk of RA 21-fold compared with non-smokers carrying no SE genes [20]. Epidemiological and genetic studies of RA have also demonstrated significant differences in subsets of patients with and

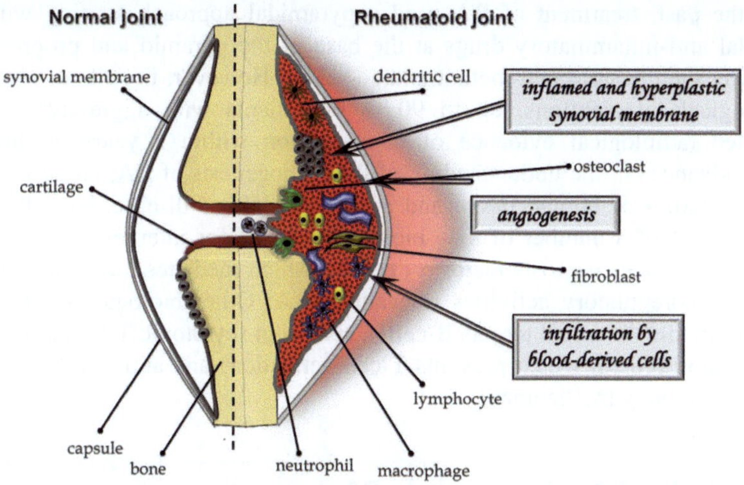

Fig. 16.1 Mechanisms of RA pathogenesis. The normal synovium is generally one to three cell layers thick but in RA becomes thickened (due to fibroblast proliferation) and infiltrated by blood-derived cells, especially lymphocytes and macrophages. Neutrophils accumulate in the synovial fluid within the joint space. This inflamed synovial tissue invades underlying bone and cartilage, particularly as a result of osteoclast activation, leading to joint destruction

without the presence of antibodies to citrullinated protein antigens (ACPA). ACPA are found in approximately 60 % of RA patients but only in 2 % of the normal population, making them highly specific for RA [22–24]. An association between smoking, HLA-DRB1 SE and antibodies to cyclic citrullinated peptides (CCP) has been demonstrated [20]. Furthermore, in anti-CCP-positive individuals, antibodies to the immunodominant citrullinated α-enolase-1 epitope were detected in approximately 40–60 % of RA patients, and this was linked to HLA-DRB1*04, suggesting that citrullinated α-enolase may be an autoantigen linking smoking to genetic risk factors in the development of RA [25, 26].

At the tissue and cellular level, RA is characterised by inflammation, hyperplasia of the synovial lining layer and marked infiltration by blood-derived cells, particularly lymphocytes and macrophages (Fig. 16.1). The normal synovium is generally one to three cell layers thick and is composed of loosely associated macrophage- and fibroblast-like cells, as well as vascular endothelial cells. In RA, the synovium is altered to a thickened tissue several cell layers thick, which covers and erodes the adjacent cartilage, bone and tendon. Histologically, the inflamed synovium shows pronounced angiogenesis, cellular hyperplasia and influx of cells, in particular T-cells, macrophages and dendritic cells. The invasive and destructive synovium is responsible for the erosions observed in RA, in that differentiation of monocyte/macrophage cells leads to formation of osteoclasts, which resorb bone matrix. Progressive destruction of the cartilage and bone produces the deformities characteristic of long-standing RA and results in functional deterioration and disability.

In the past, treatment of RA used a pyramidal approach starting with non-steroidal anti-inflammatory drugs at the base of the pyramid and progressing to DMARD such as gold and methotrexate (MTX). However, in spite of such pharmacological interventions, up to 90 % of patients with aggressive synovitis exhibited radiological evidence of bone erosion within 2 years of diagnosis. Major advances in the understanding of the pathogenesis of RA, based on bench-bedside studies of human tissue and animal models of disease, have led to the identification of a number of new biological targets for intervention, the first of which was tumour necrosis factor α (TNFα), which mediates many inflammatory and immunoregulatory activities relevant to RA. Other biologicals used in RA include rituximab (which targets B-cells), abatacept (cytotoxic T-lymphocyte antigen 4 fusion protein which prevents T-cell activation) and anti-interleukin (IL)-6 receptor antibody tocilizumab.

16.2 Role of Angiogenesis in RA

RA synovium is a site of active angiogenesis, due to the expression of numerous angiogenic factors, leading to the formation of new blood vessels. Blood vessels fulfil an important role in RA, fuelling synovial expansion and infiltration by cells from the blood, by supplying oxygen and nutrients necessary for cell metabolism and division, as well as by bringing in leukocytes and signalling mediators such as cytokines and growth factors [27–33]. As the synovium expands, more blood vessels are needed to supply poorly perfused and hypoxic areas distant from the pre-existing blood vessels with oxygen and nutrients (Fig. 16.2). The number of synovial blood vessels has been found to correlate with synovial cell hyperplasia and indices of joint tenderness [34]. When affected joints of RA patients were examined using power Doppler ultrasonography, increased synovial blood flow was observed in 81 % of RA patients, but only in less than 10 % of healthy controls, with a correlation between intra-articular microvascular power Doppler flow and clinical synovitis in RA [35]. A number of subsequent studies have confirmed these findings and have further shown that ultrasonographic measures correlate with disease severity in RA and may be useful as a marker of response to therapy [36–39]. Endothelial cells lining blood vessels within RA synovium have been shown to express cell cycle-associated antigens [40], and endothelial proliferation was shown to be increased in synovium from patients with RA [41]. This results in altered synovial blood vessel density [42, 43], with blood vessels of different sizes observed both in areas of diffuse synovitis and in regions of large leukocytic infiltrates with germinal centre-like structures [42, 44]. Despite the formation of blood vessels, their morphology appears similar to that seen in post-capillary venules of lymphoid tissue, and it is thought that there may be a failure to form functional mature vessels. RA synovium contains a significant fraction of neoangiogenic, immature and leaky blood vessels which may be observed from early stages of RA. Comparison of the staining patterns for CD31 and the pericyte marker α-smooth muscle cell actin revealed a significant fraction of CD31-positive

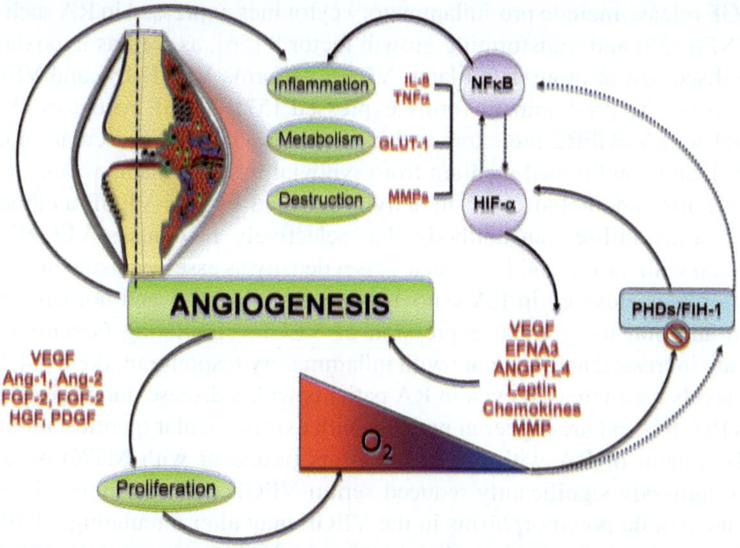

Fig. 16.2 Central role of angiogenesis in RA. Angiogenesis plays a key role in perpetuating disease in RA, by fuelling synovial hyperplasia. As cells such as FLS proliferate and become increasingly distant from pre-existing blood vessels, the synovium becomes hypoxic. Reduced oxygen tension and activation of HIF signalling drives expression of angiogenic factors such as VEGF, ANGPTL4, leptin and EFNA3, further promoting angiogenesis. Hypoxia also promotes expression of inflammatory cytokines, chemokines, molecules involved in metabolism (such as the glucose transporter GLUT-1) and MMPs and thus maintains inflammation and synovial destruction. Potential interactions between HIF and NFκB signalling, in part mediated by inhibition of PHD/FIH-1 leading to downstream NFκB activation, also contributes to the destructive and inflammatory synovial milieu

but α-smooth muscle cell actin-negative cells in RA tissue when compared to osteoarthritis (OA) or control tissue [45, 46]. The presence or density of immature vessels is increased in patients with longer disease duration, higher activity and greater cell infiltration [45].

16.2.1 Angiogenic Factors Expressed in RA: A Key Role for VEGF

A range of different factors can promote angiogenesis, directly or indirectly, and many of these factors have been reported to be expressed in RA [47, 48]. Over 30 years ago it was reported that synovial fluids from patients with RA, as well as from those with OA, contained a low molecular weight angiogenesis factor, now termed ESAF (endothelial cell stimulating angiogenesis factor) [49]. In 1994, the groups of Koch and Fava almost simultaneously reported vascular endothelial growth factor (VEGF) expression in RA synovial fluids and tissue [50, 51]. VEGF expression is localised to synovial macrophages, neutrophils and fibroblasts [52, 53], and stimuli for VEGF release include pro-inflammatory cytokines expressed in RA such as IL-1 [54], TNFα [55] and transforming growth factor β [56], as well as hypoxia, which will be discussed in more detail later. VEGF isoforms VEGF-165 and VEGF-121 appear to be the predominant forms expressed [57]. VEGF receptors (VEGFR) VEGFR1 and VEGFR2 are expressed by RA synovial microvascular endothelial cells [51], and conditioned medium from synovial tissue explants was shown to be mitogenic for endothelial cells, an activity reduced by anti-VEGF antibody [50]. Another study utilised an antibody that selectively recognises VEGFR2 when complexed with VEGF and found that vessel density as assessed using this antibody expression was elevated in RA synovium compared to OA and normal synovium [58]. In addition to synovial expression of VEGF, circulating (serum) levels of VEGF are increased and correlate with inflammatory response markers [54, 59–62]. VEGF levels are increased even in RA patients with a disease duration of less than 2 years [63, 64] and are higher in patients with extra-articular manifestations of RA [65]. Treatment of RA with TNFα inhibitors (alone or with MTX) or anti-IL-6 receptor antibody significantly reduced serum VEGF concentrations [54, 66–70]. Single nucleotide polymorphisms in the VEGF may alter circulating VEGF levels and have been suggested to correlate in RA with both disease activity [71, 72] and onset age [73], although contradictory effects of VEGF polymorphisms on the risk of cardiovascular disease in RA have been reported [74, 75]. In addition to members of the VEGF family, expression of angiopoietin (Ang)-1 and Ang-2 [76, 77] and angiopoietin receptors Tie (Tyrosine kinase with immunoglobulin-like and epidermal growth factor-like domains)-1 and Tie-2 [78–80] in RA has been described. Additionally, fibroblast growth factor (FGF)-1 and FGF-2 have been detected in RA synovial tissue [81, 82], together with platelet-derived growth factor (PDGF) [83, 84] and hepatocyte growth factor (HGF) [85].

While VEGF is traditionally considered for its role in angiogenesis, there may be another important function for this growth factor in terms of blood vessel development in RA. In RA, CD133/CD34/VEGFR2-positive endothelial progenitor cells were found close to RA synovial blood vessels [86]. These cells were generated at a higher rate from bone marrow samples taken from RA patients, compared to normal subjects. Furthermore, the capacity of bone marrow-derived cells from RA patients to progress into endothelial cells correlated with synovial microvessel density [87]. Circulating endothelial progenitor cells in patients with active RA have been reported to be lower than in individuals with inactive disease or in healthy controls [88]. Interestingly, a recent study assessed late-outgrowth endothelial progenitor cells and actually found these to be enhanced in RA and suggested that these cells contribute to synovitis perpetuation by promoting blood vessel formation through vasculogenesis [89]. It therefore seems likely that increased synovial vessel density in RA is not just due to angiogenesis but also results from post-natal vasculogenesis, due to increased mobilisation of endothelial progenitors from the bone marrow driven by VEGF and increased endothelial progenitor cell homing to the synovium.

Finally, there is evidence that angiogenesis may be a feature of other joint diseases. In OA, expression of VEGF has been reported, together with chemokines,

HGF and members of the Ang-Tie family [90–93]. Expression of Ang-2 and VEGF is higher in synovium of patients with psoriatic arthritis, relative to RA, whereas Ang-1 levels were more comparable. Psoriatic arthritis and RA exhibited different features in terms of vascular morphology, in that blood vessels in psoriatic synovium were highly tortuous in appearance, compared to the straight and branching vessels seen in RA, suggesting that the balance between Ang and VEGF may affect vessel growth and maturation in arthritic synovium [44].

In summary, a strong pro-angiogenic drive appears to exist in RA. However, increased blood-vessel formation might not necessarily result in improved perfusion and oxygenation, due to the formation of immature vessels with abnormal morphology, and might explain the apparent paradox that although synovial blood vessel density is increased, hypoxia is nevertheless a feature of RA. VEGF is a key pro-angiogenic factor expressed in RA, although its function as a vascular permeability factor may contribute to the enhanced vessel leakiness which is a feature of RA. VEGF may also contribute to synovial vasculogenesis in RA, which may impact detrimentally on vasculogenesis in other organs and could be linked to the increased cardiovascular disease seen in RA patients.

16.2.2 Regulation of Angiogenesis by Hypoxia in RA

Oxygen is a key molecular ingredient in life, and hence hypoxia—disruption in O_2 homeostasis caused when oxygen demand exceeds oxygen supply—leads to a number of adaptive changes in cellular responses. The alterations in gene expression induced by hypoxia underlie various physiological processes, including normal embryonic development [94–97] and adaptation to exercise and/or high altitudes [98]. Hypoxia is a feature of certain diseases, in particular solid tumours, but also, with relevance to this chapter, hypoxia plays a key role in RA [33]. The objective of this section is to describe the role of hypoxia in the setting of RA, particularly in terms of angiogenesis, and the potential crosstalk between hypoxia-mediated pathways and inflammatory signalling cascades. As macrophages represent important cellular players in RA, their specific responses to the hypoxic synovial microenvironment will also be considered.

The molecular basis for cellular sensing and adaptation to oxygen-depleted conditions has been extensively described, revealing key roles for members of the family of transcription factors termed hypoxia-inducible factors (HIFs), frequently termed the 'master regulators' of the response to alterations in oxygen tension [99]. HIF is a heterodimeric transcription factor composed of two subunits, namely HIF-α (regulated by oxygen levels and post-translational modifications that are sensitive to oxygen levels) and HIF-β (expressed constitutively in the nucleus) [99–101]. HIF-α accumulates in the cytoplasm, followed by translocation into the nucleus, where it dimerises with HIF-β and binds HIF co-activators, before binding hypoxia-response elements (HRE) in target genes to initiate transcription. The main regulators of HIF-α post-translational modifications are dioxygenases requiring O_2, 2-oxoglutarate, ferrous iron and ascorbic acid, named HIF prolyl hydroxylase domain-containing enzymes (PHDs) and factor inhibiting HIF-1 (FIH-1). The PHD enzymes hydroxylate proline residues within HIF-α, thus making HIF-α recognisable by the von Hippel Lindau tumour suppressor, which leads to polyubiquitination and proteasomal degradation [102]. In contrast, hydroxylation of asparagine residues by FIH-1 prevents recruitment of co-activators p300/CBP (CREB-binding protein) [103]. Under conditions of oxygen deprivation, PHD and FIH-1 are inactive, allowing HIF nuclear translocation and induction of transcription of HRE-containing genes.

As mentioned above, hypoxia is a feature of some diseases, particularly when tissue expansion and cellular metabolism result in oxygen demand exceeding supply, as is the case in RA synovium. Both HIF-α isoforms (HIF-1α, HIF-2α) are reported to be expressed in the setting of human RA synovium. In particular, HIF-1α was expressed abundantly by macrophages in most rheumatoid synovia, predominantly close to the intimal layer but also in sub-intimal zone. Of note there was markedly lower expression of HIF-1α in OA synovia, and HIF-1α was absent from healthy synovium [104, 105]. In a subsequent study, using a microelectrode technique in patients having elective hand surgery for RA tendon disease, synovial hypoxia and expression of HIF-2α were shown in inflammatory infiltrates [106]. Synovial hypoxia is probably the main driving force behind angiogenesis in RA. HIF-α expression in human RA synovium correlates with angiogenesis onset [105] and has been proved to control the expression of a wide range of genes with various roles. VEGF is one of the best characterised HIF-regulated genes, and increased levels of VEGF and VEGFR in RA have been reported in many studies, as mentioned earlier [50, 51, 54, 107]. Importantly, hypoxia increased the angiogenic potential of RA synovium-derived cells, as demonstrated by enhanced blood vessel formation using in vitro angiogenesis assay, together with enhanced synovial cell invasiveness [108]. Hypoxia also drives the expression of other pro-angiogenic players including the chemokines IL-8 (also known as CXC chemokine ligand 8 or CXCL8) [109], CC chemokine ligand 20 [110] and SDF-1 (stromal cell-derived factor 1, CXCL12) [111–113]. We and others have also reported that hypoxia increases expression of angiogenic genes other than VEGF, including angiopoietin-like 4 (ANGPTL4), ephrin A3 (EFNA3) and leptin [114, 115]. Increased levels of pro-inflammatory cytokines such as IL-6 and matrix metalloproteinases (MMP) MMP-1 and MMP-3 [109] have also been reported in response to hypoxia. Synovial hypoxia is, therefore, likely to contribute to RA by promoting inflammation, angiogenesis, cellular infiltration and cartilage degradation.

16.2.3 Interplay Between Hypoxia, Angiogenesis and Inflammation

As previously mentioned, low oxygen levels are associated with accumulation of inflammatory cells, enhanced expression of HIF and hypoxia-responsive genes and increased levels of pro-inflammatory cytokines, formally establishing a link between hypoxia and inflammation in the context of many diseases including RA (Fig. 16.2) [116]. Indications that HIF-α could have an important role in inflammation came from seminal studies utilising mice bearing myeloid-specific deletion of the *Hif-α* gene showing that HIF-1α allows myeloid cells to generate ATP in oxygen-deprived inflamed tissues, thereby operating energy-requiring processes including aggregation, motility, invasiveness and bactericidal activity [117, 118]. Indeed, the fact that hypoxia and the HIF pathway influence many inflammatory and immune responses relevant to synovitis, including monocyte–-macrophage responses, is now well recognised. Pro- and anti-inflammatory cytokines are capable of directly affecting the HIF pathway, thus highlighting the convergence between inflammatory and oxygen-dependent signalling pathways. For instance, studies on gingival fibroblasts and RA fibroblast-like synoviocytes (FLS) have demonstrated that IL-1β and TNFα are capable of activating HIF in normoxia, an effect requiring both mitogen-activated protein kinase and phosphoinositide 3-kinase (PI3K) pathways [119, 120]. In contrast to HIF-α protein stabilisation that occurs under hypoxic conditions due to inhibition of PHDs and FIH-1, cytokines have been reported to stimulate HIF-α gene transcription in macrophages and RA FLS [120, 121]. Importantly, it has been demonstrated that T helper cell-type (TH) 1 cytokines can synergise with hypoxia to induce HIF-1 in various cells, including RA FLS [120, 122, 123]. We have also shown that TH1 cytokines in combination with hypoxia are not sufficient to induce angiogenic activity by RA FLS despite inducing HIF-1 and VEGF. In contrast, TH2 cytokines induce pro-angiogenic activity in normoxia and hypoxia, despite their inability to activate HIF-1 in FLS, highlighting the complex relationships between hypoxia, angiogenesis and inflammation in RA [114]. These observations might go some way to explain the apparent paradox of the concurrent presence of hypoxia and angiogenesis in RA synovium, since TH1 cytokines and hypoxia may not lead to formation of fully functional blood vessels, which have been shown to be present in RA [45, 46].

As well as FLS, monocytes and/or macrophages are considered key players within the context of RA, and thus hypoxia inevitably influences critical aspects of their behaviour, namely the transcriptional programme, polarisation status and metabolism, which might determine the perpetuation or resolution of the inflammatory disease. HIF-1 and HIF-2 were shown to be transcriptional effectors regulating macrophage responses to hypoxia [121, 124–126], and hypoxia upregulates a number of HRE-containing genes, including VEGF and ANGPTL4 [125, 127]. With regard to polarisation status, Mantovani and colleagues [128] presented a classification system for macrophage activation in which macrophages are divided into two groups: M1-type macrophages, characterised by enhanced ability of killing intracellular microorganisms and exhibiting pro-inflammatory properties, and M2-type macrophages, which exert anti-inflammatory effects [128]. So-called 'classically activated' M1 macrophages stimulate delayed-type hypersensitivity responses and increase surface levels of major histocompatibility complex (MHC) class II molecules, expression of markers such as inducible nitric oxide synthase (iNOS) and secretion of pro-inflammatory cytokines (IL-1β, TNFα, IL-6, IL-12, IL-15). M2 ('alternatively activated') macrophages are activated in response to IL-4 or IL-13, synthesise anti-inflammatory cytokines (such as IL-10) and facilitate TH2 responses. The relevance of HIF factors to macrophage polarisation ability comes from studies involving a specific macrophage subset, primarily expressed in tumour regions, called tumour associated macrophages (TAMs) and exhibiting a M1/M2 mixed phenotype with M2 being slightly more prominent. Werno et al., utilising a tumour-spheroid model, showed that HIF-1α-deficient macrophages developed a more prominent TAM marker profile (M2-skewed) together with reduced cytotoxicity and also displayed reduced angiogenic potential [129]. The precise role of HIF-2α during macrophage-mediated inflammatory responses has also been investigated. It was shown that mice lacking HIF-2α in myeloid cells are resistant to lipopolysaccharide (LPS)-induced endotoxemia and display an inability to mount inflammatory responses to cutaneous and peritoneal irritants. This phenotype was also associated with reduced TAM infiltration in murine hepatocellular and colitis-associated colon carcinoma models, and reduced tumour cell proliferation and progression, probably due to reduced expression of the macrophage-colony stimulating factor receptor and the CXC chemokine receptor 4 [126]. Chemical stabilisation of HIF following treatment with the HIF hydroxylase inhibitor dimethyloxaloylglycine (DMOG) has also been described to promote LPS-induced tolerance via expansion of the M2 macrophage population and activation of nuclear factor κ B (NFκB) signalling [130]. Hypoxia-mediated angiogenesis via enhanced blood vessel formation facilitates tissue oxygen delivery, thus contributing to repair of injured tissues, which is a characteristic function of M2 macrophages. On the other hand, hypoxia/HIF induces the expression of pro-angiogenic molecules such as VEGF. A recently published study investigated the differential roles of macrophage-released HIF-1α and HIF-2α with regard to angiogenesis progression. In this setting, it was shown that HIF-1α exhibits pro-angiogenic behaviour via its effects on VEGF, but that HIF-2α displays anti-angiogenic behaviour through production of the angiogenesis inhibitor soluble VEGFR1 [131]. HIF hydroxylases have also been ascribed different roles in macrophage polarisation. PHD-2 haplodeficient (*Phd-2+/−*) mice displayed preformed collateral arteries that preserved limb perfusion and prevented tissue necrosis in ischemia. Improved arteriogenesis in *Phd-2+/−* mice was due to expansion of tissue-resident, M2-like macrophages, with higher expression of M2-type genes. Conversely, several pro-inflammatory or anti-angiogenic (M1-type) molecules were downregulated, including IL-6, iNOS and IL-12 [132]. On the other hand, PHD-3 was found to be preferentially expressed by M1 macrophages [133]. Modulation of the HIF-PHD-NFκB axis might thus represent a novel approach for targeting inflammatory diseases where a macrophage polarisation switch may be a contributory factor.

The presence of hypoxia and HIF expression against a background of inflammation has prompted the question of whether hypoxia can activate signalling pathways other than the HIF cascade. Of particular relevance in inflammation is the dynamic interplay between the HIF and NFκB pathways. The NFκB pathway is activated in many different cells, including macrophages, dendritic cells, fibroblasts and endothelial cells, important cellular components in the establishment and progression of RA. Activation of this key transcription factor can be induced by a large number of stimuli, including bacterial products, viral molecules and pro-inflammatory cytokines. Hypoxia has been shown to drive NFκB activation and thereby regulate the release of critical pro-inflammatory mediators [134, 135]. The very first indication that HIF induces NFκB activation came from Walmsley et al. providing data showing that hypoxia promotes human neutrophil survival, an effect requiring HIFα-dependent regulation of inhibitor of NF-κB (IκB) α and NFκB [136]. Furthermore, the participation of PHD HIF hydroxylases in NFκB activation has been proposed [103, 137–140]. In the canonical pathway of NFκB activation, activity is controlled by IκB kinases (IKK), which mediate serine phosphorylation and degradation of IκBα and allow nuclear accumulation of NFκB. IKK-2 contains an evolutionarily conserved amino acid consensus motif which can potentially be hydroxylated by PHD. Mimicking hypoxia by using small interfering RNA against PHD-1 or PHD-2 resulted in NFκB activation in HeLa cells [137]. Similar studies in other cells also suggest that PHD may negatively regulate IKK via prolyl hydroxylation [138, 139]. Potential candidates for NFκB activation in the setting of RA include pro-inflammatory cytokines, which are likely to cause a vicious cycle of signals that result in chronic inflammation [141]. Activation of NFκB and HIF by converging hypoxic and inflammatory signalling pathways thus probably leads to pathological changes associated with RA, such as inflammation, angiogenesis and bone or cartilage destruction.

16.3 Angiogenesis Inhibition as a Therapeutic Option in RA

The principle underlying angiogenesis blockade for treatment of cancer is to 'starve' the tumour of nutrients and oxygen. There is now increasing appreciation that RA can also be classified as an angiogenesis-dependent disease, due to the many parallels with solid tumours, such as hypoxia, inflammation and altered vascularity. We and others have extensively described potential therapeutic applications aiming at inhibiting angiogenesis in RA [27, 29, 31–33, 142]. Therefore in this chapter we will focus on VEGF as a target in RA and review more extensively new potential angiogenesis inhibitors for treatment of RA (Table 16.1).

Over the last decade anti-VEGF biologicals have been approved for the treatment of angiogenesis-related diseases. The best examples are anti-VEGF antibody bevacizumab, used in different forms of cancer including metastatic colon cancer, and ranibizumab and pegaptanib sodium (anti-VEGF antibody and VEGF aptamer respectively) for age-related macular degeneration. The promising clinical outcomes of the use of such biologicals suggested that such treatments may also be an option to treat other angiogenic disorders in which VEGF is over-expressed including RA. Murine collagen-induced arthritis (CIA) is a model widely used for the testing of potential therapeutics for RA and was utilised to develop TNFα inhibitors as a therapeutic modality for RA. In CIA, expression of VEGF and VEGFR has been demonstrated [143–146]. Anti-VEGF antibody in this model delayed disease onset but appeared less effective when administered during the chronic phase of disease [144]. In another study, anti-VEGF inhibited synovitis in

Table 16.1 Angiogenesis inhibition in models of RA

Target	Inhibitor	Effect	References
VEGF	Anti-VEGF antibody	Delayed onset and reduced disease in mouse model(s)	[144, 147]
	Soluble VEGFR1	Delayed onset and reduced disease in mouse model(s)	[148, 149]
	Anti-VEGFR1 antibody	Delayed onset and reduced disease in mouse model(s)	[150–152]
	Anti-NRP-1 antibody	Reduced disease in mouse model(s)	[145, 152]
IL-17	Anti-CXCL5 antibody	Reduced disease in mouse model(s)	[155]
	Anti-IL-17 antibody	Reduced disease in mouse model(s)	[154]
Ang	Soluble Tie-2	Reduced disease in mouse model(s)	[159]
	Tie-1 splice variant	Reduced disease in mouse model(s)	[160, 161]
Endothelial proliferation	Inhibitors of methionine aminopeptidase 2	Reduced disease in mouse model(s)	[162–164]
	6ODS-LHbD	Reduced disease in mouse model(s)	[165]
	Human plasminogen-related protein B	Reduced disease in mouse model(s)	[168]
Unknown	ICM0301B	Reduced disease in mouse model(s)	[167]

Table summarises angiogenesis inhibitors that have been investigated in in vivo models of RA

CIA, as indicated by a reduction in clinical score and paw swelling relative to untreated mice [147]. A soluble form of VEGFR1 has also been shown to significantly suppress established arthritis [148, 149]. A different strategy to limit angiogenesis via the VEGF pathway was to directly target VEGFR. In a spontaneous model of arthritis, De Bandt et al. observed that treatment with anti-VEGFR1 antibody abrogated bone and cartilage destruction. The antibody delayed the onset of arthritis and attenuated the severity of disease [150]. The group of Carmeliet also demonstrated that treatment with antibody against VEGFR1 reduced the incidence of joint disease, whereas antibody specific for VEGFR2 appeared ineffective [151]. More recently, Kong et al. reported that inhibition of VEGF via neuropilin-1 (NRP-1), an alternative VEGFR, significantly inhibited the survival, adhesion and migration of FLS. The anti-NRP-1 peptide also inhibited the proliferation, capillary tube formation and migration of endothelial cells in vitro, and neovascularisation in vivo, and suppressed experimentally induced arthritis in mice by inhibiting hyperplasia and angiogenesis in the arthritic joints, suggesting that anti-NRP-1 may offer a new approach for the treatment of RA [152]. To provide a more profound understanding of arthritis-associated angiogenesis, we recently evaluated the expression of angiogenesis modulating genes in CIA. One of the important findings of our study was that NRP-1 was a key player in the pathogenesis of CIA. Treatment with anti-NRP-1 antibody significantly reduced disease severity and joint destruction in CIA [145]. Kong et al. further developed a previously investigated hexapeptide which prevents VEGF binding to VEGFR1 by generating a stereochemical D-form of the peptide which was conjugated to PEG to prolong its half-life and found that the new peptide was much more beneficial in the treatment of angiogenesis in RA than the unmodified peptide. The effect was due to increased stability and delivery, suggesting that further development of current strategies could be applied to treat angiogenesis and RA in humans [153].

In addition to VEGF, pro-inflammatory cytokines have been demonstrated to be important in promoting angiogenesis. In 2010 Pickens et al. documented a novel role for IL-17A (also known as IL-17) in mediating angiogenesis, showing that IL-17 concentrations similar to those found in RA joints were capable of inducing endothelial cell migration and tube formation through the PI3K/AKT pathway. Furthermore, local expression of IL-17 in mouse ankles was able to induce joint inflammation and vascularity and demonstrated that IL-17 was angiogenic and promoted blood vessel formation in mice [154]. In continuation of their previous work the same group has further demonstrated that antibody blockade of CXCL5 ameliorates the angiogenic effect of IL-17 [155]. The ability of IL-17 to stimulate endothelial tube formation and invasiveness was also demonstrated by another recent study, highlighting the potential of targeting IL-17 in treatment of RA [156]. In an endothelial and synovial cell co-culture system, it was shown that IL-6 can also induce angiogenesis [157]. A follow-up study reported that IL-6 stimulation induced endothelial cell growth, by decreasing Ang-1 and increasing VEGF and Ang-2, and destabilised angiogenesis in RA, suggesting that targeting IL-6 signalling in RA (e.g. using tocilizumab) could inhibit angiogenesis [158]. Other approaches have targeted the Ang-Tie pathway [159–161] and the methionine aminopeptidase enzyme involved in endothelial proliferation [162–164].

Chemically modified heparins have also been developed as possible candidates for inhibiting angiogenesis; however, heparin therapy is limited by poor oral bioavailability. A newly synthesised low molecular weight heparin and deoxycholic acid conjugate (6ODS-LHbD) has been described to possess high oral bioavailability in rats. In a murine anti-collagen antibody-induced arthritis model this compound inhibited joint neovascularisation and disease development, suggesting that 6ODS-LHbD may be a promising candidate as an orally active angiogenesis inhibitor for treatment of RA [165]. In another recent study Kumagai and colleagues examined the anti-arthritic effect of ICM0301B, an angiogenesis inhibitor first described in 2004 [166]. They found that in the mouse CIA model ICM0301B inhibited disease development with an efficacy comparable to that of a non-steroidal anti-inflammatory drug, indomethacin, and suggested that the observed anti-arthritic effect of ICM0301B might be partially attributed to its anti-angiogenic activity [167]. Another novel inhibitor of angiogenesis in RA was described by Tanaka in 2011, who tested the effect of human plasminogen-related protein B in mouse CIA and found a significant reduction of disease development, VEGF expression and new blood vessel formation [168]. Chang and colleagues examined the role of macrophage CCAAT/enhancer binding protein delta (CEBPD) in CIA using *Cebpd*$^{-/-}$ mice. In the absence of CEBPD, disease development was significantly inhibited, with a decrease in the number of affected paws and reduced angiogenesis. Activation of CEBPD in macrophages was involved in promoting tube formation by endothelial cells and the migration and proliferation of rat synoviocytes [169]. Finally, another study aimed to examine whether angiogenesis induced by inflammatory agents is mediated via monocyte chemotactic protein-induced protein (MCPIP). Inflammatory agents, including TNFα, IL-1β and IL-8, were found to induce endothelial tube formation via MCPIP, and moreover MCPIP was shown to stimulate angiogenesis [170].

Conclusions

Many new therapies have been developed for treatment of RA, directed at biological responses which play a role in disease pathogenesis, such as B-cell and T-cell responses, or cascades driven by inflammatory cytokines, including IL-6, IL-1 and, notably, TNFα. These important advances offered a targeted strategy, unlike conventional but non-specific traditional treatments such as DMARD. However, in addition to side effects such as injection/infusion site reactions, the most significant side-effect of these therapies is an increase in the risk of all types of infections, including tuberculosis. Indeed, there is a significantly increased risk of reactivation of latent tuberculosis when using TNF inhibitors, and while pre-treatment screening has allowed identification of at-risk individuals; nonetheless, there is a drive to seek alternative treatment approaches. In this regard, angiogenesis represents an attractive option. VEGF is expressed in RA and VEGF inhibitors are already in clinical use for oncology applications. However, up to now VEGF blockade has been intensively used only in animal models of arthritis and no clinical trials of VEGF inhibition have been undertaken for RA. In addition adverse effects of VEGF-targeted biologicals, such as hypertension and gastro-intestinal perforation, coupled with the high cost of such therapies, are major concerns. Nevertheless, there is a valid hypothesis that such therapies can be beneficial in treating angiogenesis in RA and some groups are considering other approaches established in cancer for treatment of RA [171]. Moreover, the presence of immature blood vessels in RA synovium suggests that simply inhibiting VEGF may not be the best option. Interestingly, treatment of RA patients with TNF inhibitors selectively depleted immature vessels, without affecting the mature vasculature [45]. Haplodeficiency of PHD-2, using an in vivo tumour model in *Phd-2+/−* mice, actually resulted in vessel maturation, leading to improved tumour perfusion and oxygenation, and reduced tumour cell invasion, intravasation and metastasis [172]. In a model of colitis, inhibition of HIF hydroxylases in vivo using DMOG reduced disease severity. The possible underlying mechanism may be due to expansion of macrophages [130, 173]. These findings suggest that the best tactic to modulate angiogenesis in RA is at present unclear, and further in vitro and in vivo studies are still needed.

References

1. Bruce TO (2008) Comorbid depression in rheumatoid arthritis: pathophysiology and clinical implications. Curr Psychiatry Rep 10(3):258–264
2. Rathbun AM, Reed GW, Harrold LR (2013) The temporal relationship between depression and rheumatoid arthritis disease activity, treatment persistence and response: a systematic review. Rheumatology (Oxford) 52:1785–1794
3. Meenan RF, Gertman PM, Mason JH (1980) Measuring health status in arthritis. The arthritis impact measurement scales. Arthritis Rheum 23(2):146–152
4. Gossec L, Paternotte S, Aanerud GJ, Balanescu A, Boumpas DT, Carmona L, de Wit M, Dijkmans BA, Dougados M, Englbrecht M, Gogus F, Heiberg T, Hernandez C, Kirwan JR, Mola EM, Cerinic MM, Otsa K, Schett G, Scholte-Voshaar M, Sokka T, von Krause G, Wells GA, Kvien TK (2011) Finalisation and validation of the rheumatoid arthritis impact of disease score, a patient-derived composite measure of impact of rheumatoid arthritis: a EULAR initiative. Ann Rheum Dis 70(6):935–942
5. Barrett EM, Scott DG, Wiles NJ, Symmons DP (2000) The impact of rheumatoid arthritis on employment status in the early years of disease: a UK community-based study. Rheumatology (Oxford) 39(12):1403–1409
6. Sokka T, Kautiainen H, Pincus T, Verstappen SM, Aggarwal A, Alten R, Andersone D, Badsha H, Baecklund E, Belmonte M, Craig-Muller J, da Mota LM, Dimic A, Fathi NA, Ferraccioli G, Fukuda W, Geher P, Gogus F, Hajjaj-Hassouni N, Hamoud H, Haugeberg G, Henrohn D, Horslev-Petersen K, Ionescu R, Karateew D, Kuuse R, Laurindo IM, Lazovskis J, Luukkainen R, Mofti A, Murphy E, Nakajima A, Oyoo O, Pandya SC, Pohl C, Predeteanu D, Rexhepi M, Rexhepi S, Sharma B, Shono E, Sibilia J, Sierakowski S, Skopouli FN, Stropuviene S, Toloza S, Valter I, Woolf A, Yamanaka H (2010) Work disability remains a major problem in rheumatoid arthritis in the 2000s: data from 32 countries in the QUEST-RA Study. Arthritis Res Ther 12(2):R42
7. Sihvonen S, Korpela M, Laippala P, Mustonen J, Pasternack A (2004) Death rates and causes of death in patients with rheumatoid arthritis: a population-based study. Scand J Rheumatol 33(4):221–227
8. Kaplan MJ (2006) Cardiovascular disease in rheumatoid arthritis. Curr Opin Rheumatol 18 (3):289–297
9. Van Doornum S, Brand C, King B, Sundararajan V (2006) Increased case fatality rates following a first acute cardiovascular event in patients with rheumatoid arthritis. Arthritis Rheum 54(7):2061–2068
10. Gkaliagkousi E, Gavriilaki E, Doumas M, Petidis K, Aslanidis S, Stella D (2012) Cardiovascular risk in rheumatoid arthritis: pathogenesis, diagnosis, and management. J Clin Rheumatol 18(8):422–430
11. Nadareishvili Z, Michaud K, Hallenbeck JM, Wolfe F (2008) Cardiovascular, rheumatologic, and pharmacologic predictors of stroke in patients with rheumatoid arthritis: a nested, case–control study. Arthritis Rheum 59(8):1090–1096
12. Ong KL, Wu BJ, Cheung BM, Barter PJ, Rye KA (2013) Arthritis: its prevalence, risk factors, and association with cardiovascular diseases in the United States, 1999 to 2008. Ann Epidemiol 23(2):80–86
13. Solomon DH, Karlson EW, Rimm EB, Cannuscio CC, Mandl LA, Manson JE, Stampfer MJ, Curhan GC (2003) Cardiovascular morbidity and mortality in women diagnosed with rheumatoid arthritis. Circulation 107(9):1303–1307
14. Kremers HM, Crowson CS, Therneau TM, Roger VL, Gabriel SE (2008) High ten-year risk of cardiovascular disease in newly diagnosed rheumatoid arthritis patients: a population-based cohort study. Arthritis Rheum 58(8):2268–2274
15. Solomon DH, Goodson NJ, Katz JN, Weinblatt ME, Avorn J, Setoguchi S, Canning C, Schneeweiss S (2006) Patterns of cardiovascular risk in rheumatoid arthritis. Ann Rheum Dis 65(12):1608–1612

16. Koivuniemi R, Paimela L, Suomalainen R, Leirisalo-Repo M (2013) Cardiovascular diseases in patients with rheumatoid arthritis. Scand J Rheumatol 42(2):131–135
17. Firestein GS (2005) Pathogenesis of rheumatoid arthritis: how early is early? Arthritis Res Ther 7(4):157–159
18. Andersson AK, Li C, Brennan FM (2008) Recent developments in the immunobiology of rheumatoid arthritis. Arthritis Res Ther 10(2):204
19. Orozco G, Rueda B, Martin J (2006) Genetic basis of rheumatoid arthritis. Biomed Pharmacother 60(10):656–662
20. Klareskog L, Stolt P, Lundberg K, Kallberg H, Bengtsson C, Grunewald J, Ronnelid J, Harris HE, Ulfgren AK, Rantapaa-Dahlqvist S, Eklund A, Padyukov L, Alfredsson L (2006) A new model for an etiology of rheumatoid arthritis: smoking may trigger HLA-DR (shared epitope)-restricted immune reactions to autoantigens modified by citrullination. Arthritis Rheum 54(1):38–46
21. Silman AJ, Newman J, MacGregor AJ (1996) Cigarette smoking increases the risk of rheumatoid arthritis. Results from a nationwide study of disease-discordant twins. Arthritis Rheum 39(5):732–735
22. Schellekens GA, Visser H, de Jong BA, van den Hoogen FH, Hazes JM, Breedveld FC, van Venrooij WJ (2000) The diagnostic properties of rheumatoid arthritis antibodies recognizing a cyclic citrullinated peptide. Arthritis Rheum 43(1):155–163
23. Vincent C, Nogueira L, Sebbag M, Chapuy-Regaud S, Arnaud M, Letourneur O, Rolland D, Fournie B, Cantagrel A, Jolivet M, Serre G (2002) Detection of antibodies to deiminated recombinant rat filaggrin by enzyme-linked immunosorbent assay: a highly effective test for the diagnosis of rheumatoid arthritis. Arthritis Rheum 46(8):2051–2058
24. van Gaalen FA, Linn-Rasker SP, van Venrooij WJ, de Jong BA, Breedveld FC, Verweij CL, Toes RE, Huizinga TW (2004) Autoantibodies to cyclic citrullinated peptides predict progression to rheumatoid arthritis in patients with undifferentiated arthritis: a prospective cohort study. Arthritis Rheum 50(3):709–715
25. Lundberg K, Kinloch A, Fisher BA, Wegner N, Wait R, Charles P, Mikuls TR, Venables PJ (2008) Antibodies to citrullinated alpha-enolase peptide 1 are specific for rheumatoid arthritis and cross-react with bacterial enolase. Arthritis Rheum 58(10):3009–3019
26. Mahdi H, Fisher BA, Kallberg H, Plant D, Malmstrom V, Ronnelid J, Charles P, Ding B, Alfredsson L, Padyukov L, Symmons DP, Venables PJ, Klareskog L, Lundberg K (2009) Specific interaction between genotype, smoking and autoimmunity to citrullinated alpha-enolase in the etiology of rheumatoid arthritis. Nat Genet 41(12):1319–1324
27. Szekanecz Z, Koch AE (2008) Targeting angiogenesis in rheumatoid arthritis. Curr Rheumatol Rev 4(4):298–303
28. Paleolog EM (2009) The vasculature in rheumatoid arthritis. Int J Exp Pathol 90(3):249–261
29. Szekanecz Z, Besenyei T, Paragh G, Koch AE (2009) Angiogenesis in rheumatoid arthritis. Autoimmunity 42(7):563–573
30. Paleolog EM (2010) Angiogenesis in joint disease: the need for clinical data. Int J Clin Rheumatol 5(4):439–449
31. Szekanecz Z, Besenyei T, Szentpetery A, Koch AE (2010) Angiogenesis and vasculogenesis in rheumatoid arthritis. Curr Opin Rheumatol 22(3):299–306
32. Thairu N, Kiriakidis S, Dawson P, Paleolog E (2011) Angiogenesis as a therapeutic target in arthritis in 2011: learning the lessons of the colorectal cancer experience. Angiogenesis 14 (3):223–234
33. Konisti S, Kiriakidis S, Paleolog EM (2012) Hypoxia: a key regulator of angiogenesis and inflammation in rheumatoid arthritis. Nat Rev Rheumatol 8(3):153–162
34. Rooney M, Condell D, Quinlan W, Daly L, Whelan A, Feighery C, Bresnihan B (1988) Analysis of the histologic variation of synovitis in rheumatoid arthritis. Arthritis Rheum 31 (8):956–963
35. Strunk J, Heinemann E, Neeck G, Schmidt KL, Lange U (2004) A new approach to studying angiogenesis in rheumatoid arthritis by means of power Doppler ultrasonography and

measurement of serum vascular endothelial growth factor. Rheumatology (Oxford) 43 (12):1480–1483
36. Taylor PC, Steuer A, Gruber J, Cosgrove DO, Blomley MJ, Marsters PA, Wagner CL, McClinton C, Maini RN (2004) Comparison of ultrasonographic assessment of synovitis and joint vascularity with radiographic evaluation in a randomized, placebo-controlled study of infliximab therapy in early rheumatoid arthritis. Arthritis Rheum 50(4):1107–1116
37. Taylor PC, Steuer A, Gruber J, McClinton C, Cosgrove DO, Blomley MJ, Marsters PA, Wagner CL, Maini RN (2006) Ultrasonographic and radiographic results from a two-year controlled trial of immediate or one-year-delayed addition of infliximab to ongoing methotrexate therapy in patients with erosive early rheumatoid arthritis. Arthritis Rheum 54 (1):47–53
38. Strunk J, Klingenberger P, Strube K, Bachmann G, Muller-Ladner U, Kluge A (2006) Three-dimensional Doppler sonographic vascular imaging in regions with increased MR enhancement in inflamed wrists of patients with rheumatoid arthritis. Joint Bone Spine 73(5):518–522
39. Larche MJ, Seymour M, Lim A, Eckersley RJ, Petavy F, Chiesa F, Rioja I, Lukey PT, Binks M, McClinton C, Dolan K, Taylor PC (2010) Quantitative power Doppler ultrasonography is a sensitive measure of metacarpophalangeal joint synovial vascularity in rheumatoid arthritis and declines significantly following a 2-week course of oral low-dose corticosteroids. J Rheumatol 37(12):2493–2501
40. Ceponis A, Konttinen YT, Imai S, Tamulaitiene M, Li TF, Xu JW, Hietanen J, Santavirta S, Fassbender HG (1998) Synovial lining, endothelial and inflammatory mononuclear cell proliferation in synovial membranes in psoriatic and reactive arthritis: a comparative quantitative morphometric study. Br J Rheumatol 37(2):170–178
41. Walsh DA, Wade M, Mapp PI, Blake DR (1998) Focally regulated endothelial proliferation and cell death in human synovium. Am J Pathol 152(3):691–702
42. FitzGerald O, Soden M, Yanni G, Robinson R, Bresnihan B (1991) Morphometric analysis of blood vessels in synovial membranes obtained from clinically affected and unaffected knee joints of patients with rheumatoid arthritis. Ann Rheum Dis 50(11):792–796
43. Stevens CR, Blake DR, Merry P, Revell PA, Levick JR (1991) A comparative study by morphometry of the microvasculature in normal and rheumatoid synovium. Arthritis Rheum 34(12):1508–1513
44. Fearon U, Griosios K, Fraser A, Reece R, Emery P, Jones PF, Veale DJ (2003) Angiopoietins, growth factors, and vascular morphology in early arthritis. J Rheumatol 30(2):260–268
45. Izquierdo E, Canete JD, Celis R, Santiago B, Usategui A, Sanmarti R, Del Rey MJ, Pablos JL (2009) Immature blood vessels in rheumatoid synovium are selectively depleted in response to anti-TNF therapy. PLoS One 4(12):e8131
46. Kennedy A, Ng CT, Biniecka M, Saber T, Taylor C, O'Sullivan J, Veale DJ, Fearon U (2010) Angiogenesis and blood vessel stability in inflammatory arthritis. Arthritis Rheum 62 (3):711–721
47. Olsson AK, Dimberg A, Kreuger J, Claesson-Welsh L (2006) VEGF receptor signalling – in control of vascular function. Nat Rev Mol Cell Biol 7(5):359–371
48. Bainbridge J, Sivakumar B, Paleolog E (2006) Angiogenesis as a therapeutic target in arthritis: lessons from oncology. Curr Pharm Des 12(21):2631–2644
49. Brown RA, Weiss JB, Tomlinson IW, Phillips P, Kumar S (1980) Angiogenic factor from synovial fluid resembling that from tumours. Lancet 1(8170):682–685
50. Koch AE, Harlow LA, Haines GK, Amento EP, Unemori EN, Wong WL, Pope RM, Ferrara N (1994) Vascular endothelial growth factor. A cytokine modulating endothelial function in rheumatoid arthritis. J Immunol 152(8):4149–4156
51. Fava RA, Olsen NJ, Spencer-Green G, Yeo KT, Yeo TK, Berse B, Jackman RW, Senger DR, Dvorak HF, Brown LF (1994) Vascular permeability factor/endothelial growth factor (VPF/VEGF): accumulation and expression in human synovial fluids and rheumatoid synovial tissue. J Exp Med 180(1):341–346

52. Kasama T, Kobayashi K, Yajima N, Shiozawa F, Yoda Y, Takeuchi HT, Mori Y, Negishi M, Ide H, Adachi M (2000) Expression of vascular endothelial growth factor by synovial fluid neutrophils in rheumatoid arthritis (RA). Clin Exp Immunol 121(3):533–538
53. Taichman NS, Young S, Cruchley AT, Taylor P, Paleolog E (1997) Human neutrophils secrete vascular endothelial growth factor. J Leukoc Biol 62(3):397–400
54. Paleolog EM, Young S, Stark AC, McCloskey RV, Feldmann M, Maini RN (1998) Modulation of angiogenic vascular endothelial growth factor by tumor necrosis factor alpha and interleukin-1 in rheumatoid arthritis. Arthritis Rheum 41(7):1258–1265
55. Bottomley MJ, Webb NJ, Watson CJ, Holt PJ, Freemont AJ, Brenchley PE (1999) Peripheral blood mononuclear cells from patients with rheumatoid arthritis spontaneously secrete vascular endothelial growth factor (VEGF): specific up-regulation by tumour necrosis factor-alpha (TNF-alpha) in synovial fluid. Clin Exp Immunol 117(1):171–176
56. Berse B, Hunt JA, Diegel RJ, Morganelli P, Yeo K, Brown F, Fava RA (1999) Hypoxia augments cytokine (transforming growth factor-beta (TGF-beta) and IL-1)-induced vascular endothelial growth factor secretion by human synovial fibroblasts. Clin Exp Immunol 115 (1):176–182
57. Pufe T, Petersen W, Tillmann B, Mentlein R (2001) Splice variants VEGF121 and VEGF165 of the angiogenic peptide vascular endothelial cell growth factor are expressed in the synovial tissue of patients with rheumatoid arthritis. J Rheumatol 28(7):1482–1485
58. Giatromanolaki A, Sivridis E, Athanassou N, Zois E, Thorpe PE, Brekken RA, Gatter KC, Harris AL, Koukourakis IM, Koukourakis MI (2001) The angiogenic pathway "vascular endothelial growth factor/flk-1(KDR)-receptor" in rheumatoid arthritis and osteoarthritis. J Pathol 194(1):101–108
59. Harada M, Mitsuyama K, Yoshida H, Sakisaka S, Taniguchi E, Kawaguchi T, Ariyoshi M, Saiki T, Sakamoto M, Nagata K, Sata M, Matsuo K, Tanikawa K (1998) Vascular endothelial growth factor in patients with rheumatoid arthritis. Scand J Rheumatol 27(5):377–380
60. Kikuchi K, Kubo M, Kadono T, Yazawa N, Ihn H, Tamaki K (1998) Serum concentrations of vascular endothelial growth factor in collagen diseases. Br J Dermatol 139(6):1049–1051
61. Lee SS, Joo YS, Kim WU, Min DJ, Min JK, Park SH, Cho CS, Kim HY (2001) Vascular endothelial growth factor levels in the serum and synovial fluid of patients with rheumatoid arthritis. Clin Exp Rheumatol 19(3):321–324
62. Sone H, Sakauchi M, Takahashi A, Suzuki H, Inoue N, Iida K, Shimano H, Toyoshima H, Kawakami Y, Okuda Y, Matsuo K, Yamada N (2001) Elevated levels of vascular endothelial growth factor in the sera of patients with rheumatoid arthritis correlation with disease activity. Life Sci 69(16):1861–1869
63. Latour F, Zabraniecki L, Dromer C, Brouchet A, Durroux R, Fournie B (2001) Does vascular endothelial growth factor in the rheumatoid synovium predict joint destruction? A clinical, radiological, and pathological study in 12 patients monitored for 10 years. Joint Bone Spine 68(6):493–498
64. Clavel G, Bessis N, Lemeiter D, Fardellone P, Mejjad O, Menard JF, Pouplin S, Boumier P, Vittecoq O, Le Loet X, Boissier MC (2007) Angiogenesis markers (VEGF, soluble receptor of VEGF and angiopoietin-1) in very early arthritis and their association with inflammation and joint destruction. Clin Immunol 124(2):158–164
65. Kuryliszyn-Moskal A, Klimiuk PA, Sierakowski S, Ciolkiewicz M (2006) A study on vascular endothelial growth factor and endothelin-1 in patients with extra-articular involvement of rheumatoid arthritis. Clin Rheumatol 25(3):314–319
66. Ballara SC, Taylor PC, Reusch P, Marmé D, Feldmann M, Maini RN, Paleolog EM (2001) Raised serum vascular endothelial growth factor levels are associated with destructive change in inflammatory arthritis. Arthritis Rheum 44(9):2055–2064
67. Nakahara H, Song J, Sugimoto M, Hagihara K, Kishimoto T, Yoshizaki K, Nishimoto N (2003) Anti-interleukin-6 receptor antibody therapy reduces vascular endothelial growth factor production in rheumatoid arthritis. Arthritis Rheum 48(6):1521–1529

68. Aggarwal A, Panda S, Misra R (2004) Effect of etanercept on matrix metalloproteinases and angiogenic vascular endothelial growth factor: a time kinetic study. Ann Rheum Dis 63 (7):891–892
69. Macias I, Garcia-Perez S, Ruiz-Tudela M, Medina F, Chozas N, Giron-Gonzalez JA (2005) Modification of pro- and antiinflammatory cytokines and vascular-related molecules by tumor necrosis factor-a blockade in patients with rheumatoid arthritis. J Rheumatol 32 (11):2102–2108
70. Nagashima M, Wauke K, Hirano D, Ishigami S, Aono H, Takai M, Sasano M, Yoshino S (2000) Effects of combinations of anti-rheumatic drugs on the production of vascular endothelial growth factor and basic fibroblast growth factor in cultured synoviocytes and patients with rheumatoid arthritis. Rheumatology (Oxford) 39(11):1255–1262
71. Han SW, Kim GW, Seo JS, Kim SJ, Sa KH, Park JY, Lee J, Kim SY, Goronzy JJ, Weyand CM, Kang YM (2004) VEGF gene polymorphisms and susceptibility to rheumatoid arthritis. Rheumatology (Oxford) 43(9):1173–1177
72. Chen Y, Dawes PT, Mattey DL (2012) Polymorphism in the vascular endothelial growth factor A (VEGFA) gene is associated with serum VEGF-A level and disease activity in rheumatoid arthritis: differential effect of cigarette smoking. Cytokine 58(3):390–397
73. Chen Y, Mattey DL (2012) Age at onset of rheumatoid arthritis: association with polymorphisms in the vascular endothelial growth factor A(VEGFA) gene and an intergenic locus between matrix metalloproteinase (MMP) 1 and 3 genes. Clin Exp Rheumatol 30 (6):894–898
74. Chen Y, Dawes PT, Packham JC, Mattey DL (2011) Interaction between smoking and polymorphism in the promoter region of the VEGFA gene is associated with ischemic heart disease and myocardial infarction in rheumatoid arthritis. J Rheumatol 38(5):802–809
75. Rodriguez-Rodriguez L, Garcia-Bermudez M, Gonzalez-Juanatey C, Vazquez-Rodriguez TR, Miranda-Filloy JA, Fernandez-Gutierrez B, Llorca J, Martin J, Gonzalez-Gay MA (2011) Vascular endothelial growth factor A and cardiovascular disease in rheumatoid arthritis patients. Tissue Antigens 77(4):291–297
76. Scott BB, Zaratin PF, Colombo A, Hansbury MJ, Winkler JD, Jackson JR (2002) Constitutive expression of angiopoietin-1 and -2 and modulation of their expression by inflammatory cytokines in rheumatoid arthritis synovial fibroblasts. J Rheumatol 29(2):230–239
77. Gravallese EM, Pettit AR, Lee R, Madore R, Manning C, Tsay A, Gaspar J, Goldring MB, Goldring SR, Oettgen P (2003) Angiopoietin-1 is expressed in the synovium of patients with rheumatoid arthritis and is induced by tumour necrosis factor alpha. Ann Rheum Dis 62 (2):100–107
78. DeBusk LM, Chen Y, Nishishita T, Chen J, Thomas JW, Lin PC (2003) Tie2 receptor tyrosine kinase, a major mediator of tumor necrosis factor alpha-induced angiogenesis in rheumatoid arthritis. Arthritis Rheum 48(9):2461–2471
79. Shahrara S, Volin MV, Connors MA, Haines GK, Koch AE (2002) Differential expression of the angiogenic Tie receptor family in arthritic and normal synovial tissue. Arthritis Res 4 (3):201–208
80. Uchida T, Nakashima M, Hirota Y, Miyazaki Y, Tsukazaki T, Shindo H (2000) Immunohistochemical localisation of protein tyrosine kinase receptors Tie-1 and Tie-2 in synovial tissue of rheumatoid arthritis: correlation with angiogenesis and synovial proliferation. Ann Rheum Dis 59(8):607–614
81. Sano H, Forough R, Maier JA, Case JP, Jackson A, Engleka K, Maciag T, Wilder RL (1990) Detection of high levels of heparin binding growth factor-1 (acidic fibroblast growth factor) in inflammatory arthritic joints. J Cell Biol 110(4):1417–1426
82. Nakashima M, Eguchi K, Aoyagi T, Yamashita I, Ida H, Sakai M, Shimada H, Kawabe Y, Nagataki S, Koji T, Nakane PK (1994) Expression of basic fibroblast growth factor in synovial tissues from patients with rheumatoid arthritis: detection by immunohistological staining and in situ hybridisation. Ann Rheum Dis 53(1):45–50

83. Sano H, Engleka K, Mathern P, Hla T, Crofford LJ, Remmers EF, Jelsema CL, Goldmuntz E, Maciag T, Wilder RL (1993) Coexpression of phosphotyrosine-containing proteins, platelet-derived growth factor-B, and fibroblast growth factor-1 *in situ* in synovial tissues of patients with rheumatoid arthritis and Lewis rats with adjuvant or streptococcal cell wall arthritis. J Clin Invest 91(2):553–565
84. Remmers EF, Sano H, Lafyatis R, Case JP, Kumkumian GK, Hla T, Maciag T, Wilder RL (1991) Production of platelet derived growth factor B chain (PDGF-B/c-sis) mRNA and immunoreactive PDGF B-like polypeptide by rheumatoid synovium: coexpression with heparin binding acidic fibroblast growth factor-1. J Rheumatol 18(1):7–13
85. Koch AE, Halloran MM, Hosaka S, Shah MR, Haskell CJ, Baker SK, Panos RJ, Haines GK, Bennett GL, Pope RM, Ferrara N (1996) Hepatocyte growth factor. A cytokine mediating endothelial migration in inflammatory arthritis. Arthritis Rheum 39(9):1566–1575
86. Ruger B, Giurea A, Wanivenhaus AH, Zehetgruber H, Hollemann D, Yanagida G, Groger M, Petzelbauer P, Smolen JS, Hoecker P, Fischer MB (2004) Endothelial precursor cells in the synovial tissue of patients with rheumatoid arthritis and osteoarthritis. Arthritis Rheum 50 (7):2157–2166
87. Hirohata S, Yanagida T, Nampei A, Kunugiza Y, Hashimoto H, Tomita T, Yoshikawa H, Ochi T (2004) Enhanced generation of endothelial cells from CD34+ cells of the bone marrow in rheumatoid arthritis: possible role in synovial neovascularization. Arthritis Rheum 50(12):3888–3896
88. Grisar J, Aletaha D, Steiner CW, Kapral T, Steiner S, Seidinger D, Weigel G, Schwarzinger I, Wolozcszuk W, Steiner G, Smolen JS (2005) Depletion of endothelial progenitor cells in the peripheral blood of patients with rheumatoid arthritis. Circulation 111(2):204–211
89. Jodon de Villeroche V, Avouac J, Ponceau A, Ruiz B, Kahan A, Boileau C, Uzan G, Allanore Y (2010) Enhanced late-outgrowth circulating endothelial progenitor cell levels in rheumatoid arthritis and correlation with disease activity. Arthritis Res Ther 12(1):R27
90. Bonnet CS, Walsh DA (2005) Osteoarthritis, angiogenesis and inflammation. Rheumatology (Oxford) 44(1):7–16
91. Haywood L, McWilliams DF, Pearson CI, Gill SE, Ganesan A, Wilson D, Walsh DA (2003) Inflammation and angiogenesis in osteoarthritis. Arthritis Rheum 48(8):2173–2177
92. Mapp PI, Avery PS, McWilliams DF, Bowyer J, Day C, Moores S, Webster R, Walsh DA (2008) Angiogenesis in two animal models of osteoarthritis. Osteoarthritis Cartilage 16 (1):61–69
93. Mapp PI, Walsh DA, Bowyer J, Maciewicz RA (2010) Effects of a metalloproteinase inhibitor on osteochondral angiogenesis, chondropathy and pain behavior in a rat model of osteoarthritis. Osteoarthritis Cartilage 18(4):593–600
94. Iyer NV, Kotch LE, Agani F, Leung SW, Laughner E, Wenger RH, Gassmann M, Gearhart JD, Lawler AM, Yu AY, Semenza GL (1998) Cellular and developmental control of O_2 homeostasis by hypoxia-inducible factor 1 alpha. Genes Dev 12(2):149–162
95. Ryan HE, Lo J, Johnson RS (1998) HIF-1 alpha is required for solid tumor formation and embryonic vascularization. EMBO J 17(11):3005–3015
96. Peng J, Zhang L, Drysdale L, Fong GH (2000) The transcription factor EPAS-1/hypoxia-inducible factor 2alpha plays an important role in vascular remodeling. Proc Natl Acad Sci USA 97(15):8386–8391
97. Tian H, Hammer RE, Matsumoto AM, Russell DW, McKnight SL (1998) The hypoxia-responsive transcription factor EPAS1 is essential for catecholamine homeostasis and protection against heart failure during embryonic development. Genes Dev 12(21):3320–3324
98. Pagel H, Engel A, Jelkmann W (1992) Erythropoietin induction by hypoxia. A comparison of in vitro and in vivo experiments. Adv Exp Med Biol 317:515–519
99. Semenza GL (2009) Regulation of oxygen homeostasis by hypoxia-inducible factor 1. Physiology (Bethesda) 24:97–106
100. Rey S, Semenza GL (2010) Hypoxia-inducible factor-1-dependent mechanisms of vascularization and vascular remodelling. Cardiovasc Res 86(2):236–242

101. Semenza GL (2010) Defining the role of hypoxia-inducible factor 1 in cancer biology and therapeutics. Oncogene 29(5):625–634
102. Maxwell PH, Wiesener MS, Chang GW, Clifford SC, Vaux EC, Cockman ME, Wykoff CC, Pugh CW, Maher ER, Ratcliffe PJ (1999) The tumour suppressor protein VHL targets hypoxia-inducible factors for oxygen-dependent proteolysis. Nature 399(6733):271–275
103. Cockman ME, Lancaster DE, Stolze IP, Hewitson KS, McDonough MA, Coleman ML, Coles CH, Yu X, Hay RT, Ley SC, Pugh CW, Oldham NJ, Masson N, Schofield CJ, Ratcliffe PJ (2006) Posttranslational hydroxylation of ankyrin repeats in IkappaB proteins by the hypoxia-inducible factor (HIF) asparaginyl hydroxylase, factor inhibiting HIF (FIH). Proc Natl Acad Sci USA 103(40):14767–14772
104. Hollander AP, Corke KP, Freemont AJ, Lewis CE (2001) Expression of hypoxia-inducible factor 1alpha by macrophages in the rheumatoid synovium: implications for targeting of therapeutic genes to the inflamed joint. Arthritis Rheum 44(7):1540–1544
105. Brouwer E, Gouw AS, Posthumus MD, van Leeuwen MA, Boerboom AL, Bijzet J, Bos R, Limburg PC, Kallenberg CG, Westra J (2009) Hypoxia inducible factor-1-alpha (HIF-1alpha) is related to both angiogenesis and inflammation in rheumatoid arthritis. Clin Exp Rheumatol 27(6):945–951
106. Sivakumar B, Akhavani MA, Winlove CP, Taylor PC, Paleolog EM, Kang N (2008) Synovial hypoxia as a cause of tendon rupture in rheumatoid arthritis. J Hand Surg [Am] 33(1):49–58
107. Kurosaka D, Hirai K, Nishioka M, Miyamoto Y, Yoshida K, Noda K, Ukichi T, Yanagimachi M, Furuya K, Takahashi E, Kingetsu I, Fukuda K, Yamada A (2010) Clinical significance of serum levels of vascular endothelial growth factor, angiopoietin-1, and angiopoietin-2 in patients with rheumatoid arthritis. J Rheumatol 37(6):1121–1128
108. Akhavani MA, Madden L, Buysschaert I, Sivakumar B, Kang N, Paleolog EM (2009) Hypoxia upregulates angiogenesis and synovial cell migration in rheumatoid arthritis. Arthritis Res Ther 11(3):R64
109. Ahn JK, Koh EM, Cha HS, Lee YS, Kim J, Bae EK, Ahn KS (2008) Role of hypoxia-inducible factor-1alpha in hypoxia-induced expressions of IL-8, MMP-1 and MMP-3 in rheumatoid fibroblast-like synoviocytes. Rheumatology (Oxford) 47(6):834–839
110. Bosco MC, Delfino S, Ferlito F, Battaglia F, Puppo M, Gregorio A, Gambini C, Gattorno M, Martini A, Varesio L (2008) Hypoxic synovial environment and expression of macrophage inflammatory protein 3gamma/CCL20 in juvenile idiopathic arthritis. Arthritis Rheum 58 (6):1833–1838
111. Hitchon C, Wong K, Ma G, Reed J, Lyttle D, El-Gabalawy H (2002) Hypoxia-induced production of stromal cell-derived factor 1 (CXCL12) and vascular endothelial growth factor by synovial fibroblasts. Arthritis Rheum 46(10):2587–2597
112. del Rey MJ, Izquierdo E, Caja S, Usategui A, Santiago B, Galindo M, Pablos JL (2009) Human inflammatory synovial fibroblasts induce enhanced myeloid cell recruitment and angiogenesis through a hypoxia-inducible transcription factor 1alpha/vascular endothelial growth factor-mediated pathway in immunodeficient mice. Arthritis Rheum 60 (10):2926–2934
113. Santiago B, Calonge E, Del Rey MJ, Gutierrez-Canas I, Izquierdo E, Usategui A, Galindo M, Alcami J, Pablos JL (2011) CXCL12 gene expression is upregulated by hypoxia and growth arrest but not by inflammatory cytokines in rheumatoid synovial fibroblasts. Cytokine 53 (2):184–190
114. Larsen H, Muz B, Khong TL, Feldmann M, Paleolog EM (2012) Differential effects of Th1 versus Th2 cytokines in combination with hypoxia on HIFs and angiogenesis in RA. Arthritis Res Ther 14(4):R180
115. Muz B, Larsen H, Madden L, Kiriakidis S, Paleolog EM (2012) Prolyl hydroxylase domain enzyme 2 is the major player in regulating hypoxic responses in rheumatoid arthritis. Arthritis Rheum 64(9):2856–2867

116. Giatromanolaki A, Sivridis E, Maltezos E, Athanassou N, Papazoglou D, Gatter KC, Harris AL, Koukourakis MI (2003) Upregulated hypoxia inducible factor-1alpha and -2alpha pathway in rheumatoid arthritis and osteoarthritis. Arthritis Res Ther 5(4):R193–R201
117. Cramer T, Yamanishi Y, Clausen BE, Forster I, Pawlinski R, Mackman N, Haase VH, Jaenisch R, Corr M, Nizet V, Firestein GS, Gerber HP, Ferrara N, Johnson RS (2003) HIF-1alpha is essential for myeloid cell-mediated inflammation. Cell 112(5):645–657
118. Peyssonnaux C, Datta V, Cramer T, Doedens A, Theodorakis EA, Gallo RL, Hurtado-Ziola N, Nizet V, Johnson RS (2005) HIF-1alpha expression regulates the bactericidal capacity of phagocytes. J Clin Invest 115(7):1806–1815
119. Thornton RD, Lane P, Borghaei RC, Pease EA, Caro J, Mochan E (2000) Interleukin 1 induces hypoxia-inducible factor 1 in human gingival and synovial fibroblasts. Biochem J 350(Pt 1):307–312
120. Westra J, Brouwer E, Bos R, Posthumus MD, Doornbos-van der Meer B, Kallenberg CG, Limburg PC (2007) Regulation of cytokine-induced HIF-1alpha expression in rheumatoid synovial fibroblasts. Ann N Y Acad Sci 1108:340–348
121. Takeda N, O'Dea EL, Doedens A, Kim JW, Weidemann A, Stockmann C, Asagiri M, Simon MC, Hoffmann A, Johnson RS (2010) Differential activation and antagonistic function of HIF-{alpha} isoforms in macrophages are essential for NO homeostasis. Genes Dev 24 (5):491–501
122. Hellwig-Burgel T, Rutkowski K, Metzen E, Fandrey J, Jelkmann W (1999) Interleukin-1beta and tumor necrosis factor-alpha stimulate DNA binding of hypoxia-inducible factor-1. Blood 94(5):1561–1567
123. Westra J, Brouwer E, Bouwman E, Doornbos-van der Meer B, Posthumus MD, van Leeuwen MA, Limburg PC, Ueda Y, Kallenberg CG (2009) Role for CaMKII inhibition in rheumatoid arthritis: effects on HIF-1-induced VEGF production by rheumatoid synovial fibroblasts. Ann N Y Acad Sci 1173:706–711
124. Talks KL, Turley H, Gatter KC, Maxwell PH, Pugh CW, Ratcliffe PJ, Harris AL (2000) The expression and distribution of the hypoxia-inducible factors HIF-1alpha and HIF-2alpha in normal human tissues, cancers, and tumor-associated macrophages. Am J Pathol 157 (2):411–421
125. Fang HY, Hughes R, Murdoch C, Coffelt SB, Biswas SK, Harris AL, Johnson RS, Imityaz HZ, Simon MC, Fredlund E, Greten FR, Rius J, Lewis CE (2009) Hypoxia-inducible factors 1 and 2 are important transcriptional effectors in primary macrophages experiencing hypoxia. Blood 114(4):844–859
126. Imtiyaz HZ, Williams EP, Hickey MM, Patel SA, Durham AC, Yuan LJ, Hammond R, Gimotty PA, Keith B, Simon MC (2010) Hypoxia-inducible factor 2alpha regulates macrophage function in mouse models of acute and tumor inflammation. J Clin Invest 120 (8):2699–2714
127. White JR, Harris RA, Lee SR, Craigon MH, Binley K, Price T, Beard GL, Mundy CR, Naylor S (2004) Genetic amplification of the transcriptional response to hypoxia as a novel means of identifying regulators of angiogenesis. Genomics 83(1):1–8
128. Sica A, Schioppa T, Mantovani A, Allavena P (2006) Tumour-associated macrophages are a distinct M2 polarised population promoting tumour progression: potential targets of anticancer therapy. Eur J Cancer 42(6):717–727
129. Werno C, Menrad H, Weigert A, Dehne N, Goerdt S, Schledzewski K, Kzhyshkowska J, Brune B (2010) Knockout of HIF-1alpha in tumor-associated macrophages enhances M2 polarization and attenuates their pro-angiogenic responses. Carcinogenesis 31 (10):1863–1872
130. Hams E, Saunders SP, Cummins EP, O'Connor A, Tambuwala MT, Gallagher WM, Byrne A, Campos-Torres A, Moynagh PM, Jobin C, Taylor CT, Fallon PG (2011) The hydroxylase inhibitor dimethyloxallyl glycine attenuates endotoxic shock via alternative activation of macrophages and IL-10 production by B1 cells. Shock 36(3):295–302

131. Eubank TD, Roda JM, Liu H, O'Neil T, Marsh CB (2011) Opposing roles for HIF-1alpha and HIF-2alpha in the regulation of angiogenesis by mononuclear phagocytes. Blood 117 (1):323–332
132. Takeda Y, Costa S, Delamarre E, Roncal C, Leite de Oliveira R, Squadrito ML, Finisguerra V, Deschoemaeker S, Bruyere F, Wenes M, Hamm A, Serneels J, Magat J, Bhattacharyya T, Anisimov A, Jordan BF, Alitalo K, Maxwell P, Gallez B, Zhuang ZW, Saito Y, Simons M, De Palma M, Mazzone M (2011) Macrophage skewing by Phd2 haplodeficiency prevents ischaemia by inducing arteriogenesis. Nature 479(7371):122–126
133. Escribese MM, Sierra-Filardi E, Nieto C, Samaniego R, Sanchez-Torres C, Matsuyama T, Calderon-Gomez E, Vega MA, Salas A, Sanchez-Mateos P, Corbi AL (2012) The prolyl hydroxylase PHD3 identifies proinflammatory macrophages and its expression is regulated by activin A. J Immunol 189(4):1946–1954
134. Taylor CT (2008) Interdependent roles for hypoxia inducible factor and nuclear factor-kappaB in hypoxic inflammation. J Physiol 586(Pt 17):4055–4059
135. Oliver KM, Taylor CT, Cummins EP (2009) Hypoxia. Regulation of NFkappaB signalling during inflammation: the role of hydroxylases. Arthritis Res Ther 11(1):215
136. Walmsley SR, Print C, Farahi N, Peyssonnaux C, Johnson RS, Cramer T, Sobolewski A, Condliffe AM, Cowburn AS, Johnson N, Chilvers ER (2005) Hypoxia-induced neutrophil survival is mediated by HIF-1alpha-dependent NF-kappaB activity. J Exp Med 201 (1):105–115
137. Cummins EP, Berra E, Comerford KM, Ginouves A, Fitzgerald KT, Seeballuck F, Godson C, Nielsen JE, Moynagh P, Pouyssegur J, Taylor CT (2006) Prolyl hydroxylase-1 negatively regulates IkappaB kinase-beta, giving insight into hypoxia-induced NFkappaB activity. Proc Natl Acad Sci USA 103(48):18154–18159
138. Fu J, Taubman MB (2010) Prolyl hydroxylase EGLN3 regulates skeletal myoblast differentiation through an NF-kappaB-dependent pathway. J Biol Chem 285(12):8927–8935
139. Xue J, Li X, Jiao S, Wei Y, Wu G, Fang J (2010) Prolyl hydroxylase-3 is down-regulated in colorectal cancer cells and inhibits IKKbeta independent of hydroxylase activity. Gastroenterology 138(2):606–615
140. Ferguson JE III, Wu Y, Smith K, Charles P, Powers K, Wang H, Patterson C (2007) ASB4 is a hydroxylation substrate of FIH and promotes vascular differentiation via an oxygen-dependent mechanism. Mol Cell Biol 27(18):6407–6419
141. Simmonds RE, Foxwell BM (2008) Signalling, inflammation and arthritis: NF-kappaB and its relevance to arthritis and inflammation. Rheumatology (Oxford) 47(5):584–590
142. Muz B, Khan MN, Kiriakidis S, Paleolog EM (2009) The role of hypoxia and HIF-dependent signalling events in rheumatoid arthritis. Arthritis Res Ther 11(1):201
143. Hermann LM, Pinkerton M, Jennings K, Yang L, Grom A, Sowders D, Kersten S, Witte DP, Hirsch R, Thornton S (2005) Angiopoietin-like-4 is a potential angiogenic mediator in arthritis. Clin Immunol 115(1):93–101
144. Lu J, Kasama T, Kobayashi K, Yoda Y, Shiozawa F, Hanyuda M, Negishi M, Ide H, Adachi M (2000) Vascular endothelial growth factor expression and regulation of murine collagen-induced arthritis. J Immunol 164(11):5922–5927
145. Raatz Y, Ibrahim S, Feldmann M, Paleolog EM (2012) Gene expression profiling and functional analysis of angiogenic markers in murine collagen-induced arthritis. Arthritis Res Ther 14(4):R169
146. Thornton S, Sowders D, Aronow B, Witte DP, Brunner HI, Giannini EH, Hirsch R (2002) DNA microarray analysis reveals novel gene expression profiles in collagen-induced arthritis. Clin Immunol 105(2):155–168
147. Sone H, Kawakami Y, Sakauchi M, Nakamura Y, Takahashi A, Shimano H, Okuda Y, Segawa T, Suzuki H, Yamada N (2001) Neutralization of vascular endothelial growth factor prevents collagen-induced arthritis and ameliorates established disease in mice. Biochem Biophys Res Commun 281(2):562–568

148. Miotla J, Maciewicz R, Kendrew J, Feldmann M, Paleolog E (2000) Treatment with soluble VEGF receptor reduces disease severity in murine collagen-induced arthritis. Lab Invest 80 (8):1195–1205
149. Afuwape AO, Feldmann M, Paleolog EM (2003) Adenoviral delivery of soluble VEGF receptor 1 (sFlt-1) abrogates disease activity in murine collagen-induced arthritis. Gene Ther 10(23):1950–1960
150. de Bandt M, Ben Mahdi MH, Ollivier V, Grossin M, Dupuis M, Gaudry M, Bohlen P, Lipson KE, Rice A, Wu Y, Gougerot-Pocidalo MA, Pasquier C (2003) Blockade of vascular endothelial growth factor receptor I (VEGF-RI), but not VEGF-RII, suppresses joint destruction in the K/BxN model of rheumatoid arthritis. J Immunol 171(9):4853–4859
151. Luttun A, Tjwa M, Moons L, Wu Y, Angelillo-Scherrer A, Liao F, Nagy JA, Hooper A, Priller J, De Klerck B, Compernolle V, Daci E, Bohlen P, Dewerchin M, Herbert JM, Fava R, Matthys P, Carmeliet G, Collen D, Dvorak HF, Hicklin DJ, Carmeliet P (2002) Revascularization of ischemic tissues by PlGF treatment, and inhibition of tumor angiogenesis, arthritis and atherosclerosis by anti-Flt1. Nat Med 8(8):831–840
152. Kong JS, Yoo SA, Kim JW, Yang SP, Chae CB, Tarallo V, De Falco S, Ryu SH, Cho CS, Kim WU (2010) Anti-neuropilin-1 peptide inhibition of synoviocyte survival, angiogenesis, and experimental arthritis. Arthritis Rheum 62(1):179–190
153. Kong JS, Yoo SA, Kang JH, Ko W, Jeon S, Chae CB, Cho CS, Kim WU (2011) Suppression of neovascularization and experimental arthritis by D-form of anti-flt-1 peptide conjugated with mini-PEG™. Angiogenesis 14(4):431–442
154. Pickens SR, Volin MV, Mandelin AM II, Kolls JK, Pope RM, Shahrara S (2010) IL-17 contributes to angiogenesis in rheumatoid arthritis. J Immunol 184(6):3233–3241
155. Pickens SR, Chamberlain ND, Volin MV, Gonzalez M, Pope RM, Mandelin AM II, Kolls JK, Shahrara S (2011) Anti-CXCL5 therapy ameliorates IL-17-induced arthritis by decreasing joint vascularization. Angiogenesis 14(4):443–455
156. Moran EM, Connolly M, Gao W, McCormick J, Fearon U, Veale DJ (2011) Interleukin-17A induction of angiogenesis, cell migration, and cytoskeletal rearrangement. Arthritis Rheum 63(11):3263–3273
157. Hashizume M, Hayakawa N, Suzuki M, Mihara M (2009) IL-6/sIL-6R trans-signalling, but not TNF-alpha induced angiogenesis in a HUVEC and synovial cell co-culture system. Rheumatol Int 29(12):1449–1454
158. Kayakabe K, Kuroiwa T, Sakurai N, Ikeuchi H, Kadiombo AT, Sakairi T, Matsumoto T, Maeshima A, Hiromura K, Nojima Y (2012) Interleukin-6 promotes destabilized angiogenesis by modulating angiopoietin expression in rheumatoid arthritis. Rheumatology (Oxford) 51 (9):1571–1579
159. Chen Y, Donnelly E, Kobayashi H, Debusk LM, Lin PC (2005) Gene therapy targeting the Tie2 function ameliorates collagen-induced arthritis and protects against bone destruction. Arthritis Rheum 52(5):1585–1594
160. Jin P, Zhang J, Sumariwalla PF, Ni I, Jorgensen B, Crawford D, Phillips S, Feldmann M, Shepard HM, Paleolog EM (2008) Novel splice variants derived from the receptor tyrosine kinase superfamily are potential therapeutics for rheumatoid arthritis. Arthritis Res Ther 10 (4):R73
161. Malik NM, Jin P, Raatz Y, Sumariwalla PF, Kiriakidis S, Shepard M, Feldmann M, Paleolog EM (2010) Regulation of the Angiopoietin-Tie ligand-receptor system with a novel splice variant of Tie1 reduces the severity of murine arthritis. Rheumatology 49(10):1828–1839
162. Arsenault AL, Lhotak S, Hunter WL, Banquerigo ML, Brahn E (1998) Taxol involution of collagen-induced arthritis: ultrastructural correlation with the inhibition of synovitis and neovascularization. Clin Immunol Immunopathol 86(3):280–289
163. Bainbridge J, Madden L, Essex D, Binks M, Malhotra R, Paleolog EM (2007) Methionine aminopeptidase-2 blockade reduces chronic collagen-induced arthritis: potential role for angiogenesis inhibition. Arthritis Res Ther 9(6):R127

164. Bernier SG, Lazarus DD, Clark E, Doyle B, Labenski MT, Thompson CD, Westlin WF, Hannig G (2004) A methionine aminopeptidase-2 inhibitor, PPI-2458, for the treatment of rheumatoid arthritis. Proc Natl Acad Sci USA 101(29):10768–10773
165. Hwang SR, Seo DH, Al-Hilal TA, Jeon OC, Kang JH, Kim SH, Kim HS, Chang YT, Kang YM, Yang VC, Byun Y (2012) Orally active desulfated low molecular weight heparin and deoxycholic acid conjugate, 6ODS-LHbD, suppresses neovascularization and bone destruction in arthritis. J Control Release 163(3):374–384
166. Someno T, Kumagai H, Ohba S, Amemiya M, Naganawa H, Ishizuka M, Ikeda D (2004) ICM0301s, new angiogenesis inhibitors from Aspergillus sp. F-1491. II. Physico-chemical properties and structure elucidation. J Antibiot (Tokyo) 57(2):104–109
167. Kumagai H, Masuda T, Ohba SI, Ikeda D (2013) Suppression of type II collagen-induced arthritis by ICM0301B, a new angiogenesis inhibitor. J Antibiot (Tokyo) 66(4):243–246
168. Tanaka K, Morii T, Weissbach L, Horiuchi K, Takeuchi K, Toyama Y, Morioka H (2011) Treatment of collagen-induced arthritis with recombinant plasminogen-related protein B: a novel inhibitor of angiogenesis. J Orthop Sci 16(4):443–450
169. Chang LH, Huang HS, Wu PT, Jou IM, Pan MH, Chang WC, Wang DD, Wang JM (2012) Role of macrophage CCAAT/enhancer binding protein delta in the pathogenesis of rheumatoid arthritis in collagen-induced arthritic mice. PLoS One 7(9):e45378
170. Roy A, Kolattukudy PE (2012) Monocyte chemotactic protein-induced protein (MCPIP) promotes inflammatory angiogenesis via sequential induction of oxidative stress, endoplasmic reticulum stress and autophagy. Cell Signal 24(11):2123–2131
171. Sheikh A, Naqvi SH, Sheikh K (2012) Itraconazole: its possible role in inhibiting angiogenesis in rheumatoid arthritis. Med Hypotheses 79(3):313–314
172. Mazzone M, Dettori D, Leite de Oliveira R, Loges S, Schmidt T, Jonckx B, Tian YM, Lanahan AA, Pollard P, Ruiz de Almodovar C, De Smet F, Vinckier S, Aragones J, Debackere K, Luttun A, Wyns S, Jordan B, Pisacane A, Gallez B, Lampugnani MG, Dejana E, Simons M, Ratcliffe P, Maxwell P, Carmeliet P (2009) Heterozygous deficiency of PHD2 restores tumor oxygenation and inhibits metastasis via endothelial normalization. Cell 136 (5):839–851
173. Cummins EP, Seeballuck F, Keely SJ, Mangan NE, Callanan JJ, Fallon PG, Taylor CT (2008) The hydroxylase inhibitor dimethyloxalylglycine is protective in a murine model of colitis. Gastroenterology 134(1):156–165

Angiogenic Signaling and Structural Abnormalities in Tumors

17

Magdalena Tertil, Klaudia Skrzypek, and Agnieszka Łoboda

Contents

17.1 Angiogenic Signaling in Tumor Cells 368
17.2 Stromal Cells in Tumor Angiogenesis 372
17.3 Heterogeneity of Tumor Vasculature 376
17.3.1 Tumor Vasculature Is Unique 376
17.3.2 Alternative Mechanisms of Neovascularization 378
References 380

Abstract

Growing tumor needs to be supplied with oxygen and nutrients; hence, the mechanisms responsible for development of new blood vessels are crucial for tumor progression. Enhanced expression of proangiogenic factors enables development of tumor vasculature and subsequent invasion of tumor cells. The key step in these events is decreased oxygen tension within the growing tumor due to limited oxygen diffusion within the tissue. Apart from canonical hypoxic signaling,

M. Tertil
Department of Medical Biotechnology, Faculty of Biochemistry, Biophysics and Biotechnology, Jagiellonian University, Gronostajowa 7, 30-387 Kraków, Poland

Department of Molecular Neuropharmacology, Institute of Pharmacology, Polish Academy of Sciences, Kraków, Poland

K. Skrzypek
Department of Medical Biotechnology, Faculty of Biochemistry, Biophysics and Biotechnology, Jagiellonian University, Gronostajowa 7, 30-387 Kraków, Poland

Department of Transplantation, Polish-American Institute of Pediatrics, Jagiellonian University Medical College, Kraków, Poland

A. Łoboda (✉)
Department of Medical Biotechnology, Faculty of Biochemistry, Biophysics and Biotechnology, Jagiellonian University, Gronostajowa 7, 30-387 Kraków, Poland
e-mail: agnieszka.loboda@uj.edu.pl

J. Dulak et al. (eds.), *Angiogenesis and Vascularisation*,
DOI 10.1007/978-3-7091-1428-5_17, © Springer-Verlag Wien 2013

there are numerous molecular pathways that may modulate angiogenic secretome of cancer cells, such as the action of angiogenic enzymes and microRNAs. Besides tumor cells, also other cellular components of tumor microenvironment play an important role in stimulation of endothelium, out of which the key players are different populations of bone marrow-derived cells of myeloid origin.

Overstimulation of endothelial cells leads to development of abnormal vasculature that may be further disorganized by adaptation of alternative mechanisms of vascularization such as vessel co-option, intussusceptive microvascular growth or glomeruloid angiogenesis. Several tumor types are also capable of forming functional vessel-like structures lined with cancer cells by means of vasculogenic mimicry.

The understanding of the complexity and the diversity of the factors leading to tumor development as well as the unique structural adaptation of tumor microvessels to form functional vasculature may be helpful for the establishing potent antitumor therapies.

Keywords
Neovascularization • Hypoxia • Tumor-associated macrophages • Chemokines • microRNA • miR-378

17.1 Angiogenic Signaling in Tumor Cells

Although tumor cells generally evade regulation by normal growth-controlling mechanisms, cancer development and progression is critically depending on supply of nutrients, oxygen, and other host-derived factors through blood, as pointed out for the first time by Judah Folkman in 1971 [1]. Hence, induction of tumor angiogenesis, the so-called "angiogenic switch" through the involvement of great range of angiogenic mediators and their downstream effectors (Fig. 17.1), has been established as one of the hallmarks of cancer [2]. Accordingly, several activated oncogenes were demonstrated to influence production of vascular endothelial growth factor (VEGF), the main stimulator of blood vessel formation [3, 4]. Mutant K-ras was linked with increased production of this angiogenic factor in colorectal carcinoma cells [5], and Myc overexpression led to a tenfold induction of VEGF in a B-cell line [6]. Interestingly, recent studies in non-small cell lung carcinoma (NSCLC) models demonstrated a direct significant upregulation of IL-8 by activation of epidermal growth factor receptor (EGFR) oncogenic signaling [7]. Nevertheless, the major molecular mechanisms contributing to tumor blood vessel development are driven by responses to hypoxia.

Decreased oxygen tension within the growing tumor due to limited oxygen diffusion in the tissue leads to activation of canonical hypoxia-responsive transcription factors: hypoxia-inducible factors (HIF-1 and HIF-2), which are considered as key players in tumorigenicity and tumor angiogenesis. HIF-1α/HIF-1β heterodimers bind to the

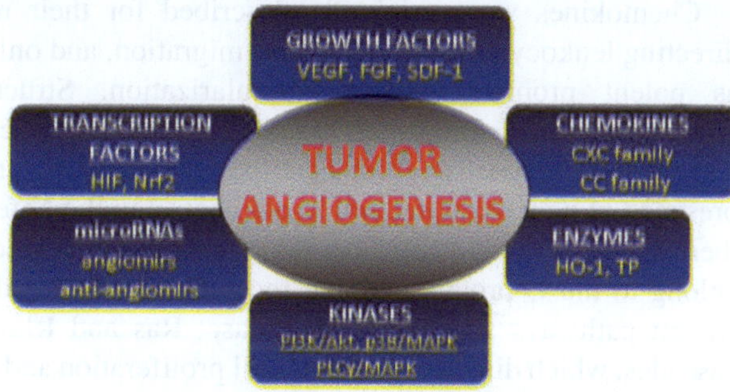

Fig. 17.1 The major groups of angiogenic mediators and their representatives important for tumor angiogenesis

hypoxia response elements (HRE) and thereby initiate expression of a broad array of targets including numerous pro-angiogenic genes [8], out of which VEGF has been most extensively studied. VEGF binding to VEGFR-2 on endothelium triggers downstream signaling cascades including PI3K/Akt, p38/MAPK, and PLCγ/MAPK to mediate canonical cellular effects of VEGF in angiogenesis by induction of endothelial cells (ECs) survival, proliferation, migration, and secretion of proteases and increased microvascular permeability [9]. Upon angiogenic switch, cancer cells usually produce high amounts of VEGF and thereby both plasma levels of VEGF as well as its expression within the tumor were found to correlate with shortened progression-free and overall survival of patients with different types of tumors [10, 11]. Another canonical target of HIF-1 is angiopoietin-2. Angiopoietins (Ang-1, Ang-2) were identified as ligands for Tie-2 tyrosine kinase receptor on ECs. Ang-1 promotes ECs survival through PI3K/Akt activation, reduces vascular permeability by tightening cell junctions, thus counteracting the action of VEGF and inflammatory cytokines, and promotes vessel maturation by increasing endothelial cell–pericyte interaction [12]. Ang-1 overexpression modulates vascular endothelium, facilitating tumor cell dissemination and metastasis [13]. Oppositely, Ang-2 antagonizes the action of Ang-1 by competitive binding to Tie-2. Autocrine Ang-2-Tie-2 axis promotes angiogenesis in activated endothelium by destabilizing the vessels and sensitizing ECs to mitogenic signals mediated by other proangiogenic factors such as VEGF [14]. In the absence of VEGF, however, increased Ang-2 production leads to pericyte detachment, ECs apoptosis, and even vessel regression [15]. Host-derived Ang-2 plays an important role in early stages of tumor development and vessel maturation, but it may be dispensable for later stages of tumor growth [16].

Other families of growth factors, which may induce angiogenesis independently of VEGF pathway, are fibroblast growth factors (FGFs) and chemokines [17]. Out of the family comprising 18 ligands binding to FGF receptors (FGFRs), FGF-1 (acidic FGF, aFGF) and FGF-2 (basic FGF, bFGF) are best characterized for the potent angiogenic activity [18, 19]. bFGF is upregulated in multiple tumor types and has mitogenic and migratory effects on many cell types, including endothelial cells [20, 21].

Chemokines were originally described for their role in immune responses, directing leukocyte proliferation and migration, and only later have been identified as potent promoters of neovascularization. Structurally, four families of chemokines are distinguished based on the position of conserved cysteine residues. The most numerous subgroups are the CXC and CC families, characterized by the presence of nonconserved amino acid between the N-terminal cysteines or lack of thereof, respectively [22]. Chemokines signal through chemokine receptors which belong to the G-protein coupled receptors family, leading to activation of downstream pathways such as MAP kinase, Ras and Rho GTPases, and PI3 kinase cascades, which directly modulate cell proliferation and migration [23]. Endothelial cells were found to express CXCR-1 and -2 as well as CCR-1 and -2 receptors which mediate direct angiogenic response to chemokines CXCL-1, -2, -3, -5, -6, -7, -8 and CCL-2 (MCP-1), -11, and -16 [24–26].

Interestingly, latest research on the molecular mechanisms underlying hypoxic HIF-1 activation shows that, paradoxically, it involves reactive oxygen species (ROS) produced by mitochondria under limited oxygenation [27]. The proposed action of ROS in HIF-1 regulation relies on the inhibition of activity of prolyl hydroxylases through oxidation of ferrous to ferric iron and/or depletion of ascorbate [27, 28]. Consequently, proteins involved in antioxidant response are being recognized as important players in regulation of HIF-1 signaling and tumor angiogenesis. Nrf2 [nuclear factor (erythroid-derived 2)-like 2, NFE2L2] transcription factor is a master regulator of cellular response to oxidative and electrophilic stress driving the expression of a broad array of cytoprotective genes [29]. Kim et al. found Nrf2 downregulation led to attenuation of tumor angiogenesis through suppression of hypoxia-induced HIF-1α stabilization and VEGF expression in colon carcinoma [30].

Nrf2 may be also involved in promotion of HIF-1 independent angiogenesis (see also Chap. 10). Our and others' work shows that its direct transcriptional target is proangiogenic chemokine interleukin-8 (IL-8, CXCL8) [31–33]. In human microvascular endothelial cells, IL-8 acts through both CXCR-1 and -2, inducing cell proliferation, migration, and capillary-like structure formation [34]. Importantly, IL-8 was demonstrated to preserve angiogenic response of HIF-1-deficient cancer cells [35].

Another target gene of Nrf2 is proangiogenic enzyme heme oxygenase-1 (HO-1), converting heme into equimolar amounts of three biologically active products: carbon monoxide (CO), ferrous iron ions (Fe^{2+}), and biliverdin [36]. HO-1 may act upstream and downstream of other important angiogenic factors such as VEGF or stromal cell-derived factor-1 (SDF-1), thus playing also a substantial role in the processes of neovascularization. The molecular mechanisms responsible for proangiogenic activities of HO-1 are well described (see also Chap. 10). Moreover, an increasing number of papers show the importance of HO-1 action in tumorigenesis. For instance, in human melanoma, HO-1 level in tumor infiltrating macrophages is accompanied by enhanced vascular density inside tumor [37] and in human glioma HO-1 expression correlates with vascular density [38]. Oppositely,

HO-1 overexpression diminished angiogenesis and distal metastasis of NSCLC subcutaneously implanted to immunodeficient mice [39].

Another interesting proangiogenic enzyme is thymidine phosphorylase (TP). TP catalyzes the reversible phosphorolysis of thymidine and 2′-deoxyuridine into their respective bases and 2′-α-D-deoxyribose-1-phosphate (dRP), which can be further dephosphorylated into 2′-α-D-deoxyribose (dR) by ubiquitous phosphatases [40]. In fact, TP was alternatively discovered as "platelet-derived endothelial cell growth factor" [41]. TP induced neovascularization in a chicken embryo chorioallantoic membrane (CAM) assay [42], in gelatin sponges implanted subcutaneously in mice [43], as well as in rat corneal and mouse dorsal air sack models in vivo [44]. In vitro, TP enhanced tube formation and migration of endothelial cells [45, 46]. The major mediators of proangiogenic action of TP are the products of its enzymatic activity, dRP and dR. The sugar products were demonstrated to induce endothelial cell migration in vitro and neovascularization in different in vivo models [47, 48]. The proangiogenic actions of TP, however, are not restricted to the direct effects on endothelium. In many tumor types, including breast, bladder, gastric, colorectal, cervical, esophageal, and lung carcinomas, expression of TP is elevated in comparison to healthy tissues [49]. In most cases it is also associated with higher microvessel density [49]. Indeed, studies have shown that cancer cells modified to overexpress the enzyme form better vascularized and larger tumors in experimental animals [50]. This effect may result not only from the formation of chemoattractant gradient of dRP/dR but also from the modulation of production of "canonical" angiogenic factors such as VEGF, bFGF, or IL-8. Accordingly, human bladder carcinoma cells RT112TP overexpressing TP secreted higher amounts of VEGF and IL-8 in the presence of excess of TP substrate [51]. Moreover, these cells displayed enhanced activation of HIF-1α under hypoxia and consequently produced more VEGF also under reduced oxygen pressure [52]. Importantly, significant correlation of expression of TP and VEGF was observed in clinical samples from cancer patients [49], including NSCLC specimens [53]. In colon cancer cells, on the other hand, TP overexpression leads to upregulation of bFGF, while VEGF level was unchanged [54]. In human cervical cancer cell line Yumoto and KB epidermoid carcinoma cells TP overexpression caused significant induction of IL-8, which was attributed to enhanced ROS production in cells with high TP levels [55].

Recently, the growing number of evidence links the alterations in the level of small noncoding RNAs—microRNA (miRNAs) and tumor development. The first evidence for it was reported in 2002 in B cell chronic lymphocytic leukemia [56] and then in 2003 in colorectal cancers [57]. It soon turned out that miRNAs can be differentially expressed in various tumor types, either benign or malignant [58], and their effect on tumor progression might be connected with the modulation of development of new blood vessels, both via promoting angiogenesis (angiomirs) or through its inhibition (anti-angiomirs) [59]. Anti-angiomirs, including miR-15a, miR-16, miR-125b, miR-199a, or miR-200b target proangiogenic VEGF and their expression could be decreased in different tumor types including lymphomas, multiple myelomas, lung, colorectal, hepatocellular, ovarian, oral, and breast cancer [60–62]. VEGF may be not only inhibited by miRNAs, but it can also be activated by them. miR-378 can indirectly promote expression of proangiogenic VEGF, because it competes with miR-125a for the same seed region in the VEGF 3′-UTR [63]. However, proangiogenic actions of miR-378 may be also explained by direct regulation of Fus-1 and Sufu [64], which functions as a negative regulator of Shh signaling. Shh induces expression of VEGF and angiopoietins (Ang-1, Ang-2) [65] and thereby it enhances formation of large-diameter vessels. Nude mice injected with miR-378 overexpressing glioblastoma cells formed tumors with larger blood vessels compared to control cells [64]. Similarly, lung adenocarcinoma cells overexpressing miR-378 implanted to nude mice were better vascularized [66]. Our study showed that media harvested from NSCLC overexpressing miR-378 enhanced formation of tubule-like structures from endothelial cells on Matrigel, and consequently NSCLC xenografts overexpressing miR-378 in immunodeficient mice were better vascularized and oxygenated and displayed enhanced distal metastasis [39]. There are many more other possible miRNAs and their targets involved in regulation of tumor angiogenesis, including HIF pathway or matrix metalloproteinases (MMPs), the subject described in detail in the Chap. 9.

Activity of tumor cell-derived factors in modulation of cancer neovascularization is not by any chance limited to direct stimulation of endothelium. VEGF, IL-8, and other cytokines and chemokines attract various circulating and surrounding cells to tumor site, which in turn may provide further proangiogenic stimuli for ECs.

17.2 Stromal Cells in Tumor Angiogenesis

Tumor vascularization is a result of an intricate interplay between the constituents of tumor microenvironment, which may differ according to tumor type and stage of progression. Solid tumors have to be considered as complex organs comprising heterogenous cellular components that shape the extracellular milieu. Besides cancer cells and endothelium, various stromal and circulating host cells recruited to the tumor contribute to the regulation of blood vessel formation in neoplastic tissues (Fig. 17.2).

The major players in inducing and sustaining the proangiogenic stimuli in tumor microenvironment are cells of myeloid origin that exhibit multifaceted functional phenotypes [67, 68]. Tumor-associated macrophages (TAMs) differentiate from tumor infiltrating monocytes primarily because of the presence of macrophage-colony stimulating factor (M-CSF) produced by cancer cells (Fig. 17.3). They adopt the phenotype biased towards an M2 [69] according to the established concept of macrophage polarization which distinguishes two extreme phenotypes. M2 (alternative activation pathway) differentiate in milieu, are rich in Th2 cytokines (IL-4, IL-13), have high scavenging capacities, and produce growth factors promoting tissue repair and suppressing adaptive immunity. M1 (classical activation) are stimulated by bacterial products and Th1 cytokines (e.g., IFNγ). These cells are characterized by high levels of MHCII expression, IL-12 and TNFα secretion, and production of ROS and nitrogen species [70]. TAMs produce an array of trophic

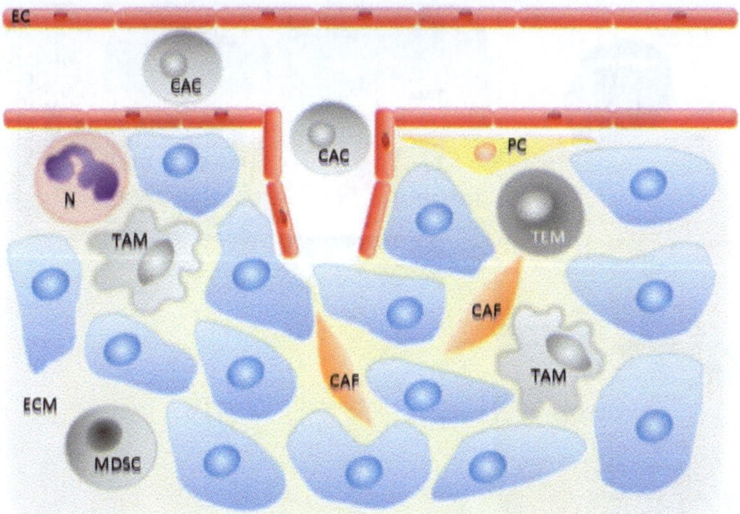

Fig. 17.2 Stromal cells in tumor angiogenesis. *EC* endothelial cells, *PC* pericytes, *TAM* tumor-associated macrophages, *TEM* Tie-2-expressing monocytes, *MDSC* myeloid-derived suppressor cells, *CAC* circulating angiogenic cells, *N* neutrophils, *CAF* cancer-associated fibroblasts, *ECM* extracellular matrix

and activating factors for both tumor and endothelial cells, including platelet-derived growth factor (PDGF), transforming growth factorβ (TGFβ). VEGF, FGF and EGF. Their crucial role in angiogenesis was illustrated by an elegant study showing that in mice devoid of M-CSF, mammary tumorigenesis was inhibited due to attenuated tumor vascularization [71]. As major producers of inflammatory mediators, TAMs are also key promoters of tumor-associated inflammation, which is considered another hallmark of cancer [2]. Several studies have shown that activation of transcription factor NFκB plays a central role in tumor promotion by TAM [72]. In response to factors released by necrotic tissues, cytokines produced by tumor and/or stromal cells or signaling triggered through HIF-1α by hypoxia, activation of NFκB leads to production of cytokines (TNFα, IL-6) and chemokines amplifying the inflammatory cascade. TNFα has been associated with increased production of chemokines such as CCL2, IL-8, and CXCL12 [73]. IL-6 activates STAT3 (signal transducer and activator of transcription 3) transcription factor which promotes cell cycle progression and resistance to apoptosis [74].

Apart from the TAMs, a distinct population of Tie-2-expressing monocytes (TEMs) contributes significantly to paracrine stimulation of angiogenesis in cancer [75]. TEMs exhibit a proangiogenic gene expression signature already in circulation, but it could be further enhanced by stimulation with Ang-2 upon recruitment to the tumor [76]. Besides releasing factors that directly stimulate ECs (e.g., VEGF and bFGF), TEMs are also important sources of factors modulating the neovascularization and metastasis in an indirect fashion, namely the proteolytic enzymes such as cathepsins and MMPs [76, 77]. Recently, it has been recognized that also CD11b

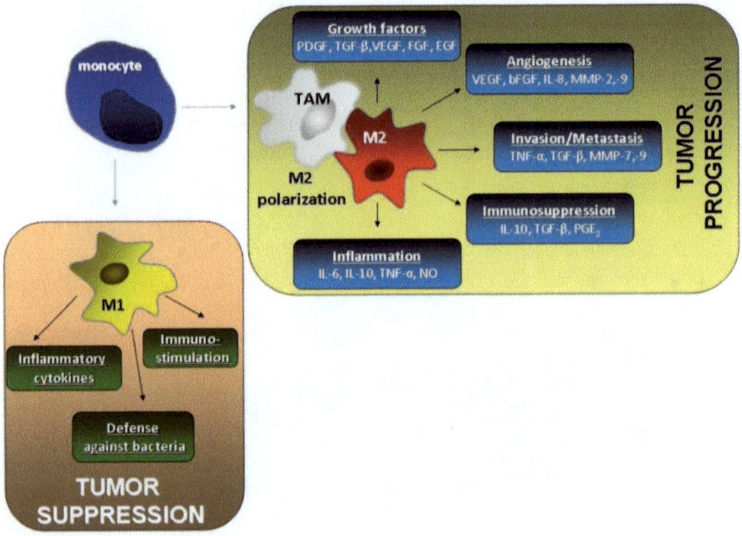

Fig. 17.3 Polarization of macrophage function. M2 cells, differentiated tumor-associated macrophages (TAMs), play a pivotal role in tumor growth. TAMs regulate tumor angiogenesis, inflammation, tumor cell invasion, and metastasis as well as immunosuppression. On the other hand, M1 macrophages display cytotoxic activity against microorganisms and neoplastic cells and may activate the immune response

+Gr1+ bone marrow-derived precursor cells known as myeloid-derived suppressor cells (MDSCs) which may constitute up to 5 % of tumor population play important role in cancer neovascularization [67]. MDSCs coinjected with cancer cells lead to formation of tumors with increased vascular density and better vessel maturation, which is abolished by deletion of MMP-9 [78], a protease crucial for increasing the bioavailability of VEGF and FGF-2 in tumor microenvironment [79]. Interestingly, there is another potent leukocytic source of proangiogenic MMP-9 within tumor stroma—cancer-infiltrating neutrophils that produce the proenzyme which is devoid of tissue inhibitor of metalloproteinases (TIMP). In contrast to other cells, including various tumor cells which secrete proMMP-9 in a complex with inhibitory TIMP-1, neutrophil TIMP-free proMMP-9 can be efficiently activated and then it acts as an angiogenic inducer [80, 81]. Cells of MDSC phenotype were also reported to incorporate directly into tumor vessels acquiring endothelial properties in tumor microenvironment [78].

The notion of contribution of bone marrow-derived vascular progenitors, so-called endothelial progenitor cells (EPC), to tumor neovascularization through vasculogenesis has been strongly argued over the years. A term EPC was introduced by Asahara et al., who described a population of naïve circulating blood cells expressing VEGFR2 and enriched with stem/endothelial marker CD34 [82]. Since then numerous experimental models indicated that the cells of the EPC phenotype may directly incorporate into tumor vasculature [83, 84] while others denied it [85]. Today it is generally accepted that these cells rather do not present endothelial precursors but are in fact of myeloid origin [86]. Nevertheless, they constitute a truly proangiogenic pool, now widely termed circulating angiogenic cells (CAC) which act mostly via paracrine stimulation of endothelium [87]. Indeed, blocking EPC mobilization was clearly associated with inhibition of tumor angiogenesis and metastasis [88]. In the clinic, CAC counts are elevated in the blood of cancer patients and may be a negative prognostic factor [89]. One of the major axis for mobilization of the bone marrow-derived pro-angiogenic cells is SDF1/CXCL12 chemokine inducible by hypoxia and VEGF acting through CXCR4 receptor [90]. It is involved in the recruitment of Tie-2-expressing monocytes, which have been already mentioned as key effectors in triggering the angiogenic switch [79], as well as in homing and engraftment of mesenchymal stromal cells to tumors [91].

Mesenchymal cells form another constituent of angiogenesis-modulating tumor microenvironment. Mesenchymal progenitors have been described to differentiate to pericytes and secrete VEGF in a model of pancreatic carcinoma [92]. Heterogenous population comprising bone marrow-derived mesenchymal cells together with local stromal fibroblasts and/or tumor cells which underwent epithelial–mesenchymal transition is considered as cancer-associated fibroblasts (CAFs) [93]. CAFs are one of the major sources of MMPs within the tumor. For example, it is MMP-2 expression in stromal fibroblasts and not in tumor cells that is a negative prognostic factor in NSCLC [94]. CAFs may secrete high amounts of SDF-1 [95] which apart from its key role in attracting CAC, may act on tumor cells to enhance their migration and adhesion to stromal cells, which in turn provide them with further signals for growth and metastasis [96–98]. Finally, fibroblasts are key producers of another factor crucial for tumor progression—hepatocyte growth factor (HGF) [99] that acts as a potent inducer of neovascularization [100]. Signaling through its tyrosine kinase receptor MET, HGF enhances endothelial cell growth in formation of new blood vessels as well as lymphatic vessels [101, 102]. MET and VEGFR-2 cooperate in promotion of angiogenesis by synergistic activation of downstream signaling pathways involving Erk-MAPK, AKT and FAK kinases [103]. Activation of HGF-MET axis can also lead to induction of VEGF and suppression of expression of antiangiogenic thrombospondin-1 [104]. Accordingly, treatment of experimental tumors with MET inhibitors results in attenuation of angiogenesis and tumor growth [105].

Cells within tumor microenvironment may communicate to modulate blood vessel formation not only by secretion of proteins and small-molecule mediators or direct cell–cell interactions but also by exchanging information through microvesicles. Alternative names help to distinguish the origin of microvesicles: those released from activated platelets are called microparticles, neutrophil-derived microvesicles are called ectosomes, whereas exosomes are microvesicles secreted by tumors [106]. However, in nomenclature the term exosomes describes also vesicles originating from endocytosis. Their size ranges from 50 nm to 1 μm, and they are secreted by healthy and damaged cells in response to external stimuli. Their role involves mediation of communication between cells, as they contain miRNAs and mRNA inside, which is called exosomal shuttle RNA (esRNA).

Vesicles released from human and murine mast cell lines contain over 1,200 mRNAs and approximately 121 miRNAs molecules [107]. For example, it has been demonstrated that genetic information can be transmitted from tumor cells through miRNAs transfer by exosomes and in that way reprogram endothelial cells to a proangiogenic phenotype [108]. During tumor growth and spreading microvesicles are actively released and they can play a role as vehicles for information transfer. For example, microvesicles derived from tumors can transfer mRNA to monocytes to activate these cells in tumor microenvironment, increasing production of cytokines to fasten tumor growth and diminish the immune response [109]. It has been also demonstrated that microvesicles released from human renal cancer stem cells stimulate angiogenesis and formation of lung pre-metastatic niche [110]. Exosomes circulating in blood can be derived also from platelets, and they can play a role in hemostatic processes induced during clotting. Moreover, they may exert a role in vascularization and metastasis of many tumors including lung cancer [111]. There are several examples of miRNAs, which can be transferred from cancer cells to endothelial cells, such as miR-210. It can be released by metastatic cancer cells and transported to endothelial cells, where it inhibits expression of proangiogenic target genes [112]. Similarly, miR-9 can be transferred from tumor cells to endothelial cells by microvesicles. Then miR-9 reduces suppressor of cytokine signaling 5 (SOCS5) level and thereby activates JAK-STAT pathway, enhancing tumor angiogenesis [113].

17.3 Heterogeneity of Tumor Vasculature

17.3.1 Tumor Vasculature Is Unique

Tumor microenvironment is characterized by uncontrolled, continuous overproduction of angiogenic factors. Paradoxically, such extreme stimulation of endothelium leads to development of immature, functionally, and structurally abnormal neovasculature (Fig. 17.4) [114]. The architecture of tumor blood vessel network is irregular, disorganized, and tortuous [115]. As a result, areas of increased vascularization and pockets of reduced vascular density are formed [116]. Endothelial cell layer is highly dysfunctional—loose connections, wider junctions, and impaired coverage with mural cells cause substantial leakiness of the vessels [117] and some studies even report cytogenetic anomalies within tumor ECs [118]. Intercellular gaps and transendothelial holes favor the escape of cancer cells to the bloodstream. Moreover, hyperpermeability of the vasculature leading to elevation of interstitial fluid pressure combined with heterogeneity of vessel distribution cause spatial and temporal variations in blood flow [119]. Such impaired perfusion not only hampers delivery of therapeutic drugs to the tumor but most importantly creates areas of hypoxic and acidic microenvironment [120].

Hypoxia drives secretion of angiogenic factors and recruitment of bone marrow-derived proangiogenic cells, thus fueling the vicious cycle of nonproductive neovascularization. What is more, it leads to pro-malignant reprogramming of

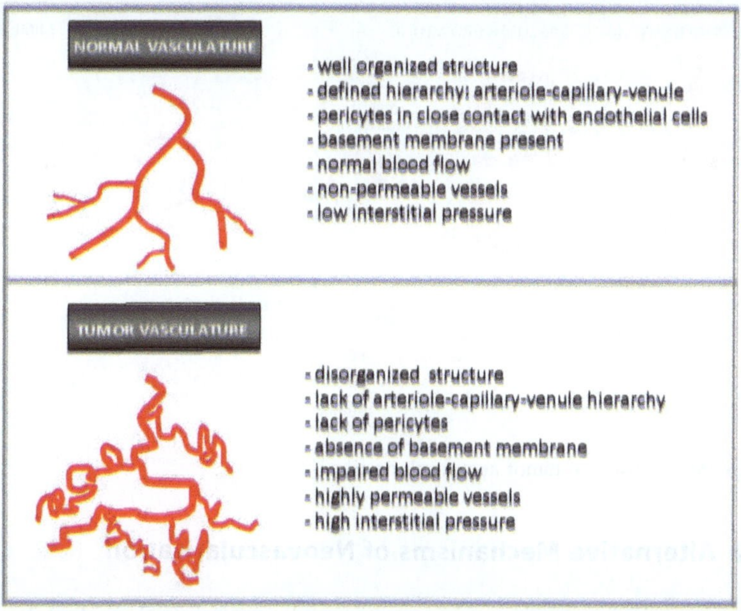

Fig. 17.4 The major differences between normal and tumor vasculature

tumor gene expression and selection of genotypes which favor resistance to hypoxia-reoxygenation injury, such as *TP53* tumor suppressor mutations [121]. Low oxygen concentration downregulates DNA repair mechanisms, thus potentiating the genomic instability in cancer cells [122]. Gene expression changes evoked by hypoxia promote switch to anaerobic metabolism [123], enhanced autophagy [124], inhibition of apoptosis [121], and increased invasiveness [125] and metastasis [126]. Hypoxia also alters receptor tyrosine kinase-mediated signaling [127] and epithelial-to-mesenchymal transition (EMT) in cancer cells [128]. Finally, oxygen deprivation may lead to suppression of antitumor immune responses and contributes to the failure of radiotherapy [129]. Not surprisingly, recent research has unraveled yet one more face of pro-tumoral action of hypoxia—its role in the maintenance of a small population of cancer stem-like cells, which are believed to drive tumorigenesis and disease recurrence [130].

Hence, hypoxia emerges as a major therapeutic target for angiogenesis-related anticancer treatment (see Chap. 18). Surprisingly, however, recent data suggest also that activation of HIFs due to the inhibition of prolyl hydroxylase-2 (PHD-2), an enzyme responsible for HIFs protein degradation in physiological conditions, can paradoxically improve the effectiveness of cancer therapy due to tumor vessel normalization, increasing the delivery of chemotherapeutics [131]. Thus, the initial view that anti-angiogenic therapies should destroy the extensive tumor blood network, depriving the tumor of oxygen and nutrients, is replaced by understanding that disregulated, chaotic vascular network has to be normalized to increase the effectiveness of chemotherapy.

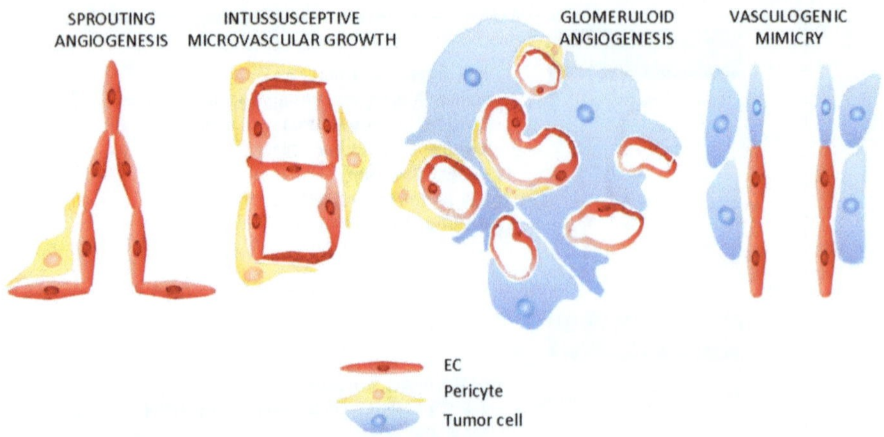

Fig. 17.5 Mechanisms of tumor neovascularization

17.3.2 Alternative Mechanisms of Neovascularization

Besides canonical sprouting angiogenesis and controversial postnatal vasculogenesis, tumor neovascularization may involve also alternative mechanisms such as vessel co-option, intussusceptive microvascular growth (IMG), glomeruloid angiogenesis, and tumor cell vasculogenic mimicry (Fig. 17.5) that have been associated with the most aggressive tumor types [132].

Vessel co-option refers to the avascular growth of primary or metastatic lesions along preexisting vessels that occurs usually in well-vascularized tissues such as lung or brain [133, 134]. This phenomenon was described to occur rather at initial stages of tumor development. In a model of glioma Holash et al. found that the coopted vessels underwent regression after 4 weeks of tumor growth which could be interpreted as a host defense mechanism [135]. The best candidate for a major molecular regulator of this process is Ang-2. Its expression raises in coopted vessels leading to destabilization of capillary walls and vessel regression at early tumor growth stages when high amounts of VEGF are not yet present [135]. Upon subsequent “angiogenic switch”, Ang-2 and VEGF cooperate to promote sprouting neovascularization. Besides glioma and lung carcinoma, vessel co-option has now been described in several other types of tumors, including also human melanoma and Kaposi sarcoma [136].

Expansion of capillary network by intussusceptive microvascular growth (IMG) involves partitioning of preexisting vessel lumen by insertion of connective tissue columns called tissue pillars and their subsequent growth [132]. Such phenomenon can occur within hours or even minutes and requires little energy as it does not depend on ECs proliferation and extracellular matrix remodeling. Instead, endothelial cells increase their volume and become thinner. Although molecular mechanisms driving IMG are not yet fully understood, it has been noted that the process could be stimulated by alterations of blood flow dynamics [137], and the best candidates for its regulation are proteins involved in mediating interaction between ECs and endothelium and mural cells, namely the PDGF-BB, angiopoietins, TGF-β, monocyte chemoattractant protein-1 (MCP-1), and ephrins together with Eph-B receptors [138]. IMG was observed in melanoma, colorectal, and mammary tumors [132].

Glomeruloid bodies (GBs) formation, another phenomenon associated with cancer vascularization, is best characterized for glial tumors. Glioblastoma multiforme models showed that tumor cells may pull microvesicular plexus into the tumor cell nest leading to appearance of coiled vascular structures which further evolve into chaotic and tortuous vascular aggregates [139]. In other cancers, including breast, endometrial, and prostate carcinoma, focal proliferative buddings of endothelial cells resembling a renal glomerulus were observed [140]. Glomeruloid microvascular proliferations are associated with poor prognosis in human malignancies [141], and VEGF was shown to play essential role in their induction and maintenance in experimental tumor models [142, 143].

Finally, the most aggressive tumors present ability to form perfusable, matrix-rich vasculogenic-like networks de novo from tumor cells, a phenomenon termed vasculogenic or vascular mimicry (VM) [144, 145]. VM structures can be in close connection with endothelial-lined capillaries and/or form chimeric vasculature providing functional perfusion pathway for growing tumor independent from angiogenic factors [146, 147]. Blood circulation in VM tubes was confirmed in humans using laser scanning confocal angiography in patients with choroidal melanoma [148]. VM is a remarkable example of tumor cell plasticity indicative of a genetic reversion of aggressive cells to an embryonic stem cell-like multipotent phenotype [149]. Extensive research on the mechanisms of VM revealed numerous genes of vascular-, stem cell-, and hypoxia-related signaling pathways to be involved in the process (reviewed in [150]). As in the case of glomeruloid microvascular proliferation, vasculogenic mimicry is a strong negative prognostic factor in the clinic [145].

Conclusions

Although research over several decades has clearly shown marked abnormalities in the tumor vasculature, further understanding of the molecular pathways responsible for tumor development as well as structural heterogeneity in tumor vessels is required. Similarly, great progress in understanding the molecular aspects of tumor angiogenesis has been made and new cross talks between signaling pathways have been revealed. Still, however, the detailed molecular events responsible for neovascularization need to be explained.

Acknowledgments AL is supported by the Foundation for Polish Science—PARENT-BRIDGE Programme cofinanced by the European Union within European Regional Development Fund (POMOST/2010-2/8) and she is the recipient of L'Oreal Poland for Women in Science Scholarship. The Faculty of Biochemistry, Biophysics, and Biotechnology of the Jagiellonian University is a beneficiary of the structural funds from the European Union and the Polish Ministry of Science and Higher Education (grants No: POIG.02.01.00-12 064/08, POIG 01.01.02-00-109/09, POIG.02.02.00-014/08 and 01.01.02-00-069/09).

References

1. Folkman J (1971) Tumor angiogenesis: therapeutic implications. N Engl J Med 285(21): 1182–1186
2. Hanahan D, Weinberg RA (2011) Hallmarks of cancer: the next generation. Cell 144(5): 646–674
3. Rak J, Yu JL, Klement G, Kerbel RS (2000) Oncogenes and angiogenesis: signaling three-dimensional tumor growth. J Investig Dermatol Symp Proc 5(1):24–33
4. Rak J, Mitsuhashi Y, Sheehan C, Tamir A, Viloria-Petit A, Filmus J, Mansour SJ, Ahn NG, Kerbel RS (2000) Oncogenes and tumor angiogenesis: differential modes of vascular endothelial growth factor up-regulation in ras-transformed epithelial cells and fibroblasts. Cancer Res 60(2):490–498
5. Okada F, Rak JW, Croix BS, Lieubeau B, Kaya M, Roncari L, Shirasawa S, Sasazuki T, Kerbel RS (1998) Impact of oncogenes in tumor angiogenesis: mutant K-ras up-regulation of vascular endothelial growth factor/vascular permeability factor is necessary, but not sufficient for tumorigenicity of human colorectal carcinoma cells. Proc Natl Acad Sci USA 95(7): 3609–3614
6. Mezquita P, Parghi SS, Brandvold KA, Ruddell A (2005) Myc regulates VEGF production in B cells by stimulating initiation of VEGF mRNA translation. Oncogene 24(5):889–901
7. Zhang Y, Wang L, Zhang M, Jin M, Bai C, Wang X (2012) Potential mechanism of interleukin-8 production from lung cancer cells: an involvement of EGF-EGFR-PI3K-Akt-Erk pathway. J Cell Physiol 227(1):35–43
8. Semenza GL (1999) Regulation of mammalian O_2 homeostasis by hypoxia-inducible factor 1. Annu Rev Cell Dev Biol 15:551–578
9. Ferrara N, Gerber HP, LeCouter J (2003) The biology of VEGF and its receptors. Nat Med 9 (6):669–676
10. Nakashima T, Huang CL, Liu D, Kameyama K, Masuya D, Ueno M, Haba R, Yokomise H (2004) Expression of vascular endothelial growth factor-A and vascular endothelial growth factor-C as prognostic factors for non-small cell lung cancer. Med Sci Monit 10(6): BR157–BR165
11. Yuan A, Yu CJ, Kuo SH, Chen WJ, Lin FY, Luh KT, Yang PC, Lee YC (2001) Vascular endothelial growth factor 189 mRNA isoform expression specifically correlates with tumor angiogenesis, patient survival, and postoperative relapse in non-small-cell lung cancer. J Clin Oncol 19(2):432–441
12. Augustin HG, Koh GY, Thurston G, Alitalo K (2009) Control of vascular morphogenesis and homeostasis through the angiopoietin-Tie system. Nat Rev Mol Cell Biol 10(3):165–177
13. Holopainen T, Huang H, Chen C, Kim KE, Zhang L, Zhou F, Han W, Li C, Yu J, Wu J, Koh GY, Alitalo K, He Y (2009) Angiopoietin-1 overexpression modulates vascular endothelium to facilitate tumor cell dissemination and metastasis establishment. Cancer Res 69 (11):4656–4664
14. Saharinen P, Bry M, Alitalo K (2010) How do angiopoietins Tie in with vascular endothelial growth factors? Curr Opin Hematol 17(3):198–205
15. Saharinen P, Eklund L, Pulkki K, Bono P, Alitalo K (2011) VEGF and angiopoietin signaling in tumor angiogenesis and metastasis. Trends Mol Med 17(7):347–362
16. Nasarre P, Thomas M, Kruse K, Helfrich I, Wolter V, Deppermann C, Schadendorf D, Thurston G, Fiedler U, Augustin HG (2009) Host-derived angiopoietin-2 affects early stages of tumor development and vessel maturation but is dispensable for later stages of tumor growth. Cancer Res 69(4):1324–1333
17. Cao Y, Cao R, Hedlund EM (2008) Regulation of tumor angiogenesis and metastasis by FGF and PDGF signaling pathways. J Mol Med (Berl) 86(7):785–789
18. Presta M, Dell'Era P, Mitola S, Moroni E, Ronca R, Rusnati M (2005) Fibroblast growth factor/fibroblast growth factor receptor system in angiogenesis. Cytokine Growth Factor Rev 16(2):159–178

19. Lindner V, Majack RA, Reidy MA (1990) Basic fibroblast growth factor stimulates endothelial regrowth and proliferation in denuded arteries. J Clin Invest 85(6):2004–2008
20. Korc M, Friesel RE (2009) The role of fibroblast growth factors in tumor growth. Curr Cancer Drug Targets 9(5):639–651
21. Beenken A, Mohammadi M (2009) The FGF family: biology, pathophysiology and therapy. Nat Rev Drug Discov 8(3):235–253
22. Zlotnik A, Yoshie O (2000) Chemokines: a new classification system and their role in immunity. Immunity 12(2):121–127
23. Thelen M, Stein JV (2008) How chemokines invite leukocytes to dance. Nat Immunol 9(9): 953–959
24. Keeley EC, Mehrad B, Strieter RM (2011) Chemokines as mediators of tumor angiogenesis and neovascularization. Exp Cell Res 317(5):685–690
25. Strieter RM, Burdick MD, Gomperts BN, Belperio JA, Keane MP (2005) CXC chemokines in angiogenesis. Cytokine Growth Factor Rev 16(6):593–609
26. Salcedo R, Ponce ML, Young HA, Wasserman K, Ward JM, Kleinman HK, Oppenheim JJ, Murphy WJ (2000) Human endothelial cells express CCR2 and respond to MCP-1: direct role of MCP-1 in angiogenesis and tumor progression. Blood 96(1):34–40
27. Klimova T, Chandel NS (2008) Mitochondrial complex III regulates hypoxic activation of HIF. Cell Death Differ 15(4):660–666
28. Page EL, Chan DA, Giaccia AJ, Levine M, Richard DE (2008) Hypoxia-inducible factor-1alpha stabilization in nonhypoxic conditions: role of oxidation and intracellular ascorbate depletion. Mol Biol Cell 19(1):86–94
29. Baird L, Dinkova-Kostova AT (2011) The cytoprotective role of the Keap1-Nrf2 pathway. Arch Toxicol 85(4):241–272
30. Kim TH, Hur EG, Kang SJ, Kim JA, Thapa D, Lee YM, Ku SK, Jung Y, Kwak MK (2011) NRF2 blockade suppresses colon tumor angiogenesis by inhibiting hypoxia-induced activation of HIF-1alpha. Cancer Res 71(6):2260–2275
31. Zhang X, Chen X, Song H, Chen HZ, Rovin BH (2005) Activation of the Nrf2/antioxidant response pathway increases IL-8 expression. Eur J Immunol 35(11):3258–3267
32. Loboda A, Stachurska A, Florczyk U, Rudnicka D, Jazwa A, Wegrzyn J, Kozakowska M, Stalinska K, Poellinger L, Levonen AL, Yla-Herttuala S, Jozkowicz A, Dulak J (2009) HIF-1 induction attenuates Nrf2-dependent IL-8 expression in human endothelial cells. Antioxid Redox Signal 11(7):1501–1517
33. Florczyk U, Czauderna S, Stachurska A, Tertil M, Nowak W, Kozakowska M, Poellinger L, Jozkowicz A, Loboda A, Dulak J (2011) Opposite effects of HIF-1alpha and HIF-2alpha on the regulation of IL-8 expression in endothelial cells. Free Radic Biol Med 51(10):1882–1892
34. Singh S, Wu S, Varney M, Singh AP, Singh RK (2011) CXCR1 and CXCR2 silencing modulates CXCL8-dependent endothelial cell proliferation, migration and capillary-like structure formation. Microvasc Res 82(3):318–325
35. Mizukami Y, Jo WS, Duerr EM, Gala M, Li J, Zhang X, Zimmer MA, Iliopoulos O, Zukerberg LR, Kohgo Y, Lynch MP, Rueda BR, Chung DC (2005) Induction of interleukin-8 preserves the angiogenic response in HIF-1alpha-deficient colon cancer cells. Nat Med 11(9):992–997
36. Zhou S, Ye W, Zhang M, Liang J (2012) The effects of nrf2 on tumor angiogenesis: a review of the possible mechanisms of action. Crit Rev Eukaryot Gene Expr 22(2):149–160
37. Torisu-Itakura H, Furue M, Kuwano M, Ono M (2000) Co-expression of thymidine phosphorylase and heme oxygenase-1 in macrophages in human malignant vertical growth melanomas. Jpn J Cancer Res 91(9):906–910
38. Nishie A, Ono M, Shono T, Fukushi J, Otsubo M, Onoue H, Ito Y, Inamura T, Ikezaki K, Fukui M, Iwaki T, Kuwano M (1999) Macrophage infiltration and heme oxygenase-1 expression correlate with angiogenesis in human gliomas. Clin Cancer Res 5(5):1107–1113
39. Skrzypek KTM, Golda S, Ciesla M, Weglarczyk K, Collet G, Guichard A, Kozakowska M, Boczkowski J, Was H, Gil T, Kuzdzal J, Muchova L, Vitek L, Loboda A, Jozkowicz A,

Kieda C, Dulak J (2013) Interplay between heme oxygenase-1 and miR-378 affects non-small cell lung carcinoma growth, vascularization and metastasis. Antioxid Redox Signal 19(7): 644–660
40. Friedkin M, Roberts D (1954) The enzymatic synthesis of nucleosides. I. Thymidine phosphorylase in mammalian tissue. J Biol Chem 207(1):245–256
41. Miyazono K, Okabe T, Urabe A, Takaku F, Heldin CH (1987) Purification and properties of an endothelial cell growth factor from human platelets. J Biol Chem 262(9):4098–4103
42. Liekens S, Bilsen F, De Clercq E, Priego EM, Camarasa MJ, Perez-Perez MJ, Balzarini J (2002) Anti-angiogenic activity of a novel multi-substrate analogue inhibitor of thymidine phosphorylase. FEBS Lett 510(1–2):83–88
43. Miyadera K, Sumizawa T, Haraguchi M, Yoshida H, Konstanty W, Yamada Y, Akiyama S (1995) Role of thymidine phosphorylase activity in the angiogenic effect of platelet derived endothelial cell growth factor/thymidine phosphorylase. Cancer Res 55(8):1687–1690
44. Uchimiya H, Furukawa T, Okamoto M, Nakajima Y, Matsushita S, Ikeda R, Gotanda T, Haraguchi M, Sumizawa T, Ono M, Kuwano M, Kanzaki T, Akiyama S (2002) Suppression of thymidine phosphorylase-mediated angiogenesis and tumor growth by 2-deoxy-L-ribose. Cancer Res 62(10):2834–2839
45. Moghaddam A, Zhang HT, Fan TP, Hu DE, Lees VC, Turley H, Fox SB, Gatter KC, Harris AL, Bicknell R (1995) Thymidine phosphorylase is angiogenic and promotes tumor growth. Proc Natl Acad Sci USA 92(4):998–1002
46. Haraguchi M, Miyadera K, Uemura K, Sumizawa T, Furukawa T, Yamada K, Akiyama S, Yamada Y (1994) Angiogenic activity of enzymes. Nature 368(6468):198
47. Sengupta S, Sellers LA, Matheson HB, Fan TP (2003) Thymidine phosphorylase induces angiogenesis in vivo and in vitro: an evaluation of possible mechanisms. Br J Pharmacol 139 (2):219–231
48. Stevenson DP, Collins WP, Farzaneh F, Hata K, Miyazaki K (1998) Thymidine phosphorylase activity and prodrug effects in a three-dimensional model of angiogenesis: implications for the treatment of ovarian cancer. Am J Pathol 153(5):1573–1578
49. Bronckaers A, Gago F, Balzarini J, Liekens S (2009) The dual role of thymidine phosphorylase in cancer development and chemotherapy. Med Res Rev 29(6):903–953
50. Akiyama S, Furukawa T, Sumizawa T, Takebayashi Y, Nakajima Y, Shimaoka S, Haraguchi M (2004) The role of thymidine phosphorylase, an angiogenic enzyme, in tumor progression. Cancer Sci 95(11):851–857
51. Brown NS, Jones A, Fujiyama C, Harris AL, Bicknell R (2000) Thymidine phosphorylase induces carcinoma cell oxidative stress and promotes secretion of angiogenic factors. Cancer Res 60(22):6298–6302
52. Brown NS, Streeter EH, Jones A, Harris AL, Bicknell R (2005) Cooperative stimulation of vascular endothelial growth factor expression by hypoxia and reactive oxygen species: the effect of targeting vascular endothelial growth factor and oxidative stress in an orthotopic xenograft model of bladder carcinoma. Br J Cancer 92(9):1696–1701
53. Volm M, Mattern J, Koomagi R (1998) Expression of platelet-derived endothelial cell growth factor in non-small cell lung carcinomas: relationship to various biological factors. Int J Oncol 13(5):975–979
54. Bijnsdorp IV, Capriotti F, Kruyt FA, Losekoot N, Fukushima M, Griffioen AW, Thijssen VL, Peters GJ (2011) Thymidine phosphorylase in cancer cells stimulates human endothelial cell migration and invasion by the secretion of angiogenic factors. Br J Cancer 104(7):1185–1192
55. Tabata S, Ikeda R, Yamamoto M, Furukawa T, Kuramoto T, Takeda Y, Yamada K, Haraguchi M, Nishioka Y, Sone S, Akiyama S (2012) Thymidine phosphorylase enhances reactive oxygen species generation and interleukin-8 expression in human cancer cells. Oncol Rep 28(3):895–902
56. Calin GA, Dumitru CD, Shimizu M, Bichi R, Zupo S, Noch E, Aldler H, Rattan S, Keating M, Rai K, Rassenti L, Kipps T, Negrini M, Bullrich F, Croce CM (2002) Frequent deletions and

down-regulation of micro- RNA genes miR15 and miR16 at 13q14 in chronic lymphocytic leukemia. Proc Natl Acad Sci USA 99(24):15524–15529
57. Michael MZ, SM OC, van Holst Pellekaan NG, Young GP, James RJ (2003) Reduced accumulation of specific microRNAs in colorectal neoplasia. Mol Cancer Res 1(12):882–891
58. Ciesla M, Skrzypek K, Kozakowska M, Loboda A, Jozkowicz A, Dulak J (2011) MicroRNAs as biomarkers of disease onset. Anal Bioanal Chem 401(7):2051–2061
59. Wang S, Olson EN (2009) AngiomiRs – key regulators of angiogenesis. Curr Opin Genet Dev 19(3):205–211
60. Sun CY, She XM, Qin Y, Chu ZB, Chen L, Ai LS, Zhang L, Hu Y (2013) miR-15a and miR-16 affect the angiogenesis of multiple myeloma by targeting VEGF. Carcinogenesis 34(2): 426–435
61. He J, Jing Y, Li W, Qian X, Xu Q, Li FS, Liu LZ, Jiang BH, Jiang Y (2013) Roles and mechanism of miR-199a and miR-125b in tumor angiogenesis. PLoS One 8(2):e56647
62. Choi YC, Yoon S, Jeong Y, Yoon J, Baek K (2011) Regulation of vascular endothelial growth factor signaling by miR-200b. Mol Cells 32(1):77–82
63. Hua Z, Lv Q, Ye W, Wong CK, Cai G, Gu D, Ji Y, Zhao C, Wang J, Yang BB, Zhang Y (2006) MiRNA-directed regulation of VEGF and other angiogenic factors under hypoxia. PLoS One 1:e116
64. Lee DY, Deng Z, Wang CH, Yang BB (2007) MicroRNA-378 promotes cell survival, tumor growth, and angiogenesis by targeting SuFu and Fus-1 expression. Proc Natl Acad Sci USA 104(51):20350–20355
65. Pola R, Ling LE, Silver M, Corbley MJ, Kearney M, Blake Pepinsky R, Shapiro R, Taylor FR, Baker DP, Asahara T, Isner JM (2001) The morphogen Sonic hedgehog is an indirect angiogenic agent upregulating two families of angiogenic growth factors. Nat Med 7(6): 706–711
66. Chen LT, Xu SD, Xu H, Zhang JF, Ning JF, Wang SF (2012) MicroRNA-378 is associated with non-small cell lung cancer brain metastasis by promoting cell migration, invasion and tumor angiogenesis. Med Oncol 29(3):1673–1680
67. Murdoch C, Muthana M, Coffelt SB, Lewis CE (2008) The role of myeloid cells in the promotion of tumour angiogenesis. Nat Rev Cancer 8(8):618–631
68. Coffelt SB, Lewis CE, Naldini L, Brown JM, Ferrara N, De Palma M (2010) Elusive identities and overlapping phenotypes of proangiogenic myeloid cells in tumors. Am J Pathol 176(4):1564–1576
69. Mantovani A, Sozzani S, Locati M, Allavena P, Sica A (2002) Macrophage polarization: tumor-associated macrophages as a paradigm for polarized M2 mononuclear phagocytes. Trends Immunol 23(11):549–555
70. Allavena P, Mantovani A (2012) Immunology in the clinic review series; focus on cancer: tumour-associated macrophages: undisputed stars of the inflammatory tumour microenvironment. Clin Exp Immunol 167(2):195–205
71. Lin EY, Nguyen AV, Russell RG, Pollard JW (2001) Colony-stimulating factor 1 promotes progression of mammary tumors to malignancy. J Exp Med 193(6):727–740
72. Biswas SK, Lewis CE (2010) NF-kappaB as a central regulator of macrophage function in tumors. J Leukoc Biol 88(5):877–884
73. Balkwill F (2009) Tumour necrosis factor and cancer. Nat Rev Cancer 9(5):361–371
74. Li N, Grivennikov SI, Karin M (2011) The unholy trinity: inflammation, cytokines, and STAT3 shape the cancer microenvironment. Cancer Cell 19(4):429–431
75. De Palma M, Venneri MA, Galli R, Sergi Sergi L, Politi LS, Sampaolesi M, Naldini L (2005) Tie2 identifies a hematopoietic lineage of proangiogenic monocytes required for tumor vessel formation and a mesenchymal population of pericyte progenitors. Cancer Cell 8(3):211–226
76. Coffelt SB, Tal AO, Scholz A, De Palma M, Patel S, Urbich C, Biswas SK, Murdoch C, Plate KH, Reiss Y, Lewis CE (2010) Angiopoietin-2 regulates gene expression in TIE2-expressing monocytes and augments their inherent proangiogenic functions. Cancer Res 70 (13): 5270–5280

77. Kostoulas G, Lang A, Nagase H, Baici A (1999) Stimulation of angiogenesis through cathepsin B inactivation of the tissue inhibitors of matrix metalloproteinases. FEBS Lett 455(3):286–290
78. Yang L, DeBusk LM, Fukuda K, Fingleton B, Green-Jarvis B, Shyr Y, Matrisian LM, Carbone DP, Lin PC (2004) Expansion of myeloid immune suppressor Gr + CD11b + cells in tumor-bearing host directly promotes tumor angiogenesis. Cancer Cell 6(4):409–421
79. Du R, Lu KV, Petritsch C, Liu P, Ganss R, Passegue E, Song H, Vandenberg S, Johnson RS, Werb Z, Bergers G (2008) HIF1alpha induces the recruitment of bone marrow-derived vascular modulatory cells to regulate tumor angiogenesis and invasion. Cancer Cell 13(3): 206–220
80. Bekes EM, Schweighofer B, Kupriyanova TA, Zajac E, Ardi VC, Quigley JP, Deryugina EI (2011) Tumor-recruited neutrophils and neutrophil TIMP-free MMP-9 regulate coordinately the levels of tumor angiogenesis and efficiency of malignant cell intravasation. Am J Pathol 179(3):1455–1470
81. Ardi VC, Van den Steen PE, Opdenakker G, Schweighofer B, Deryugina EI, Quigley JP (2009) Neutrophil MMP-9 proenzyme, unencumbered by TIMP-1, undergoes efficient activation in vivo and catalytically induces angiogenesis via a basic fibroblast growth factor (FGF-2)/FGFR-2 pathway. J Biol Chem 284(38):25854–25866
82. Asahara T, Murohara T, Sullivan A, Silver M, van der Zee R, Li T, Witzenbichler B, Schatteman G, Isner JM (1997) Isolation of putative progenitor endothelial cells for angiogenesis. Science 275(5302):964–967
83. Jung SY, Choi JH, Kwon SM, Masuda H, Asahara T, Lee YM (2012) Decursin inhibits vasculogenesis in early tumor progression by suppression of endothelial progenitor cell differentiation and function. J Cell Biochem 113(5):1478–1487
84. Liang PH, Tian F, Lu Y, Duan B, Stolz DB, Li LY (2011) Vascular endothelial growth inhibitor (VEGI; TNFSF15) inhibits bone marrow-derived endothelial progenitor cell incorporation into Lewis lung carcinoma tumors. Angiogenesis 14(1):61–68
85. Purhonen S, Palm J, Rossi D, Kaskenpaa N, Rajantie I, Yla-Herttuala S, Alitalo K, Weissman IL, Salven P (2008) Bone marrow-derived circulating endothelial precursors do not contribute to vascular endothelium and are not needed for tumor growth. Proc Natl Acad Sci USA 105(18): 6620–6625
86. Yoder MC (2009) Defining human endothelial progenitor cells. J Thromb Haemost 7(Suppl 1): 49–52
87. Li Calzi S, Neu MB, Shaw LC, Kielczewski JL, Moldovan NI, Grant MB (2010) EPCs and pathological angiogenesis: when good cells go bad. Microvasc Res 79(3):207–216
88. Gao D, Nolan DJ, Mellick AS, Bambino K, McDonnell K, Mittal V (2008) Endothelial progenitor cells control the angiogenic switch in mouse lung metastasis. Science 319(5860): 195–198
89. Fleitas T, Martinez-Sales V, Gomez-Codina J, Martin M, Reynes G (2010) Circulating endothelial and endothelial progenitor cells in non-small-cell lung cancer. Clin Transl Oncol 12(8):521–525
90. Domanska UM, Kruizinga RC, Nagengast WB, Timmer-Bosscha H, Huls G, de Vries EG, Walenkamp AM (2013) A review on CXCR4/CXCL12 axis in oncology: no place to hide. Eur J Cancer 49(1):219–230
91. Song C, Li G (2011) CXCR4 and matrix metalloproteinase-2 are involved in mesenchymal stromal cell homing and engraftment to tumors. Cytotherapy 13(5):549–561
92. Beckermann BM, Kallifatidis G, Groth A, Frommhold D, Apel A, Mattern J, Salnikov AV, Moldenhauer G, Wagner W, Diehlmann A, Saffrich R, Schubert M, Ho AD, Giese N, Buchler MW, Friess H, Buchler P, Herr I (2008) VEGF expression by mesenchymal stem cells contributes to angiogenesis in pancreatic carcinoma. Br J Cancer 99(4):622–631
93. Allen M, Louise Jones J (2010) Jekyll and Hyde: the role of the microenvironment on the progression of cancer. J Pathol 223(2):162–176

94. Ishikawa S, Takenaka K, Yanagihara K, Miyahara R, Kawano Y, Otake Y, Hasegawa S, Wada H, Tanaka F (2004) Matrix metalloproteinase-2 status in stromal fibroblasts, not in tumor cells, is a significant prognostic factor in non-small-cell lung cancer. Clin Cancer Res 10(19):6579–6585
95. Mishra P, Banerjee D, Ben-Baruch A (2011) Chemokines at the crossroads of tumor-fibroblast interactions that promote malignancy. J Leukoc Biol 89(1):31–39
96. Burger JA, Stewart DJ, Wald O, Peled A (2011) Potential of CXCR4 antagonists for the treatment of metastatic lung cancer. Expert Rev Anticancer Ther 11(4):621–630
97. Wald O, Izhar U, Amir G, Kirshberg S, Shlomai Z, Zamir G, Peled A, Shapira OM (2011) Interaction between neoplastic cells and cancer-associated fibroblasts through the CXCL12/CXCR4 axis: role in non-small cell lung cancer tumor proliferation. J Thorac Cardiovasc Surg 141(6):1503–1512
98. Imai H, Sunaga N, Shimizu Y, Kakegawa S, Shimizu K, Sano T, Ishizuka T, Oyama T, Saito R, Minna JD, Mori M (2010) Clinicopathological and therapeutic significance of CXCL12 expression in lung cancer. Int J Immunopathol Pharmacol 23(1):153–164
99. Matsumoto K, Nakamura T (2006) Hepatocyte growth factor and the Met system as a mediator of tumor-stromal interactions. Int J Cancer 119(3):477–483
100. Kaga T, Kawano H, Sakaguchi M, Nakazawa T, Taniyama Y, Morishita R (2012) Hepatocyte growth factor stimulated angiogenesis without inflammation: differential actions between hepatocyte growth factor, vascular endothelial growth factor and basic fibroblast growth factor. Vascul Pharmacol 57(1):3–9
101. Grant DS, Kleinman HK, Goldberg ID, Bhargava MM, Nickoloff BJ, Kinsella JL, Polverini P, Rosen EM (1993) Scatter factor induces blood vessel formation in vivo. Proc Natl Acad Sci USA 90(5):1937–1941
102. Gherardi E, Birchmeier W, Birchmeier C, Vande Woude G (2012) Targeting MET in cancer: rationale and progress. Nat Rev Cancer 12(2):89–103
103. Sulpice E, Ding S, Muscatelli-Groux B, Berge M, Han ZC, Plouet J, Tobelem G, Merkulova-Rainon T (2009) Cross-talk between the VEGF-A and HGF signalling pathways in endothelial cells. Biol Cell 101(9):525–539
104. Zhang YW, Su Y, Volpert OV, Vande Woude GF (2003) Hepatocyte growth factor/scatter factor mediates angiogenesis through positive VEGF and negative thrombospondin 1 regulation. Proc Natl Acad Sci USA 100(22):12718–12723
105. Puri N, Khramtsov A, Ahmed S, Nallasura V, Hetzel JT, Jagadeeswaran R, Karczmar G, Salgia R (2007) A selective small molecule inhibitor of c-Met, PHA665752, inhibits tumorigenicity and angiogenesis in mouse lung cancer xenografts. Cancer Res 67(8):3529–3534
106. Ratajczak J, Wysoczynski M, Hayek F, Janowska-Wieczorek A, Ratajczak MZ (2006) Membrane-derived microvesicles: important and underappreciated mediators of cell-to-cell communication. Leukemia 20(9):1487–1495
107. Valadi H, Ekstrom K, Bossios A, Sjostrand M, Lee JJ, Lotvall JO (2007) Exosome-mediated transfer of mRNAs and microRNAs is a novel mechanism of genetic exchange between cells. Nat Cell Biol 9(6):654–659
108. Bussolati B, Grange C, Camussi G (2011) Tumor exploits alternative strategies to achieve vascularization. FASEB J 25(9):2874–2882
109. Baj-Krzyworzeka M, Szatanek R, Weglarczyk K, Baran J, Zembala M (2007) Tumour-derived microvesicles modulate biological activity of human monocytes. Immunol Lett 113 (2):76–82
110. Grange C, Tapparo M, Collino F, Vitillo L, Damasco C, Deregibus MC, Tetta C, Bussolati B, Camussi G (2011) Microvesicles released from human renal cancer stem cells stimulate angiogenesis and formation of lung premetastatic niche. Cancer Res 71(15):5346–5356
111. Janowska-Wieczorek A, Wysoczynski M, Kijowski J, Marquez-Curtis L, Machalinski B, Ratajczak J, Ratajczak MZ (2005) Microvesicles derived from activated platelets induce metastasis and angiogenesis in lung cancer. Int J Cancer 113(5):752–760

112. Kosaka N, Iguchi H, Hagiwara K, Yoshioka Y, Takeshita F, Ochiya T (2013) Neutral sphingomyelinase 2 (nSMase2)-dependent exosomal transfer of angiogenic microRNAs regulate cancer cell metastasis. J Biol Chem 288(15):10849–10859
113. Zhuang G, Wu X, Jiang Z, Kasman I, Yao J, Guan Y, Oeh J, Modrusan Z, Bais C, Sampath D, Ferrara N (2012) Tumour-secreted miR-9 promotes endothelial cell migration and angiogenesis by activating the JAK-STAT pathway. EMBO J 31(17):3513–3523
114. De Bock K, Cauwenberghs S, Carmeliet P (2011) Vessel abnormalization: another hallmark of cancer? Molecular mechanisms and therapeutic implications. Curr Opin Genet Dev 21(1): 73–79
115. Ribatti D, Nico B, Crivellato E, Vacca A (2007) The structure of the vascular network of tumors. Cancer Lett 248(1):18–23
116. Less JR, Posner MC, Skalak TC, Wolmark N, Jain RK (1997) Geometric resistance and microvascular network architecture of human colorectal carcinoma. Microcirculation 4(1): 25–33
117. Dvorak HF, Nagy JA, Dvorak JT, Dvorak AM (1988) Identification and characterization of the blood vessels of solid tumors that are leaky to circulating macromolecules. Am J Pathol 133(1):95–109
118. Hida K, Hida Y, Amin DN, Flint AF, Panigrahy D, Morton CC, Klagsbrun M (2004) Tumor-associated endothelial cells with cytogenetic abnormalities. Cancer Res 64(22):8249–8255
119. Jain RK (1988) Determinants of tumor blood flow: a review. Cancer Res 48(10):2641–2658
120. Lunt SJ, Chaudary N, Hill RP (2009) The tumor microenvironment and metastatic disease. Clin Exp Metastasis 26(1):19–34
121. Graeber TG, Osmanian C, Jacks T, Housman DE, Koch CJ, Lowe SW, Giaccia AJ (1996) Hypoxia-mediated selection of cells with diminished apoptotic potential in solid tumours. Nature 379(6560):88–91
122. Bristow RG, Hill RP (2008) Hypoxia and metabolism. Hypoxia, DNA repair and genetic instability. Nat Rev Cancer 8(3):180–192
123. Cairns RA, Harris IS, Mak TW (2011) Regulation of cancer cell metabolism. Nat Rev Cancer 11(2):85–95
124. Rouschop KM, van den Beucken T, Dubois L, Niessen H, Bussink J, Savelkouls K, Keulers T, Mujcic H, Landuyt W, Voncken JW, Lambin P, van der Kogel AJ, Koritzinsky M, Wouters BG (2010) The unfolded protein response protects human tumor cells during hypoxia through regulation of the autophagy genes MAP1LC3B and ATG5. J Clin Invest 120(1):127–141
125. Pennacchietti S, Michieli P, Galluzzo M, Mazzone M, Giordano S, Comoglio PM (2003) Hypoxia promotes invasive growth by transcriptional activation of the met protooncogene. Cancer Cell 3(4):347–361
126. Chang Q, Jurisica I, Do T, Hedley DW (2011) Hypoxia predicts aggressive growth and spontaneous metastasis formation from orthotopically grown primary xenografts of human pancreatic cancer. Cancer Res 71(8):3110–3120
127. Wang Y, Ohh M (2010) Oxygen-mediated endocytosis in cancer. J Cell Mol Med 14(3): 496–503
128. Jiang J, Tang YL, Liang XH (2011) EMT: a new vision of hypoxia promoting cancer progression. Cancer Biol Ther 11(8):714–723
129. Yotnda P, Wu D, Swanson AM (2010) Hypoxic tumors and their effect on immune cells and cancer therapy. Methods Mol Biol 651:1–29
130. Alison MR, Lim SM, Nicholson LJ (2010) Cancer stem cells: problems for therapy? J Pathol 223(2):147–161
131. Leite de Oliveira R, Deschoemaeker S, Henze AT, Debackere K, Finisguerra V, Takeda Y, Roncal C, Dettori D, Tack E, Jonsson Y, Veschini L, Peeters A, Anisimov A, Hofmann M, Alitalo K, Baes M, D'Hooge J, Carmeliet P, Mazzone M (2012) Gene-targeting of Phd2 improves tumor response to chemotherapy and prevents side-toxicity. Cancer Cell 22(2): 263–277

132. Dome B, Hendrix MJ, Paku S, Tovari J, Timar J (2007) Alternative vascularization mechanisms in cancer: pathology and therapeutic implications. Am J Pathol 170(1):1–15
133. Holmgren L, O'Reilly MS, Folkman J (1995) Dormancy of micrometastases: balanced proliferation and apoptosis in the presence of angiogenesis suppression. Nat Med 1(2): 149–153
134. Pezzella F, Pastorino U, Tagliabue E, Andreola S, Sozzi G, Gasparini G, Menard S, Gatter KC, Harris AL, Fox S, Buyse M, Pilotti S, Pierotti M, Rilke F (1997) Non-small-cell lung carcinoma tumor growth without morphological evidence of neo-angiogenesis. Am J Pathol 151(5):1417–1423
135. Holash J, Maisonpierre PC, Compton D, Boland P, Alexander CR, Zagzag D, Yancopoulos GD, Wiegand SJ (1999) Vessel cooption, regression, and growth in tumors mediated by angiopoietins and VEGF. Science 284(5422):1994–1998
136. Hillen F, Griffioen AW (2007) Tumour vascularization: sprouting angiogenesis and beyond. Cancer Metastasis Rev 26(3–4):489–502
137. Djonov VG, Kurz H, Burri PH (2002) Optimality in the developing vascular system: branching remodeling by means of intussusception as an efficient adaptation mechanism. Dev Dyn 224(4):391–402
138. Kurz H, Burri PH, Djonov VG (2003) Angiogenesis and vascular remodeling by intussusception: from form to function. News Physiol Sci 18:65–70
139. Dome B, Timar J, Paku S (2003) A novel concept of glomeruloid body formation in experimental cerebral metastases. J Neuropathol Exp Neurol 62(6):655–661
140. Straume O, Akslen LA (2003) Increased expression of VEGF-receptors (FLT-1, KDR, NRP-1) and thrombospondin-1 is associated with glomeruloid microvascular proliferation, an aggressive angiogenic phenotype, in malignant melanoma. Angiogenesis 6(4):295–301
141. Straume O, Chappuis PO, Salvesen HB, Halvorsen OJ, Haukaas SA, Goffin JR, Begin LR, Foulkes WD, Akslen LA (2002) Prognostic importance of glomeruloid microvascular proliferation indicates an aggressive angiogenic phenotype in human cancers. Cancer Res 62(23): 6808–6811
142. Nagy JA, Dvorak HF (2012) Heterogeneity of the tumor vasculature: the need for new tumor blood vessel type-specific targets. Clin Exp Metastasis 29(7):657–662
143. Nagy JA, Chang SH, Shih SC, Dvorak AM, Dvorak HF (2010) Heterogeneity of the tumor vasculature. Semin Thromb Hemost 36(3):321–331
144. Maniotis AJ, Folberg R, Hess A, Seftor EA, Gardner LM, Pe'er J, Trent JM, Meltzer PS, Hendrix MJ (1999) Vascular channel formation by human melanoma cells in vivo and in vitro: vasculogenic mimicry. Am J Pathol 155(3):739–752
145. Seftor RE, Hess AR, Seftor EA, Kirschmann DA, Hardy KM, Margaryan NV, Hendrix MJ (2012) Tumor cell vasculogenic mimicry: from controversy to therapeutic promise. Am J Pathol 181(4):1115–1125
146. Ruf W, Seftor EA, Petrovan RJ, Weiss RM, Gruman LM, Margaryan NV, Seftor RE, Miyagi Y, Hendrix MJ (2003) Differential role of tissue factor pathway inhibitors 1 and 2 in melanoma vasculogenic mimicry. Cancer Res 63(17):5381–5389
147. Hendrix MJ, Seftor RE, Seftor EA, Gruman LM, Lee LM, Nickoloff BJ, Miele L, Sheriff DD, Schatteman GC (2002) Transendothelial function of human metastatic melanoma cells: role of the microenvironment in cell-fate determination. Cancer Res 62(3):665–668
148. Frenkel S, Barzel I, Levy J, Lin AY, Bartsch DU, Majumdar D, Folberg R, Pe'er J (2008) Demonstrating circulation in vasculogenic mimicry patterns of uveal melanoma by confocal indocyanine green angiography. Eye (Lond) 22(7):948–952
149. Hendrix MJ, Seftor EA, Hess AR, Seftor RE (2003) Vasculogenic mimicry and tumour-cell plasticity: lessons from melanoma. Nat Rev Cancer 3(6):411–421
150. Kirschmann DA, Seftor EA, Hardy KM, Seftor RE, Hendrix MJ (2012) Molecular pathways: vasculogenic mimicry in tumor cells: diagnostic and therapeutic implications. Clin Cancer Res 18(10):2726–2732

Endothelium in Pathologic Angiogenesis and Angiogenesis-Mediated Therapies

18

Maria Paprocka, Catherine Grillon, Danuta Duś, and Claudine Kieda

Contents

18.1 Introduction 390
18.2 Endothelium Organ-Specificity: Contribution to New Insights in the Microenvironment-Based Therapies 391
18.2.1 Organ-Specificity: Why Should Endothelium Be Considered as Organ Specific? 391
18.2.2 Endothelial Cells as Diagnostic Markers and Disease Reporters 395
18.3 Endothelium in Pathologies and Pathologic Endothelium 398
18.3.1 Targeting the Pathologic Endothelial Cells 398
18.3.2 Why Normalization Could Bring New Means for Tumor Therapy? 400
References 403

Abstract

This chapter describes a short historical overview of the progress in endothelium research and point the importance of organ-selective characteristics according to the present knowledge about endothelium biology. Uncovering the advantages that the endothelial cell properties and characteristics provide for the development of future targeted therapies, the review describes why mature endothelial cells due to their organ-specificity can be useful to target diseased organs.

In the same line, endothelium properties will be exploited to make the endothelial cells a disease marker, e.g., in diabetes, stroke, cancer, inflammation, or ischemia and to provide a potential diagnostic indicator for the estimation of metastatic progression. New perspectives are thus opened by endothelial cells that can be considered both as a reporter and a target. These features can be

M. Paprocka • D. Duś
Ludwik Hirszfeld Institute of Immunology, and Experimental Therapy, PAN, 53114 Wrocław, Poland

C. Grillon • C. Kieda (✉)
Centre for Molecular Biophysics, UPR 4301 CNRS, 45071 Orléans, France
e-mail: claudine.kieda@cnrs-orleans.fr

J. Dulak et al. (eds.), *Angiogenesis and Vascularisation*,
DOI 10.1007/978-3-7091-1428-5_18, © Springer-Verlag Wien 2013

combined with new cell-mediated and cell-targeted therapeutics designed to correct angiogenesis. Examples of such possible applications are detailed in the repair of tumor angiogenesis with help of endothelial cell precursors through their ability to target the pathologic angiogenesis and participate to normalization of the pathologic vasculature. The hypothesis that normalized angiogenesis may provide an efficient treatment, working as adjuvant to classical therapies, is being developed. The objective is to reach a mechanical stabilization that should result in an advantageous change of the tumor microenvironment.

Keywords
Tumor • Chemokines • Down syndrome • VEGF • Glycosaminoglycans • Circulating endothelial cells • Endothelial progenitor cells

18.1 Introduction

In 2007 a review of endothelium research history by Hwa and Aird [1] indicated that the concept of site-specific phenotype of endothelium got to be accepted quite recently in the scientific literature. Since the idea that endothelium can serve as a marker and a reporter for pathologies is now being exploited, one can measure the gap that have to be closed to progress in the knowledge of the biology of this long ignored tissue for its unique properties and consequently, in recognizing the endothelial cells as its representative entities.

The distinction among endothelial cells shows up first at the level of the type of vessel that the endothelium is lining. From arteries to veins, capillaries, post-capillary venules, and lymphatics, the diversity appears as early as at the embryonic state to produce cells that are phenotypically and functionally different. Clearly distinct from one vessel type to another, endothelial cells are deeply involved in regulating the metabolism of the organ they belong to, as well as the activities of the surrounding cells they interact with, both in the vessel lumen and in the underlying tissue. As such, the endothelium fulfills fundamental activities:

- Blood to tissue transport of nutrients and cells, functional relationships between blood and lymph
- Selection and recruitment of cells and molecules from blood and lymph to tissue, and reciprocally

Endothelial cells imprint their specificity. This property is fundamental to achieve not only intercellular interactions that govern immune and inflammatory responses but also pathologic autoimmune reactions, allergic response and, in cancer, tumor cell spreading and establishment of metastases. Vascular bed-specific gene expression can also arise in genetic diseases like the Down syndrome, where changes in endothelial cell activation might explain immune response defects. Moreover, the species specificity adds to the organ-specificity and contributes to explain variation in the molecular mechanisms of infectious processes [2].

Proving the distinct characteristics of the endothelial cells, the presently available methods for gene expression analysis have also allowed demonstrating that endothelial reactivity is highly sensitive to the microenvironmental conditions. This represents a rationale for the diagnostic potential and further therapeutic developments, based on endothelial cells serving as tools for cell and gene therapies.

Among the main characteristics of endothelial cells is their ability to make vessels through angiogenesis. Often occurring for repair purposes, angiogenesis may also promote pathologic processes during inflammation or tumor stroma development, in response to the common microenvironmental factor, such as hypoxia.

This review will focus on the possibilities offered by the organ-specificity of endothelial cells and take advantage of their particular features to understand specific mechanisms such as aberrant leukocyte homing or selection leading to inadequate immune response. This knowledge will ultimately serve to design new therapeutic strategies directed against pathologies such as solid tumors, whose development depends on the hypoxia-induced angiogenesis.

18.2 Endothelium Organ-Specificity: Contribution to New Insights in the Microenvironment-Based Therapies

Progress in research on endothelial cells, especially on their organ-specificity and disease-specific changes, raises the following questions (1) can endothelium be considered as a disease reporter like it has been evoked for diabetes, stroke, cancer, inflammation, and ischemia? And, if yes, (2) can the organ-specific endothelium be used for diagnostic purposes? On the other hand, endothelial cells play an important role in mediating the neoplasm metastasis and other invasive diseases. Mature organ-specific endothelial cells could be used for the design of strategies to target the diseased organs. This is a promising perspective that permits to plan new therapeutics by formatting the endothelial precursors to reach pathological sites and participate in the pathological angiogenesis and/or the endothelial repair mechanisms.

18.2.1 Organ-Specificity: Why Should Endothelium Be Considered as Organ Specific?

18.2.1.1 Endothelium Implies Site Selectivity in Physiological and Pathological Conditions

The first description that endothelium may be something else than an inert cover layer of similar cells, brought a breakthrough into research although it should have appeared earlier as a logically admitted concept. This had not been the case partly because most in vitro experiments were conducted solely with the human umbilical vein endothelial cells (HUVEC).

The elaborated level of organ-specificity displayed by endothelial cells can be convincingly illustrated by quite different endothelial functions: maintaining the blood brain barrier on the one hand and, on the other, the activity and selectivity of post-capillary venules in the lymph nodes. As a consequence, the organ-specificity is a parameter that should be taken into account to understand the endothelium biology and subsequently its reactions and mechanisms of action [3–5].

As described by Butcher et al. [6], the organ-specific migration of lymphocytes during their homing illustrates the selective role of endothelial cells and, more precisely, of the adhesion molecules they express. This suggested the need for a clear-cut description of the endothelial signature in terms of cell surface molecules. We had shown the presence of membrane lectins in lymphocytes [7], and their significant influence on cell recognition and adhesion [8–11]. It was further demonstrated that organ specificity of lymphocyte migration is determined by the selective interaction of lymphocytes with specialized endothelial cells as described by the famous work of Stamper and Woodruff [12]. Then, Butcher and coworkers showed that the organ-restricted endothelial cell determinants mediate the antigen-independent organ specificity of lymphocyte migration [6]. This model strongly suggested the existence of defined pairs of complementary receptors/ligands on the lymphocytes and endothelial cells, which can separately mediate lymphocyte-high endothelial venule adherence in Peyer's patches and in lymph nodes [6].

We contributed to bring a clearer demonstration of this selectivity by a simple approach based on the isolation of murine endothelial cells from different organs and analyzing their distinct biochemical and biological behavior in terms of lymphocyte adhesion and cell selection [13, 14]. Moreover, we could evidence the micro-environmental modulation of the endothelial addressins' composition and their posttranslational modification [15]. This allowed further extensive studies on selectins and their ligands, as well as permitted to decipher the exquisite definition of lymphocyte homing according to immunologically defined populations.

Indeed, intercellular interactions are also modulated by endothelial cell-expressed selectins and leukocyte-expressed ligands. Such a double sugar-protein recognition, first described in the case of cross talk between tumor and endothelial cells [9], can be generalized. Indeed, selectins (which are lectins) and their addressin ligands dictate the first level of interactions that govern the rolling and arrest of leukocytes on the endothelium of a given organ [16]. Further elucidation of this mechanism could provide the means to control the extravasation processes. Promoting or inhibiting the leukocyte recruitment requires the knowledge on fine-tuning of molecular signal transduction cascades which underlie the cell interactions, including the role of chemokines.

18.2.1.2 Chemokines Are Acting as Regionally Expressed Chemo-attracting Molecules

Chemokines are adding a degree to the molecular mechanisms that explain the spatiotemporal characteristics of leukocyte homing. Their expression is highly inducible as well as the molecules that allow their cell surface presentation. Namely, the glycosaminoglycans (GAGs) are differently expressed both quantitatively and qualitatively on the endothelial cell surface according to the site and to the biological context [17]. This confers a higher degree of organ-specificity to the endothelium. Moreover, GAGs and chemokines expression or presentation indicates how endothelial cell phenotype may reflect the biological state of a tissue.

As reviewed by Ribatti et al. [5], endothelial cells constitute a heterogeneous population. The endothelium which is covering the entire inner surface of blood vessels shows various properties depending on the state of the tissue and the function the endothelial cells must achieve. Although the clearest morphological and functional differences are visible between large and small vessels, organ-specificity appears also between cells derived from various microvascular endothelial beds. The second degree of distinction results from the microenvironment and from the role of the extracellular matrix, influencing the endothelial phenotype and affecting the distinct response to growth factors. Consequently, the concept of organ-specificity implies a sum of posttranslational modifications but combined with the expression of unique genes that are controlled by organ-characteristic regulatory elements. Finally, pathological conditions like tumor growth are accompanied by the development of a pathologic tumor vasculature and tumor stroma, regulated by the endothelial cells and their interaction with tumor cells. Nevertheless, for the time being the organ-specificity of endothelium remained to be directly proven.

18.2.1.3 Endothelial Cell Lines as Tools to Demonstrate Organ-Specificity

To analyze in details the putative organ-specificity of endothelium, we isolated and established a set of lines originating from human endothelial cells [4]. Their immortalization, once being standardized, allowed comparing the gene expression profiles. These analyses have proven that microvascular endothelial cells differ significantly in terms of phenotype and gene expression [18] according to their organ of origin. Functional significance of these differences was illustrated in human cells during the specific chemokine-directed homing [17] and during the recruitment of lymphocytes in allergic pathologies as asthma [19].

The endothelial organ-specificity, together with species-specificity, is evident in some infectious diseases and their organ-specific propagation. Our work on cat scratch disease revealed the intrinsic differences between human and feline endothelial cell susceptibility to *Bartonella henselae* infection. While no effect was observed on the feline cells upon bacteria exposure, the human cells displayed accelerated angiogenesis and wound healing. Noticeable differences were demonstrated between human microvasculature- and macrovasculature-derived endothelial cells. *B. henselae* could stimulate the activation of vascular endothelial growth factor (VEGF) receptor-2 (VEGFR-2) and trigger VEGF production by human skin microvascular endothelial cells only, but not HUVEC or any feline endothelial cells tested. These results may explain the decreased pathogenic potential of *B. henselae* infection for cats as compared to humans and suggest that secretion of VEGF by human skin endothelial cells might induce their growth and ultimately lead to bacillary angiomatosis formation [2].

Endothelial phenotype can reflect pathology and can influence the immune cell selection, as illustrated by experiments performed in murine model of Down

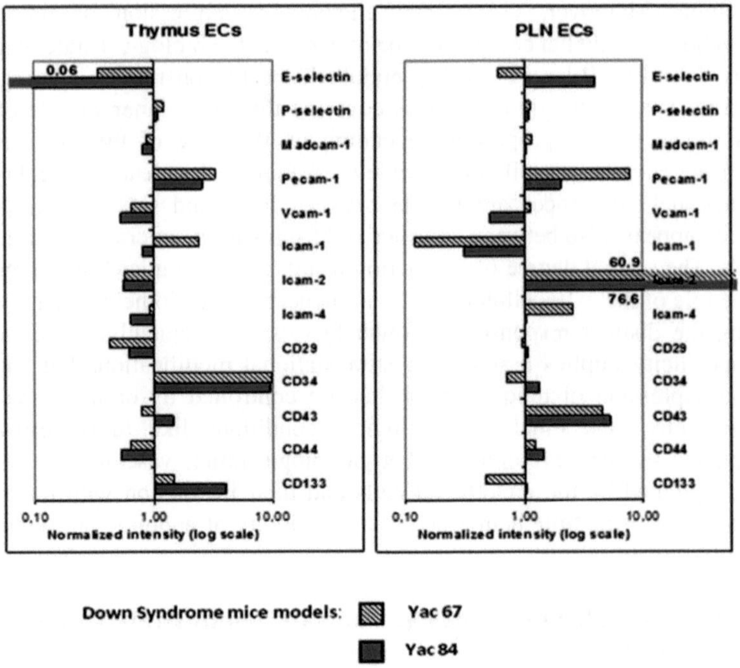

Fig. 18.1 Differential expression of characteristic cell surface adhesion molecules and endothelial markers in thymus and peripheral lymph nodes derived endothelial cells from trisomic Down syndrome mouse models as compared to normal mice. Yac67 and Yac84 mice bear two and three extracopies of the GIRK2 gene, respectively (see text)

syndrome. In the trisomic mice, obtained by the yeast artificial chromosome method and expressing double and triple copies of the GIRK2 gene (G protein-activated inward rectifier potassium channel 2, critical in the Down syndrome region of chromosome 16 [20]), it was possible to isolate endothelial cells and establish cell lines that represent their organ-specific features. In such endothelial cells we showed different expression pattern of adhesion molecules, when compared to their non-trisomic counterparts (Fig. 18.1). Moreover, we demonstrated that this differential expression strongly influences the mobilization and thymic export of lymphocytes (M. Tomczynska, doctoral thesis, Orleans, 2006), confirming the functional significance of the endothelial cell phenotype. The differential expression of adhesion molecules can explain the differences in the selection of circulating lymphocytes both on their release from the thymus and their homing into lymph nodes. Hence, E-selectin which is an efficient early adhesion molecule for lymphocyte rolling and adhesion, is downregulated in the thymic endothelial cells from trisomic mice. It is also decreased in the peripheral lymph node-derived endothelial cells of YAC 67 (one GIRK2 extra copy) mice, while it is over-expressed in YAC 84 (two GIRK2 extra copies). Similarly, the trisomic cells over express ICAM-2; this can explain the strong differences in lymphocyte adhesion (not shown).

Generally, organ-specificity of endothelium is more pronounced when cells react to the microenvironment changes, especially when cells were transferred from normoxic conditions to physiologic oxygen tension (physioxia) or to pathologic hypoxia [21].

18.2.2 Endothelial Cells as Diagnostic Markers and Disease Reporters

Circulating endothelial cells were searched for their possible detachment from damaged sites and, as such, for their ability to carry the characteristics of the tissue they come from and the conditions of the milieu they were in. The work by Dignat-George et al. [22, 23] has brought valuable information but still the major difficulties remain in the yield and viability of the recovered cells [24]. This makes it difficult for circulating endothelial cells to serve as a diagnostic marker. An easier approach is the detection of endothelial cell-associated factors that can be considered as markers of diseases. Among them the VEGFs and the soluble forms of VEGF receptors are considered as markers of ischemia-related disorders. Later on, the circulating endothelial precursors were checked as potential markers of diseases. Their presence in the peripheral blood is most likely a consequence of their mobilization from the bone marrow in response to signals from damaged or hypoxic tissues [25].

Repair mechanisms are among the main possibilities offered by the normally functioning endothelial cells. As such, the circulating mature endothelial cells are specifically recruited to the injured sites. Thus, their increased number could be suggested as potential marker of diseases. Importantly, it has been shown that endothelial cells, when extracted from a given organ, are able to return preferentially into the same organ. As indicated by the data presented (Fig. 18.2) on the localization of the endothelial cells from bone marrow and peripheral lymph nodes derived cell lines, 96 h after intravenous injection and recirculation from the primary "filters" like spleen and lungs, their relative localization appears to be preferential into the bone marrow and into the peripheral lymph nodes, respectively. Moreover, we have observed that conditioning of nonspecialized endothelial cells like precursor cells, with a proper organ-representative medium [15] allowed to increase the efficacy of the specific tissue targeting by the conditioned endothelial cells. This was estimated by detection and quantification of the preferential homing upon intravenous injection, recirculation, and settling of the cells. Consequently, these results indicate that circulating endothelial cells could be considered as diagnostic markers on the one hand but, on the other hand, could be potentially used to target given organs.

During the intravasation process, after the initial contact, the tightened adhesion and further extravasation of the adhered cell into a tissue is facilitated by a whole set of chemo-attracting molecules, i.e., chemokines and their receptors. In this process one should take into account the glycobiological mechanisms of cell recognitions. Extensive studies performed on endogenous lectins have confirmed that the

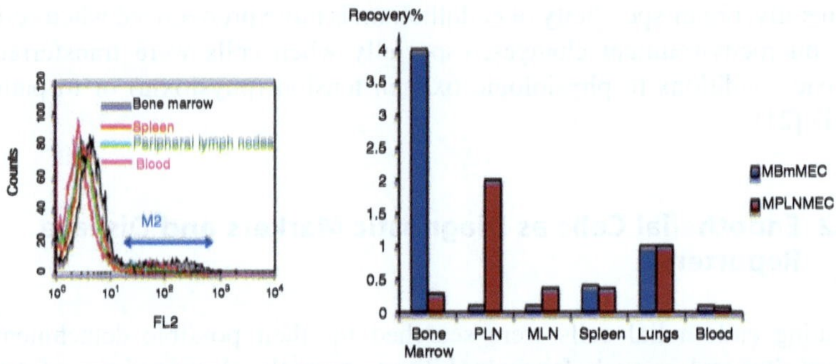

Fig. 18.2 Quantification by flow cytometry of endothelial cells preferential homing and distribution upon intravenous injection and settling for 96 h in vivo. Fluorescently labeled bone marrow (BM)- and peripheral lymph nodes (PLN)-derived endothelial cell from established lines (10^6 in 50 μL) were injected into mice and their distribution quantified by flow cytometry (M2 markers indicates the numbers of positive cells) according to the example shown on the *left panel*. The *right panel* reports the relative distribution of labeled endothelial cells injected intravenously and detected 96 h later. Data are percent of injected cells relatively to 1 as the amount of labeled endothelial cells found par 10^9 lung cells. % ± SD = 20 %

selectivity comes from adhesins, and a higher degree of sophistication is brought by glycoconjugates that decorate the adhesion molecules and their ligand counterpart. Consequently, the endothelial organ selectivity is most probably provided by the interplay of lectins and receptors together with interactions of chemokines and receptors that are differently activated and expressed depending on the site where the posttranscriptional modifications of the glycanic parts play a decisive part. As such, glycans and lectins are highly modified according to the physiological state of the tissue. This can be detected on the endothelium [4, 15, 26].

Glycans and mainly glycosaminoglycans contribute to this selectivity by their role in chemokine binding and presentation. Indeed, chemokines are not only expressed differently in the selected organs, but they insure the cell recruitment by being presented at the endothelial surface by glycanic structures, that are also representative of the endothelium site [16, 27]. This is namely the case for the 6CKine (CCL21)-CCR7 axis that rules the lymphocyte homing into lymph nodes upon 6Ckine presentation on the endothelial cell surface through specific GAGs [17]. Similar mechanisms are fundamental for specific localization of tumor cells during the establishment of metastases. Indeed, the same chemokine-mediated pathway is valid for tumor cell dissemination and site-specific homing on the one hand, and for the tumor-mediated recruitment of bone marrow-derived cells into the tumor site, on the other hand [25]. Both effects are crucial for tumor growth and invasion.

18.2.2.1 Tumor Cell Dissemination and Endothelial Chemokine Specificity Presentation

Tumor cell dissemination is quite selective for the secondary organ where cells localize to make metastases. This depends, among other molecular mechanisms, on the chemokine/chemokine receptor interactions. The pre-metastatic niche is an example of a selective interplay of the primary tumor signaling towards distant organs to induce a, so-called, pre-metastatic state in these secondary sites. The local secondary micro-environmental state is prepared by factors synthesized upon hypoxic conditions at the local primary tumor site that act in a paracrine manner to precondition the secondary metastatic site to host the escaped circulating tumor cells [28].

18.2.2.2 Bone Marrow-Derived Cell Homing into the Tumor and Endothelial Chemokine-Mediated Recruitment

Endothelial cells within the tumor vasculature may play a role in homing of the bone marrow-derived cells. First, they are conditioned by the hypoxic stromal microenvironment, which may result, for example, in an enhanced production of stromal cell-derived growth factor-1 (SDF-1, CXCL12). The recruitment of bone marrow-derived cells is highly dependent on this SDF1/CXCR4 axis. Then the recruited cells, including circulating endothelial precursors, can participate in formation of tumor stroma, which promotes the tumor development. Because the recruitment is a tumor chemokine-dependent process, the number of $CXCR4^+$ circulating cells can be a criterion of tumor evolution [25, 29]. Consequently, the interplay between the whole organism and the tumor is characterized by the primary influence of hypoxia to set the tumor stromal composition and then to affect the external response of the organism. This response is largely dependent on the signals given from the tumor site.

From this point of view, the recruited cell populations can be considered a selective response to the pathology. In search for the possibility of characterizing disease by analysis of circulating endothelial cells, a large body of data has led to the indication that this population may reflect the response of an organism to the tumor growth [30] and response of the tumor to therapy. Additionally, recruitment of endothelial precursors may participate in neovascularization of metastases [31]. This population can also be an indicator in other diseases as cardiac failure [32], hemophilia [33, 34], diabetes [35, 36], and more generally, ischemia-linked processes [37]. The phenotype definition of the circulating endothelial cells versus endothelial precursors is still a matter of vigorous debate, although the use of endothelial progenitors is currently considered as a promising approach for cardiovascular diseases [38]. The doubts come also from the low number of endothelial precursors found in the diseased sites, despite the active and selective homing.

We have isolated and immortalized the human cord blood endothelial precursors [39] and established a comparative model of endothelial progenitors in mice. These cell lines allowed for studies of interactions between tumor and endothelial cells and for building of in vitro 3D models that permit the quantitative assessment of cell recruitment kinetics as a function of tumor stromal composition. We have shown that these cell lines represent early precursors, which can get matured in vitro upon a proper treatment with chemokines. Moreover, these cells should reach tumor sites when intravenously injected, according to their ability to respond to VEGF-A chemo-attracting properties.

18.3 Endothelium in Pathologies and Pathologic Endothelium

Endothelium may reflect pathologies because it very often participates in the disease pathogenesis. In cancer it is well documented that progression depends on the angiogenic switch. The latter results, in turn, from the response of endothelial cells to hypoxia, which is the common character of the tumor milieu. The pathologic endothelium can be described in terms of common but abnormal features and the differences they make as compared to normal endothelium. Such distinctions will be taken into account for therapeutic designs.

Hypoxia, both in normal processes, like the embryonic development, as well as during abnormal ones, is the main signal to which endothelial cells react. It potently influences the activity of endothelial cells into pathologic vessels. Rendering the tumor angiogenesis abnormal, hypoxia affects the tumor progression. It is remarkable that, because of their ability to respond to hypoxia, endothelial cells make new vessels in various ways to reestablish the oxygen delivery inside the tumor.

The actively formed abnormal new vessels are a hallmark of cancer [40]. Their malfunction results mainly from the activation of signaling cascades regulated by the hypoxia inducible factor-1 (HIF-1). In the absence of oxygen and inhibition of the prolyl hydroxylases (PHD)-mediated oxygenation, HIF-1α is stabilized and no longer removed by the von Hippel-Lindau (VHL)-dependent ubiquitinylation into the proteasome. Transcriptional activity of the HIF-1α/β heterodimers, after binding to the hypoxia responsive element (HRE) in the target genes, leads to production of proangiogenic factors, mainly VEGFs. VEGF is a major mitogen and survival factor for endothelial cells. Usually, within hypoxic tumors it is overproduced and its activity is not effectively balanced by other molecules that regulate angiogenesis in physiological conditions.

In addition to VEGFs, hypoxia upregulates expression of angiopoietin-2 (Ang2) or interleukin-8 (IL-8); these cooperate to render the tumor angiogenesis strongly abnormal. Their action can be, to some extent, hindered by so-called "normalizing" factors. Among them is angiopoietin-1 (Ang1) that competes with Ang2 for their common receptor Tie-2. Similarly, the platelet-derived growth factor-β (PDGF-β) would compete with VEGFs for PDGF-β receptor, as they inhibit PDGF-β signaling. These molecules are well known to be differentially expressed in endothelial cells located in a pathologic site as compared to endothelial cells situated in the corresponding normal tissues [40, 41].

18.3.1 Targeting the Pathologic Endothelial Cells

Based on differences between endothelial cells in healthy and diseased tissues we have designed an approach to target the endothelium inside the tumor. The rationale lays in the possibility to reduce the abnormalities in endothelial cell functions. The demonstration that natural killer (NK) cells can actively kill activated endothelial cells in tumor-like milieu [42] was validated in vivo [43] and the mechanism of endothelial cell-mediated antigen presentation could be demonstrated [44] opening a way to kill selectively the abnormal endothelial cells inside a tumor.

Interestingly, when NK cells were recruited in vivo and could penetrate the tumor to kill the tumor cells, increased NO production derived from inducible nitric oxide synthase (iNOS) was detected at the level of endothelium [43]. It was further shown in vitro that the endothelial reaction to NO can downregulate the CD31 expression at the junction of vicinal endothelial cells and, consequently, influence the cohesion of the endothelial barrier [45].

The concept of a positive effect of endothelial cell-based treatments on the antitumor reaction is supported by the results brought by the angiogenesis-targeted strategies. As numerous works corroborate the fact that inhibition of angiogenesis should no longer be considered the main objective, the molecular mechanisms of those strategies had to be deciphered. Because the efficient anti-angiogenic treatments lead to total destruction of the vessels, the anoxia is reached, but the expected eradication of the tumor is not. On the contrary, such a strategy often leads to the selection of tumor-initiating cells [46], which are the most resistant and able to reproduce the whole tumor diversity. In fact, their eradication is the actual challenge in tumor therapy. From this point of view, it is necessary to modulate the anti-angiogenic strategies and make them evolve towards controlled and long lasting normalization of tumor blood vessels.

Since Folkman had indeed described the tumor angiogenesis as pathologic [47] and according to Jain's hypothesis that pathologic vessels should be normalized [48] instead of being destroyed, all angiogenesis targeted therapies were analyzed for their ability to normalize the vasculature [49]. Remarkably, all anti-angiogenic treatments were shown to undergo a step of vessel normalization accompanied by the concomitant and transient raise in the local partial oxygen pressure (pO_2) inside the tumor.

Increase in pO_2 is an adjuvant to radiotherapy or chemotherapy and is expected to counteract the effects of hypoxia. It is hypothesized that an increase in the oxygen tension inside the tumor stroma would deeply change the microenvironment both in terms of cellular as well as molecular balance [50, 51]. Namely, it should locally downregulate the expression of hypoxia-induced genes. Consequently, the normalization process is expected, resulting in modifications of the molecular and cellular composition of the milieu, and bringing a deep change in the types and mechanisms of cell recruitment into the tumor.

The main downregulated proteins by treatments that compensate hypoxia are: HIF-1α, PHDs, and VHL [29]. Table 18.1 lists some of the genes affected by oxygen tension normalization. These genes that regulate the production of VEGFs are indeed strongly modulated. This directly affects the vessel permeability and thus the endothelial cell reorganization with a strong impact on intra-tumoral vessel structure (CD31 increase), pericytes recruitment, and cytokine balance. Noticeably, osteopontin, the molecule that has a deep effect on tumor cell escape, is strongly downregulated.

Table 18.1 Effect of hypoxia compensation on tumor metabolism and phenotype, qRT-PCR analysis of hypoxia/oxygen sensing genes

Hypoxia compensation in tumor: effect on gene expression	
Downregulated genes	Upregulated genes
Osteopontin	CD31
VHL, HIF-1, HIF-2, HO-1	CD34, VEGF-R1, R2
PHD-1, PHD-2, PHD-3	

18.3.2 Why Normalization Could Bring New Means for Tumor Therapy?

Shortcuts to the normalization strategy often come from the difficult identification of the therapeutic windows. Decision concerning the proper time course for the therapy is a great challenge. Visualization techniques may be helpful but their applications to human treatment still needs improvements in the definition and the speed of the response. Normalization window should put in accordance the parameters of efficacy in modulation of size and structure of the vessels, leading to proper pruning together with avoiding toxicity towards normal tissue as well as precluding excessive pruning. The normalization appears to be an adjuvant strategy that permits to increase the efficacy of the treatments.

The biochemical markers that usually serve as decision parameters correspond to the proteins expressed under the influence of hypoxia, mainly HIF-1α, which is closely related to tumor progression [52]. Among them is the glucose transporter 1 (GLUT-1), which is overexpressed in response to hypoxia and low pH, when cells switch on the anaerobic glycolysis [53], and the carbonic anhydrase IX that contributes to maintaining the pH and which is elevated in hypoxia [54]. A very interesting marker appears to be the osteopontin as the key mediator of cell movement which is considerably increased in hypoxia; it is a key molecule for tumor cell escape and metastases [55]. Similarly, the lysyl oxydase (LOX) is highly dependent on pO_2 and controls cell motility through its action on collagen fibers [56].

The main methods to measure pO_2 as directly as possible, inside a tumor, are based on the application of oxygen electrodes that are, unfortunately, limited in their use by the invasiveness of measurements [57]. Reduced invasiveness of the probes makes them extremely useful for experimental purposes, especially when it comes to the phosphorescence quenching-based ones [58]. Methods based on electron paramagnetic resonance [59] and magnetic resonance spectroscopy are still at the prefeasibility stage [60].

The applied techniques are aimed at developing the better sensitivity and resolution in imaging. The ultrasounds are contrast-enhanced to image small vessels [61], where the degree of oxygenation is given by the near infrared spectroscopy that provides the Hb/HbO_2 ratio. Most of the methods require, however, the magnetic resonance imaging from dynamic contrast enhancement with gadolinium derivatives as contrast agents. In this line the blood oxygen level dependent (BOLD) magnetic resonance imaging (MRI) is very sensitive to measure the time related changes in blood oxygenation [62].

Specific sites of hypoxia are currently chemically detected by the pimonidazole derivatives (PIMO) that make adducts with reduced proteins. These adducts can be visualized either histochemically or monitored by dynamic imaging [63] like positron emission tomography (PET) for ^{18}F-fluoromisonidazole (^{18}F-MISO) [64] or ^{18}F-fluoroazomycinarabinofuranoside (^{18}F-FAZA) and derivatives [65]. Together with those derivatives, ^{18}F-fluorodeoxyglucose (^{18}F-FDG) is used for PET assessment of hypoxia but appears to detect also sites of high metabolic activity in cells which, indeed, may also be related to a boost of O_2. This explains the lack of correlation with copper-diacetyl-bis-N4-methylthiosemicarbazone (^{60}Cu-ATSM) labeling [66]. Detection of accumulations of reduced forms of such compounds by the PET-based imaging is the method of choice for noninvasive assessment of hypoxia in a tumor. In the necrotic parts the non-metabolically active cells are not labeled, which is an additional advantage. The main limitations come from the use of short-lived radioactive elements as ^{18}F.

All these methods have proven to be very valuable in clinical applications of chemotherapeutics and in radiotherapy. For example, the intensity-modulated radiation strategy improved the efficacy of radiotherapy protocols allowing for the hypoxia imaging to monitor the doses required [67]. Such a progress shows the need for an increased or safer assessment of the therapeutic windows on the basis of the vasculature normalization to allow better treatment. The ultimate aim is to elaborate better carriers of chemotherapeutic drugs, allowing an efficient penetration inside the tumor mass, and more effective inducers of a higher blood flow, promoting the vessels to be less permeable for the spreading tumor cells, thus exploiting the main effect of vasculature normalization, i.e., limitation/prevention of metastases.

As described by Jain [68], normalization of endothelial cells in the tumor vasculature has direct effect on the tumor treatment outcome. Because pO_2 modulation also directly affects the chemokine composition of the microenvironment, it plays a direct role in the chemoattraction of the circulating cells and their mobilization from the bone marrow. As we have shown, when the tumor pO_2 is increased by hypoxia compensation, the numbers of the bone marrow-derived $CXCR4^+$ cells recruited inside the tumor [29] are considerably lowered. This can be meaningful for the selection of immune cell populations and their antitumor reactivity. Illustrated in Fig. 18.3 the relative recruitment of immune cells as NKs and Tregs is considerably modified upon hypoxia compensation in the tumor site. This reflects a change in the chemokines production balance as NKs express the CCR7 for CCL21 while the reduced Tregs express CCR10+, the receptor for CCL28 [69].

Considering such impact of the endothelial cells on the tumor microenvironment, they can be regarded as a potential tool to reach a long-term normalization of the vessels. It has been shown that signaling through tumor vasculature [70] can bring a deep improvement to treatment, a process quite independent of the tumor type but dependent on the endothelial reaction. This opens potentially new types of adjuvant treatments which would increase antitumor efficacy, provided the normalization is stabilized [71]. Indeed, long-lasting normalization could be reached if endothelial cell activation is controlled. The PI3K/AKT/mTOR pathway is very intensely investigated, as it can downregulate the response to an excessive amount

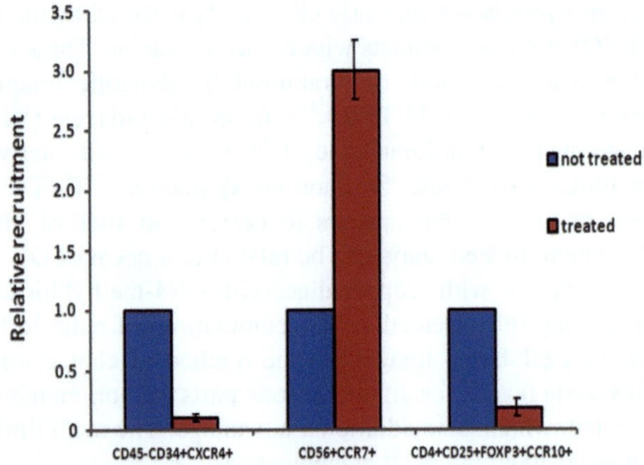

Fig. 18.3 Tumor microenvironment-dependent cell recruitment and escape regulation by chemokines and pO_2. Numbers of CXCR4+CD34+CD45− bone marrow-derived cells among tumor cells appear to be reduced upon vessel normalization treatment; CD4+CD25+FOXP3 +CCR10+ Tregs cells are strongly reduced and CD56+CCR7+ NK cells are increased. Data report one typical experiment from $N > 10$ animals

of VEGF-A produced by the tumor cells. By making the treatment independent of the tumor cells themselves, but addressed to endothelial cells only, this adjuvant strategy is an example of microenvironment-targeted approach. While oxygen tension restoration is a switch able to start the reversal of the tumor-type angiogenesis (response to VEGFs, NO, and Ang2), the stabilization of a close-to-normal phenotype needs the long-term activity of molecules able to control the endothelial cells. As we have shown, this can be achieved through the activation of the phosphatase and tensin homologue (PTEN), which controls the cell proliferation and survival growth pathways at the level of PI2P production from PI3P, thus balancing the AKT activation.

To recapitulate the general approaches we proposed to reach tumor endothelial cell normalization and stabilization. Such strategies have been already validated in melanoma and mammary carcinoma models [29].

Conclusion

Endothelium has gained a full appreciation of its significance after evidencing its roles in pathologies. This review has been focusing on the function of the endothelial cells within tumors, because of the critical role of angiogenesis in tumor progression. This phenomenon, which is necessary not only for tumor development but also for tumor cells dissemination, illustrates particularly how important it is to know the properties of endothelial cells for their ability to control the tumor microenvironment.

We underlined the multiple biological effects dependent on endothelial cells in diseases and the proposed ways to control them, introducing also the concept of endothelium organ-specificity. It helps understanding the infectious disease mechanisms, especially their propagation and species-specificity, combined with organ-specificity. Endothelial cells may insure selection of lymphocytes and their homing to lymphoid organs or inflamed tissues. Moreover, endothelial gene expression profile is deeply perturbed in some genetic diseases as Down syndrome. In such cases, patients display unbalanced circulating lymphocyte populations and a resulting disconcerted immune response possibly leading to a higher susceptibility to leukemia as opposed to resistance to solid tumors.

The strong impact of angiogenesis on ischemia-related diseases, as documented by pathologic vessel formation within tumors, brings a rationale for targeting endothelial cells to modulate the tumor microenvironment and modify, in a controlled way, the tumor stroma. Endothelial cells inside the tumor, influenced by the tumor milieu, might be a direct therapeutic target or serve as markers of tumor progression, providing a new diagnostic tool. Newest data on endothelial cells interactions with other cell types suggest an important part played by exosomes-carried regulatory molecules, such as microRNAs [50, 72], thus opening to new strategies and tools to achieve cell-targeted therapies.

Acknowledgments This work was supported by the Ligue Nationale contre le Cancer, Fondation Jérôme Lejeune, the Institut National du Cancer (grant N° 347/N-INCA/2008/0), the French National Research Agency (ANR 3Sens), and by Normoxys Inc. and the CG45.

References

1. Hwa C, Aird WC (2007) The history of the capillary wall: doctors, discoveries, and debates. Am J Physiol Heart Circ Physiol 293:H2667–H2679
2. Berrich M, Kieda C, Grillon C et al (2011) Differential effects of *Bartonella henselae* on human and feline macro- and micro-vascular endothelial cells. PLoS One 6:e20204
3. Auerbach R (1991) Vascular endothelial cell differentiation: organ-specificity and selective affinities as the basis for developing anti-cancer strategies. Int J Radiat Biol 60:1–10
4. Kieda C, Paprocka M, Krawczenko A et al (2002) New human microvascular endothelial cell lines with specific adhesion molecules phenotypes. Endothelium 9:247–261
5. Ribatti D, Nico B, Vacca A et al (2002) Endothelial cell heterogeneity and organ specificity. J Hematother Stem Cell Res 11:81–90
6. Butcher EC, Scollay RG, Weissman IL (1980) Organ specificity of lymphocyte migration: mediation by highly selective lymphocyte interaction with organ-specific determinants on high endothelial venules. Eur J Immunol 10:556–561
7. Kieda CM, Bowles DJ, Ravid A et al (1978) Lectins in lymphocyte membranes. FEBS Lett 94:391–396
8. Grosse E, Kieda C, Nicolau C (1984) Flow cytofluorometric investigation of the uptake by hepatocytes and spleen cells of targeted and untargeted liposomes injected intravenously into mice. Biochim Biophys Acta 805:354–361
9. Kieda C, Monsigny M (1986) Involvement of membrane sugar receptors and membrane glycoconjugates in the adhesion of 3LL cell subpopulations to cultured pulmonary cells. Invasion Metastasis 6:347–366
10. Kieda C, Roche AC, Delmotte F et al (1979) Lymphocyte membrane lectins. Direct visualization by the use of fluoresceinyl-glycosylated cytochemical markers. FEBS Lett 99:329–332

11. Monsigny M, Roche AC, Kieda C et al (1988) Characterization and biological implications of membrane lectins in tumor, lymphoid and myeloid cells. Biochimie 70:1633–1649
12. Stamper HB Jr, Woodruff JJ (1976) Lymphocyte homing into lymph nodes: in vitro demonstration of the selective affinity of recirculating lymphocytes for high-endothelial venules. J Exp Med 144:828–833
13. Bizouarne N, Denis V, Legrand A et al (1993) A SV-40 immortalized murine endothelial cell line from peripheral lymph node high endothelium expresses a new alpha-L-fucose binding protein. Biol Cell 79:209–218
14. Bizouarne N, Mitterrand M, Monsigny M et al (1993) Characterization of membrane sugar-specific receptors in cultured high endothelial cells from mouse peripheral lymph nodes. Biol Cell 79:27–35
15. Denis V, Dupuis P, Bizouarne N et al (1996) Selective induction of peripheral and mucosal endothelial cell addressins with peripheral lymph nodes and Peyer's patch cell-conditioned media. J Leukoc Biol 60:744–752
16. Salmi M, Jalkanen S (1999) Molecules controlling lymphocyte migration to the gut. Gut 45:148–153
17. Crola Da Silva C, Lamerant-Fayel N, Paprocka M et al (2009) Selective human endothelial cell activation by chemokines as a guide to cell homing. Immunology 126:394–404
18. Lamerant N, Kieda C (2005) Adhesion properties of adhesion-regulating molecule 1 protein on endothelial cells. FEBS J 272:1833–1844
19. Lewandowicz-Uszynska A, Paprocka M, Kieda C et al (2004) Efficiency of lymphocytes adhesion to endothelial cells of distinct tissue origin from children with asthma. Pol Merkur Lekarski 16:104–107
20. Harashima C, Jacobowitz DM, Stoffel M et al (2006) Elevated expression of the G-protein-activated inwardly rectifying potassium channel 2 (GIRK2) in cerebellar unipolar brush cells of a Down syndrome mouse model. Cell Mol Neurobiol 26:719–734
21. Carreau A, El Hafny-Rahbi B, Matejuk A et al (2011) Why is the partial oxygen pressure of human tissues a crucial parameter? Small molecules and hypoxia. J Cell Mol Med 15:1239–1253
22. Camoin-Jau L, Kone-Paut I, Chabrol B et al (2000) Circulating endothelial cells in Behcet's disease with cerebral thrombophlebitis. Thromb Haemost 83:631–632
23. Dignat-George F, Sampol J (2000) Circulating endothelial cells in vascular disorders: new insights into an old concept. Eur J Haematol 65:215–220
24. Blann AD, Woywodt A, Bertolini F et al (2005) Circulating endothelial cells. Biomarker of vascular disease. Thromb Haemost 93:228–235
25. Chouaib S, Kieda C, Benlalam H et al (2010) Endothelial cells as key determinants of the tumor microenvironment: interaction with tumor cells, extracellular matrix and immune killer cells. Crit Rev Immunol 30:529–545
26. Opolski A, Wietrzyk J, Dus D et al (1998) Metastatic potential and saccharide antigens expression of human colon cancer cells xenotransplanted into athymic nude mice. Folia Microbiol (Praha) 43:507–510
27. Szczepanek K, Kieda C, Cichy J (2008) Differential binding of hyaluronan on the surface of tissue-specific endothelial cell lines. Acta Biochim Pol 55:35–42
28. Sceneay J, Smyth MJ, Moller A (2013) The pre-metastatic niche: finding common ground. Cancer Metastasis Rev 32:449–464
29. Kieda C, El Hafny-Rahbi B, Collet G et al (2013) Stable tumor vessel normalization with pO increase and endothelial PTEN activation by inositol trispyrophosphate brings novel tumor treatment. J Mol Med (Berl) 91:883–899
30. Mancuso P, Calleri A, Bertolini F (2012) Circulating endothelial cells and circulating endothelial progenitors. Recent Results Cancer Res 195:163–170
31. Li CX, Shao Y, Ng KT et al (2012) FTY720 suppresses liver tumor metastasis by reducing the population of circulating endothelial progenitor cells. PLoS One 7:e32380
32. Singh N, Van Craeyveld E, Tjwa M et al (2012) Circulating apoptotic endothelial cells and apoptotic endothelial microparticles independently predict the presence of cardiac allograft vasculopathy. J Am Coll Cardiol 60:324–331

33. Biguzzi E, Mancuso P, Franchi F et al (2009) Circulating endothelial cells (CECs) and progenitors (CEPs) in severe haemophiliacs with different clinical phenotype. Br J Haematol 144:803–805
34. Mancuso P, Bertolini F (2010) Circulating endothelial cells as biomarkers in clinical oncology. Microvasc Res 79:224–228
35. Lombardo MF, Iacopino P, Cuzzola M et al (2012) Type 2 diabetes mellitus impairs the maturation of endothelial progenitor cells and increases the number of circulating endothelial cells in peripheral blood. Cytometry A 81:856–864
36. Palombo C, Kozakova M, Morizzo C et al (2011) Circulating endothelial progenitor cells and large artery structure and function in young subjects with uncomplicated type 1 diabetes. Cardiovasc Diabetol 10:88
37. Sun CK, Leu S, Sheu JJ et al (2013) Paradoxical impairment of angiogenesis, endothelial function and circulating number of endothelial progenitor cells in DPP4-deficient rat after critical limb ischemia. Stem Cell Res Ther 4:31
38. Lavergne M, Vanneaux V, Delmau C et al (2011) Cord blood-circulating endothelial progenitors for treatment of vascular diseases. Cell Prolif 44(Suppl 1):44–47
39. Paprocka M, Krawczenko A, Dus D et al (2011) CD133 positive progenitor endothelial cell lines from human cord blood. Cytometry A 79:594–602
40. De Bock K, Cauwenberghs S, Carmeliet P (2011) Vessel abnormalization: another hallmark of cancer? Molecular mechanisms and therapeutic implications. Curr Opin Genet Dev 21:73–79
41. Carmeliet P, Jain RK (2011) Principles and mechanisms of vessel normalization for cancer and other angiogenic diseases. Nat Rev Drug Discov 10:417–427
42. Bielawska-Pohl A, Crola C, Caignard A et al (2005) Human NK cells lyse organ-specific endothelial cells: analysis of adhesion and cytotoxic mechanisms. J Immunol 174:5573–5582
43. Bielawska-Pohl A, Blesson S, Benlalam H et al (2010) The anti-angiogenic activity of IL-12 is increased in iNOS−/− mice and involves NK cells. J Mol Med 88:775–784
44. Benlalam H, Jalil A, Hasmim M et al (2009) Gap junction communication between autologous endothelial and tumor cells induce cross-recognition and elimination by specific CTL. J Immunol 182:2654–2664
45. Carreau A, Kieda C, Grillon C (2011) Nitric oxide modulates the expression of endothelial cell adhesion molecules involved in angiogenesis and leukocyte recruitment. Exp Cell Res 317:29–41
46. De Bock K, Mazzone M, Carmeliet P (2011) Antiangiogenic therapy, hypoxia, and metastasis: risky liaisons, or not? Nature reviews. Clin Oncol 8:393–404
47. Kieran MW, Folkman J, Heymach J (2003) Angiogenesis inhibitors and hypoxia. Nat Med 9:1104, author reply 1104–1105
48. Fukumura D, Jain RK (2007) Tumor microvasculature and microenvironment: targets for anti-angiogenesis and normalization. Microvasc Res 74:72–84
49. Goel S, Duda DG, Xu L et al (2011) Normalization of the vasculature for treatment of cancer and other diseases. Physiol Rev 91:1071–1121
50. Collet G, Skrzypek K, Grillon C et al (2012) Hypoxia control to normalize pathologic angiogenesis: potential role for endothelial precursor cells and miRNAs regulation. Vascul Pharmacol 56:252–261
51. Matejuk A, Collet G, Nadim M et al (2013) MicroRNAs and tumor vasculature normalization: impact on anti-tumor immune response. Arch Immunol Ther Exp (Warsz) 61:285–299
52. Semenza GL (2002) Involvement of hypoxia-inducible factor 1 in human cancer. Int Med 41:79–83
53. Medina RA, Owen GI (2002) Glucose transporters: expression, regulation and cancer. Biol Res 35:9–26
54. Loncaster JA, Harris AL, Davidson SE et al (2001) Carbonic anhydrase (CA IX) expression, a potential new intrinsic marker of hypoxia: correlations with tumor oxygen measurements and prognosis in locally advanced carcinoma of the cervix. Cancer Res 61:6394–6399
55. Hui EP, Sung FL, Yu BK et al (2008) Plasma osteopontin, hypoxia, and response to radiotherapy in nasopharyngeal cancer. Clin Cancer Res 14:7080–7087

56. Erler JT, Bennewith KL, Nicolau M et al (2006) Lysyl oxidase is essential for hypoxia-induced metastasis. Nature 440:1222–1226
57. Vaupel P, Hockel M, Mayer A (2007) Detection and characterization of tumor hypoxia using pO_2 histography. Antioxid Redox Signal 9:1221–1235
58. Griffiths JR, Robinson SP (1999) The OxyLite: a fibre-optic oxygen sensor. Br J Radiol 72:627–630
59. Gallez B, Baudelet C, Jordan BF (2004) Assessment of tumor oxygenation by electron paramagnetic resonance: principles and applications. NMR Biomed 17:240–262
60. Lee CP, Payne GS, Oregioni A et al (2009) A phase I study of the nitroimidazole hypoxia marker SR4554 using 19F magnetic resonance spectroscopy. Br J Cancer 101:1860–1868
61. Di Martino EF, Gagel B, Schramm O et al (2005) Evaluation of tumor oxygenation by color duplex sonography: a new approach. Otolaryngol Head Neck Surg 132:765–769
62. Taylor NJ, Baddeley H, Goodchild KA et al (2001) BOLD MRI of human tumor oxygenation during carbogen breathing. J Magn Reson Imaging 14:156–163
63. Raleigh JA, Chou SC, Arteel GE et al (1999) Comparisons among pimonidazole binding, oxygen electrode measurements, and radiation response in C3H mouse tumors. Radiat Res 151:580–589
64. Rasey JS, Grunbaum Z, Magee S et al (1987) Characterization of radiolabeled fluoromisonidazole as a probe for hypoxic cells. Radiat Res 111:292–304
65. Kumar P, Ohkura K, Beiki D et al (2003) Synthesis of 1-beta-D-(5-deoxy-5-iodoarabinofuranosyl)-2-nitroimidazole (beta-IAZA): a novel marker of tissue hypoxia. Chem Pharm Bull 51:399–403
66. Bowen SR, Van Der Kogel AJ, Nordsmark M et al (2011) Characterization of positron emission tomography hypoxia tracer uptake and tissue oxygenation via electrochemical modeling. Nucl Med Biol 38:771–780
67. Chao KS, Bosch WR, Mutic S et al (2001) A novel approach to overcome hypoxic tumor resistance: Cu-ATSM-guided intensity-modulated radiation therapy. Int J Radiat Oncol Biol Phys 49:1171–1182
68. Jain RK (2005) Normalization of tumor vasculature: an emerging concept in antiangiogenic therapy. Science 307:58–62
69. Facciabene A, Peng X, Hagemann IS et al (2011) Tumour hypoxia promotes tolerance and angiogenesis via CCL28 and T(reg) cells. Nature 475:226–230
70. Qayum N, Muschel RJ, Im JH et al (2009) Tumor vascular changes mediated by inhibition of oncogenic signaling. Cancer Res 69:6347–6354
71. Sato Y (2012) The vasohibin family: novel regulators of angiogenesis. Vascul Pharmacol 56:262–266
72. Skrzypek K, Tertil M, Golda S et al (2013) Interplay between heme oxygenase-1 and miR-378 affects non-small cell lung carcinoma growth, vascularization and metastasis. Antioxid Redox Signal 19:644–660

Index

A
α1-Acid glycoprotein (AGP), 105
Activator protein-1 (AP-1), 126, 149–151, 254
ADMA/DDAH pathway
  actin dynamics regulation
    DDAH1 interaction, 252
    gene expression modulation, 252–253
    NO signalling, 252
    oxidative stress and inflammation, 253–254
    RhoA, 251
    Rho GTPases, 251
  cell–cell adhesion, 250
  cell migration, 250
  endothelial cells, effects of, 248, 249
  gap junctional communication, 250
  proliferation and apoptosis, 249–250
  signalling mediators, angiogenesis regulation, 250–251
Adrenomedullin (AM) signaling, 38
AKT1 serine/threonine protein kinase, 41
Angiogenesis
  arteriogenesis, 13
  endothelial cells
    activation, 11
    extracellular matrix, 11–13
  hypoxia-inducible factor-1α, 11
  inhibitors (*See* Vasohibin (VASH) family)
  intracellular/intercellular model, of lumen formation, 13
  matrix metalloproteinases, 12
  proteases, 12–13
  stages of, 11, 12
  stalk cells, 12
Angiogenic signaling pathways
  blood vessel formation, HO-1 function
    proangiogenic function, 217
    SDF-1 signaling interactions, 219–220
    VEGF signaling interactions, 217–219
  Nrf2 transcription factor, 226–229
  tumor cells
    angiogenic switch, 368
    angiopoietins, 369
    anti-angiomirs, 371–372
    chemokines, 370
    fibroblast growth factors, 369
    heme oxygenase-1, 370–371
    hypoxia-inducible factors, 368–369
    mediators, 369
    Nrf2, 370
    thymidine phosphorylase, 371
    VEGF, 372
Angiogenic switch, 294, 368
Angiopoietins (Ang), 40–41, 369
Angiotensinogen (AGT), 110
Anti-angiogenic therapies, 301–303
Anti-lymphangiogenic therapy
  in cancer, 53–54
  in graft rejection, 52
  in inflammatory disorders, 51
Antithrombin, 111
Antitrypsin, 102
Antitumor angiogenic drug screening
  drug screening, 304–305
  drugs targeting tumor angiogenesis in vivo, 303–304
  target validation, 306–307
  toxicity assay, 306
Apolipoprotein E (Apo E), 203
Apolipoprotein L domain containing 1 (APOLD1), 155
Arterial endothelial cell, 15
Arterial–venous–lymphatic identity, 9–11
Arteriogenesis, 13, 246, 285, 318, 328, 330, 332, 350
Arteriolar endothelial cell, 15

J. Dulak et al. (eds.), *Angiogenesis and Vascularisation*,
DOI 10.1007/978-3-7091-1428-5, © Springer-Verlag Wien 2013

Asymmetric-di-methyl arginine (ADMA). *See also* ADMA/DDAH pathway
metabolism dysregulation
cancer, 247
diabetes mellitus, 247–248
hypercholesterolaemia, 246
ischemia, 246
renal disease, 247
proteolytic degradation, methylated proteins, 243–244

B

Basic fibroblast growth factor (bFGF), 4–5
Biogenesis, of microRNA, 188, 189
Blood vessel formation
angiogenic signaling cascades, HO-1 function
proangiogenic function, 217
SDF-1 signaling interactions, 219–220
VEGF signaling interactions, 217–219
angiopoietin-1, 5
basic fibroblast growth factor, 4–5
HO-1 knockout, overall effect, 216
therapeutic angiogenesis and progenitor cells
damaged tissue regeneration, 220–222
proangiogenic cells, 222–223
tumor angiogenesis and growth, 224–225
vascular endothelial growth factor, 5

C

Cancer. *See also* Vasohibin (VASH) family
ADMA metabolism dysregulation, 247
anti-lymphangiogenic therapy, 53–54
hypoxia, tumor progression, 272, 275–276
in lower vertebrates, zebrafish model, 300–301
Carbonyl stress, 316–317
Caveolae, 14, 16, 17
CCAAT/enhancer-binding protein delta (CEBP-delta), 152
CCN2. *See* Connective tissue growth factor (CTGF)
Cerebral ischemia, 329–330. *See also* Ischemia
Chemokines
and bone marrow-derived cells, 397
endothelial cells, 392–393
escape regulation, 402
production, 401
tumor cell dissemination, 396–397
Circulating endothelial cells, 395, 397. *See also* Endothelial cells
c-Myc, 151, 176
Collagen and calcium-binding EGF domain-containing protein 1 (CCBE1), 42
Connective tissue growth factor (CTGF), 153
Continuous endothelium, 14
Continuous, non-fenestrated endothelium, 14
Coronary artery disease (CAD), 328
Cyclooxygenase-2 (COX-2), 153
Cysteine cathepsins, 89
Cysteine-rich angiogenic inducer 61 (CYR61), 154

D

Danio rerio. *See* Zebrafish model
Developmental angiogenesis, microRNA
delta-like 4, 191–192
dicer deficiency, 190–191
miR-10, 191
miR-30, 191, 192
miR-126, 191
schematic illustration, 192
VEGF, 191, 192
Diabetes mellitus
ADMA metabolism dysregulation, 247–248
definition, 315–316
hyperglycemia, 315
macroangiopathy
cerebral ischemia, 329–330
coronary artery disease, 328
description, 326
neovascularization, 326–327
peripheral arterial disease, 329
microangiopathy
description, 318
nephropathy, 321–323
neuropathy, 323–324
retinopathy, 319–321
wound healing, 324–326
pathway imbalance
carbonyl stress, 316–317
diacylglycerol/protein kinase C pathway, 316–317
endothelium, 318
hexosamine pathway, 316–317
oxidative stress, 317–318
polyol pathway, 316–317
vascular complications, 314–315, 330–331

Dimethylaminohydrolase (DDAH) expression, genetic manipulation, 244, 245. *See also* ADMA/DDAH pathway
Down syndrome, 390, 394
Drug screening. *See* Antitumor angiogenic drug screening
Dual specificity protein phosphatase 1 (DUSP1), 153

E
Early growth response (Egr) family, 148, 149, 175
EGF-like domain-containing protein 7 (EGFL7), 154
Embryonic hemangioblasts, 5
Emilin-1, 42
Endothelial cells
  angiogenic mechanisms, 295–296
  arterial, 15
  arterial-venous-lymphatic identity, 9–11
  arteriolar, 15
  arteriovenous cell identity, 9, 10
  characteristics, 390–391
  of circulatory system, 4
  continuous, non-fenestrated endothelium, 14
  description, 13–14
  diagnostic markers and disease reporters
    bone marrow-derived cells, 397
    tumor cell dissemination, 396–397
  discontinuous, 15
  ETS transcription factors
    ETS target gene, 127–129
    vascular development, 129–131
    in vivo models of angiogenesis, 131, 132
  fenestrated, 14–15
  functions of
    adaptive and innate immune response, 18
    anti-inflammatory function, 17
    E-selectin, 19
    fibrin and fibrinolysis, 19, 20
    platelet activation and aggregation, 19
    prostacyclin, 19, 20
    semipermeability, 17
    thrombin, 19
    vasomotor tone and blood pressure, regulators of, 20
    von Willebrand factor, 19
  glomerular, 16
  hemogenic, 7, 9
  hepatic sinusoidal, 16
  lymphatic, 16
  Notch signalling, 10
  organ-specificity
    cell lines as tool, 393–395
    chemokines, 392–393
    site selectivity, 391–392
  origin and differentiation, 5–9
  pathologies and pathologic endothelium
    normalization strategy, 400–402
    targeting of, 398–400
  pulmonary capillary, 16
  sinusoidal, 15
  venous, 15–16
Endothelial progenitor cells (EPCs), vi, 7, 9, 67, 71, 73, 95, 325, 346, 374
  ADMA/DDAH effects, 248
  HO-1, 216, 223
  proangiogenic cells, HO-1, 222
  SDF-1 signaling, 220
Endothelial side population cells
  EC–SP cells, 75–76
  identification of, 74
  isolation of, 73–75
  therapeutic potential of, 76–80
    angiogenic growth factors, 77
    antiangiogenic approach, 78
    heterogeneous angiogenesis model, 79, 80
    hind limb ischemia model, 77–78
    vascular regenerative medicine, 76–77
Endothelin-1 (ET-1), 20
EPCs. *See* Endothelial progenitor cells (EPCs)
EphrinB2, 41
3,6-di(2,3-epoxypropoxy)xanthone (EPOX), 155
Erg (ETS-related gene), 131, 133–135
Erg p55, 126
Ets-1, 130–133
ETS transcription factors
  alternative splicing mechanism, 125
  AP-1 transcription complex, 126
  for basic cellular functions, 122
  characterization, 123
  DNA-binding properties, 123
  in endothelial cells
    ETS target gene, 127–129
    vascular development, 129–131
    in vivo models of angiogenesis, 131, 132
  endothelial homeostasis, 129, 133, 134
  PNT domain, 123–124
  structural and functional domains of, 123, 124

ETS transcription factors (*cont.*)
  targets of signal transduction pathways, 125–126
  transcriptional networks, 126–127
  transcriptional repressors, 124–125
Etv-2 (Etrsp/Er71), 130
Extracellular matrix (ECM), 105, 106, 108

F
Factor-inhibiting HIF-1 (FIH-1), 167–168
Fasting-induced adipose factor (Fiaf), 39
Fenestrated endothelia, 14–15
Fibrinogen, 93–94
Fibrins, 93, 94
Fluorescence-activated cell sorting (FACS), 69, 70, 72, 74, 75
FOXC2, 40

G
Glass catfish model, hypoxia
  characteristics, 285–286
  drawbacks, 274–275
  lymphatic vessels, 287–288
  tolerance, 287
  transparency, 287
  vascular permeability, 286–287
Glomerular endothelial cell, 16
Glomeruloid bodies (GBs) formation, 379
Glucose transporter 3 (GLUT3), 154–155
Glycosaminoglycans (GAGs), 392, 396
Graft rejection, 52

H
Haematogenous metastasis, 46
Hemangioblasts, 5–7, 9
Hematopoietic stem cells (HSCs), surface marker schemes for, 69
Heme oxygenase-1 (HO-1) function. *See also* NF-E2-related factor 2 (Nrf2) transcription factor
  angiogenic signaling, tumor cells, 370
  blood vessel formation
    angiogenic signaling cascades, 217–220
    HO-1 knockout, overall effect, 216
    therapeutic angiogenesis and progenitor cells, 220–223
    tumor angiogenesis and growth, 224–225
  against vascular injury, 215–216
Hemogenic endothelium, 7, 9
Hepatic sinusoidal endothelial cell, 16
Hepatoma-derived growth factor (HDGF), 201
Heterogeneous angiogenesis model, 79, 80
High mobility group protein 1-like 2 (HMG-1 L2). *See* Hepatoma-derived growth factor (HDGF)
High molecular weight (HMW) fibrinogen, 93, 94
Histone deacetylases (HDACs), 170
HO-1. *See* Heme oxygenase-1 (HO-1) function
Hoechst–SP method, 70, 73
Hypercholesterolaemia, 246
HypoxamiRs, 173
  miR-16, 195
  miR-210, 193–194
  miR-503, 195
  miR-15a/b, 195
  miR-200b, 195–196
  miR-23~27~24 cluster, 194–195
Hypoxia
  angiogenesis regulation, 295
  embryo zebrafish metastasis and tumor model, 277–280
  in glass catfish
    characteristics, 285–286
    lymphatic vessels, 287–288
    tolerance, 287
    transparency, 287
    vascular permeability, 286–287
  intratumoral endothelial cells, 276
  metastasis, 276
  retinopathy, in adult zebrafish model
    age-related macular degeneration, 281
    characteristics, 280–281
    diabetic retinopathy, 281
    disease-progression, 281
    Notch signaling pathway, 284
    retinal vasculature, 282–283
    retinal vessels, 284
  rheumatoid arthritis, angiogenesis role, 347–348
  tumor progression, 275–276
  zebrafish, versatile model, 274
Hypoxia-induced angiogenesis
  microRNA, 173
  nitric oxide, 171–172
Hypoxia inducible factors (HIFs)
  angiogenic signaling pathways, 368–369
  deacetylation of, 169–170
  HIF-α hydroxylation
    factor-inhibiting HIF-1, 167–168
    HIF-α isoforms, 166

hypoxia response element, 165
prolyl hydroxylases, 167
stabilization during hypoxia, 167
HIF-1 and HIF-2, 164, 176–177
HIF-dependent genes, 164, 165
and nuclear factor κB, 170–171
reactive oxygen species-mediated stabilization, 168
transcriptional VEGF regulation, 174–175
Hypoxia-regulated angiogenic inhibitors, 177–178
Hypoxia response element (HRE), 165
Hypoxia-responsive miRs. *See* HypoxamiRs

I

Id1 (DNA-binding protein inhibitor), 151–152
Immediate-early genes (IEGs)
angiogenesis
anti-apoptotic proteins, 155
AP-1 family, 149–151
ATF3, 151
CEBP-delta, 152
connective tissue growth factor, 153
COX-2, 153
Cyr61, 154
cytokines, 152–153
cytoskeletal proteins, 155
DNA-binding protein inhibitor Id1, 151–152
DUSP1, 153
early growth response proteins, 148, 149
EGFL7/ZNeu1, 154
Ets family, 151
gene-silencing technologies, 146
glucose transporter 3, 154–155
mechanisms of action, 146, 147
Myc, 151
NR4A family, 149
serum response factor, 152
TNFAIP3, 154
urokinase-type plasminogen activator, 154
description, 146
Inflammatory disorders
lymphangiogenesis
absorption of tissue edema, 43
acute and chronic inflammation, 43
extranodal lymphangiogenesis, 45
intranodal lymphangiogenesis, 45–46
lymphatic vessels, 43–44
VEGF-A, 44
VEGF-C-VEGFR-3 axis, 44
pro-and anti-lymphangiogenic therapy, 51–52
rheumatoid arthritis, 348–349
$\alpha 9\beta 1$ Integrins, 41, 42
Intercellular model, of lumen formation, 13
Interleukin-6 (IL-6), 152
Interleukin-8 (IL-8), 152–153, 175, 176
Intracellular model, of lumen formation, 13
Intussusceptive microvascular growth (IMG), 378–379
Ischemia
ADMA metabolism dysregulation, 246
HO-1, 217, 220–221, 223
hypoxia, 273
Nrf2, 215–216, 229–230
Ischemia-associated miRs, 197–198

K

Kallistatin, 108–109

L

Lipid-lowering drugs, 331
Low molecular weight (LMW) fibrinogen, 93, 94
Lymphangiogenesis
description, 28
embryonic development, of mammalian lymphatic vasculature, 30, 36
in human diseases
in cancer, 46–50
in hypertension, 49
inflammatory lymphangiogenesis, 43–46
lymphatic endothelial cells, 30
Coup-tf2 deficiency, 37
vs. extracellular matrix interactions, 41–42
LYVE-1, 30
proliferation and migration, 37–39
PROX1, 30
source of, 29
SOX18, 30, 37
transdifferentiation of macrophages, 29
lymphatic vascular system
in humans embryos, 28–29
in mice, 28
lymphatic vessels
postnatal maturation of, 43
remodeling and maturation of, 40–41
lymphovascular separation process, 39

Lymphangiogenesis (*cont.*)
molecular factors, 30–35
pathological, 28
physiological, 28
therapeutic modulation of
animal models, 49, 51
anti-lymphangiogenic therapy, 51–54
pro-lymphangiogenic therapy, 51–52
Lymphatic endothelial cells, 16, 30
Coup-tf2 deficiency, 37
vs. extracellular matrix interactions, 41–42
LYVE-1, 30
proliferation and migration, 37–39
PROX1, 30
source of, 29
SOX18, 30, 37
transdifferentiation of macrophages, 29
Lymphatic vascular system
in humans embryos, 28–29
in mice, 28
Lymphatic vessel endothelial hyaluronian 1 (Lyve-1), 11, 16
Lymphatic vessels
postnatal maturation of, 43
remodeling and maturation of, 40–41
Lymphedema-distichiasis (LD) syndrome, 40
Lymph node (LN) metastasis, 46, 47
Lymphovascular separation process, 39

M
Macroangiopathy, diabetes mellitus
cerebral ischemia, 329–330
coronary artery disease, 328
description, 326
neovascularization, 326–327
peripheral arterial disease, 329
Mammalian lymphatic vasculature, embryonic development of, 30, 36
Markers and disease reporters, endothelium
bone marrow-derived cells, 397
tumor cell dissemination, 396–397
Maspin, 110–111
Matrikines, 93
Matrix metalloproteinases (MMPs), 88–89
Matrix proteins, proteolytic modification of, 93–94
Mcl-1 (induced myeloid leukaemia cell differentiation protein), 155
Membrane-type MMPs, 88, 91, 92
Methylarginines
ADMA/DDAH pathway
actin dynamics regulation, 251–254
cell–cell adhesion, 250
cell migration, 250
endothelial cells, effects of, 248, 249
gap junctional communication, 250
proliferation and apoptosis, 249–250
signalling mediators, angiogenesis regulation, 250–251
angiogenesis, 242
arginine methylation, 243
DDAH expression, genetic manipulation, 244, 245
definition, 242–243
elimination, 244
generation and metabolism, 242–243
regulatory role, angiogenesis
ADMA metabolism dysregulation, 246–248
cultured cells and knockout animals, 244, 246
EPCs, ADMA/DDAH effects, 248
Microangiopathy, diabetes mellitus
description, 318
nephropathy, 321–323
neuropathy, 323–324
retinopathy, 319–321
wound healing, 324–326
MicroRNA (miR) regulation
angiogenesis regulation, 296
biogenesis of, 188, 189
in developmental angiogenesis
delta-like 4, 191–192
dicer deficiency, 190–191
miR-10, 191
miR-30, 191, 192
miR-126, 191
schematic illustration, 192
VEGF, 191, 192
hypoxia-induced angiogenesis, 173
premature miRs, 189
primary miR, 189
tissue regeneration angiogenesis
developmental angiomiRs, 193
HIF signaling, 196–197
hypoxia-responsive miRs, 193–196
ischemia-associated miRs, 197–198
tumor angiogenesis
metastatic angiogenesis, 202–204
secretory factors, 201–202
VEGF signaling, 198–200

zebrafish model, 297–299
MMPs. *See* Matrix metalloproteinases (MMPs)
Multidrug resistance transporter 1 (MDR1), 69

N

Neovascularization
HO-1 function, blood vessel formation, 220–222
Nrf2 transcription factor, 229–230
tumor vasculature heterogeneity
capillary network expansion, 378–379
glomeruloid bodies formation, 379
intussusceptive microvascular growth, 378
vasculogenic/vascular mimicry, 379
vessel co-option, 378
vascular complications, macroangiopathy, 326–327
Nephropathy, 321–323
Nerve growth factor IB (NGFIB). *See* Nur77
Neuropathy, 323–324
Neuropilin-1 (Nrp1), 38
Neuropilin-2 (Nrp2), 38
Nexin-1 (PN-1), 102, 108–111
NF-E2-related factor 2 (Nrf2) transcription factor. *See also* Heme oxygenase-1 (HO-1) function
angiogenic signaling, tumor cells, 370
proangiogenic function
angiogenic signaling pathways, 226–229
damaged tissue, regenerative neovascularization, 229–230
Nrf2 knockout effect, 225
tumor angiogenesis, 231
against vascular injury, 215–216
Nitric oxide (NO)
HIF stabilization, 171–172
signalling, 252
Notch-1, 88, 90
Nrf2. *See* NF-E2-related factor 2 (Nrf2) transcription factor
Nuclear factor κB and HIFs, 170–171
Nuclear receptor subfamily 4 group A (NR4A), 149
Nur77, 149

O

Organ-specificity, endothelium
cell lines as tool, 393–395
chemokines, 392–393
site selectivity, 391–392
Oxidative stress, 317–318

P

PAI-1. *See* Plasminogen activator inhibitor type 1 (PAI-1)
Pathological lymphangiogenesis, 28
Peripheral arterial disease (PAD), 329
Physiological lymphangiogenesis, 28
Pigment epithelial-derived factor (PEDF), 112
Plasmin-mediated proteolysis, 106, 107
Plasminogen activator inhibitor type 1 (PAI-1)
α1-acid glycoprotein, 105
active and latent forms, 105
in angiogenesis and tumor invasion, 106–107
antiangiogenic activity, 108
endothelial cell apoptosis, 107–108
functional flexibility, 104
latency transition, 105
at low and high concentrations, 108
proangiogenic activity, 107
pro-apoptotic cell signaling, 107
proteolytic ECM fragments, 108
self-inactivation mechanism, 104
structure, 104
vitronectin-binding site, 105
Podoplanin function, 39
Pointed/PNT domain, 123–124
Proangiogenic function
gene regulation, in hypoxic conditions, 173–175
HO-1 function, blood vessel formation, 217
Nrf2 transcription factor
angiogenic signaling pathways, 226–229
Nrf2 knockout effect, 225
regenerative neovascularization, damaged tissue, 229–230
tumor angiogenesis, 231
Prolyl hydroxylases (PHDs), 167
Pro-lymphangiogenic therapy
in inflammatory disorders, 51–52
in secondary lymphedema, 52
Prostaglandin-endoperoxide synthase 2. *See* Cyclooxygenase-2 (COX-2)
Protease(s), in angiogenesis, 94–95
activity for cardiovascular development, 86
anastomosis formation, 92
cysteine cathepsins, 89
matrix metalloproteinases, 88–89

Protease(s) (*cont.*)
  migration and invasive cell growth, 91–92
  pericellular proteolytic activity, 91–92
  proteolytic modification
    of cellular receptors, 93
    of growth factors, 92–93
    of matrix proteins, 93–94
  serine proteases, 87–88
  tread-milling model, 92
  in VEGF-induced sprouting, 89–91
Protein C inhibitor (PCI), 109, 110
Proteolytic modification
  of cellular receptors, 93
  of growth factors, 92–93
  of matrix proteins, 93–94
Pulmonary capillary endothelial cell, 16

R
Reactive oxygen species-mediated HIFs stabilization, 168
Regulator of G protein signaling 5 (RGS5), 178
Related transcription enhancer factor-1 (RTEF-1), 175
Renal disease, 247
Response gene to complement-32 (RGC-32), 178
Retinopathy
  hypoxia, in adult zebrafish model
    age-related macular degeneration, 281
    characteristics, 280–281
    diabetic retinopathy, 281
    disease-progression, 281
    Notch signaling pathway, 284
    retinal vasculature, 282–283
    retinal vessels, 284
  vascular complications, microangiopathy, 319–321
Rheumatoid arthritis (RA)
  angiogenesis role
    chemical stabilization, HIF, 350
    FLS, 349
    HIF pathway, 349
    hypoxia, 347–348, 350
    IκB kinases, 351
    inflammatory cells accumulation, 348–349
    synovium expansion and infiltration, 344–345
    TNFα, 349
    tumour associated macrophages, 350
    VEGF, 345–347
  characteristics, 341, 343
  pathogenesis, 343
  symptoms, 341
  therapeutic option, angiogenesis inhibition
    anti-NRP-1 antibody, 352
    anti-VEGF antibody, 351
    cancer treatment, 351
    CCAAT/enhancer binding protein delta, 353
    heparins, 353
    in models of RA, 352
    neuropilin-1, 352
    pro-inflammatory cytokines, 353

S
Secondary lymphedema, pro-lymphangiogenic therapy, 52
Secreted matrix metalloproteinases, 88
Self-renewal, 79
Serine protease inhibitors (Serpins)
  in angiogenesis regulation, 108, 109
  angiotensinogen, 110
  antithrombin, 111
  antitrypsin, 102
  features of, 103
  human serpins, 102
  inhibition mechanism, 103–104
  kallistatin, 108–109
  maspin, 110–111
  nexin-1, 111
  PAI-1 (*see* Plasminogen activator inhibitor type 1 (PAI-1))
  pigment epithelial-derived factor, 112
  protein C inhibitor, 109
  structure of, 103
Serine proteases, 87–88
SerpinA1. *See* Antitrypsin
Serum response factor (SRF), 152
Side population (SP) cells
  cardiac, 70
  endothelial SP
    identification of, 74
    isolation of, 73–75
    therapeutic potential of, 76–80
  Hoechst 33342 exclusion, 69
  purification approach, advantage of, 70
Sinusoidal endothelium, 15
SIRT1, 169, 197, 264
SIRT3, 168, 169
Sirtuins, 168–171
Solute carrier family 2, facilitated glucose transporter member 3 (SLC2A3). *See* Glucose transporter 3 (GLUT3)

Sonic hedgehog (Shh) protein, 10
SP cells. *See* Side population (SP) cells
Sphingosine-1-phosphate (S1P), 39, 44, 88, 296
Stem cell homing, 193
Stem/progenitor cells, in peripheral vascular system, 71
  endothelial progenitor cells, 73
  human aortic endothelial cells, 72
  human umbilical vein endothelial cells, 72
  tunica adventitia, 72
  tunica intima, 71–72
  tunica media, 72
Stroke. *See* Cerebral ischemia

T

Tertiary lymphoid organs, 45
Thrombospondin 1 (TSP-1), 177, 201
Thymidine phosphorylase (TP), 371
Tie receptors, 40
Tissue regeneration angiogenesis, miR regulation
  developmental angiomiRs, 193
  HIF signaling, 196–197
  hypoxia-responsive miRs, 193–196
  ischemia-associated miRs, 197–198
Tissue repair angiogenesis, 197, 202, 204
  miR-210, 193
  miR-92a, 196
  miR-dependent pathways, 198
Tissue stem cells, 69
TNFAIP3. *See* Tumour necrosis factor alpha-induced protein 3 (TNFAIP3)
Transcription factors (TF), 122. *See also* ETS transcription factors
Transforming growth factor β receptor 2 (TGFBR2), 201
Tumor angiogenesis, 203. *See also* Vasohibin (VASH) family
  angiogenic mechanisms, in endothelial cells, 295–296
  antitumor angiogenic drug screening, in zebrafish
    drug screening, 304–305
    drugs targeting tumor angiogenesis in vivo, 303–304
    target validation, 306–307
    toxicity assay, 306
  blood vessel formation, HO-1 function, 224–225
  cell dissemination, 396–397
  description, 294–295
  metastatic angiogenesis, miR regulation, 202–204
  nrf2 transcription factor, 231, 370
  secretory factors, miR regulation
    angiogenin, 201
    HDGF, 201
    miR-9, 202
    miR-210, 202
    TGFBR2, 201
    TSP-1, 201
  stromal cells
    bone marrow-derived vascular progenitors, 374–375
    cells within tumor microenvironment, 375–376
    exosomes, 376
    macrophage function polarization, 372–374
    mesenchymal cells, 375
    Tie-2-expressing monocytes, 373
    tumor-associated macrophages, 372–373
    tumor vascularization, 372–373
  VEGF signaling, miR regulation, 198–200
  zebrafish model
    cancer in lower vertebrates, 300–301
    genetic engineering, 299–300
    and microRNA, 297–299
    vessel formation and homeostasis, 296–297
    xenotransplants models, anti-angiogenic therapies, 301–303
Tumor-associated macrophages (TAMs), 372–373
Tumor vasculature, heterogeneity
  characteristics, 376–377
  neovascularization, alternative mechanisms, 378–379
Tumour necrosis factor alpha-induced protein 3 (TNFAIP3), 154
Tyrosine kinase inhibitors, of VEGF-A pathway, 53–54

U

Urokinase-type plasminogen activator (uPA), 154

V

Vascular complications
  diabetes (*see* Diabetes mellitus)
  lipid-lowering drugs, 331
Vascular early response gene (Verge), 155
Vascular endothelial growth factor (VEGF)
  blood vessel formation, 5

Vascular endothelial growth factor (VEGF) (*cont.*)
diagnostic markers, 395
importance of, 5
microRNA regulation
developmental angiogenesis, 191, 192
tumor angiogenesis, 198–200
mRNA translation, 175
organ-specificity, endothelium cell lines, 393
proteases in VEGF-induced sprouting, 89–91
regulation by HIFs, 173–175
rheumatoid arthritis, angiogenesis role, 345–347
vascular endothelial growth factor-A (VEGF-A), 49, 54
inflammatory lymphangiogenesis, 44, 45
in tumors, 47
tyrosine kinase inhibitors, 53
vascular endothelial growth factor-C (VEGF-C), 37–38
vascular endothelial growth factor-D (VEGF-D), 37–38
vascular endothelial growth factor receptor 3 (VEGFR-3), 37–38
Vascular-resident endothelial side population cells. *See* Side population cells
Vasculogenesis, 4
Vasculogenic/vascular mimicry (VM), 368, 378, 379
Vasohibin (VASH) family
vasohibin-1
ECs, survival/death, 264
isolation, 262
pathophysiological conditions, 262–264
vasohibin-2
isolation, 264–265
opposing role of, angiogenesis, 265
pathophysiological conditions, 265
VEGF. *See* Vascular endothelial growth factor (VEGF); Vasohibin (VASH) family
Venous endothelial cell, 15–16
Verge. *See* Vascular early response gene (Verge)
Vesiculo-vacuolar organelles (VVOs), 14, 17
Vitronectin-binding site, PAI-1, 105

W

Wound healing, vascular complications, 324–326

X

Xenograft, zebrafish model, 301–303, 308

Z

Zebrafish model
antitumor angiogenic drug screening
drug screening, 304–305
drugs targeting tumor angiogenesis in vivo, 303–304
target validation, 306–307
toxicity assay, 306
in cancer progression, 308
embryo metastasis, hypoxia
angiogenic effects, 279
migratory properties, specific cell line, 279
transparency, 274, 277
tumor cell dissemination, 278
tumor cell invasion, 277
vasculature, 278, 280
hypoxia
drawbacks, 274–275
hypoxia-induced angiogenic retinopathy, 280–285
hypoxia, versatile model, 274
setup, 279, 280
in pharmaceutical companies, 308
tumor angiogenesis
cancer in lower vertebrates, 300–301
genetic engineering, 299–300
and microRNA, 297–299
vessel formation and homeostasis, 296–297
xenotransplants models, anti-angiogenic therapies, 301–303